CONTENTS

Preface ix
About the Authors .xi

Chapter 1   The Poultry Industry—An Overview

Domestication and Early Use of Chickens 1
Transformation of the American Poultry
    Industry 3
Domestication and Early Use of Turkeys 4
Development of American Poultry
    Production 5
Poultry Products: Basics for Good
    Nutrition 8
Health Issues and Poultry Products 9
Poultry Products to Feed the Hungry 10
World Poultry Distribution 12
U.S. Poultry Industry 13

Chapter 2   Poultry Biology

Evolution and Classification of Birds 22
Integument 23
Structural Systems 25
Circulatory System 26
Respiratory System .27
Digestive System 28
Excretory System 33
Reproductive System 33
Integration of Body Processes 39

Chapter 3   Incubation

Poultry Reproduction 46
Incubation 47
Brooding 57

Chapter 4   Poultry Genetics and Breeding

Genetics 58
Breeding of Chickens 65
Breeds and Breeding Turkeys 72

Chapter 5   Fundamentals of Poultry Nutrition

Nutrient Composition of Poultry and Eggs . . .75
Classification of Nutrients 76
Functions of Nutrients 76
Nutrients 78

Chapter 6   Feeds and Additives

Feeds for Poultry 97
Evaluating Poultry Feeds 112
Buying Feeds 114
Feed Substitutions 116
Feed Preparation 116
Presence of Substances Affecting Product
    Quality 119
Toxic Levels of Inorganic Elements for
    Poultry 119

Chapter 7   Poultry Feeding Standards–Diet Formulation and Feeding Programs

Factors Involved in Formulating Poultry
    Diets 122
Nutrient Requirement Determination 122
Feeding Standards 122
How to Balance Diets 123
Feeding Programs 126

Feeding Turkeys 129
Feeding Ducks 130
Feeding Geese 130
Feeding Ostriches 131
Feeding Pigeons .131
Feeding Game Birds 131

Chapter 8   Poultry Management

The Poultry Enterprise 133
Management 134
Improving Efficiency 135
Management of Poultry 136
Caponizing 136
Cannibalism 137
Troubleshooting 137
Computers 139
Communication System 139
Management of Light 139
Brooding 140
Save Energy, Save Costs 141
Control Pests 142
Rodents 142
Litter 144

Chapter 9   Animal Waste and Other Environmental Issues

Animal (Poultry) Waste 147

Chapter 10   Poultry Behavior, Stress, and Welfare

Animal Behavior 153
Stress 159
Welfare 159

Chapter 11   Poultry Health

Normal Mortality and Morbidity Losses 162
Monitoring Health 162
Causes of Avian Diseases 163
Pathogenic or Infectious Diseases 163
Parasites 172
Immunity and Other Defenses Against
    Disease 179
Infectious Disease Prevention and Control . .180
Metabolic Diseases 186
Environmental Stress 187
Behavioral Problems 188

Chapter 12   Food Safety

Government and Food Safety 191
Pathogens and Microorganisms and
    Food Safety 193
Food Safety Aspects of Eggs and Poultry
    Meat 193
Role of USDA's Food Safety and Inspection
    Service 196
Irradiation 197
Antibiotics 198

Chapter 13   Poultry Houses and Equipment

Location 199
Space Requirements of Buildings and
    Equipment 200
Other Poultry Facilities and Equipment 212
Specialized Buildings and Equipment for
    Specialized Purposes 213
Housing and Equipment for Layers 213
Brooder Houses and Equipment for
    Chicks 215
Housing and Equipment for Replacement
    (Started) Pullets 215
Houses and Equipment for Broilers 216
Houses and Equipment for Turkeys 216

Chapter 14   Eggs and Layers

The Composition of the Egg 218
Biology of the Production of the Egg 223
Commercial Production of Eggs–Layers 223
Management of Egg Production 234
Marketing Eggs 240

Chapter 15   Chicken Meat and Its Production (Broilers)

Marketing of Chicken Meat 242
Composition of Chicken Meat 245
The Biology of Meat Production 246

POULTRY SCIENCE

FOURTH EDITION

Colin G. Scanes
Iowa State University

George Brant
Iowa State University

M. E. Ensminger

PEARSON

Prentice
Hall

Upper Saddle River, New Jersey 07458

Library of Congress Cataloging-in-Publication Data
Scanes, C. G.
    Poultry science / Colin G. Scanes, George Brant, M.E. Ensminger.—4th ed.
      p. cm.
    Rev. ed. of: Poultry science / M. Eugene Ensminger. 3rd ed. 1992.
    Includes bibliographical references (p.).
    ISBN 0-13-113375-6
    1. Poultry. I. Brant, George. II. Ensminger, M. Eugene. III. Ensminger, M. Eugene.
Poultry science. IV. Title.

SF487.E59 2004
636.5—dc21
                                        2003049858

Publisher: Stephen Helba
Executive Editor: Debbie Yarnell
Associate Editor: Kimberly Yehle
Production Editor: Lori Dalberg, Carlisle Publishers Services
Production Liaison: Janice Stangel
Director of Manufacturing and Production: Bruce Johnson
Managing Editor: Mary Carnis
Manufacturing Buyer: Cathleen Petersen
Creative Director: Cheryl Asherman
Cover Design Coordinator: Miguel Ortiz
Cover Designer: Amy Rosen
Cover Illustration: Brooding chicks. ISU photo courtesy of Bob Elbert.
Marketing Manager: Jimmy Stephens
Formatting: Carlisle Communications, Ltd.

Pearson Education LTD.
Pearson Education Singapore, Pte. Ltd.
Pearson Education Canada, Ltd.
Pearson Education—Japan
Pearson Education Australia PTY. Limited
Pearson Education North Asia Ltd.
Pearson Educación de Mexico, S. A. de C.V.
Pearson Education Malaysia Pte. Ltd.

ISBN 0-13-113375-6

DEDICATION

To the giants of the previous generation who achieved so much and to the present and next generation of whom so much is expected.

Commercial Production of Broilers 247
Broiler Management 255
Broiler Breeders and Reproduction 257
Harvesting and Processing 260
Roasters 263
Capons 264

Chapter 16　Turkeys and Turkey Meat

Marketing of Turkey Meat 265
Composition of Turkey Meat 267
The Biology of Meat Production 267
Commercial Production of Turkeys 267
Breeding and Genetics 268
Nutrition and Feeding 272
Turkey Management 274
Processing Turkeys 278
Turkey Breeders (Turkey Poult
　　Production) 280

Chapter 17　Marketing and Grading

Poultry and Egg Consumption and Prices . . .288
Market Value of Poultry and Eggs 293
Factors Affecting the Price of Poultry
　　Products 293
Market Changes 294
Market Channels and Selling
　　Arrangements 294
Marketing Eggs 296
Grading and Sizing of Eggs 299
Marketing Poultry Meat 301
Current and Emerging Issues in Marketing
　　(Consumer Trends) 305
Trends Ahead in Marketing 307

Chapter 18　Poultry Industry and Business

Vertical Coordination of the Poultry
　　Industry 309
Vertical Integration 309
World Trade in Poultry and Eggs 311
International Competitiveness 313
Poultry Business Issues 315
Records and Databases 316
Allied Poultry Industries 317
Poultry Research 317
Careers for Poultry Science Graduates 318

Chapter 19　Other Poultry–Ducks, Geese, Pigeons, Ratites (Ostriches Plus Emus and Rheas), Game, and Ornamental Birds

Ducks and Geese 319
The Ratites–Ostriches, Emus, and Rheas . . .327
Pigeons 332
Game and Ornamental Birds 333

Appendices

Appendix I　Business Suggestions for
　　Small- and Moderate-Sized Poultry
　　Producers 337
Appendix II　Raising Ducks and Geese
　　in Free-Range Conditions 346
Appendix III　Raising Ostriches 354
Appendix IV　Raising Pigeons 355
Appendix V　Raising Game and
　　Ornamental Birds 356

Index 363

PREFACE

Just as poultry production throughout the world has undergone tremendous changes and growth, this book has been radically revised to reflect this. Building on the strengths of the series of books by M. E. Ensminger, we have concentrated in this extensive revision on accurately reflecting both the science on which poultry production is based and the fundamentals of the industry. This is an extensive revision and updating of the third edition of *Poultry Science* (published in 1992). Many new figures and photographs are included to illustrate formation and to reflect today's poultry production and the diversity of poultry. The book includes the various facets of poultry production ranging from broiler chickens, eggs, turkeys, ducks, geese, ostriches, and other ratites and game birds. It includes the industry (U.S. and world), poultry biology, development, genetics, production, processing, and marketing. New chapter/sections have been added on food safety, welfare, and environmental issues together with emerging issues including organic production. Increased attention is focused on the structure of the industry, the wide variety of poultry products on the market, new developments (including social and ethical issues), and broiler breeders.

The book is intended as an introduction to poultry science courses, particularly units in poultry science at colleges in North America and throughout the world. It should also provide a useful overview of the poultry industry for business and engineering and other graduates entering the poultry industry. To aid updating, reference is made to pertinent websites. It is hoped the volume will be useful to small- and medium-sized producers, for poultry and game birds, poultry farmers, and future farmers. The appendices specifically address issues of special interest to these types of producers. The volume will provide a useful resource for extension educators and instructors in general courses in animal sciences.

Although written in the United States, the book attempts to be useful to students, faculty, and the industry throughout North America together with South America, Europe, Asia, and Africa. Units are provided in both the U.S. and metric systems. Issues particularly appropriate to countries in the European Union are now included. Recognizing the role of small- and medium-sized producers and critical niche markets (e.g., organic and free range), pertinent issues are addressed in the text together with a series of appendices specifically directed toward the small- and medium-sized producer.

We recognize the invaluable role of our mentors throughout our careers and lives including in date order: the late John Phillips, Sir Brian Follett, Frank Cunningham, Arthur Chadwick, Mike Forbes, Paul Griminger, Hans Fisher, the late Ed Oleski, David Topel, John Kozak, Paul Siegel, and many others (for Colin Scanes) and J. C. Gilbreath, Rollin Thayer, Sergio Brambila, F. W. Hill, Emmitt Haynes, and Rex Thomas, among others (for George Brant).

We would like to thank the following reviewers for their valuable input: Kevin Downs—Middle Tennessee State University; Richard A. Barczewski—Delaware State University; Alan Sams—Texas A&M; and Jason Emmert—University of Arkansas.

We are indebted to our colleagues and particularly Susan Lamont and David Topel of Iowa State University's Department of Animal Science for their advice and wisdom. The tremendous support from our wives and families during the writing is gratefully acknowledged.

Colin G. Scanes and George Brant
Ames, Iowa, 2003

Colin G. Scanes is a professor of Animal Science at Iowa State University. He was educated in the UK with a B.S. from Hull University and Ph.D. from the University of Wales. He was formerly on the faculty at the University of Leeds and Rutgers—the State University of New Jersey where he was department chair. He has published extensively in poultry and animal science with 10 books and over 500 papers. He has received numerous awards, including election as a fellow of the Poultry Science Association and Honorary Professor at the Agricultural University of Ukraine.

George Brant is a Professor of Animal Science at Iowa State University. He received his undergraduate education at Oklahoma State University (graduating in 1963). His graduate education was at Oklahoma State University (Master of Science, 1965) and University of California–Davis (Doctor of Philosophy, 1971). He is a longtime member of the Poultry Science Association. Dr. Brant teaches extensively in animal science: basic concepts of animal science, domestic animal physiology, poultry science, companion animal science, companion animal systems management, agriculture travel course preparation, and agriculture travel course (AnS 496).

M. E. Ensminger (1908–1998) grew up on a Missouri farm. He was educated at the University of Missouri (B.S. and M.S.) and the University of Minnesota (Ph.D.). He was on the staff of the University of Massachusetts, then Minnesota, and finally Washington State University (serving as department head). He was president of the Agriservices Foundation.

Dr. Ensminger authored 22 books on animal, dairy, beef, and sheep science and nutrition (some are translated into several languages) and numerous articles. His international recognition stemmed from his service to the animal industry, and his book, together with his International Stockman's Schools and his International Ag-Tech Schools. He received numerous awards.

1

The Poultry Industry—
An Overview

Objectives

After studying this chapter, you should be able to:

1. Define the term *poultry*.
2. Describe the domestication of the chicken.
3. Describe the domestication of the turkey.
4. Discuss the development of the American poultry industry.
5. Understand the changes that have occurred in the poultry industry.
6. Discuss the numbers of broilers, layers, and turkeys raised in the United States.
7. List reasons why eggs and poultry meat are desirable in a diet.
8. List nonfood uses of eggs and poultry.
9. List the top states in egg, broiler, and turkey production.
10. Understand the importance of poultry in the United States.
11. Understand the importance of poultry in the world.

There has been tremendous growth of poultry and egg production in North America, Europe, South America, and Asia. Poultry meat and eggs meet consumer demand with their good eating properties; their nutritious, high-quality protein and low fat; and their low price (see Figures 1.1 and 1.2). World production of poultry and eggs is shown in Table 1.1. U.S. production of poultry meat and eggs is shown in Table 1.2. Poultry meat production in the major producing countries is shown in Table 1.3. The price in the United States (in nominal dollars) is the same today as 50 years ago. Average per capita consumption of chicken meat surpassed pork in 1985 and beef in 1992 in the United States.

Up until the 1920s, chicken meat was for special occasions. Candidates ran for office calling for "a chicken in every pot," a description of economic prosperity. For the etymology (origin of words) of poultry-related words, see Table 1.4. Chicken meat was a by-product of egg production by dual-purpose backyard flocks. The production of chicken meat resulted from roosters that were seasonally raised for meat and the "spring chicken," together with cull old hens.

The term *poultry* applies to domesticated birds that are normally slaughtered and prepared for market. Poultry include chickens, turkeys, ducks, geese, swans, pigeons, peafowl, guinea fowl, pheasants, quail, and other game birds, together with the ratites such as ostriches and emus. There are about 10,000 species of birds in the world. The study of birds is known as ornithology, while the study of poultry is poultry science or poultry biology.

DOMESTICATION AND EARLY USE OF CHICKENS

Our understanding of domestication has greatly expanded through a synergy between archaeology and the application of the techniques of molecular biology to existing populations of chickens. Chickens were likely domesticated for food (eggs and meat), for religious or ceremonial purposes, and for cockfighting.

Figure 1.1 Turkey confinement unit. *(Courtesy, Gretta Irwin, Iowa Turkey Federation)*

Figure 1.2 Poultry products.

TABLE 1.1 Poultry and Egg Production in the World in 2001

	Production (billion lb)	Production (million metric tons)
Chicken meat	132	59.8
Eggs	125	56.6
Turkey meat	11	5.1
Duck meat	6.4	2.9
Goose meat	4.6	2.1
Pigeon and other	0.04	0.02

Source: FAO.

TABLE 1.2 U.S. Poultry and Egg Production in 2001

	Production (million lb)	Production (thousand metric tons)
Chicken meat	31,333	14,210
Eggs	11,201	5,080
Turkey meat	5,479	2,485
Duck meat	116	52.6

Source: FAO.

TABLE 1.3 World Poultry Meat Production in 2000

Country	2000 Production (million tons)
1. United States	16.4
2. China (including Hong Kong)	11.5
3. Brazil	6.0
4. France	2.2
5. United Kingdom	1.5
6. Japan	1.2
7. Thailand	1.1
8. Italy	1.1
9. Canada	1.1
10. Spain	1.0
11. Argentina	0.9
12. Germany	0.7
13. Netherlands	0.7
14. Indonesia	0.7
15. Taiwan	0.7
16. India	0.7
17. Malaysia	0.7
18. Turkey	0.7
19. Colombia	0.7
20. Australia	0.6
Poland	0.6
Egypt	0.6
World	58.0

Source: USDA—National Agricultural Statistics Service.

The ancestral stock for the chicken is the red jungle fowl (*Gallus gallus*) from Southeast Asia. Other species of jungle fowl include the Java or green jungle fowl, Ceylon or Lafayette's jungle fowl, and the grey jungle fowl. Domestication of chickens is thought to have occurred in East Asia (most likely northeast China). This occurred about 7,500 years ago, based on radiocarbon dating of chicken bones at archaeological sites. Chickens then spread from northeast China to the Fertile Crescent and Europe (probably via the Silk Route) and to Africa. Colonists took chickens to the Americas. The first recorded chickens were on the island of Cuba in 1495. From their introduction until the advent of large-scale production techniques, chickens were raised essentially as scavengers in backyards and farmyards. There is still considerable potential for the continuation and expansion of this small-scale approach in many communities in developing countries. The ad-

TABLE 1.4   Word Origins Related to Poultry

The English language derives from a mixture of Anglo-Saxon (a Germanic language), Old English, and Middle French (a Romance language derived from Latin). Both the Germanic and Romance languages are members of the Indo-European group of languages that date back at least 5,000 years.

Chick (young chicken)—abbreviation of chicken

Chicken (a young fowl)—from Old English *cicen* and ancestral language that gave rise to all Germanic languages *Kiukinam*

Cock (adult male chicken)—from Old English *kok* and ancestral language that gave rise to all Germanic languages *Keuk* (possibly imitative of the sound the bird makes)

Cockerel (young cock)—diminutive form

Duck—from Old English *duce* and ancestral language that gave rise to all Germanic languages *ducan* (to dive)

Emu—from *ema* (Portuguese)

Fowl—from Old English *fugal* akin to Old High German *fogol* and ancestral language that gave rise to all Germanic languages *fogal*

Goose (plural geese)—from Old English *gos (ges)* and ancestral language that gave rise to all Germanic languages *gans* akin to *anser* in Latin and hence Indo-European root language

Hen (adult female chicken)—from Old English *henna* akin to Old High German *henna* and ancestral language that gave rise to all Germanic languages *henna*

Ostrich—from Middle English *ostrice* from Old French *ostrice*

Pigeon—from Middle English *pijon* from Middle French *pijon* from Late Latin *pipion* (a young bird) from Latin *pipire* (to chirp), which is imitative of the sound the bird makes

Poultry, Pullet, Poult—from Old French *poult* (female chicken) from Latin *pullus* (a small animal)

Quail—from Old French *caille*

Rooster (adult male chicken)—from American English (bird that roosts)

---

dition of egg and meat protein to protein-deficient diets, together with cash income, can make small-scale poultry production very appealing.

TRANSFORMATION OF THE AMERICAN POULTRY INDUSTRY

The American poultry industry had its humble beginnings when chickens were first brought to the continent by the early European settlers. Small home flocks were started at the time of the establishment of the first permanent settlement at Jamestown in 1607. For many years thereafter, chickens were tenderly cared for by the farmer's wife, who fed them on table scraps and the unaccounted-for grain from the crib.

As villages and towns were established and increased in size, the farm flocks were also increased. Surplus eggs and meat were sold or bartered for groceries and other supplies in the nearby towns. Eventually, grain production to the West, the development of transportation facilities, the use of refrigeration, and artificial incubation further stimulated poultry production in the latter part of the 1800s.

Until the late 1920s, chicken meat was viewed essentially as a by-product of egg production. The production of chicken meat transitioned from by-product to dedicated broiler chicken via dual-purpose breeds. Dedicated meat chickens or broilers were initially based on Barred Plymouth Rock and New Hampshire breeds, but later they also included Cornish and White Plymouth Rock backgrounds. Breeding for meat production started after the advent of broiler chickens. The resulting synthetic breeds needed to have white plumage because of processor and consumer preferences.

Since World War II, changes in poultry and egg production and processing have paced the whole field of agriculture. Practices in all phases of poultry production such as breeding and genetics, feeding, management, housing, marketing, and processing have become highly specialized. The net result is that more products have been made available to consumers at favorable prices, comparatively speaking, and per capita consumption and value of production have soared. Note that the combined value of production from broilers, eggs, turkeys, and chickens in 2000 was $21.2 billion. Of the combined total, 66% was from broilers, 20% from eggs, 13% from turkeys, and 1% from other chickens.

Significant events important in the development of the U.S. poultry industry are shown in Boxes 1.1, 1.2, and 1.3. Box 1.1 shows events contributing to increased consumption of poultry products. Box 1.2 shows important events resulting in improved nutrition for poultry. Box 1.3 relates events resulting in improved management practices associated with the care and production of poultry.

BOX 1.1   Events Important in the Development of Poultry Research and Consumption in the United States

Year	Event
1862	Morrill Agricultural Land Grant Act passed, leading to establishment of Colleges of Agriculture.
1862	Congress authorized creation of USDA.
1875	Connecticut established the first state agricultural experiment station.
1878	First commercial egg-drying operation established.
1887	Hatch Act passed, establishing agricultural experiment stations in all states.
1899	Frozen eggs first marketed.
1900	First egg-breaking operation established.
1906	First U.S. county agent appointed.
1914	Smith-Lever Agricultural Extension Act passed by Congress, establishing Cooperative Extension Service.
1918	First USDA federal-state grading programs for poultry established.
1918	U.S. Post Office shipped chicks by mail.
1926	USDA inaugurated Federal Poultry Inspection Service.
1928	USDA began inspection of poultry products for wholesomeness.
1934	USDA issued standards for eggs, with legal status.
1940	Mechanical poultry dressing initiated.
1946	Chicken-of-Tomorrow contest initiated by the Great Atlantic and Pacific Tea Company (A&P).
1956	Col. Harland Sanders began franchise operations of Kentucky Fried Chicken™.
1957	U.S. Congress authorized the Poultry Products Inspection Act.
1964	Buffalo Chicken Wings introduced.
1969	Col. Sanders' Kentucky Fried Chicken™ began integrating, acquiring a broiler producer and a processor.
1973	Egg McMuffin™ introduced.
1983	Chicken McNuggets™ introduced in the United States.
2003	Americans ate more broilers and turkeys than ever before because of (1) lower prices in relation to other meats, and (2) new and further processed products.

Source: Adapted from American Poultry History, 1823–1973, American Poultry Historical Society, Inc., Chapter 20, "Chronology."

DOMESTICATION AND EARLY USE OF TURKEYS

Turkeys (*Meleagris gallopavo*) were domesticated in the New World by the pre-Columbian civilizations of present-day Mexico and Central America at least 2,200 years ago and possibly as much as 4,500 years ago. The native turkey domesticated was of the subspecies *Meleagris gallopavo gallopavo*. It is possible that other populations of turkeys (also of different subspecies) were domesticated by the American Indians of the Southwest of what is now the United States. The Spanish colonists transported domesticated turkeys to Spain as documented in 1501. Later, these were distributed to the Mediterranean countries, to other countries in Europe, and by the colonists to what became the United States. The original turkeys for the Pilgrims' "thanksgiving" celebration in 1621 were wild turkeys. There is a misperception about that first thanksgiving. It was really a harvest feast with thanksgiving being a day of prayer and fasting!

There has been some confusion in the naming of the turkey. Domesticated turkeys were taken back to Spain by the Spanish Conquistadors. In Europe, the turkey was confused with the guinea fowl (called a turkey-cock). Another mistake was in thinking that the guinea fowl came from Turkey when, in fact, it comes from West Africa! Domesticated turkeys were reared on a small scale by farmers and fanciers particularly in Europe and North America.

Breeding programs to improve the turkey, preceding those of broiler chickens, were initiated in England in the 1920s. Improved turkeys were imported to the United States and crossed with North American wild turkeys to create the "Broad-Breasted Bronze." It is possible that the "Broad-Breasted Bronze" derived in part also from the wild turkey populations (of the subspecies *Meleagris gallopavo silvestris*) on the Eastern Seaboard of the United States. In the 1950s, breeders succeeded in developing turkeys with white plumage.

BOX 1.2   Nutritional Events Important in the Development of the Poultry Industry

Year	Event
1912	Casimar Funk proposed the "vitamine theory" to describe certain feed components essential for animal life.
1914	Vitamin A discovered.
1922	First soybean meal processed.
1929	Pelleting of poultry feed initiated.
1933	Riboflavin isolated, with milk established as a valuable source of it.
1935	Pyridoxine discovered.
1935	First sulfa drug announced.
1936	Vitamin E isolated.
1937	Vitamin A, which was first identified in 1914, synthesized.
1948	Crystalline B_{12} isolated.
1949	New nutritional developments discovered, including animal protein factor, vitamin B-12, antibiotics, hormones, and surfactants.
1974	Selenium was approved by the Food and Drug Administration for use at the rate of 0.1 parts per million (ppm) in complete feeds for growing chickens to 16 weeks of age, for breeder hens producing hatching eggs, and at the rate of 0.2 ppm in complete rations for turkeys.
1987	FDA increased the maximum allowance of selenium in complete feeds for chickens, turkeys, and ducks from 0.1 ppm to 0.3 ppm.
1994	NRC Nutrient Requirements of Poultry, Ninth Revised Edition published.

Source: Adapted from *American Poultry History, 1823–1973*, American Poultry Historical Society, Inc., Chapter 20, "Chronology."

DEVELOPMENT OF AMERICAN POULTRY PRODUCTION

As noted earlier, the history of the poultry industry is given in Boxes 1.1, 1.2, and 1.3. A discussion of some of the most important changes in the poultry industry follows:

1. Changes in breeding methods. Standard breeds of chickens decreased as modern breeding methods were applied. The poultry geneticist used family, as well as individual bird records, to develop high egg production. From this base, breeders created strains for high egg production and feed efficiency. When breeding for broiler production, hybrid vigor is obtained by systematic matings that may involve crossing different breeds, different strains of the same breed, or the crossing of inbred lines. Many of the strains used as sires trace their ancestry to the broad-breasted Cornish breed. The main objective is the improvement of broiler growth rate to 6 weeks of age, although improvement in other economic factors such as feed efficiency is also sought.

2. Changes in hatcheries. In the beginning, hatching was done according to nature's way; by a setting hen hovering over eggs. Then came the first incubator, patented in 1844. Hatcheries became larger in size and fewer in number. In 1934, 0.50 billion chicks were hatched in 11,000 hatcheries. In 2002,

there were 322 chicken hatcheries, including both broiler and egg type, each with an average capacity of 2.7 million eggs; and 65 turkey hatcheries, each with an average capacity of 736,000 eggs. In 2001, 9.0 billion broiler chicks, 452 million egg-type chicks, and 302 million turkey poults were hatched.

3. Changes in egg production. A hundred years ago, a hen produced about 100 eggs per year. In 2000, the U.S. average was 257 eggs per hen (see Figures 1.3 and 1.4 for comparison of the best). After rising continuously for many years, average egg production fell very slightly in 2001 to 256 eggs per hen. Eggs used to be sold largely on an ungraded basis. Today, eggs are candled for interior quality, weighed, placed into cartons, and sold according to size and quality. In modern egg-grading plants, efficient, power-operated weighing machines speed grading and move the eggs through the marketing system with dispatch (see Figure 1.5). Beginning in the early 1930s, three important changes took place in relation to egg production:

   a. The greater emphasis on commercial-size flocks allowed the light breeds and strains of chickens to gradually replace general or dual-purpose breeds for egg production.

   b. The technique for "sexing" chicks became perfected, and only the female chicks of egg-type breeds were sold by the hatcheries to layer operations.

BOX 1.3   Genetic Events Important in the Development of the Poultry Industry

Year	Event
1607	Small flocks were established in the Jamestown Colony.
1828	First Single-Comb White Leghorns imported into the United States.
1830	Breeding programs involving red chickens began around Narragansett Bay in Rhode Island.
1840	First census of poultry taken in the United States. Birds from China began to enter the United States.
1844	First recorded American incubator.
1850	Large-scale raising of ducks began on Long Island, New York.
1869	First trapnest patent granted.
1870	Toe punch introduced for identification in breeding programs.
1870	First commercial hatchery established in the United States.
1872	A commodity exchange for the trading of cash eggs was chartered.
1873	Pekin ducks first imported into the United States.
1874	First American Standard of Excellence issued, known as Standard of Perfection.
1874	Chicken wire first introduced.
1889	Artificial light first used to stimulate egg production.
1892	First long-distance express shipment of baby chicks (Stockton, New Jersey to Chicago, Illinois), marking the beginning of the commercial hatchery industry.
1895	Commercial feed industry began in Chicago.
1898	Battery fattening introduced by Swift and Company.
1901	Dry mash feeding first recommended to the public.
1903	Cornell gasoline brooder developed by Professor James Rice of Cornell University.
1904	First official egg-laying contest started in America.
1905	First chickens raised in batteries.
1905	First battery brooder developed by Charles Cyphers.
1909	Electric candler developed.
1911	First U.S. egg-laying contests established at Storrs, Connecticut and Mountain Grove, Missouri.
1913	First oil brooder marketed.
1923	Electrically heated incubator developed by Ira Petersime. USDA issued first tentative classes, standards, and grades of eggs.

    c. The feeding, breeding, management, and disease-control practices were improved so that more eggs were produced per layer, thereby requiring fewer layers to provide the eggs necessary to supply the market demands.

4. Changes in egg consumption. In recent years, per capita egg consumption in the United States has begun to increase after declining for many years.

5. Changes in chicken meat production. Prior to 1930, chicken meat was mainly the by-product of egg production. Birds, no longer producing eggs at a satisfactory rate, were sold for meat purposes, mainly in the fall of the year. Males raised with the pullets were disposed of as fryers or roasters at weights of 3 to 8 lb (1.4 to 3.6 kg).

    a. In 1934, 34 million broilers were produced in the United States.

    b. In 2000, 8.3 billion broilers were produced in the United States.

Modern broiler production is concentrated and highly commercialized. Very little land is required beyond the space necessary for a broiler house and a driveway. Chicks, feed, and other items used in production are purchased or are obtained from another division of an integrated operation of which the broiler production is a part. The operation is highly specialized and mechanized. Similar progress has been made in the processing of poultry. Poultry processing is a highly industrialized, large-scale operation using modern mechanical equipment and sanitary methods.

6. Changes in turkey production. As with chicken production, the initial production of turkeys was mostly a small sideline enterprise until 1910. At that time, 870,000 farmers raised 3.7 million turkeys, or an average of four turkeys per farm. In 2001, turkey production totaled 272 million birds and 7.15 billion pounds. The value of the turkeys produced was $4.44 billion.

BOX 1.3   Continued

Year	Event
1923	English investigator reported sex linkage and sex determination of chicks at hatching.
1923	Commercial broiler industry started on the Delmarva Peninsula near Ocean View in Sussex County, Delaware.
1928	Sex-linked crosses advertised.
1929	Layers kept in individual cages.
1931	All-night lights for layers used.
1932	Forced molting for commercial egg production started in Washington.
1935	Artificial insemination of poultry introduced.
1935	First successful incrossbred-hybrid chickens produced for egg production.
1940	Hy-Line Poultry Farms, Des Moines, Iowa, marketed its first Hy-Line layers, thereby applying inbred-hybrid corn breeding principle to egg production.
1940	USDA announced Beltsville Small White turkey.
1941	Newcastle disease reported in the United States.
1942	Beak trimmer developed.
1947	Antibiotics first used in treatment of poultry diseases.
1954	USDA discovered parthenogenesis (reproduction without male fertilization) in the fowl.
1956	Integration and contracting began in the egg business.
1962	Corporate egg farms became a viable part of the poultry industry.
1969	University of Delaware began growing broilers in wire cages.
1970	Economists estimated that vertical integration, a form of contract production, involved 9% of the nation's broilers, 85% of the turkeys, and 30% of the table eggs.
1971	Exotic Newcastle disease discovered in southern California requiring destruction of more than 3,000,000 birds.
1972	Marek's vaccine approved.
1981	For the first time in the United States, the cash receipts from marketing poultry exceeded the cash receipts from marketing hogs.
2002	Production and marketing contracts and vertical integration account for more than 90% of the production of eggs, broilers, and turkeys. The companies supplied the chicks/poults, feed, litter, fuel, and medication, and the contract growers provided housing and labor for the care of the birds.

Source: Adapted from American Poultry History, 1823–1973, American Poultry Historical Society, Inc., Chapter 20, "Chronology."

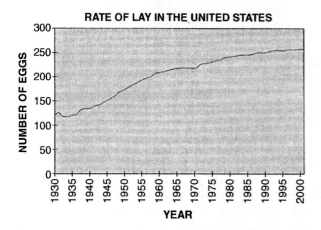

Figure 1.3   Changes in egg production per hen.

7. Changes in the number and size of poultry farms.

  a. In 1910, more than 5.5 million farms in the United States (88% of the 6.4 million farms in the nation) kept chickens. The average size flock in the United States numbered 50 laying hens.

  b. Today the average egg farm has about 175,000 hens.

  c. In 1959, the average number of broilers raised per farm was about 34,000 birds.

  d. Today it is almost 200,000.

  e. During this same period, turkey numbers rose sharply with bigger and bigger turkey operations.

8. Changes in ownership and organization. As poultry operations grew in size and efficiency, they vertically integrated (chiefly with feed companies and processors) to secure more credit, and contractual production became commonplace. Today, 95% of the broilers are produced under contract. Also, it is estimated that about 90% of the turkeys and eggs are produced under some kind of integrated or contract arrangement.

9. Changes in feed efficiency. There has been a marked lowering of feed required to produce a unit

Figure 1.4 "Superlayer" perched on the eggs that she laid every day for 448 days. *(Courtesy, University of Missouri)*

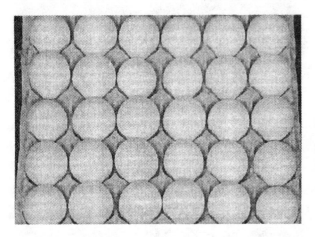

Figure 1.5 White eggs. *(Courtesy, Hy-Line International)*

of eggs, turkeys, and broilers since 1940. In 1940, it required 7.4 units (lb) of feed to produce a dozen eggs; in 2000, it took only 3.8 units of feed. In 1940, it required 4.7 units (lb or kg) of feed to produce 1 unit (lb or kg) of weight gain of broilers; in 2000, it took only 1.9 units of feed. In 1940, it required 4.5 units (lb or kg) of feed to produce 1 unit of weight gain of turkeys; in 2000, it took only 2.4 units of feed.

TABLE 1.5 Poultry Labor Requirements

| Type of Poultry | Work-Hours/Hundredweight Production[1] | |
	1940	2000
	(hours)	*(hours)*
Layers	1.7/100 eggs	≤0.2/100 eggs
Broilers	8.5	≤ 0.1
Turkeys	23.7	≤ 0.2

[1] *Based on "work-hours/hundredweight production."*
Source: USDA.

10. Changes in labor requirements. Poultry producers have achieved remarkable efficiency, primarily through increased confinement production and mechanization. In the late 1930s, it required 8.5 work-hours to produce 100 lb of broiler meat. By the 1990s, only 0.1 work-hour was required. In this same time, the labor requirements to produce 100 lb of turkey meat were lowered from 23.7 work-hours to 0.2 work-hour (Table 1.5). Much less progress has been made in reducing the labor requirements for the production of red meats, with the result that poultry producers have achieved a very real economic advantage.

11. Changes in processing. In 1947, New York dressed poultry (only the blood and feathers removed) accounted for 80% of the chickens and 75% of the turkeys marketed. Today, practically all chickens and turkeys are eviscerated then marketed chilled on ice or frozen. Not only that, much of the meat is cut up or further processed—changes designed to lessen labor in the kitchen and to add variety to menus. Also, there is a big trend toward ready-to-eat items and fast food.

POULTRY PRODUCTS: BASICS FOR GOOD NUTRITION

Most nutrition programs can be improved by the inclusion of eggs and/or poultry meat. Poultry products are of the highest nutritional quality and are marketed at prices that in developed countries even the poor can afford.

Eggs

Of all the foods available to people, the egg most nearly approaches a perfect balance of all the nutrients. This is evidenced by the fact that the egg is the total source of nutrition for the developing embryo. Unlike mammals, the chick embryo cannot secure needed nutrients from the reserves of the mother; rather, it lives in a closed system that must contain all the food needed for development. If a single nutrient were to be lacking, the developing embryo would die.

Eggs contain an abundance of proteins, vitamins, and minerals. The protein fraction of eggs is

TABLE 1.6 Examples of Essential Nutrients in Two Medium Eggs (0.24 lb or 108 g) without Shell

Nutrient	Function	Percentage of Daily Dietary Recommendations
		(%)
Protein	Required to build muscles and body tissues	16.3
Iron	Required for red blood cells and helps promote good health	17.3
Vitamin A	Required for normal vision and clear healthy skin together with embryonic and fetal development	22.0
Thiamin	Required as a coenzyme for some enzyme functions; promotes normal growth, good appetite, and a healthy nervous system	8.3
Riboflavin	Required as a coenzyme for some enzyme functions; promotes growth and good health	14.8
Vitamin B_{12}	Required for some enzyme functions; prevents pernicious anemia	1.0 mcg
Vitamin D	Required for calcium absorption and building bones and teeth	25.0

Source: Poultry and Egg National Board.

highly digestible and of high quality, having a biological value of 94 on a scale of 100—the highest rating of any food. The reason for this high-quality protein is that egg proteins are complete proteins; that is, they contain all the essential amino acids required to maintain life and promote growth and health. Additionally, eggs are a rich source of iron, phosphorus, trace minerals, vitamins A, E, and K, and all the B vitamins, including vitamin B_{12}. Eggs are second only to fish-liver oils as a natural source of vitamin D. Eggs are moderate from the standpoint of calorie content, with a medium-size egg containing about 77 calories. Table 1.6 illustrates the nutritive value of eggs.

Notwithstanding the high nutritive value of eggs, millions of Americans routinely eat eggs for breakfast each morning simply because they like them. Additionally, eggs are used in numerous baked and processed foods to enhance flavor and texture.

Per capita egg consumption by Americans declined steadily for many years after World War II. In 1945, the final year of World War II, the U.S. per capita average exceeded 421 eggs. Per capita egg consumption declined, reaching a low of 234 eggs in 1991. Since then it has steadily increased and was at 253 in 2003.

Poultry Meat

Poultry meat is supplied chiefly by chickens and turkeys, although other fowl contribute as well. Poultry meat is economical and quick and easy to prepare and serve. Also, it has a number of desirable nutritional properties.

Nutritionally, people eat poultry meat for its high content of high-quality protein and its low-fat content. Turkey and chicken meat is slightly higher in protein and slightly lower in fat than beef and other red meats (see Table 1.7). Additionally, the protein is a rich source of all the essential amino acids, as shown in

Table 1.8. The close resemblance of the amino acid content of poultry meat to the amino acid profiles of milk and eggs serves to emphasize the latter point.

HEALTH ISSUES AND POULTRY PRODUCTS

Americans are becoming more and more health and diet conscious.

Allergies

Occasionally, a child exhibits an allergic sensitivity to eggs. In most cases, the white of the egg is the portion that creates the reaction, and the yolk is generally readily tolerated. In highly sensitive individuals, the diet must be totally devoid of eggs; hence reading food labels is critically important. The presence of egg products in foods with inaccurate labeling has led to the foods being recalled by the USDA.

Cholesterol, Animal Fat, and Heart Disease

Coronary heart disease is a major cause of death, particularly in developed countries. It involves narrowing of blood vessels, including those leading to the heart due to deposition of cholesterol-containing plaque in blood vessels, which is called atherosclerosis. This is characterized by clogged blood vessels, which reduce the absorption of oxygen and nutrients in tissues. Evidence from epidemiology (study of diseases in populations) is that high serum cholesterol and low-density lipoprotein (LDL) cholesterol (the so-called bad cholesterol) are risk factors for coronary heart disease. Other risk factors include heredity, hypertension (high blood pressure), diabetes mellitus, smoking, and lack of exercise. Causes for high cholesterol and particularly high LDL cholesterol are heredity, lack of exercise, being overweight or obese, stress,

TABLE 1.7   Comparison of Composition of Cooked Turkey, Chicken, and Beef

Kind of Meat	Protein	Fat	Moisture	Food Energy Calories
	(%)	(%)	(%)	(per 3 1/2 oz or 100 g)[1]
Turkey (roasted and boned):				
Breast (white meat)	32.9	3.9	62.1	176
Leg (dark meat)	30.0	8.4	60.5	204
Chicken (roasted and boned):				
Breast (white meat)	32.4	5.0	61.3	182
Leg (dark meat)	29.2	6.5	62.7	185
Beef (cooked and boned):				
Round steak	31.3	6.4	61.2	183
Rump roast	29.6	7.1	62.0	190
Hamburger	27.4	11.3	60.0	219

[1]Standard portion size.
Source: Foods and Nutrition Encyclopedia, Ensminger and colleagues.

TABLE 1.8   Comparison of Amino Acid Composition of Animal Foods

Amino Acid	Percentage of the Food						
	Turkey	Chicken[1]	Eggs	Beef	Pork	Lamb	Milk
Arginine	1.31	1.47	0.86	1.12	1.09	1.17	0.10
Cystine	0.20	0.40	0.28	0.24	0.22	0.22	0.03
Histidine	0.60	0.44	0.34	0.49	0.54	0.46	0.09
Isoleucine	1.00	0.90	0.90	0.87	0.83	0.82	0.18
Leucine	1.53	1.45	1.09	1.43	1.28	1.25	0.33
Lysine	1.81	1.65	0.87	1.43	1.32	1.31	0.27
Methionine	0.52	0.40	0.42	0.39	0.42	0.39	0.08
Phenylalanine	0.74	0.88	0.69	0.68	0.70	0.66	0.16
Threonine	0.80	0.88	0.70	0.68	0.87	0.83	0.15
Tryptophan	0.18	0.18	0.24	0.19	0.24	0.22	0.05
Valine	1.02	1.47	1.05	0.97	0.85	0.85	0.23

[1]Average of values for whole chicken (white meat and dark meat).

Source: Adapted from Protein Resources and Technology: Status and Research Needs, *edited by M. Miner, N. S. Scrimshaw, and D. I. C. Wang. The Avi Publishing Company, Inc., Westport, CT, 1978.*

age, and diet (high dietary levels of either saturated fatty acids or trans fatty acids). Cholesterol can come from the diet or is produced by the liver.

The United States' National Institute of Health (National Heart, Lung, and Blood Institute) has developed a National Cholesterol Education Program: "Live healthier, live longer." Among the recommendations are reducing fat and saturated fatty acids and cholesterol from the diet. The American Heart Association recommends using egg substitutes and egg whites, and limiting intake of egg yolks because of both cholesterol and saturated fatty acids.

When evaluating the risk of heart disease posed by the consumption of animal products, we must weigh the benefits against the hazards. A well-planned diet, along with exercise and a minimum of stress, constitutes the best prevention against heart disease. The nutrients supplied by eggs and poultry meat provide well-balanced nutrition. For heart healthy diets the American Heart Association recommends boneless and skinless chicken breasts, turkey breast fabricated into chops and tenderloins, and ground turkey breast.

POULTRY PRODUCTS TO FEED THE HUNGRY

World population continues to increase (see Table 1.9). The United Nation's Food and Agriculture Organization provides a startling picture of the number of people in the world who are hungry. A numerical picture of the extent of undernourished people in the world is presented in Tables 1.10, 1.11, and 1.12.

Through the ages, eggs and poultry meat have been a basic food. In the future, they will become increasingly important in meeting the challenge for

TABLE 1.9   Population of the World

	Number of People (billion)	Number of People (billion)
	1990	*2001*
World	5.255	6.134
Developing countries	3.999	4.816
Countries in transition (e.g., former Soviet Union)	0.412	0.411
Industrialized/ developed countries	1.256	1.318

Source: FAO.

TABLE 1.10   Estimate of Number of Hungry People in the World in 1998–2000

	Number of Undernourished People (million)
World	840
Developing countries	799
Countries in transition (e.g., former Soviet Union)	30
Industrialized/developed countries	11
Together with ~2 billion people with micronutrient (e.g., vitamin) deficiencies	

Source: FAO.

TABLE 1.11   Estimate of the Increase in the Number of Hungry People in the World

Year	Number of Undernourished People (million)
1990–1992	822
1994–1996	828
1998–2000	840

Source: FAO.

TABLE 1.12   Estimate of Number of Hungry People by Region in the World

	Number of Undernourished People (million)	
	1990–1992	*1994–1996*
World	822	828
Asia	526	512
Africa and Near East	230	252
Latin America and Caribbean	64	63

Source: FAO.

TABLE 1.13   Feed Efficiency in Livestock

Species	Feed:Gain
Broiler chicken	1:1.9
Turkey	1:2.4
Pigs	1:3.15
Beef cattle	1:6–9*

*For feedlot cattle.

feeding the hungry people of the world. Nevertheless, the food versus feed argument will wax hot. Can we afford to feed grains to animals while denying them to starving people? There are many arguments for and against the practice of feeding animals to produce human food. However, poultry products have proven to be extremely valuable as a source of food for the following reasons:

1. **Poultry convert feed to food efficiently.** Table 1.13 lists the feed gain efficiency rating by animal species. Broilers have the most favorable protein efficiency ratio of any animal, and when compared to the mammalian livestock species grown for meat, they have the most favorable feed conversion ratio. Protein and feed efficiencies of turkeys and layers are also favorable when compared to the other types of livestock.

2. **The poultry industry is dynamic.** Because of the short periods required for growing and marketing, the poultry industry can adjust rapidly to a variety of economic factors (e.g., feed availability, numbers of birds on feed, costs, etc.). Other livestock enterprises, notably the cattle industry, necessitate relatively long periods to adjust to market changes because of length of time required for the animals to mature and reproduce.

3. **Poultry feeds are not commonly used for human consumption.** A large number of by-product feeds are fed to poultry; among them are blood meal, fish meal, and meat and bone meal. It may be argued that poultry do not compete to any appreciable extent with the hungry people of the world for food grains such as rice or wheat. Instead, they eat feed grains like corn/maize and grain sorghum, for which there is less demand for human consumption.

4. **Layers provide a continuous source of food.** Unlike meat animals that must be fed for a period of time before a usable product can be attained, layers produce eggs throughout the year. Thus, the layer can produce several times her weight in eggs throughout her life, while the products derived from meat animals are restricted in their final market weight.

5. **Vegetarians (but not vegans) consume eggs.** In some countries, the eating of meat is taboo by religious precept. But the consumption of eggs is acceptable and is a major source of animal protein consumed by the people of these countries.

6. **Poultry products are inexpensive.** Poultry meat and eggs are among the "best buys" in the supermarket.

WORLD POULTRY DISTRIBUTION

It is important that poultry producers have a global perspective concerning poultry. The production of poultry is worldwide. Figures 1.6 to 1.15, along with Tables 1.14 to 1.16, give pertinent details pertaining to the leading poultry-producing countries of the world.

1. The United States is the world's leading producer of chicken meat, with China second and Brazil third. Production of chicken meat has increased between 1990 and 2000 in most countries, including the United States. Phenomenal increases are seen for Argentina, the People's Republic of China, Brazil, and Mexico (see Figures 1.6 to 1.8).

2. The People's Republic of China is by far the world's leading producer of eggs, with the United States second and Japan third. Production of eggs has shown only a modest increase between 1990 and 2000 in many countries, including the United States and those of the European Union. Very large increases are seen for China, Mexico, and India (see Figures 1.9 to 1.11).

WORLD PRODUCTION OF CHICKEN MEAT (BILLION LB)

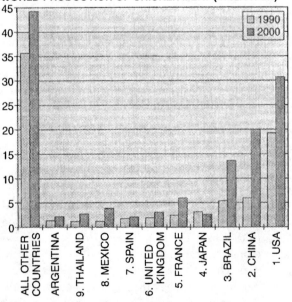

Figure 1.6 Changes in chicken meat production between 1990 and 2000.

WORLD PRODUCTION OF CHICKEN MEAT IN 1990 (%)

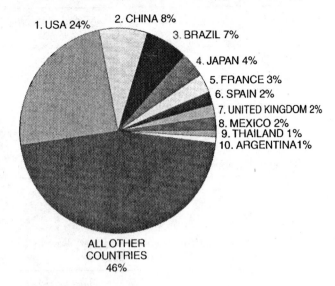

Figure 1.7 Leading countries for chicken meat production in 1990.

WORLD PRODUCTION OF CHICKEN MEAT IN 2000 (%)

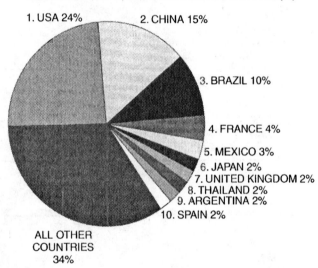

Figure 1.8 Leading countries for chicken meat production in 2000.

3. The United States is by far the world's leading producer of turkey meat, with France second and Italy third. Production of turkey meat has shown only a modest increase between 1990 and 2000 in the United States and Canada, but large increases are seen for Brazil, Germany, and France (see Figures 1.12 to 1.14).

4. The leading duck- and goose-producing country is China, which ranks first in duck and goose production by a wide margin (Table 1.17 on page 16 and Figure 1.15 on page 14). (Also see Chapter 19 for details).

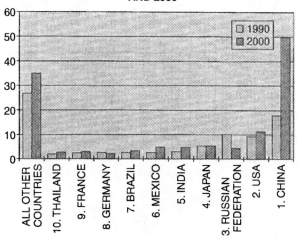

Figure 1.9 Changes in egg production between 1990 and 2000.

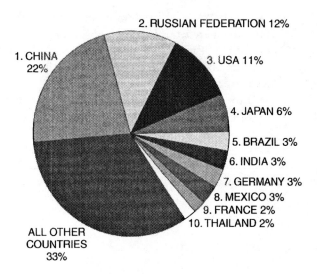

Figure 1.10 Leading countries for egg production in 1990.

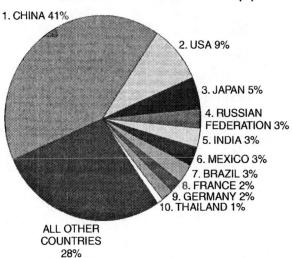

Figure 1.11 Leading countries for egg production in 2000.

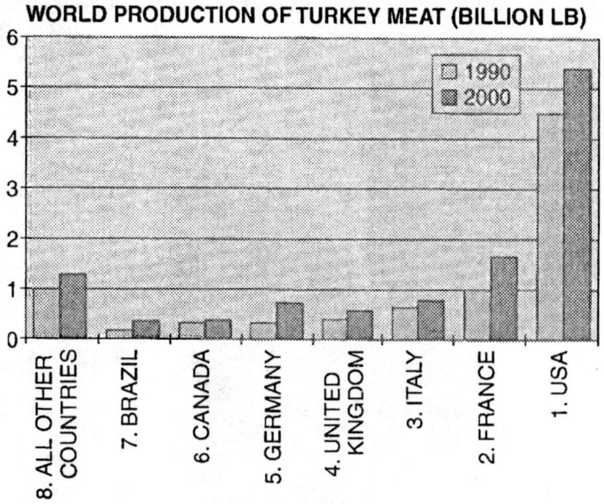

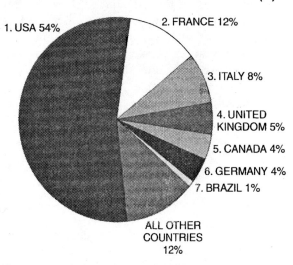

Figure 1.12 Changes in turkey production between 1990 and 2000.

WORLD PRODUCTION OF TURKEY MEAT IN 1990 (%)

Figure 1.13 Leading countries for turkey production in 1990.

U.S. POULTRY INDUSTRY

The U.S. poultry industry is concentrated geographically and in a relatively small number of companies.

Leading States

Tables 1.18, 1.19, and 1.20, together with figures 1.16, 1.17, and 1.18 on pages 16–17, show the 10 leading U.S. states in number of eggs, broilers, and turkeys produced. From these tables, the following deductions can be made:

- From an overall standpoint, the Southeast ranks high as a poultry area.
- Egg production is now predominantly focused in the Corn Belt states (see Table 1.19). The

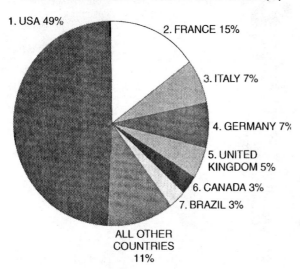

Figure 1.14 Leading countries for turkey production in 2000.

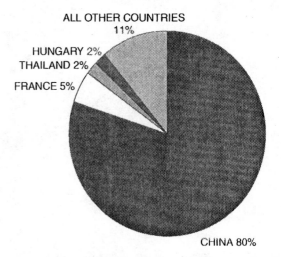

Figure 1.15 Leading countries for duck and goose meat production in 2001.

major reasons for companies to locate in these states are availability of the locally produced corn and soybeans and the elimination of transportation costs. In addition, there is a land base for manure disposal.

- Broiler production is focused in a relatively small number of states (see Table 1.18).

- Broiler chickens are predominantly produced in states in an arc from Pennsylvania and Delaware through the Eastern Seaboard and the Southern and Gulf States, together with Arkansas. This reflects the following:
  - Low costs of operation including land, labor, and costs for heating buildings

TABLE 1.14   World Production of Chicken Meat

Country	Production in Billion *lb* (million metric tons)	Production in Billion *lb* (million metric tons)	Percentage Increase
	1990	*2000*	
1. USA	19.1 (8.7)	30.6 (13.9)	+60.9%
2. China	6.0 (2.7)	19.8 (9.0)	+239%
3. Brazil	5.3 (2.4)	13.2 (6.0)	+154%
4. Mexico	1.5 (0.7)	4.0 (1.8)	+143%
5. United Kingdom	1.8 (0.8)	2.6 (1.2)	+54%
6. France	2.2 (1.0)	5.8 (1.2)	+16.5%
7. Japan	3.1 (1.4)	2.6 (1.2)	−14.1%
8. Thailand	1.1 (0.5)	2.4 (1.1)	+94%
9. Spain	1.8 (0.8)	2.0 (0.9)	+19.6%
10. Argentina	0.7 (0.3)	2.0 (0.9)	+418%
World	78.3 (35.5)	128.3 (58.2)	+64%

Source: FAO or calculated from FAO.

TABLE 1.15   World Production of Eggs

Country	Production in Billion lb (million metric tons)	Production in Billion lb (million metric tons)	Percentage Increase
1. China	18.1 (8.2)	50.3 (22.8)	+179%
2. USA	8.1 (4.0)	11.0 (5.0)	+23.9%
3. Japan	5.3 (2.4)	(2.5)	+4.5%
4. Russian Federation	10.1 (4.6)[+]	4.2 (1.9)	Not comparable
5. India	2.6 (1.2)	4.0 (1.8)	+53.6%
6. Mexico	2.2 (1.0)	4.0 (1.8)	+77.1%
7. Brazil	2.6 (1.2)	3.5 (1.6)	+23.6%
8. France	2.0 (0.9)	2.2 (1.0)	+172%
9. Germany	2.2 (1.0)	2.0 (0.9)	−9.5%
10. Thailand	1.5 (0.7)	1.8 (0.8)	+9.6%
World	82.9 (37.6)	122.2 (55.4)	+47.3%

[+]Former Soviet Union.
Source: FAO database or calculated from FAO.

TABLE 1.16   World Production of Turkey Meat

Country	Production in Billion lb (million metric tons)	Production in Billion lb (million metric tons)	Percentage Increase
1. USA	4.52 (2.05)	5.38 (2.44)	+19.2%
2. France	0.97 (0.44)	1.63 (0.74)	+68.1%
3. Italy	0.62 (0.28)	0.73 (0.33)	+18.2%
4. Germany	0.29 (0.13)	0.71 (0.30)	+131.8%
5. United Kingdom	0.37 (0.17)	0.55 (0.25)	+49.0%
6. Canada	0.29 (0.13)	0.33 (0.15)	+18.3%
7. Brazil	0.11 (0.05)	0.31 (0.14)	+158%
World	8.16 (3.70)	10.84 (4.92)	+32.7%

Source: FAO database or calculated from FAO.

- Knowledge base and human capital
- Infrastructure
- Minnesota and North Carolina have long been the top states for turkey production. Production is also found in other states (e.g., Arkansas) often where broiler chickens are produced, or in the Corn Belt (see Table 1.20).

Relative Importance of Different Species of Poultry

Tables 1.1 and 1.2 indicate the relative importance of different species of poultry in the world and the United States, respectively. Production of all poultry has increased.

TABLE 1.17   World Production of Duck and Goose Meat in 2001

Country	Production in Billion lb (million metric tons)	Production as a Percentage of World Production
China	8.6 billion (3.9 million)	77.7
France	0.53 billion (0.24 million)	4.8
Thailand	0.23 billion (0.10 million)	2.1
Hungary	0.23 billion (0.10 million)	2.1
World	10.77 billion (5.00 million)	

Source: FAO database or calculated from FAO.

TABLE 1.18   Broiler Production by U.S. States (2000)

State	Number Raised (billion)
1. Georgia	1.25
2. Arkansas	1.17
3. Alabama	1.01
4. Mississippi	0.77
5. North Carolina	0.71
6. Texas	0.57
7. Maryland	0.29
8. Virginia	0.27
9. Delaware	0.26
10. Pennsylvania	0.13

Source: USDA—National Agricultural Statistics Service.

TABLE 1.19   Egg Production by U.S. States (2000)

State	Number of Eggs (billion)
1. Ohio	8.2
2. Iowa	7.6
3. Pennsylvania	6.3
4. California	6.3
5. Indiana	6.1
6. Georgia	5.1
7. Texas	4.4
8. Arkansas	3.6
9. Minnesota	3.3
10. Nebraska	3.0

Source: USDA—National Agricultural Statistics Service.

Components of the Poultry Industry

The process of getting eggs and poultry meat from the birds to the consumer involves the coordinated efforts of many people, from production through marketing. Poultry production is highly integrated. It is the most integrated livestock industry in the

TABLE 1.20   Turkey Production by U.S. States (2000)

State	Number Raised (million)
1. Minnesota	44
2. North Carolina	43
3. Arkansas	27
4. Missouri	24
5. Virginia	24
6. California	19
7. Indiana	14
8. Pennsylvania	9.5
9. South Carolina	9.2
10. Iowa	7.1

Source: USDA—National Agricultural Statistics Service.

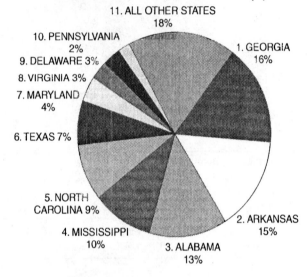

BROILER PRODUCTION BY U.S. IN 2000 (%)

- 11. ALL OTHER STATES 18%
- 10. PENNSYLVANIA 2%
- 9. DELAWARE 3%
- 8. VIRGINIA 3%
- 7. MARYLAND 4%
- 6. TEXAS 7%
- 5. NORTH CAROLINA 9%
- 4. MISSISSIPPI 10%
- 3. ALABAMA 13%
- 2. ARKANSAS 15%
- 1. GEORGIA 16%

Figure 1.16   Leading states for chicken meat (broiler) production in 2000.

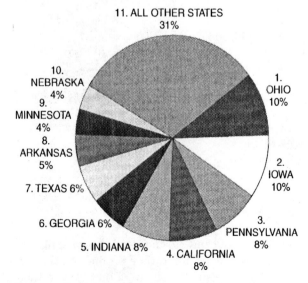

EGG PRODUCTION BY U.S. STATES IN 2000 (%)

- 11. ALL OTHER STATES 31%
- 10. NEBRASKA 4%
- 9. MINNESOTA 4%
- 8. ARKANSAS 5%
- 7. TEXAS 6%
- 6. GEORGIA 6%
- 5. INDIANA 8%
- 4. CALIFORNIA 8%
- 3. PENNSYLVANIA 8%
- 2. IOWA 10%
- 1. OHIO 10%

Figure 1.17   Leading states for egg production in 2000.

TURKEY PRODUCTION BY U.S. STATES IN 2000 (%)

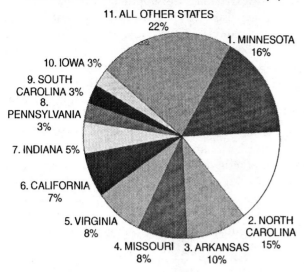

Figure 1.18 Leading states for turkey production in 2000.

Figure 1.19 Mature adult male turkey (tom turkey). *(Courtesy, USDA)*

United States and elsewhere. Firms are organized to control every level of production, from the laying of the egg or the production of the meat through the ultimate promotion and marketing of the finished product—be it an egg or a loaf of turkey pastrami. However, the chain of events in producing the poultry product and getting it to the consumer can be broken down into three primary divisions or stages: (1) producers, (2) processors, and (3) market specialists.

Producers

Because poultry operations are highly integrated, the average poultry production division is very specialized. In general, poultry production operations can be broken down as follows:

1. **Breeding flocks.** Breeding flocks provide the eggs necessary to supply industry demand for layers and young meat birds (for instance see Figure 1.19). Fertility and rate of lay are extremely important factors in evaluating breeding stock. The nutrient requirements for these birds are high and exacting.
2. **Hatchery.** Eggs produced by breeders are transported to hatcheries where they are incubated and hatched. The newly hatched birds are then shipped to grow-out operations.
3. **Grow-out operations.** When the newly hatched birds are to be used as replacement breeders, the management protocol becomes more involved, necessitating carefully planned feeding, vaccinating, and lighting programs. In cases where the newly hatched birds are from meat-

type strains, grow-out operations feed a high-quality ration to promote growth as quickly and as inexpensively as possible.
4. **Market egg-producing operations.** Eggs are produced by caged layer hens. The eggs collected in these operations are cleaned, graded, and packaged on the farm, enabling the marketed product to go directly from the farm to the consumer.
5. **Game bird, sport, and hobby operations.** In general, these types of operations are small and represent sidelines of the farm operator. The market for game birds raised for game reserves (e.g., quail, pheasant, chukar, etc.) has been increasing in recent years as more and more people have the money and time for recreation. There has always been a large number of poultry hobbyists who raise birds for a number of purposes such as show (see Figure 1.20), fighting (which is not legal in many places), or sport flying (e.g., pigeons).

Figure 1.20 Domestic geese. *(Photo by Nicholas Stein)*

Processors

The second division of the poultry industry involves the processing of the various poultry products. The branches of this division are:

1. **Egg-breaking industry.** Most of the market quality eggs are processed at the farm level. However, there is an increasing demand for shell-less eggs for the food industry. Eggs (e.g., small) are taken to egg-breaking plants where the broken eggs are marketed in frozen, liquid, or dry form to such industries as the baking industry and fast-food franchises.
2. **Processing industry.** The processing industry utilizes almost every part of a slaughtered bird. Inedible or condemned parts are processed and sold as fertilizer, livestock feed, and so on. Poultry meat is now being used in such processed meats as hot dogs, sausages, turkey bacon, cold cuts, Buffalo wings, and in fast-food restaurants (known in the industry as quick service), for example, McDonald's Mc-Nuggets™, and ready-to-eat meats.

Marketing Specialists

Once poultry products have been processed for marketing (see Figure 1.21), they must be packaged, distributed, and promoted in such a way as to achieve maximum consumer demand. Marketing of poultry products is a diverse field with each aspect requiring the services of a specialist in that particular area. The following market specialists are involved in the marketing of poultry products:

1. **Market forecasters.** These specialists accumulate a large volume of data from which to predict future trends in supply and demand.

Figure 1.21 Marketing eggs. *(Courtesy, Hy-Line International)*

2. **Research and development scientists.** These specialists (e.g., in universities and commercial companies) improve the quality and acceptability of poultry products and develop new products with consumer appeal.
3. **Advertisers.** These specialists study the wants and desires of consumers and tailor the marketing of poultry products to meet these needs.
4. **Public relations and education experts.** These specialists are playing an ever-increasing role in marketing. Their primary function is to promote the poultry industry as a whole rather than the individual products. Their main responsibility is to educate the public relative to the nutritional and economic merits of poultry products.

Per Capita Consumption of Poultry Products

Poultry meat and eggs are used chiefly for human food. In 2000, the U.S. per capita consumption of chicken and turkey on a retail basis totaled 82.3 and 17.7 lb, respectively. Additionally, in that same year 258 eggs were consumed per person.

Table 1.21 illustrates the trends of per capita consumption of poultry meat. As noted, the consumption of broilers and turkeys on a per capita basis has increased sharply while the consumption of certain other types of meat, notably beef, has declined. One trend, the steady decline in the per capita use of eggs, reversed in the early 1990s. Egg consumption increased from 236 in 1995 to 251 in 2003.

Food Safety

Poultry meat and eggs are a significant source of food-borne disease with disease-causing organisms found in both. It should be emphasized that food-borne disease is associated with other foods including beef, pork, seafood, vegetables, and fruits. Government (federal and state) agencies have strict regulations to reduce the incidence of food-borne diseases.

Food-borne disease occurs particularly where the food is uncooked or undercooked. For instance, raw eggs are used in many dishes. Incorrect handling of poultry meat can also lead to problems; for example, cutting up raw chicken on boards that are then used for the preparation of salads and other fresh or uncooked food. Food safety of poultry products is considered in detail in Chapter 12.

Poultry in Scientific Research

Poultry, especially chickens, have been used extensively as biological tools through which humans

TABLE 1.21   Meat Consumption (lb/capita/year) and Projections in the United States

	Year			
	1997	*2001*	*2004*	*2008*
Retail Weight				
Chicken	73.2	82.3	89.5	98.3
Turkey	17.6	17.5	16.6	16.1
Beef	66.9	61.7	61.3	58.2
Pork	48.7	52.7	53.0	54.3
Boneless Weight				
Chicken	51.8	58.3	63.4	69.6
Turkey	13.9	13.8	13.1	12.7
Beef	63.4	58.5	58.1	55.2
Pork	45.8	49.5	49.7	51.0

Source: USDA baseline projections.

have been able to learn more about themselves and other animals. Birds can be effectively used in many areas of biomedical research.

Nutrition Studies

Chicks are used as bioassays to demonstrate the need for many of the essential nutrients along with their functions. The high sensitivity of chicks (particularly their growth) to nutritional deficiencies proved of great value in vitamins, minerals, and amino acids, and in their more accurate estimation of requirements. Also, chicks have the advantage of being cheap and readily available, and large numbers of them can be hatched at the same time so as to provide the accuracy that goes with large numbers.

Much of the early work on vitamins involved chicks and/or eggs (e.g., biotin, pantothenic acid, riboflavin, thiamin, and vitamin K).

Embryonic Development

The chick embryo has been the model of choice for the study of embryonic development for over 100 years. There is a wealth of information available on its anatomical, physiological, biochemical, and endocrine development and the underlying molecular and cellular mechanisms. Some of the reasons for the use of the chick embryo by researchers include:

- The availability of a base of information to build upon and of reagents (probes and antisera).
- The fact that development occurs without influences by the mother, in contrast to the situation in mammals.
- The ease of experimental manipulations of the embryo *in ovo*.

- Fewer regulations for experimentation than is the case in studies with mammals.

Toxicology

Both wild birds and poultry may be exposed to toxicants in the environment. The adverse effects of early pesticides, such as DDT, on wild birds were described by Rachel Carson in her 1962 book *Silent Spring*. This book was one of the catalysts for the formation of the environmental movement. Chickens, particularly embryos, have been used as model systems for the effects of toxicants on birds—both wild birds and poultry. Researchers have found that the chicken is very sensitive to many toxicants, frequently much more sensitive than other birds.

The chicken embryo has been used extensively as a model for toxicant effects during development. It is particularly useful in examining teratogenic effects (teratogens are chemicals that cause abnormalities of development).

Genetics

Chickens are very good models for genetic studies. They have the following advantages:

- Short generation interval (~ 6 months)
- Availability of artificial insemination technology
- Defined closed genetic population with specific characteristics (e.g., single mutation as in the dwarf or obese chick, defined growth rate, or disease resistance)
- Availability of numerous breeds with marked genetic differences
- Populations (F_2 or greater) that contain great genetic differences due to crosses between

either domestic chickens and red jungle fowl or laying breeds or broiler breeds

Genomics

Tremendous progress has been made in genomics (the study of the DNA of animals, plants, and microorganisms). The full sequences have been reported for the human and mouse genomes together with those of the model plant, *Arabidopsis*, the fly, *Drosophila*, the nematode, *C. elegans*, and a number of microorganisms. Researchers in the United States and Europe have mapped the genome of the chicken, and the sequences or partial sequences of over 500 genes have been reported. The National Human Genome Research Institute (one of the institutes of the United States' National Institute of Health) announced plans in 2002 to fully sequence the chicken genome. Washington University in St. Louis is the lead institution for this program. The chicken was one of the six priority species; the others being the chimpanzee, various fungi, honeybee, sea urchin, and *Tetrahymena*. The rationale for choosing the chicken was its use as the model system for embryonic development.

The initiative to sequence the chicken genome will be a boon to poultry geneticists. It is likely to lead to even greater progress in improving the genetics of commercial chickens with, for instance, improved disease resistance (and hence reduced need for antibiotics), improved production traits (growth, feed efficiency, egg production), and improved product quality.

Transgenics

Transgenic animals, plants, and microorganisms have received an additional gene or genes (transgene or transgenes) from another individual (often of a different species) or a synthetic gene. Transgenic animals can have the transgene in tissues such as the liver where transmission to the next generation is not possible. This is a somatic cell transgenic. Alternatively, the transgene may be in the germ cells that develop into gametes. In these germ-line transgenics, the transgene is passed on to the next generation and can be duplicated into many animals. Another form of transgenic animal has the gene of interest deleted or inactivated. This is called a gene "knockout."

Transgenic chickens have been produced successfully. The strategy employed was to transform embryonic stem cells or many embryonic cells by placing transgene(s) into these cells. The placing of a gene is achieved by using a gene "gun" (shoot the gene attached to a particle into the cell) or using a virus to carry the gene into the cell. The resulting chick is a chimera, with cells that contain no, one, or multiple transgenes at various points in the genome. If the transgene is expressed (RNA is produced and then translated as protein) and if the transgene is passed to the next generation, we have a transgenic chicken. This is essentially a random process of gene incorporation. By selection based on determination of gene expression and use of relatively large numbers of chick embryos, transgenic chickens are being obtained.

Avian transgenic technology is being developed for pharmaceutical and agricultural applications.

Behavioral Studies

Without doubt, more behavioral studies have been conducted with birds than with any other class of animal. The pioneering work on imprinting (socialization) was done with goslings by the Austrian zoologist, Konrad Lorenz. Social order, known as "peck order," was first observed in chickens. The mating behavior in chickens and turkeys involving a chain reaction between the male and female is well known. Of all abnormal animal behaviors, cannibalism in chickens is most common. Also, the effect of prolactin (a hormone) on broodiness of poultry is of scientific and practical interest (also see Chapter 10).

Poultry in Medicine

The use of eggs for research in human and veterinary medicine and in the preparation of serums, vaccines, and pharmaceutical products is invaluable.

Bacteriological laboratories use large quantities of eggs in the preparation of culture media. The developing chick embryo is used for culturing viruses and others for the production of vaccines against a long list of human and animal diseases including encephalomyelitis (sleeping sickness), measles, smallpox, and so forth.

Poultry Products in Industry

Science and technology have teamed up to utilize poultry and eggs for many food and nonfood uses (see Box 1.4). Diluents containing egg yolk are widely used in artificial insemination.

BOX 1.4   Future of the Poultry Industry

Science and technology have been key to the tremendous expansion of the poultry industry. Added poultry production will be needed to take care of our consumer demand. Per capita consumption of poultry meat and eggs (see Table 1.21) will increase in developed and developing countries. It appears reasonable to expect that the following transitions, which are well under way, will continue in the future:

1. Production units will be larger and more commercial. In the future, it appears likely that layers, broilers, and turkeys will be produced in larger integrated companies.
2. Some specialization will evolve. Increases in specialty niche products such as organic and free-range eggs will continue.
3. Automation will increase. Higher priced labor, along with more sophisticated equipment, will make for increased mechanization along the line from production through processing and marketing.
4. Improved housing and environmental control will be realized. Engineering, behavioral sciences, and stress physiology will be combined in such a way as to bring about improved houses, brooders, incubators, and labor-saving equipment.
5. Bird density may decrease, with increasing concerns on welfare.
6. Growth rate and feed efficiency of broilers will increase. The application of molecular biology to breeding and feed ingredients will bring about greater efficiency of feed utilization.
7. Egg production per layer will increase. Some hens among the better strains now lay more than 300 eggs per year. There is increasing evidence that this trait (egg production) may be plateauing in high-producing strains. Breeders of egg-producing hens will now pay more attention to efficiency of production and product quality.
8. Livability will increase. Drugs and vaccines have helped increase livability, and they will continue to be used in the future. However, it is expected that breeding stock will continue to be selected for greater disease resistance and livability.
9. The issues of food safety (including irradiation of product), animal welfare, and removal of antibiotics from poultry feed will become even more critical.
10. More specialty meat and egg products will evolve. In particular, attention will be focused on new fast-food items, improved shelf life, and on convenience foods.
11. Improved processing will occur. Despite the strides already made in poultry processing, more improvements lie ahead. Deboning will be perfected, better methods of extending shelf life will be developed, and improved by-product utilization will be realized.
12. Marketing efficiency and exports from countries with competitive advantages (e.g., low-cost ingredients for feed) will increase. The better operations are now very efficient from the standpoint of marketing.

 USEFUL WEBSITES

Poultry and Egg Production in the World

http://apps.fao.org/page/collections?subset=agriculture

http://apps.fao.org/page/form?collection=Production.
Livestock.Stocks&Domain=Production&servlet=1&language=
EN&hostname=app

Poultry and Egg Production in the United States

http://www.usda.gov/nass/pubs/
reportname.htm#Eggs_Products

http://nhlbisupport.com/chd1/S2Tipsheets/eggs.htm

C H A P T E R

2

Poultry Biology

Objectives

After studying this chapter, you should be able to:

1. List the evolutionary ancestors of birds.
2. List the classification scheme for chickens and turkeys.
3. List the major parts and functions of the bird integument.
4. List the three major muscle types.
5. List and describe the major skeletal muscle cell types found in birds.
6. Describe the path of blood as it flows around the bird's body.
7. List the major blood cell types and give their function.
8. Describe the major differences in the respiratory systems of birds and mammals.
9. List the parts of the bird's digestive system and give the function of each part.
10. List eight major functions of the liver.
11. Describe the digestion and absorption of carbohydrate, fat, and protein.
12. Describe the absorption of vitamins and minerals.
13. List the parts of the large intestine and the functions of each part.
14. Describe the kidney and list the major nitrogen excretory compound.
15. Describe male reproduction.
16. Describe female reproduction.
17. List factors affecting egg size.
18. Identify parts of an egg.
19. Be able to list common egg abnormalities.
20. Define clutch size.

If poultry producers are to achieve maximum bird performance, they must have a basic understanding of the bird's structural makeup and of how the various systems of the body function (Figures 2.1 and 2.2). Through this knowledge, they can manipulate the body functions of the birds to attain maximum production. One example of how the knowledge of physiology has been adapted to poultry production is the manipulation of light to promote egg production. Additionally, the structure of the digestive tract is very different from that of the ruminant (e.g., sheep, cattle) or of the mammalian nonrumi-nant (e.g., horse, pig); hence, feeds and procedures must be designed accordingly.

EVOLUTION AND CLASSIFICATION OF BIRDS

From an evolutionary viewpoint, birds are warm-blooded, feathered, flying reptiles. Birds evolved from the reptilian group, the Archosaurs, during the Mesozoic era. Based on comparison of genes in existing birds and reptiles, together with the molecular clock, the common ancestor of birds and croc-

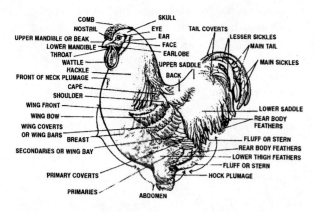

Figure 2.1 External anatomy of the male chicken. *(Original diagram from E. Ensminger)*

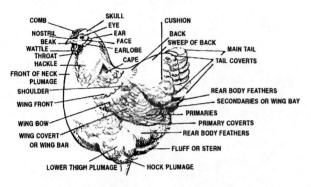

Figure 2.2 External anatomy of the female chicken. *(Original diagram from E. Ensminger)*

Figure 2.3 *Gallus gallus*—the ancestor of the domestic chicken. *(Original diagram from E. Ensminger)*

odiles (living archosaurs) originated about 254 million years ago. A fossil of a primitive bird, *Archaeopteryx*, was first discovered in 1860 in rocks of the late Jurassic period. It had wings, feathers, and also teeth. Based on its anatomy, it is thought that *Archaeopteryx* could fly. There are also fossils of earlier archosaurs with feathers. Hence, it seems that birds evolved from feathered bipedal (two-legged) reptiles probably early in the Jurassic period.

Many of the major groups of birds diverged from each other during the Cretaceous period, at the end of the Mesozoic era. Based on comparison of genes and the molecular clock, the ancestors of galliform birds (chickens, turkeys, etc.) diverged from the ratites about 92 million years ago and from the ancestors of ducks and other anseriform birds even earlier.

Birds (class *Aves*) are classified traditionally as either paleognathous (ancient-jawed) or neognathous (new-jawed). Paleognathous include many long extinct species (e.g., *Archaeopteryx*). Birds alive today are classified as neognathous in two superorders:

1. **Ratitae**—(the ratites—a group of flightless birds such as emu and ostrich)
2. **Carinatae**—(all the other living birds and their ancestors)

Classification

Domestic chickens are in the species *Gallus gallus* (see Figure 2.3) but were sometimes referred to as *Gallus domesticus*. Chickens and turkeys are classified as follows:

> Phylum: *Chordata*
> Subphylum: *Vertebrata*
> Class: *Aves*
> Superorder: *Carinatae*
> Order: *Galliformes*
> Species: *Gallus gallus* (chicken)
>   *Meleagris gallopavo* (turkey)

The classification of ducks, geese, pigeons, and ratites is considered in Chapter 19.

INTEGUMENT

In poultry, the skin and feathers collectively form the integument, that is, the outer protection of the body. They do the following:

- Protect the body from injury
- Help to maintain a relatively constant body temperature
- Aid in flight
- Act as receptors for sensory stimuli

Skin

The skin is rather thin and relatively free of secretory glands, with the exception of the uropygial or oil gland (preen gland) that is located on the upper part of the tail. Oil produced by this gland is collected in the beak and distributed on the feathers to act as water repellant. This oil is of particular importance for aquatic species.

In yellow-skinned chickens, the yellow color of the skin and shanks is due to several xanthophylls derived from the feed. These are deposited in the skin.

Several specialized structures consist of exposed areas of skin. These include the comb, wattles, snood (in turkeys), earlobes, beak, claws, and spurs (see Figures 2.1 and 2.2). The comb, wattles, and snood (in turkeys) are sensitive to the effects of sex hormones and consequently serve as indicators of secondary sex characteristics. Male hormones cause these appendages to become enlarged. The comb is the fleshy protuberance on top of the head. It is usually red and occurs in a number of shapes that are classified as (1) single comb, (2) rose comb, (3) pea comb, (4) cushion comb, (5) buttercup comb, (6) strawberry comb, or (7) V-shaped comb (see Figure 2.4 and color plate 17). The wattles,

which are usually red, are pendulous growths of flesh at either side of the base of the beak and upper throat. The snood in turkeys is a fleshy protuberance at the base of the upper beak. The earlobe is a fleshy patch of bare skin below each ear. It varies in color, depending on the breed. The beak, claws, and spurs are horny, keratinized structures. Additionally, the exposed parts of the legs and feet are covered with hard scales.

Feathers

Feathers are epidermal outgrowths that form the external covering or plumage of birds. At hatching, birds are covered by down feathers, which are soft, fine, fluffy, plumule-like feathers. These feathers are rapidly replaced by a coarser type of feather. Adult feathers can be classified into three types:

1. **Contour feathers.** These outermost feathers can be divided into four distinct parts: (a) quill, (b) shaft or rachis, (c) fluff or undercolor, and (d) web. The quill and shaft are continuous and hollow, tapering to a fine point at the distal end of the feather. The web is formed by barbs that contain small barbules. The barbules interlock with other barbs, thereupon forming a continuous, uniform series (see Figure 2.5). The fluff or undercolor is a series of barbs having no barbules. The absence of barbules causes this area of the feather to take on a scattered, downy appearance.

Figure 2.4 Heads of male chickens showing different comb types. (Note: For color photos of comb types see color plate 17) *(Original diagram from E. Ensminger)*

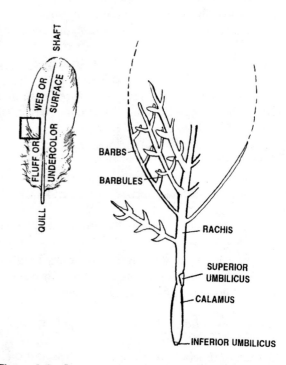

Figure 2.5 Structure of contour feathers. *(Original diagram from E. Ensminger)*

2. **Plumules.** These feathers form a soft downy undercoat. The rachis is short, and the barbs and barbules radiate freely.
3. **Filoplume.** These feathers have a short, flexible, hairlike rachis with barbs confined to the apex.

Feathers are distributed on the skin in well-defined tracts (see Figures 2.1 and 2.2). Through this ordered arrangement of tracts, flight is facilitated, body heat conserved, and abrasions reduced. In periods of cold weather, muscles attached to the feathers cause them to become erect in relation to the skin, thereupon creating a thicker and more efficient insulation.

STRUCTURAL SYSTEMS

The two physiological systems most involved with the structural integrity of the fowl are the skeletal system and the muscular system. Because flying involves a considerable expenditure of energy, it is essential that the structure of the fowl be designed so as to maximize the efficiency of energy utilization.

Skeletal System

Poultry are bipeds; that is, they stand on two legs. The skeletal system of the chicken is shown in Figure 2.6.

The basic skeletal arrangement is generally analogous to that of mammals. Yet, there are several differences. Birds possess a pair of bones in the shoulder area, called the coracoids. This pair of bones facilitates wing movement and offers additional support for the wing. Also, several morphological differences as compared to mammals are evident in the spine. The cervical vertebrae (neck bones) form an S-shaped column connecting the body to the head. When a flying bird lands, considerable pressure is exerted throughout the body, and this S-shaped conformation acts as a spring to minimize the impact on the head. Another difference from the mammalian spine is that the vertebrae along the trunk and body of the bird are fused together. This rigid conformation of the back provides considerable support for the wings.

The skeletal system is intimately connected to the respiratory system; many bones are pneumatic. Pneumatic bones are hollow and are connected to the respiratory system, thereby serving as a reservoir for air and reducing the weight of the bird for flight. The skull, humerus, keel, clavicle, and lumbar and sacral vertebrae are all part of this system. In fact, if one were to cut off the inflow of air through the trachea but open one of these bones, (e.g., the humerus), the bird would continue to breathe.

Egg laying places a great demand on the hen for calcium since eggshells consist primarily of calcium carbonate. To facilitate mobilization of cal-

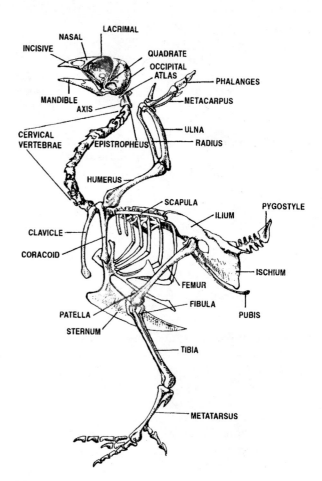

Figure 2.6 Skeletal system of the chicken. *(Original diagram from E. Ensminger)*

cium in the body for this type of production, birds have what is termed *medullary* bone. The marrow cavity of some bones is filled with interlacing spicules of bone. The spaces between the spicules are filled by red marrow and blood sinuses. In pullets, medullary bone involving the tibia, femur, pubic bone, sternum, ribs, toes, ulna, and scapula develops about 10 to 14 days prior to the laying of the first egg. At the onset of lay, the medullary bone enables the hen to mobilize calcium rapidly, so fast that if the hen is fed a diet very low in calcium she will lose 40% of the calcium in her skeleton after laying six eggs. Medullary bone is generally absent in males or nonlaying females. However, administration of estrogen, the female hormone, together with testosterone can initiate the formation of medullary bone.

Muscular System

Birds, like mammals, possess three distinct types of muscles: (1) smooth, (2) cardiac, and (3) skeletal. Smooth muscle is the type of muscle that lines many of the organs over which the bird has no voluntary control (e.g., gastrointestinal tract). Cardiac muscle is the type of muscle found in the heart. To a large degree, smooth muscle and cardiac muscle are self-regulating. That is, they need no outside (extrinsic) stimulus for the initiation of contraction. The third type of muscle, skeletal muscle, constitutes most of the muscle mass in the body. It is responsible for executing most voluntary movements.

The skeletal muscles of poultry contain five types of muscle fibers divided into two major categories based on the myofibril type:

- Fast twitch fiber
    I
    IIA
    IIB
- Slow twitch or tonic fibers
    IIIA
    IIIB

Muscles with tonic fibers predominate in white muscle. Fast twitch fibers predominate in red muscle, which also has more blood supply and hence more hemoglobin and also more myoglobin (a protein that stores oxygen) (also see Chapter 15).

Red fibers predominate in what is commonly called *dark* meat. These fibers contain large quantities of myoglobin, an iron-containing, oxygen-carrying compound very similar to hemoglobin. The white fibers, collectively forming white or pale meat, contain relatively little myoglobin.

Red fibers abound in muscles that are used continuously. They receive more blood and contain more fat and myoglobin than white fibers, thereby favoring the aerobic (with oxygen) production of energy that is conducive to prolonged activity. White fibers, however, are rich in glycogen, a sugar-rich compound that is readily broken down in anaerobic (without oxygen) conditions needed to sustain brief spurts of activity. Thus, one would expect the muscles used in flight by a good flying bird to contain more red fibers than those of a poor flier. This is, in fact, the case when the pigeon (a flying bird) is compared to the chicken (a poor flier). The pectoralis muscle (a breast muscle) of the pigeon contains about 40 times as much myoglobin as in the same muscle of the chicken.

CIRCULATORY SYSTEM

The circulatory system of birds functions much like that of mammals. The avian heart consists of four chambers: right atrium, right ventricle, left atrium, and left ventricle (see Figure 2.7). Incoming deoxygenated blood is received in the right atrium and is subsequently passed to the right ventricle.

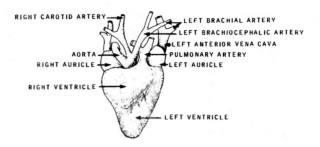

Figure 2.7 The heart of the chicken—anteroventral view. *(Original diagram from E. Ensminger)*

Contraction of the heart pushes the blood from the right ventricle to the lungs where oxygen is picked up by the blood and carbon dioxide removed. Freshly oxygenated blood travels from the lungs to the left atrium and left ventricle. Upon contraction of the left ventricle, blood is pushed through the arterial system where it eventually reaches the target cells, gives off its oxygen, and picks up waste products that will ultimately be excreted. The deoxygenated blood then returns to the heart through the venous system, and the process is repeated.

Also, when the demand for oxygen and nutrients is great, such as in flying, the heart rate increases dramatically from the rate sustained when the bird is at rest, on the order of a 5- to 14-fold increase. There is considerable variation in heart rates between species (see Table 2.1) with higher rates in smaller birds. In fact, heart rate is inversely proportional to the log of the body weight of the bird.

Poultry Blood

Poultry blood consists of a mixture of two components: plasma and cells.

1. **Plasma**—is a light yellowish fluid containing dissolved minerals, nutrients, and proteins.
2. **Blood cells**—include the following:
    - red blood cells (erythrocytes)
    - white blood cells (basophils, eosinophils, heterophils, monocytes, and lymphocytes)
    - thrombocytes

TABLE 2.1   Heart Rate in Birds

Type of Bird	Heart Rate Beats/Minute
Canary	1,000
Quail	500–600
Chicken	350–475
Turkey	200–275
Pigeon	220
Goose	200
Ostrich, young	80
Ostrich, adult	30–60

In contrast to mammals, avian erythrocytes and thrombocytes contain a nucleus.

Blood from poultry has a much higher viscosity than that from mammals. This may be due to the red blood cells (RBCs) and free hemoglobin. Avian RBCs are larger, have a nucleus, and cannot change shape as they pass through a capillary.

Poultry Species and Sex	RBC Concentration $\times 10^6$/ml	Leukocyte Concentration $\times 10^3$/ml	Thrombocyte Concentration $\times 10^3$/ml
Adult male chicken	3.6	16.6	27.6
Adult female chicken	2.8	31.3	42.8

The *life span of erythrocytes* of chickens averages from 28 to 35 days compared to about 120 days for humans. Blood volume remains fairly constant from day to day. A 5.5 lb (2.5 kg) chicken will generally have about 240 ml of blood (roughly the equivalent of 0.5 pt). Erythrocytes constitute 30 to 40% of the volume of blood, with males generally having a higher percentage of erythrocytes than females.

Poultry Species and Sex	% RBC/Erythrocyte Volume
Adult male chicken	43.5
Adult female chicken	28.6
Adult male turkey	38.5
Adult female turkey	37.2

In the 1-week-old chick, blood represents about 8.7% of the body weight. This percentage steadily decreases as the bird becomes older, declining to about 4.6% of body weight at sexual maturity.

RESPIRATORY SYSTEM

Because of the extremely heavy demands for energy in flight, birds have developed a respiratory system that permits the greatest exchange of oxygen per unit time of any animal; yet the lungs of birds are smaller than those of mammals in relation to body size. The anatomy and physiology of the avian respiratory system differ markedly from mammalian systems. The first difference is the role of the lungs. In mammals, the diaphragm muscle controls the expansion and contraction of the lungs. Birds have no diaphragm, and the lungs do not expand and contract upon inspiration and expiration, respectively. Rather, they act solely as organs in which gas exchange in the blood takes place.

Birds possess an extensive air sac system that receives air during inspiration. Most birds have nine

A

B

Figure 2.8 Respiratory system of the chicken showing (A) pathway of air during inspiration (breathing in) and (B) expiration (breathing out). *(Reprinted from Fedde, M. R.(1998). Relationship of structure and function of the avian respiratory system to disease susceptibility. Poultry Science 77, 1130–1138 with permission from Poultry Science)*

air sacs: the paired cervical sacs, median clavicular sac, and paired cranial thoracic, caudal thoracic, and abdominal sacs (see Figure 2.8). In a few birds there is one cervical sac, thereby decreasing the number of air sacs to eight.

When a bird inhales, inspiratory muscles increase the volume of the body cavity creating subatmospheric pressure in the air sacs. This draws fresh air through the lungs and into the air sacs. During exhalation, the expiratory muscles decrease the volume of the body cavity forcing air out of the air sacs, back through the lungs, and out of the body (see Figure 2.8).

DIGESTIVE SYSTEM

The digestive system of poultry differs considerably from that of nonruminant mammals. Figure 2.9 shows the gastrointestinal system of poultry. Existing birds have no teeth; hence, there is no chewing. The esophagus empties directly into the crop, where the feed is stored and soaked. From the crop, the feed passes to the proventriculus (or glandular "stomach"), the thick-walled organ immediately in front of the gizzard. Here it is stored temporarily while digestive juices are copiously secreted and mixed with it. From there, it passes to the gizzard, a very muscular organ, where with the aid of stones or grit swallowed by the bird, the feed is crushed and ground. Then the feed moves through the small intestine, the ceca, and colon to the cloaca.

Digestion in the fowl is rapid; it requires only about 3 hours in the broiler chick and laying hen for feed to pass from the mouth to the cloaca (this is called the gut transit time). During the night, the rate of passage slows with feed stored in the crop.

Process of Digestion

Digestion can be defined as the process whereby proteins, fats, and complex carbohydrates are broken down into units small enough to be absorbed. This process is accomplished primarily through the action of digestive enzymes.

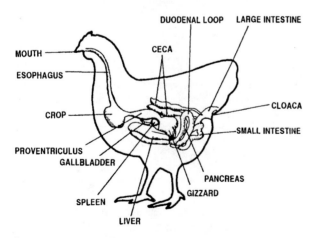

Figure 2.9 The digestive system of poultry. *(Original diagram from E. Ensminger)*

Enzymes are protein catalysts produced by cells within the body. They increase the rate of biochemical reactions at ordinary body temperatures without being used up in the process. Enzymatic activity is responsible for most of the chemical changes occurring in feeds as they move through the digestive tract. A summary of the enzymes involved in the digestive process of poultry is presented in Table 2.2.

Many of the digestive enzymes are stored in an inactive form. When they are in the inactive form, they are called zymogens or proenzymes. Once secreted into a favorable environment for digestion, generally governed by pH or other enzymes, these inactive enzymes "turn on" and perform their specific digestive functions.

Physiology of Digestion

The discussion of the physiology of digestion centers on the specific organs.

Oral Region

Two physical processes occur in the oral region of birds: prehension and initiation of deglutition.

Prehension can be defined as the act of bringing food into the mouth. *Deglutition* is the act of swallowing. The manner in which birds pick up food and swallow it depends largely on whether they have a soft palate. Birds lacking a soft palate (e.g., chicken, goose, duck, turkey) pick up feed with their beaks, mix the feed with saliva, and then raise their heads and extend their necks, thereupon allowing the feed to progress downward by gravity and negative pressure in the esophagus. Birds having a soft palate (e.g., pigeons) utilize this structure to help move the feed to the back of the mouth and force it down the esophagus. Thus, these birds can drink with their heads in a downward position.

The tongue of birds is a rigid, heavily cornified structure in contrast to the labile, soft tongue found in mammals. Taste buds are located at the base of the tongue but are few in number, a fact that explains the relative absence of taste sensitivity in birds (also see Chapter 10).

The development of the salivary glands in poultry depends on the eating habits of the particular species being studied. Birds that consume aquatic feed have poorly developed salivary glands. Conversely, birds that eat dry feed have, by necessity, well-developed salivary glands. In some species, such as the sparrow, saliva contains amylase, an enzyme that breaks down starch. However, most domesticated birds do not secrete amylase. The uses of saliva in digestion are manyfold, including the following:

TABLE 2.2   Digestive Processes in Birds

Region	Secretion (secreted by)	Enzyme	Enzyme Acts on, or Function	End Product of Digestion
Mouth	Saliva (salivary glands)	Amylase (ptyalin)	Starch, dextrins Lubricates the food	Dextrins Glucose
Crop	Mucus			
Proventriculus	Gastric juice and acids, chiefly HCl (walls of stomach) Mucus	Pepsin Lipase (in carnivores) Amylase	Protein Fat	Proteases, polypeptides, peptides Higher fatty acids and glycerol Coating of stomach lining and lubrication of food
Gizzard			Grinding	Ground foods Reduced particle size
Duodenum (small intestine)	Pancreatic juice (pancreas) Bile (liver)	Trypsin Chymotrypsin Amylase Lipase Carboxypeptidase Collagenase Cholesterol esterase	Proteins, proteases, and peptides Starch, dextrins Fats Peptides Collagen Cholesterol Fats	Peptones, peptides Amino acids Maltose, dextrins Fatty acids and monoglyceride Amino acids and peptides Peptides Cholesterol esterified with fatty acids Emulsion of fats
Small Intestine	Intestinal juice (secreted by intestinal wall)	Peptidase Sucrase (invertase) Maltase Lactase Polynucleotidase	Peptides Sucrose Maltose Lactose Nucleic acid	Amino acids and dipeptides Glucose and fructose Glucose Glucose and galactose Mononucleotides
Ceca		Microbial activity	Cellulose, polysaccharides, starches, sugars	Volatile fatty acids Microbial proteins B vitamins Vitamin K

1. **Lubricant.** These secretions act as aids in mastication, the formation of the bolus, and swallowing. Without this moisture, swallowing would be extremely difficult.
2. **Enzymatic activity.** The enzyme alpha-amylase (ptyalin) is found in the saliva of some birds, but not in chickens and turkeys. It serves to break α-1, 4 glucosidic linkages in starch and glycogen.
3. **Buffering capacity.** A large quantity of bicarbonate is secreted in saliva, thus serving as a buffer in the ingesta.
4. **Taste.** Saliva solubilizes a number of the chemicals in the feed. Once in solution the chemicals can be detected by the taste buds.
5. **Protection.** The membranes within the mouth must be kept moist in order to remain viable.

Saliva provides one means by which this is accomplished.

Oropharynx and Esophagus Regions

The oropharynx is the structure through which both air and feed pass. Unlike mammals, birds have no sharp demarcation where the mouth ends and the pharynx begins. However, when the neck is extended in the process of eating, there is a change in the position of the trachea that prevents the downward passage of food.

The esophagus is a muscular tube extending from the pharynx to the cardia of the stomach. The musculature and innervation of the esophagus are such that peristaltic waves move the bolus. Peristalsis is the coordinated contraction and relaxation of smooth muscles, creating a unidirectional movement that pushes the bolus through the digestive tract.

Crop

At the junction of the cervical segment and the thoracic segment of the esophagus in birds, there is a differentiated outpouching of the esophagus called the crop. If the bird has been starved, feed will bypass this structure and go directly to the proventriculus and gizzard. As feeding progresses, the crop begins to fill and acts as a storage organ. In the crop, there is limited digestion due to microbiological fermentation. Limited absorption of glucose and volatile fatty acids in the crop has been demonstrated in some birds. The size and shape of the crop is dependent upon the eating habits of the bird. Birds consuming large amounts of grain tend to have large bilobed crops while birds that primarily consume insects have rudimentary crops or, in some cases, no crop at all. Some birds, notably the pigeon and the dove, have the ability to produce a secretion in the crop called crop milk. Crop epithelial cells become filled with lipid under the influence of the hormone, prolactin, and are sloughed off. This cell-rich fluid is regurgitated and fed to their young.

Proventriculus and Gizzard

Gastric digestion in birds is carried out in two separate and distinct organs: the proventriculus and the gizzard.

The proventriculus is a small organ through which ingested feed passes rapidly. Its main function is that of gastric fluid secretion. The fluids secreted by the proventriculus are very similar to those in the stomach of the nonruminant, containing both pepsinogen and hydrochloric acid. Very little churning and mixing of feed occurs in this organ. The function of the gizzard is the mechanical action of mixing and grinding the feed. Since the bird has no teeth and swallows its feed whole, this muscular organ mixes in the ingesta during grinding. Grit, such as small pieces of granite, may be added to poultry rations to increase the digestibility of whole grains or grains with a minimal amount of processing. Grit stimulates motility in the gizzard as well as provides additional surface for grinding. When feed is provided in mash form, the benefits of grit are minimal. Modern commercial diets do not contain grit because they are finely ground.

Pancreas

The pancreas, an accessory organ of digestion, is a glandular structure that plays an essential role in the digestive physiology of poultry. The pancreas, being

TABLE 2.3　Composition of Pancreatic Secretions

Item	Functions
Proteolytic enzymes: Trypsinogen Chymotrypsinogen A Chymotrypsinogen B Procarboxypeptidase A Procarboxypeptidase B	Splits proteins into peptides and amino acids
Collagenase	Breakdown of collagen
Lipolytic enzymes: Pro-phospholipase A Pancreatic lipase	Breakdown of lipids
Cholesterol esterase	Esterification of cholesterol
Nucleolytic enzymes: Ribonuclease Deoxyribonuclease	Breakdown of nucleic acids
Amylase enzymes: Pancreatic amylase	Breakdown of starches
Cations: Sodium Potassium Calcium Magnesium	Buffers, cofactors, osmotic regulators
Anions: HCO^- Cl^- SO_4^- HPO_4^-	Buffers, osmotic regulators

both an endocrine and exocrine gland, serves two physiologically distinct functions. The endocrine function is that of the secretion of the hormones insulin, glucagon, somatostatin, and pancreatic polypeptide. The exocrine function deals with the production and secretion of fluids that are necessary for digestion within the small intestine.

The pancreatic duct leads from the pancreas to the small intestine. Many of the pancreatic enzymes are stored and secreted in an inactive form that becomes activated at the site of digestion. Trypsinogen is a proteolytic enzyme that is activated in the small intestine by enterokinase, an enzyme secreted from the intestinal mucosa. When activated, trypsinogen becomes trypsin. Trypsin, in turn, can then activate chymotrypsinogen to chymotrypsin.

The nucleases, lipases, and pancreatic amylase are secreted in their active form. Many of the enzymes require a specific environment before they will function. For example, amylase requires a pH of about 6.9 and the presence of inorganic ions before it will digest complex carbohydrates (see Table 2.3).

Liver

In addition to the pancreas and salivary glands, the liver is an indispensable accessory organ of the gastrointestinal tract. Closely associated with the liver are the gallbladder and bile duct.

From the gizzard and small intestine, most of the absorbed nutrients travel through the portal vein to the liver, the largest gland in the body. The liver not only plays an important part in nutrient metabolism and storage, but also forms bile, a fluid essential for lipid absorption in the small intestine. The numerous physiological functions of the liver follow:

1. Secretion of bile
2. Detoxification of harmful compounds
3. Metabolism of proteins, carbohydrates, and lipids:
    a. Regulation of blood glucose
    b. Synthesis of glucose from amino acid and other gluconeogenic precursors
    c. Fatty acid synthesis from glucose, amino acids, and volatile fatty acids (VFAs)
    d. Synthesis of cholesterol
4. Storage of vitamins
5. Storage of glucose (carbohydrate) as the polysaccharide, glycogen
6. Destruction of red blood cells
7. Formation of plasma proteins:
    a. Albumen
    b. Prothrombin and other clotting factors
    c. Carrier proteins for hormones, vitamins, and so on
8. Hormonal roles include:
    a. Production of hormones such as insulin-like growth factor-I (IGF-I) and hormone precursors (e.g., angiotensinogen)
    b. Activation of thyroxine (T_4) to triiodothyronine (T_3)
    c. Inactivation of hormones

The primary role of the liver in digestion and absorption is the production of bile. Bile facilitates the solubilization and absorption of dietary fats and also aids in the excretion of certain waste products such as cholesterol and by-products of hemoglobin degradation. The greenish color of bile is due to the end products of red blood cell destruction, biliverdin and bilirubin. Bile contains a number of salts resulting from the combination of sodium and potassium with bile acids. These salts combine with lipids in the small intestine to form micelles. Micelles are colloidal complexes of monoglycerides and insoluble fatty acids that have been emulsified and solubilized for absorption. When the micelle has been formed, the lipid can be digested and the resulting products (fatty acids and glycerol) can cross the mucosal barrier of the small intestine and enter the lymphatic system. Bile salts, however, do not travel with the lipid; rather, they are recycled into the enterohepatic circulation.

The volume of bile production is highly variable. A bird that has been starved produces little bile. Conversely, a bird fed a high-fat diet will produce substantial quantities in order to keep up with absorptive requirements. Generally, the volume of bile is dependent on (1) blood flow, (2) nutritive state of the bird, (3) type of ration being fed, and (4) the enterohepatic bile salt circulation.

In many animals, the gallbladder is the storage site for bile. Several species of birds do not have gallbladders, among them pigeons and doves.

Small Intestine

The small intestine is divided anatomically into three sections: duodenum, jejunum, and ileum. There is no histological distinction between the three sections. The first segment, the duodenum, originates at the distal end of the gizzard. Anatomists use Meckel's diverticulum as the separation point between jejunum and ileum. The length of the small intestine varies according to the eating habits of birds. Carnivorous birds (meat eaters) have substantially shorter intestines than herbivorous birds (plant eaters). This can be explained by the fact that meat products are more readily digested and absorbed than plant products.

Throughout the luminal surface of the small intestine lies an extensive network of fingerlike projections called villi (See Figures 2.10 and 2.11). Each villus contains a lymph vessel called a lacteal and a series of capillary vessels. On the surface of the villi are a great number of microvilli, which provide further surface area for absorption.

Three types of motility can be observed in the small intestine. The first type is called pendular motion. These waves do not advance down the intestine. Rather, they are merely a localized shortening and lengthening of the intestine that produces a mixing action. Segmentation contractions are the second type of intestinal motility. These intestinal movements are ringlike contractions at regular intervals that periodically relax, whereupon the area that had been previously relaxed contracts. This type of motility provides a means of mixing in addition to the pendular contractions. Peristalsis, a form of motility that has been previously discussed, is the third type of intestinal motility, providing a means for movement of chyme (intestinal contents) down the tract.

DIGESTION AND ABSORPTION IN THE SMALL INTESTINE. The small intestine is the primary organ of digestion and absorption. Specialized enzymes present in

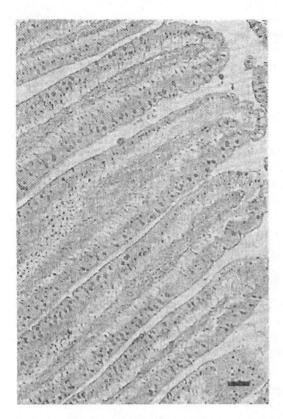

Figure 2.10 Light micrograph of cross section of the small intestine (duodenum) showing villi projecting into the lumen and providing a large surface area for the absorption of nutrients. *(Reprinted from Yaissle, J. E., Morishita, T. Y. and Lilburn, M. S. (1999). Effects of dietary protein on restrict-fed broiler pullets during a coccidial challenge.* Poultry Science *78, 1385–1390 with permission from* Poultry Science*)*

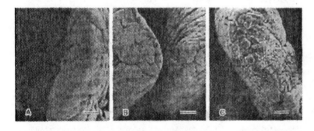

Figure 2.11 Scanning electronmicrograph of the inner surface of small intestine (duodenum) showing tips of the villi that project into the lumen and provide a large surface area for the absorption of nutrients. *(Reprinted from Shamoto, K. and Yamauchi, K. (2000). Recovery responses of chick intestinal villus morphology to different refeeding procedures.* Poultry Science *79, 717–723 with permission from* Poultry Science*)*

the various segments of this long organ provide fast, effective means of breaking down carbohydrates, lipids, and proteins for subsequent absorption.

Carbohydrates. Digestion and absorption of most carbohydrates occur in the small intestine. Here, such enzymes as sucrase and maltase split carbohydrates into monosaccharides, whereupon absorp-

tion takes place. The region of the greatest absorption of sugars is in the jejunum. Glucose and galactose are absorbed through an active transport mechanism. Sodium ion concentration within the intestinal contents has been shown to be critical in this mechanism. A high Na^+ concentration facilitates rapid absorption of these sugars while a low Na^+ concentration reduces the rate of absorption. Some pentoses and hexoses are absorbed through diffusion, a process considerably slower than that of active transport.

Lipids. Lipids are digested and absorbed primarily in the upper part of the small intestine, but considerable absorption can take place as far down as the ileum. When lipids, emulsified by bile salts, come into contact with the various lipases that are found in the duodenum, they are broken down into monoglycerides and fatty acids. Short-chain fatty acids are then absorbed directly into the mucosa of the small intestine and are transported to the portal circulation. Monoglycerides and insoluble fatty acids are emulsified by bile salts, forming micelles. By attaching to the surface of epithelial cells, the micelles enable these components to be absorbed into the mucosal cells. Once inside these cells, the long-chain fatty acids are reesterified to form triglycerides. Tryglycerides then combine with cholesterol, lipoproteins, and phospholipids to form portomicrons, minute fat droplets. These are then transported by the portal blood.

Proteins. Although protein digestion is initiated in the proventriculus and gizzard, most digestion and absorption occur in the small intestine. Numerous pancreatic and intestinal enzymes split proteins into their constituent amino acids, which are subsequently absorbed.

Amino acid absorption is not clearly understood; but an active transport mechanism involving Na^+, similar to that of glucose absorption, is implicated. Amino acids are rapidly absorbed in the duodenal and jejunal segments but are poorly absorbed in the ileum.

Minerals and Vitamins. Mineral absorption occurs throughout the small and large intestines, with the rate of absorption depending on a number of factors such as pH, carriers, and so forth. Numerous mechanisms of mineral absorption have been elucidated. Many minerals, for example, iron and sodium, require active transport systems. Others, such as calcium, utilize both carrier proteins and diffusion mechanisms.

Most of the vitamins are absorbed in the upper portion of the intestine, with the exception of vitamin B_{12}, which is absorbed in the lower intestine. Water-soluble vitamins are rapidly absorbed, but the absorption of fat-soluble vitamins relies heavily on the fat absorption mechanisms, which are generally slow.

Ceca, Colon, Large Intestine, and Cloaca

The large intestine consists of the colon and ceca. The cecum (pl. ceca) is a blind-ended tube found at the junction of the small and large intestines. In grain-eating birds, there are two large ceca, while in some other types of birds there may be only one rudimentary pouch, or none at all.

The large intestine, or colon, is extremely short in birds and is very similar in structure to the small intestine. The large intestine in birds plays a much more important role in digestion and absorption than once thought. In birds, all waste products, both urinary and fecal, empty into a structure called the cloaca, which leads to the vent. Thus, urinary and fecal waste products are mixed together.

EXCRETORY SYSTEM

The kidneys in birds are rather large and elongated and are situated along the fused backbone. Each kidney consists of three lobes (cranial lobe, middle lobe, and caudal lobe). These lobes empty into a ureter leading to the urodeum of the cloaca from which urine is excreted (see Figure 2.12). The primary functions of the kidney are twofold: (1) to filter the blood so as to remove water and waste products

therefrom, and (2) to reabsorb any nutrients (e.g., glucose or electrolytes) that might be recycled for additional use. Because the kidneys control the absorption and excretion of water and electrolytes, they are the primary control center for maintaining the proper osmotic balance of body fluids.

The urine of birds is cream colored. Much of it consists of a thick, pasty mucoid material that contains uric acid. Unlike mammals, whose urine contains urea primarily, birds excrete uric acid as the primary nitrogen metabolite. Uric acid is synthesized in the liver and is excreted via the urine. It comprises 60 to 80% of the total urinary nitrogen. Birds can, however, produce urea to a limited extent as an end product of purine metabolism and the catabolism of arginine; but the level of urea found in the urine is insignificant as compared to the uric acid content.

Because urine is transported to the cloaca, feces and urine are excreted from the body together. In fact, some urine may pass into the colon where additional water can be reabsorbed.

REPRODUCTIVE SYSTEM

The reproductive physiology of poultry is markedly different from that of mammals. The most obvious differences are that the egg is large and yolk filled, is fertilized in the infundibulum, is supplied with additional nutrients as egg white, is surrounded by a shell, and is expelled from the body in birds. In contrast, the very small, fertilized egg remains *in utero* receiving nutrients from the mother until birth in most mammals.

Male Reproductive System

The male reproductive system of birds is extremely simple, consisting of two very large testes, each having an epididymis and vas deferens that lead to the copulatory organ (see Figure 2.12; also see color plate 15). In male birds, the testes are located along the backbone within the abdominal cavity. In mammals, except for the elephant and marine mammals, the testes are located in a sac called the scrotum that hangs from the body. This sac keeps the testes cooler than the interior of the abdominal cavity because spermatogenesis cannot take place efficiently in mammals at normal body temperature. However, birds have evolved a mechanism that permits their intra-abdominal testes to be fully functional.

The process of removing the testes from young males to promote meat quality for market birds is called caponizing, and castrated birds are called capons. Because incisions must be made in the abdominal wall to remove the testes, the operation is more dangerous than the castration of mammals where the testes hang away from the body and can be easily removed. Because broilers are now marketed

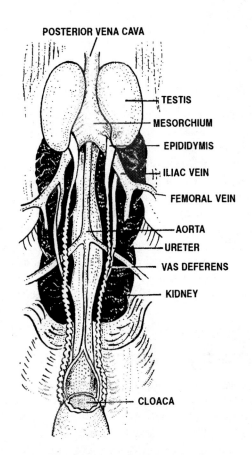

POSTERIOR VENA CAVA

— TESTIS

— MESORCHIUM

— EPIDIDYMIS

— ILIAC VEIN

— FEMORAL VEIN

— AORTA

— URETER

— VAS DEFERENS

— KIDNEY

— CLOACA

Figure 2.12 The male reproductive and urinary systems. (Note: For color photos of reproductive system see color plate 15) *(Original diagram from E. Ensminger)*

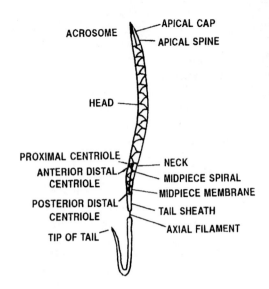

Figure 2.13 Schematic diagram showing structure of avian sperm. *(Original diagram from E. Ensminger)*

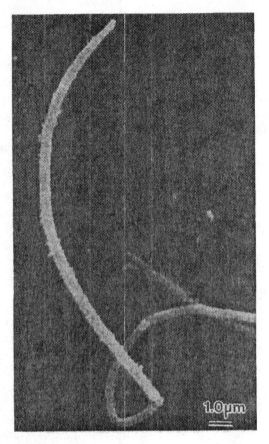

Figure 2.15 Scanning electromicrograph of quail spermatozoon. *(Reprinted from Korn, N., Thurston, R. J., Pooser, B. P. and Scott, T. R. (2000). Ultrastructure of spermatozoa from Japanese quail.* Poultry Science *79, 407–414 with permission from* Poultry Science)

Figure 2.14 Light micrograph of turkey spermatozoa (also see color plate 6). *(Reprinted from Korn, N., Thurston, R. J., Pooser, B. P. and Scott, T.R. (2000). Ultrastructure of spermatozoa from Japanese quail.* Poultry Science *79, 407–414 with permission from* Poultry Science)

at about 45 days of age, well before they reach sexual maturity, caponizing is rarely performed.

The copulatory apparatus for the turkey and chicken consists of two papillae and a rudimentary copulatory organ that is located at the vent. In ducks and geese, this organ is fairly well developed and is erectile in nature.

The gross appearance of avian sperm is somewhat different from mammalian sperm. It has a long cylindrical head with a pointed acrosome, a short midpiece, and a long tail (see Figures 2.13 to

2.16 and color plate 6). Because the fowl has no seminal vesicles or prostate gland, the volume of seminal fluid is very low. The seminal fluid for avian sperm contains no fructose, citrate, inositol, phosphorylcholine, or glyceryl phosphorylcholine compounds commonly found in mammalian semen. Chickens normally produce about 1 ml of whitish semen per ejaculate, and turkeys produce about 0.5 ml of yellowish or brownish semen per ejaculate. Although turkeys produce only about one-half the volume of semen as chickens, the sperm concentration is about twice that of chickens. Thus, chickens yield roughly the same number of sperm per ejaculate as turkeys, about 1.75 to 3 billion.

Female Reproductive System

Females of most animals have two functional ovaries. The hen has only *one functional* ovary, the left one, which is situated in the body cavity near the backbone (Figure 2.17; also see color plate 16). At the time of hatching, the female chick's left ovary contains up to approximately 3,600 to 4,000 tiny ova from which full-sized yolks may develop when the hen matures.

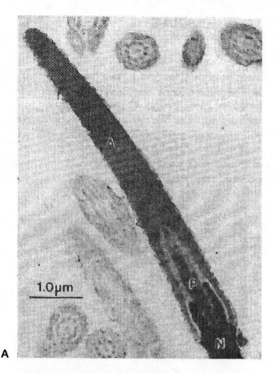

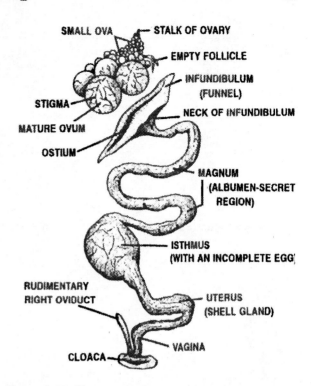

Figure 2.17 The reproductive tract of the female chicken. (Note: For color photos of reproductive system see color plate 16) *(Original diagram from E. Ensminger)*

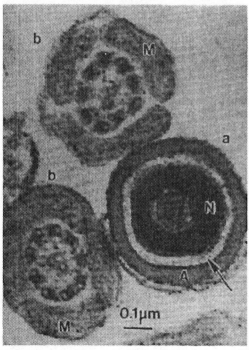

Figure 2.16 Transmission electron micrograph of spermatozoon from Japanese quail.

A is a longitudinal section of the head of the spermatozoan showing the nucleus (N) and acrosome (A).

B is a series of cross sections of the head (a) showing the acrosome (A) surrounding the nucleus (N) and mid-piece (b) showing the mitachondria (M) surrounding the microtubules with the typical 9+2 arrangement of the spermatozoan. *(Reprinted from Korn, N., Thurston, R. J., Pooser, B. P. and Scott, T. R. (2000). Ultrastructure of spermatozoa from Japanese quail. Poultry Science 79, 86–93 with permission from Poultry Science)*

Each follicle consists of the yolk-filled ovum enclosed in a thin-walled sac that is attached to the ovary by a stalk. This sac contains the vast network of blood vessels that supply the yolk precursors. When a pullet reaches sexual maturity, or comes into egg production, some of the ova develop to mature yolks. When mature, the yolk is released from the follicle by rupture of the follicle wall along a line called the stigma. Soon after its release, the yolk is picked up, or engulfed, by the funnel of the oviduct.

The oviduct is a coiled, folded tube about 25 in. (~65 cm) long, occupying a large part of the left side of the abdominal cavity. It is divided into five rather clearly defined regions, each of which plays a specific role in the completion of the whole egg. A normal hen requires about 24 hours to complete an egg. Within 30 minutes after the egg is laid, another ovum/yolk is released from the ovary for laying the following day. The functions of each of the five parts of the oviduct are described in Table 2.4.

The Egg

The bird egg is a marvel of nature. It is one of the most complete foods known, as evidenced by the excellent balance of proteins, fats, carbohydrates, minerals, and vitamins that it provides during that 21-day in-the-shell period when it serves as the developing chick's only source of food. It is the reproductive cell (ovum) of the hen. Upon fertilization by the sperm, the egg will develop into a chick when

TABLE 2.4　Functions of the Oviduct

Part	Approximate Time Egg Spends in Section	Functions
Infundibulum (funnel)	15 minutes	Picks up yolk from the body cavity after it is released from the follicle. If live sperm are present, fertilization occurs in this section.
Magnum (albumen-secreting region)	3 hours	Thick white (albumen) is deposited around the yolk. This layer later forms the chalaziferous layer—the chalaza and inner thin and thick white.
Isthmus	1 hour 15 minutes	Inner and outer shell membranes are added and some water and mineral salts. These membranes give some protection to the egg contents from outside contamination.
Uterus (shell gland)	20 hours 45 minutes	During the first part of the egg's stay in the shell gland, water and minerals pass through the white, inflating the egg and giving rise to the outer layer of white. Soon after the egg is inflated, the shell gland starts to add calcium over the shell membranes, continuing this process until just prior to laying. If the shell is going to be colored, pigment is added in this section.
Vagina	15 minutes	The egg passes into this section just prior to laying. The egg is rotated end for end 1–2 minutes before laying.
	Time from ovulation to laying is about 25½ hours.	

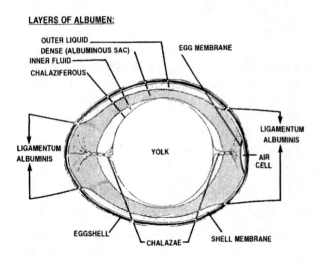

LAYERS OF ALBUMEN:

OUTER LIQUID
DENSE (ALBUMINOUS SAC)
INNER FLUID
CHALAZIFEROUS
EGG MEMBRANE
LIGAMENTUM ALBUMINIS
LIGAMENTUM ALBUMINIS
YOLK
AIR CELL
EGGSHELL
CHALAZAE
SHELL MEMBRANE

Figure 2.18　Structure of the egg. *(Original diagram from E. Ensminger)*

TABLE 2.5　Egg Production of Poultry

Species	Age of Sexual Maturity	Eggs/Year	Egg Size	
	(mo)	(no.)	(lb)	(g)
Chicken:				
Light-type	5–6	240	0.13	58
Broiler-type		170		
Turkey	7	105	0.19	85
Goose	24	15–60	0.47	215
Duck (Pekin)	7–8	110–175	0.18	80
Pheasant	8–10	40–60	0.07	32
Quail (Bobwhite)	8–10	150–200	0.02	9
Pigeon	6	12–15	0.04	17
Guinea fowl	10–12	40–60	0.09	40

incubated properly. See Figure 2.18 for structure of the egg. For more information on the composition of eggs, see Chapter 14.

FECUNDITY. The term *fecundity* is used to describe the inherent capacity of an organism to reproduce rapidly. In higher animals, reproduction is possible only after the ovum (female gamete) is fertilized or united with the spermatozoon (male gamete). In chickens, fertilization is not a necessary preliminary to egg laying. Thus, the hen can lay eggs continuously without being mated or without being stimulated by the presence of a male. This biological phenomenon has been advantageously utilized in producing infer-

tile eggs for food. Infertile eggs are of more economic value for food than fertile eggs, because there is no danger of loss through development of the embryo.

EGG SIZE. There is an enormous range in size among the eggs of different species of birds. Ostrich eggs average about 2 lb (~1 kg) (see Chapter 19), whereas hummingbird eggs weigh only 0.001 lb (0.5 g). The average size of eggs laid by a number of domestic species of birds is shown in Table 2.5.

There is also considerable variation in egg size within a species. Thus, in chickens, the eggs of Dark Brahmas are more than twice as heavy as those of Japanese Bantams. The eggs of the large meat-producing breeds are at the upper end of size and those of the smaller game birds at the lower end. Also, the size of the eggs laid by one individual may differ widely from those laid by another of the same species and breed. An understanding of the various factors that influence egg size is important because, economically, it is not unusual for the price of medium-size eggs to be 5 to 10 cents lower per dozen than the price of large eggs.

The most important factors affecting egg size are:

1. **Genetics.** Egg size is an inherited trait. Although environmental factors may result in smaller eggs, the upper limit in egg size is determined when the pullet is hatched. For more than 70% of all eggs to Grade Large or better, a strain with an average egg weight of 25 oz (709 gm) per dozen is needed.
2. **Age of bird.** When pullets start to lay, 80% or more of their eggs will be under 21 oz (396 gm) per dozen. Egg size should increase gradually until the birds are around 12 to 14 months of age, following which some decrease in egg size may be expected.
3. **Clutch order.** The order of an egg in the clutch affects its weight. The first egg of a clutch is usually the heaviest, and there is usually a progressive decrease in the weight of the rest of the eggs of the clutch.
4. **Total eggs laid in a year.** There is a tendency toward a decline in egg size with the total number of eggs laid in a year.
5. **Age at maturity and season of hatch.** Delay in maturity will usually result in larger eggs at the start of production. Also, season of hatch when birds are grown in natural daylight affects both maturity and size of eggs at the start. Late fall- and winter-hatched birds will be subjected to an increasing day length during the critical age period of 8 to 20 weeks, if no supplementary lights are used. This will stimulate early maturity and result in a prolonged period of small eggs.
6. **Temperature.** The size of eggs declines during the hot summer months. Fortunately, temperatures favoring large eggs also favor high egg production. Thus, anything that can be done in the summer to keep layers cool will benefit egg size. This includes insulation, proper ventilation, reflective paint on the roof, foggers, and plenty of cool water.
7. **Type of housing.** Caged birds lay eggs about 1/2 oz (14 gm) per dozen larger than birds on the floor, and the birds on slats lay about 10% more large eggs than birds on the floor with part litter. Virtually all table eggs produced in the United States are from caged hens.
8. **Feed and water.** Normally, feed is a minor factor in egg size so long as a well-balanced ration is fed. However, underfeeding (as a result of feeders being empty for lengthy periods, or of timid birds not getting enough to eat when feeder space is limited), a deficiency of needed minerals (especially calcium), or a very low level of protein will result in smaller eggs. The dietary fat in the ration may increase egg size slightly. If water intake is too low because the water is too cold, too hot, too unpalatable, or because water is not available, egg size will be hurt along with egg production.
9. **Disease.** Some diseases affect egg size dramatically, and in some cases the effect persists for months after the layers appear to be healthy. Both Newcastle disease and infectious bronchitis will cause a drop in egg production, with many of the eggs being small and often odd shaped. With bronchitis, especially, small egg size may continue for months.
10. **Grain fumigants.** Ethylene dibromide, an ingredient of a fumigant commonly used on seed oats, will decrease egg size. A drop in egg size to as small as 16 oz (454 gm) per dozen from an initial 23.8 oz (675 gm) average has been reported.

EGG SHAPE. Eggs differ considerably in shape. Although many are truly ovate, some are nearly spherical, whereas others are elongated. Some eggs are almost equally pointed or rounded at both ends while others taper sharply from the large end to the small end. The eggs laid by birds of the same species resemble each other in shape, but they are not identical, nor are all eggs of a particular bird alike.

Aristotle (384–322 B.C.) believed that the cock hatched from the more pointed chicken egg and the hen from the rounder type. Early 19th-century naturalists argued that egg contours indicated the general body form of the bird that would develop within. Somewhat later, natural selection advocates theorized that through adaptation, the eggs of different birds had assumed the shape most likely to ensure the survival of each species in the particular environment.

It is now generally agreed that physiological factors largely determine the diversities in the form of the egg, but that the shape may be modified by certain conditions. Among causes of the variations in egg shape within species or of a particular bird are the following:

1. **Pressure exerted by the oviduct muscles.** The walls of the oviduct contain two layers of muscles: (a) the inner circular layer, which moves the egg forward; and (b) the outer longitudinal layer, which expands the oviduct. Muscle tones and difference in the degree of coordination between these muscles likely account for many of the minor variations in the shape of the egg.
2. **Volume of albumen and size of isthmus.** It has been postulated that an elongated egg might be

formed if a large volume of albumen is secreted and then forced through a narrow isthmus.

3. **Breed and flock variation.** The eggs laid by birds of the same breed resemble each other in shape, but there may be enormous variation within a single flock.

4. **Heredity.** Chickens may be bred and selected for a specific egg shape with comparative ease; for example, they may be bred and selected for round eggs or long eggs.

5. **Commencement of laying.** The first eggs laid by a pullet are likely to be atypical in shape.

6. **Cycle of laying.** The first egg of a cycle is usually longer and narrower than the second egg.

7. **Pause in laying.** The first egg laid after a pause of 7 days is longer and narrower than the last egg preceding the pause.

EGG COLOR. The eggs of the domestic hen may be white, many shades of brown, or yellow (see color plate 10). One breed lays blue-green eggs. Sometimes very small, dark flecks are present on the shell, especially if it is brown. The vast majority of eggs sold in the United States are white.

Among domestic fowl, the color of the eggs is peculiar to the breed, although tinted eggs occasionally appear in breeds that ordinarily lay white eggs. Of the four races recognized in the United States, only the Mediterranean (comprised of Leghorns, Minorcas, Anconas, Black Spanish, and Blue Andalusian) lays white eggs. The other three races (Asiatic, English, and American) lay tinted eggs, with the exception of two or three breeds. Cochin China hens lay eggs that range from bright yellow to dark yellow speckled with small red dots. Langshans lay dark yellow eggs. Brahmas lay reddish yellow eggs. Most of the continental European breeds lay white eggs. The Araucana of South America lays light bluish green eggs. Hens normally lay eggs of the same color, but there may be considerable variation in color among hens of the same breed.

Colored eggs occur because pigment is deposited in the shell as it is formed in the uterus. The hereditary nature of such pigmentation is evidenced by the manner in which egg color is identified with species and breed. However, the mode of inheritance of shell color is difficult to determine because most varieties of fowl are the product of many years of crossbreeding. Egg color often assumes economic importance since there are numerous local prejudices in favor of certain shell tints.

STRUCTURE OF THE EGG. A schematic side view of an egg is shown in Figure 2.18, with the various parts labeled in their normal position.

TABLE 2.6 Chemical Composition of the Egg

	Whole Egg	White	Yolk	Shell
		(percent of whole egg)		
	100	*58*	*31*	*11*
	(%)	(%)	(%)	(%)
Water	65.5	88.0	48.0	—
Protein	11.8	11.0	17.5	—
Fat	11.0	.2	32.5	—
Ash	11.7	.8	2.0	96.0
Total	100.0	100.0	100.0	96.0

The protective covering, known as the shell, is composed primarily of calcium carbonate, with 6,000 to 8,000 microscopic pores permitting transfer of volatile components. The air cell, located in the large end of the egg, is formed when the cooling egg contracts and pulls the inner and outer shell membranes apart. The cordlike chalazae hold the yolk in position in the center of the egg. As shown, the yolk is surrounded by membrane, known as the vitelline membrane. The germinal disc, a normal part of every egg, is located on the surface of the yolk. Embryo formation begins here only in fertilized eggs.

COMPOSITION OF THE EGG. The chemical composition of the egg is given in Table 2.6 (for details see Chapter 14). Cholesterol is present in eggs, as it is in the bodies of birds and other animals. Cholesterol is essential to cell membranes and the synthesis of steroid hormones. However, high levels of cholesterol are associated with arterial diseases in people. Eggs have an average cholesterol content as follows: extra large eggs, 230 mg; large eggs, 213 mg; and medium-size eggs, 180 mg.

EGG ABNORMALITIES. Periodically, a deviation in the mechanics of egg laying will create abnormal eggs. Some of these abnormalities are as follows:

1. **Double-yolked eggs.** This is the result of two follicles developing and ovulating at the same time.

2. **Blood spots.** These are caused when a small blood vessel breaks during ovulation.

3. **Meat spots.** Meat spots are degenerated blood clots in the egg.

4. **Yolkless eggs.** Occasionally, foreign material may get into the oviduct and stimulate the secretion of albumen in much the same manner as the yolk.

5. **Dented eggshells.** When one egg is kept too long in the uterus, a second egg may pass down the tract and actually touch the first egg, there-

upon creating an indentation in the shell of the second egg.

6. **Soft-shelled eggs.** These eggs result when either no shell is secreted or an abnormally small amount of shell is secreted. This may be associated with calcium deficiency and some other diseases.

EGG LAYING.   After the yolk is released from the follicle, it is picked up by the infundibulum and subsequently passed into the magnum. In the magnum, the egg begins to take shape as the chalazae (a ropelike substance) and three of the four layers of albumen become attached to the yolk. The albumen deposited in the magnum first appears to be homogenous, but as the egg passes through this organ it is continually turned, which gives rise to the various layers of egg white. The egg spends a total of about 3 hours in the magnum.

From the magnum, the egg is passed into the isthmus where the inner and outer shell membranes are added along with some mineral salts. After spending about 1¼ hours in the isthmus, the egg travels to the uterus or shell gland. Here, additional water and minerals pass through the shell membranes into the white, inflating the egg and giving rise to the fourth layer of albumen (outer layer of thin white). It is also in this organ where the shell composed primarily of calcium carbonate is formed. If the eggshell is to be colored, the pigments are deposited in the uterus. The egg spends about 21 hours in the uterus.

From the uterus, the egg passes through the vagina and out the cloaca. In the vagina, a thin protein coating called bloom or cuticle is applied to the shell. When this substance dries, it seals the openings of the porous eggshell. One of the interesting physiological phenomena in the hen is the fact that, as the egg passes down the oviduct, it travels small end first; however, immediately prior to expulsion, the egg is turned 180° so that the large end is exposed first. This turning is believed to aid the muscles of the tract in expelling the egg.

INTEGRATION OF BODY PROCESSES

The bird is an extremely complex biological organism, requiring considerable information to be passed to and from its different physiological systems. This integration of messages is accomplished through two modes of transport: neural (via nerves) and *endocrine* (via blood). The nervous systems provide electrical impulses that stimulate or inhibit various body functions. Endocrine controls are provided via the endocrine glands and their secretions called hormones. A *hormone* can be defined as a chemical released by a specific area of the body that is transported to another region within the animal via the bloodstream where it elicits a physiological response.

Nervous System

The nervous system can be divided into two anatomical systems: the central nervous system (CNS) and the peripheral nervous system (PNS). The CNS consists of the brain and spinal cord (see Figure 2.19 for anatomy). The skull and vertebral column protect these. The PNS comprises the rest of the nervous system and is made up of nerves. Some of the nerves originate in the brain while the rest originate from the spinal cord. The PNS is functionally divided into the somatic nervous system and the autonomic nervous system. The somatic nervous system enables the body to adapt to stimuli from the external environment. Various stimuli, such as touch, are perceived by specialized receptors within this system, and the body responds accordingly. The autonomic system involves the maintenance of homeostasis, the internal environment of the body. This is the system that controls the gastrointestinal tract.

The autonomic system can be further divided into the sympathetic autonomic nervous system and the parasympathetic autonomic nervous system. Both the sympathetic system and the parasympathetic system are usually associated with routine integration of normal activity. When the sympathetic system is stimulated, there is a need for large amounts of blood in peripheral tissues, such as skeletal muscle. In order to accommodate this need, blood is shunted from such areas as the gastrointestinal tract. Heart rate and respiratory rate increase to accommodate the increased demand for oxygen.

Special Senses

One means by which a bird can react to the environment is through the use of its special senses—eyes, ears, and so forth. (This is discussed in detail in Chapter 10.)

Endocrine System

The endocrine system is composed of a number of glands that produce, store, and secrete hormones. Table 2.7 lists the hormones produced by the various glands and the physiological functions of the hormones. Hormones can be classified into two broad categories, according to structural properties: (1) protein and peptide

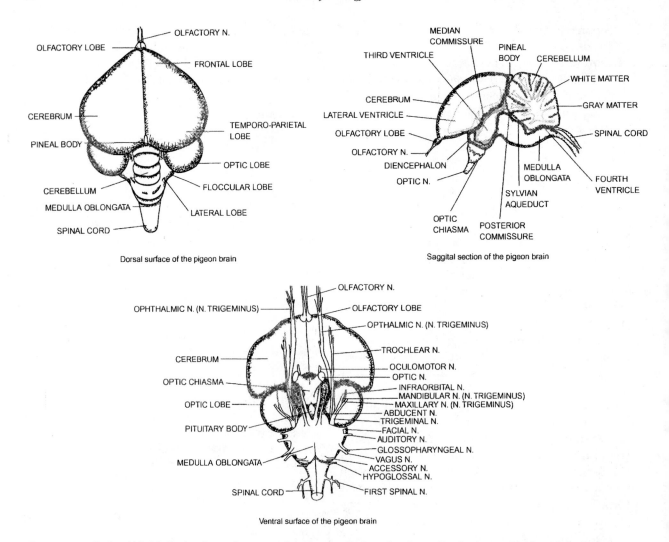

Figure 2.19   Brain of bird (left, view from above; top right, longitudinal section; immediately above, view from below).

hormones (including modified amino acids), and (2) steroid hormones.

Mesotocin and vasotocin are synthesized in the hypothalamus and are transported along nerve tracts to the posterior lobe of the pituitary where they are subsequently stored. They, like all of the hormones, are secreted into the blood when called upon to promote their physiological roles.

Hormones carry out their regulatory roles through several types of feedback systems as seen in Figure 2.20. Two types of feedback systems, negative feedback and positive feedback, are utilized.

Negative Feedback

Control of the various feedback systems depends on the circulating levels of various metabolites,

hormones, and nutrients (Figure 2.20). For example, if the level of calcium in the blood drops, the parathyroid gland secretes a hormone that mobilizes body calcium and subsequently increases the concentration of calcium in the blood. This increase in the blood calcium then decreases the secretion of parathyroid hormone. Thus, the endocrine gland initially reacts to the absence of a particular compound—a negative feedback.

Another example of negative feedback involves a three-step process. If the metabolic state of the bird calls for the secretion of corticosterone or it is stressed, the hypothalamus is stimulated to secrete the releasing factor for adrenocorticotropic hormone (ACTH). This corticotropin-releasing hormone (CRH) stimulates the anterior pituitary to

TABLE 2.7 Hormones of the Fowl

Gland	Hormone	Type of Hormone	Functions of Hormone
Pituitary gland (Hypophysis): 1. Anterior lobe (adenohypophysis)	Gonadotropic hormones 1. Follicle-stimulating hormone (FSH)	Glycoprotein	Stimulates growth of ovarian follicles in females. Maturation of sperm in males.
	2. Luteinizing hormone (LH)	Glycoprotein	Acts in ovarian granulosa cells to synthesize progesterone. Triggers ovulation in females. Acts on Leydig cells of testes to produce androgen.
	Prolactin (PRL)	Protein	Causes broodiness in turkeys. Initiates crop-sac secretion in pigeons.
	Growth hormone (GH) or somatotropin (ST)	Protein	Growth promotion via insulin-like growth factor-I.
	Adrenocorticotropic hormone (ACTH)	Peptide	Stimulation of the adrenal cortex and release of the adrenal corticoids.
	Thyrotropin (TSH)	Glycoprotein	Stimulation of the thyroid glands to (1) release thyroxine (T_4), and (2) absorb iodine to synthesize T_4.
	Melanotropin (MSH)	Peptide	Function is not known in birds.
2. Posterior lobe (neurohypophysis)	Mesotocin (storage)	Peptide	Stimulates uterine tissue.
	Vasotocin (storage)	Peptide	Antidiuretic hormone. May initiate the contraction of the uterus that begins oviposition.
Hypothalamus	Mesotocin	Peptide	See neurohypophysis.
	Vasotocin	Peptide	See neurohypophysis.
	Releasing hormone	Peptide	Stimulates the adenohypophysis to release its hormones.
	1. Gonadotropin-releasing hormone-I (GnRH-I)	Peptide	Stimulation of LH and FSH release.
	2. Thyrotropin-releasing hormone (TRH)	Modified tripeptide	Stimulation of TSH and GH release.
	3. Vasoactive intestinal peptide (VIP)	Polypeptide	Stimulation of PRL release.
	4. Corticotropin-releasing hormone (CRH)	Polypeptide	Stimulation of ACTH release.
	5. Growth hormone releasing hormone (GHRH)	Polypeptide	Stimulation of GH release.
	6. Somatostatin (SRIF)	Peptide	Inhibition of GH and TSH release.
Thyroid glands	Thyroxine (T_4) and triiodothyronine (T_3)	Modified amino acid	Affects metabolic rate and control of body temperature. Affects feather growth and color. Inhibits release of TSH.
Ultimobranchial glands (C-cells)	Calcitonin	Peptide	Calcium metabolism. May play a role in the regulation of serum phosphorus.
Parathyroid glands	Parathyroid hormone (PTH)	Peptide	Calcium mobilization and also phosphorus metabolism.
Adrenals: 1. Cortex	Aldosterone	Steroid	Electrolyte and water metabolism.
	Corticosterone	Steroid	Carbohydrate, fat, and protein metabolism.
2. Medulla	Catecholamines 1. Epinephrine (adrenaline)	Amino acid derivative	"Fight-or-flight" changes in blood flow, heart rate, and metabolism.
	2. Norepinephrine (noradrenaline)	Amino acid derivative	"Fight-or-flight" changes in blood flow, heart rate, and metabolism. Neurotransmitter.

(*continued*)

TABLE 2.7   Hormones of the Fowl (*continued*)

Gland	Hormone	Type of Hormone	Functions of Hormone
Pancreas	Glucagon	Polypeptide	Carbohydrate, fat, and protein metabolism.
	Insulin	Polypeptide	Carbohydrate, fat, and protein metabolism.
	Pancreatic polypeptide	Polypeptide	Hormone secretion.
	Somatostatin	Peptide	Hormone secretion and lipid metabolism.
Testes (male)	Testosterone	Steroid	Secondary sex characteristics. Sexual behavior. Spermatogenesis. Inhibits LH release.
	Inhibin	Glycoprotein	Inhibits FSH release.
Ovary (female)	Estrogens. estradiol (E_2), estriol, estrone	Steroid	Oviduct growth and functioning. Production of yolk precursors. Calcium absorption and retention in medullary bone synthesis. Inhibits LH secretion.
	Progesterone (P_4)	Steroid	Oviduct growth. Involved in synthesis of avidin (component of egg white). Stimulates LH release by positive feedback ovulation.
	Testosterone	Steroid	Calcium retention in medullary bone.
Pineal gland	Melatonin	Modified amino acid	Functions unclear in poultry.
Liver	Insulin-like growth factor-I (IGF-I)	Polypeptide	Stimulation of growth.
	Activation of thyroxine to triiodothyronine (T_3)	Modified amino acid	As T_4.
Kidney	1,25 dihydroxy vitamin D_3 $[1,25(OH)_2D_3]$	Steroid	Calcium absorption; bone metabolism.
	Renin	Protein	Activation of angiotensin which thereby affects blood pressure and aldosterone release.
Adipose tissue	Leptin	Protein	Affects food intake.
Gastrointestinal tract	Multiple hormones (e.g., cholecystokinin, gastrin, ghrelin.)	Polypeptides	Affects gut functioning and feed intake (some).

secrete ACTH. This ACTH travels via the bloodstream to the adrenal cortex, which is then stimulated to produce corticosterone. The corticosterone travels to both the target cells and to the hypothalamus and pituitary where it inhibits the release of CRH and ACTH.

Positive Feedback

In positive feedback mechanisms, a hormone is secreted, which promotes the production of a second compound, and the second compound then travels back to the original endocrine gland and stimulates the production of even more of the initial hor-

mone. For example, the hypothalamus is stimulated to release gonadotropin-releasing factor I, which causes the anterior pituitary to secrete luteinizing hormone (LH). LH travels to the ovary, stimulating it to produce progesterone. This in turn stimulates the hypothalamus to secrete more releasing factor (see Figure 2.16).

Regulation of Feed Intake

The hypothalamus is a structure in the ventral region of the brain. It plays a critical role in the control centers of feeding. Within the hypothalamus, certain areas can be differentiated. Two of these

NEGATIVE FEEDBACK
General scheme

Endocrine Organ → Target organ → Change in circulating metabalite/ion/hormone

Examples

1.

Parathyroid gland → Bone → Increase in plasma calcium
Parathyroid hormone
(PTH)

High calcium inhibiting PTH release

2.

or β cells of pancreas
(islets of Langerhans) → Liver → Increase in plasma glucose
Insulin

High glucose inhibiting insulin release

NEGATIVE FEEDBACK ON THE HYPOTHALAMUS AND PITUITARY GLAND

1.

Hypothalamus → Anterior Pituitary Gland → Adrenal Cortex
Cortex
CRH   ACTH
plasma   Increase in Corticosterone

High corticosterone inhibiting CRH and hence ACTH release

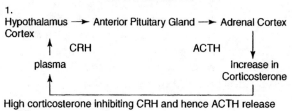

2.

Hypothalamus → Anterior Pituitary Gland → Thyroid → Increase in plasma Thyroxine → Liver
TRH   TSH
Thyrotropin
Triodothyronine (T_3)

High T_3 inhibiting TRH and TSH release

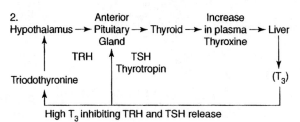

3.

Hypothalamus → Anterior Pituitary Gland → Testes → Increase in plasma testosterone
GnRH-I   LH

High testosterone inhibiting GnRH-I release

4.

Hypothalamus → Anterior Pituitary Gland → Ovary → Increase in plasma estradiol
GnRH-I   LH
Luteinizing Hormone

High estradiol inhibiting GnRH-I release

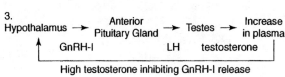

5.

Hypothalamus → Anterior Pituitary Gland → Gonads → Increase in plasma inhibin
FSH
Follicle Stimulating
Hormone

High inhibin decreasing FSH release

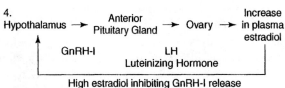

6.

SRIF
Hypothalamus → Anterior Pituitary Gland → Liver → Increase in plasma IGF-I
GHRH   GH
Growth Hormone

High IGF-I inhibiting GHRH and GH release

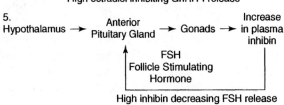

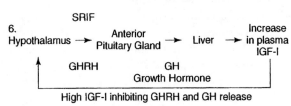

Figure 2.20 Feedback systems.

POSITIVE FEEDBACK

1.

Hypothalamus → Anterior Pituitary Gland → Ovary → Increase in plasma progesterone
GnRH-I   LH
Luteinizing Hormone

High progesterone stimulating GnRH-I release

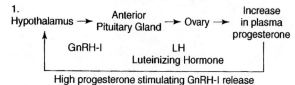

2. Ability of hypothalamus to respond to progesterone requires the presence of estrogen.

are of particular importance in the regulation of appetite. The first area is that of the lateral hypothalamus. It is commonly called the feeding center because, upon stimulation of this region, the bird commences to eat whether or not it is hungry. If this area is damaged, the bird loses all desire to eat and eventually starves. The ventromedial area of the hypothalamus functions as the satiety center. Stimulation of this region will depress appetite. If the ventromedial nuclei are destroyed, there is no inhibition of feed intake, and the bird has an uncontrollable appetite. (Also see Chapter 10).

Thermoregulation

Birds are warm-blooded animals—homeotherms. Hence, they must maintain a constant body temperature for normal physiological functions. If the bird is in a hot environment, the body must give off heat and utilize cooling mechanisms. If the bird is in a cold environment, the body must produce heat through metabolic processes and use insulatory mechanisms to keep this heat from escaping. Table 2.8 shows the deep body temperature of several species of poultry.

Adaptation to Heat Stress

Birds have no sweat glands and must, therefore, use other means of dissipating heat. The comb and wattle areas have highly vascularized and ex-

TABLE 2.8 Deep Body Temperature of Poultry at Rest and in Thermoneutral Conditions

Species	Temperature	
	(°F)	(°C)
Chicken	106.7	41.5
Turkey	106.2	41.2
Duck	107.8	42.1
Goose	105.8	41.0
Pigeon	108.0	42.2

posed skin. Additionally, respiratory mechanisms allow for some cooling, and the feathers act as a means of protecting the body from exposure to heat. The primary cooling mechanism is respiratory in nature. When air is inhaled, heat is emitted from the nasal membranes and other areas of the respiratory tract. This heat is then lost when the air is exhaled.

Heat can be transferred from the inner body to the surface through the following mechanisms:

1. **Conduction.** In this mechanism, heat is transferred from molecule to molecule in the body and is eventually lost when the heat reaches the outermost portions. Conduction is increased by the birds crouching to the ground and squashing the breast feathers to facilitate heat loss to the litter.
2. **Convection.** In this mechanism, heat is transferred from heat-producing tissues to the blood. The blood then travels to the skin, resulting in an increased skin temperature and a loss of heat to the atmosphere.

Adaptation to Cold Stress

Birds are very well adapted to cold, due primarily to their highly efficient insulation provided by feathers. When birds are exposed to cold stress, several compensatory mechanisms are used. The first line of defense is the plumage. Feathers are erected to provide more efficient protection from the environment and to conserve body heat. Birds shiver in response to cold. This activity increases the metabolic rate of the body to produce additional heat.

Effect of Light on Poultry

Reproduction in poultry is extremely sensitive to photoperiodism (length of daylight). Turkeys are more sensitive to the length of daylight than chickens.

The wavelength of the light can affect the amount of stimulation. Orange to red (664 to 740 nm) wavelengths yield the most satisfactory results. Since incandescent lightbulbs emit these wavelengths, they are entirely satisfactory for light stimulation programs.

Effect of Light on Sexual Maturity and Reproduction

When growing pullets are exposed to increasing duration of light, sexual maturity is stimulated. The increasing duration of light stimulates the hypothalamus to secrete gonadotropin-releasing factor (GnRH) that, in turn, stimulates the anterior pituitary to release LH and FSH. If egg production is started at too early an age, the eggs will be too small to be profitable. Thus, most producers utilize a growing program designed to suppress sexual maturity in pullets until they reach an age and size when marketable-size eggs can be produced.

Birds in lay require only about 0.5 to 1.0 footcandle of light. Light intensities greater than 1 footcandle are wasteful if artificial light is used. Generally, a 14- to 16-hour light program is used. The length of light exposure should never be decreased for birds in production.

Clutches

Hens lay eggs in a certain time pattern called a clutch. For example, a hen may lay an egg on each of 10 consecutive days and on the eleventh day fail to produce an egg. This pattern is repeated. Another hen may have a pattern of eggs laid for 3 successive days followed by a missed day.

Examples of clutch patterns

3-egg clutch XXX- XXX- XXX- XXX- XXX- XXX- XXX- XXX- XXX- XXX- XXX- XXX- XXX- XXX- XXX-

5-egg clutch XXXXX- XXXXX- XXXXX- XXXXX- XXXXX- XXXXX- XXXXX- XXXXX- XXXXX- XXXXX-

11-egg clutch XXXXXXXXXXX- XXXXXXXXXXX- XXXXXXXXXXX- XXXXXXXXXXX- XXXXXXXXXXX-

Generally, each successive egg of a clutch becomes smaller and smaller. The weight of the yolks remains relatively constant, but the amount of albumen decreases as the clutch progresses. Clutches are determined by the hormonal cycles in the hen and are highly variable, ranging from 1 day to as many as 200 days. Genetic selection has extended clutch length.

A hen having a three- to four-egg clutch has an ovulation cycle of about 26 hours. The 2-hour difference over the 24 hours in 1 day is called lag. The "missed day" allows the hen to get back into synchrony. As the clutch becomes progressively longer, there is a reduction of the length of the ovulatory cycle to a minimum of 24 hours, normally. Genetic selection has reduced the length of the ovulatory cycle. Most hens on a 14 hours of light to 10 hours of darkness program lay their eggs in the morning between 12 to 16 hours after the beginning of the dark period.

FURTHER READING

Physiology and Biochemistry of the Domestic Fowl. Vol. 5. Edited by B. M. Freeman. New York: Academic Press, 1984.

Reproduction in Farm Animals. Edited by B. Hafez and E. S. E. Hafez. Philadelphia: Lippincott, Williams and Wilkins, 2000.

Sturkie's Avian Physiology, 5th ed. Edited by G. C. Whittow. San Diego: Academic Press, 2000.

3

Incubation

Objectives

After studying this chapter, you should be able to:

1. Understand the history of artificial incubation.
2. Be able to define hatchability.
3. List the roles of a hatchery.
4. List the major factors for successful egg storage and incubation.
5. Understand the importance of embryonic development.
6. List the major events that occur during embryonic development.

Eggs have been incubated by artificial means for thousands of years. Aristotle (384–322 B.C.) told of Egyptians incubating eggs by the hotbed method in which decomposing manure furnished the heat. The ancient Egyptians constructed large brick incubators heated with fires. The Chinese developed artificial incubation as early as 246 B.C. They burned charcoal or used dung heaps for heating.

For many years, the application of artificial incubation principles was a "closely guarded secret," passed from one generation to the next. The hatchery operator determined the proper temperature by placing an incubating egg in the socket of his or her eye. Temperature changes were effected in the incubator by moving the eggs, by adding additional eggs to use the heat of embryological development of older eggs, and by regulating the flow of fresh air through the hatching area. Apparently humidity was no problem, for primitive incubators were located in highly humid areas, and the heat source, such as decomposing manure, often furnished moisture. Turning was done up to five times daily, after the fourth day of incubation.

The Smith incubator, patented in 1918, was the forerunner of today's efficient, large-scale incubators used for the hatching of chickens, turkeys, ducks, and other eggs. The artificial incubation of

poultry was followed, somewhat later, by artificial brooding. Natural brooding was commonplace until about 1930.

The brooding period is usually considered to extend from the time poultry hatch, or are received from the hatchery, until they are about 3 weeks old. Today, incubators with capacities of more than 500,000 eggs, equipped with sophisticated controls to maintain optimum conditions for hatchability, are rather common; and by using banks of incubators as many as one million chicks a week can be hatched (see Figures 3.1 and 3.2). Also, commercial brooding of poultry has become larger and more scientific. It is not uncommon to find houses and equipment so arranged that one employee is responsible for as many as 50,000 chicks.

POULTRY REPRODUCTION

Poultry reproduction is markedly different from the reproduction of mammals. The most obvious differences are that the egg is fertilized in the infundibulum, supplied with nutrients, surrounded by a shell, and expelled from the body, while the fertilized egg in mammals remains *in utero* until birth. Also, in higher animals, reproduction is possible only after the ovum (female gamete) is fertilized or

Figure 3.1 Newly hatched chick in an incubator. *(Courtesy USDA)*

Figure 3.2 Trays of hatching eggs in incubator. *(ISU photo by Bob Elbert)*

united with the spermatozoon (male gamete). In chickens, fertilization is not a necessary preliminary to egg laying; a hen can lay continuously without being mated. However, except for parthenogenesis, fertilization of poultry eggs is a requisite to reproduction. Table 3.1 summarizes pertinent information relative to the reproduction of poultry.

Parthenogenesis

Parthenogenesis, the development of unfertilized eggs, occurs in both chickens and turkeys—especially the latter. By genetic selection, the incidence of parthenogenesis in turkeys has been increased to over 40% in eggs of experimental flocks. Most of the parthenogenetic embryos die, but about 1% of them complete development and hatch. All of those that hatch have been diploid males, but testes weights are low and only about 20% produce semen.

INCUBATION

Incubation is the act of bringing an egg to hatching. It may be either natural or artificial. Natural incubation is with a setting hen hovering over eggs. Incubation independent of the hen is known as artificial incubation.

With chickens, turkeys, ducks, quail, pheasants, and ostriches, artificial methods have been perfected that give results superior to those achieved by natural means. One broiler breeder hen will produce about 150 offspring in a year, weighing about 750 lb (340 kg) at marketing. Without artificial incubation, it would not be possible to capitalize on this reproductive capacity. Artificial incubation has freed the breeding hen from incubating eggs.

Hatchability Determining Factors

Hatchability refers to the percentage of eggs hatched. It may be reported as either the percentage of fertile eggs hatched, or the percentage of chicks hatched from all eggs placed in incubation.

The percentage based on fertile eggs is more precise, provided adequate techniques are used to distinguish between early mortality of the embryo and infertility of the egg. However, the percentage

TABLE 3.1   Reproduction in Poultry

Species	Incubation Period	Eggs/Year	Fertility	Hatchability of Fertile Eggs	Age of Sexual Maturity
	(days)	(no.)	(%)	(%)	(mo)
Chicken					
Light-type	21	240	90	90	5–6
Broiler-type	21	170	85	81	6
Turkey	28	105	90	90	7
Goose	28–32	15–60	85	65–70	24
Duck (Pekin)	28	110–175	85	65–70	7–8
Pheasant	24	40–60	90	85	8–10
Quail (Bobwhite)	23–24	150–200	85	80	8–10
Pigeon	18	12–15	90	85	6
Guinea fowl	27	40–60	90	80	10–12

of salable chicks hatched from all eggs set is often a useful measure, because it prevents errors in estimates of hatchability that arise from inaccurate determinations of fertility.

A number of factors influence the hatchability of eggs, including the following:

1. Fertility
2. Genetics
3. Nutrition
4. Diseases
5. Egg selection
6. Handling of hatching eggs and incubation conditions

Fertility

Fertility refers to the capacity to reproduce. It is the factor that determines the number of offspring that may be obtained from a given number of eggs. If the male mates with a hen, the male sperm unites with the ova, which is found on the yolk. Unless one or both birds are sterile, such an egg will be fertile and should hatch. Factors influencing fertility include the number of females mated to one male, the age of breeders, the length of time between mating and saving eggs for hatchability, and management practices.

It is estimated that about 10% of the eggs set in the United States are infertile (see Table 3.1). Such eggs are not only a loss to the industry, but they occupy valuable incubator space and require time-consuming labor and handling.

Some eggs laid within 24 hours after mating may be fertile, but generally it requires 2 weeks after the flock is mated before satisfactory fertility may be obtained. The removal of males from a flock is followed by decline in fertility within 1 week for chickens and 2 weeks for turkeys. Few, if any, fertile eggs will be produced in chickens after 3 weeks and in turkeys after 6 weeks. If the males in the mating are changed, the eggs laid within a few days are fertilized by the new males and there is little overlapping in progeny.

Genetics

Hatchability is an inherited trait and is selected for by primary breeders. Selection is made for strains that possess high fertility and hatchability. Among the genetic factors affecting hatchability are the following:

- **Inbreeding.** Close breeding, without rigid selection for hatchability, has been shown to lower hatchability in both chickens and turkeys.
- **Crossbreeding and incrossbreeding.** Although the results of crossing pure breeds or incross breeds will depend upon the characters or genes carried by the parent stock, such crossing usually results in increased hatchability.

- **Lethal and semilethal genes.** More than 30 lethal or semilethal genes are known in poultry. These genes cause the death of the developing embryo before the end of incubation or soon after hatching.
- **Egg production.** Eggs laid by hens producing at a high rate are more fertile and possess higher hatchability than eggs laid by low producers.
- **Age.** Hatchability tends to be highest during the first laying year for both chickens and turkeys.

Nutrition

The egg must contain all the nutrients needed by the embryo at the time it is laid by the hen, since there is no further contact with the mother once the egg is completed. Therefore, breeder hens must be fed rations that supply adequate quantities of the nutrients needed for embryo development. The nutrients most likely to affect hatchability are the vitamins and trace minerals. For high hatchability and good development of young, breeder hens (chickens, turkeys, ducks, geese, and so forth) require greater amounts of vitamins A, D, E, B_{12}, riboflavin, pantothenic acid, niacin, and of the mineral manganese than birds kept for commercial laying. Birds intended as breeders should be started on special breeder rations at least a month before hatching eggs are to be saved.

Diseases

Hatching eggs from healthy flocks produce the most chicks.

Pullorum disease, caused by *S. pullorum*, once common, has been eradicated from most commercial flocks. Transmission of this disease is chiefly by the egg. Control is based on a regular testing program of breeding stock to ensure freedom from infection. Hatching eggs should be produced under MG (*Mycoplasma gallisepticum*)-negative and MS (*Mycoplasma synovial*)-negative programs.

Mycoplasma gallisepticum is the organism responsible for chronic respiratory disease (CRD), a respiratory disease affecting the air sac of chickens and turkeys. The organism is egg-transmitted, and the accepted method for control is by elimination of breeder flocks containing carrier birds, which is accomplished by the serum agglutination test. MS, which is caused by *Mycoplasma synovial*, is also a respiratory disease, but the respiratory tract is seldom involved in sickness or death. The organisms locate in the synovial fluids of the hock and joints of the footpads, and in some cases the wing joints. These areas become swollen and inflamed. The disease is also egg-transmitted. Two methods of control are available: (1) eradication in the breeders by the serum agglutination test, and (2) heat treatment (at

115°F or 46°C) of the hatching eggs before they are placed in the incubator.

Certain other diseases affect hatchability, even though the disease organisms may not pass into the egg and affect the embryo directly. Newcastle disease and infectious bronchitis, for example, may affect egg shape and shell porosity. Eggs from hens affected by these diseases usually do not hatch as well.

Egg Selection

Certain physical characteristics of eggs are related to hatchability; among them, size, shape, shell quality, and interior quality.

- **Size of egg.** The size of eggs is related to hatchability in all species of poultry. Extremely large or very small eggs do not hatch well.
- **Shape of egg.** Eggs that deviate greatly from normal shape do not hatch well. Since the shape of eggs is inherited, it follows that hatchability can be increased by selection.
- **Shell quality.** The quality of the shell is related to hatchability. Eggs possessing strong shells hatch better than eggs with thin shells. The kind of shell depends upon breeding, nutrition, and weather. Some strains or families produce eggs with thick, strong shells, whereas others lay eggs with thin, weak shells. The amount of calcium and vitamin D in the ration affects the shell. Also, eggs produced in hot weather have thinner shells than those produced when the weather is more moderate.
- **Interior quality.** There is evidence that eggs that show quality when candled before incubation (i.e., show movement of yolk and well-centered position of the yolks) hatch more chicks than eggs showing a weak or low-quality condition.

Handling Hatching Eggs

The following practices for handling eggs are recommended:

1. **Collect or gather frequently.** Generally, hatching eggs are gathered more frequently than eggs intended for table use. When temperatures are normal, eggs should be gathered at least three times per day. When temperatures are above 85°F, hatching eggs should be gathered five or more times a day. Frequent gathering reduces the likelihood of contaminating eggs from contact with nesting material and feces, and prevents chilling in winter and overheating in summer.
2. **Clean soiled eggs.** It is best to use only clean eggs for hatching purposes. Realistically, however, hatching eggs are usually valuable; hence,

soiled eggs should be cleaned if possible. The following washing procedure is recommended:
   a. Keep the washing water temperature higher than the egg temperature.
   b. Use a sanitizer-detergent at the recommended level.
   c. Use a nonrecycling washer.
   d. Cool the eggs quickly after washing.
3. **Sanitize.** For effective sanitizing, eggs should be treated within 1 to 2 hours after they are gathered. Several decontaminants may be used, including the following:
   a. **Quaternary ammonia.** It is sprayed on eggs in a lukewarm water solution containing 200 ppm. Its chief advantage is that it may be sprayed on the eggs as they are picked up from the nest.
   b. **Chlorine dioxide.** It may be used as a spray or as a dip on the eggs soon after gathering. It is noteworthy that chlorine dioxide, used as a spray or dip, gives results superior to fumigation with formaldehyde gas. Moreover, it is easier to use.
   c. **Ozone (O_3).** When generated at 100 ppm, ozone (O_3) is an effective sanitizer.
   d. **Formaldehyde gas.** While formaldehyde gas is very effective its use is restricted in many countries including the United States. Triple strength (made by mixing 0.6 g of potassium permanganate with 1.2 cc of Formalin [37.5%] for each cubic foot of space in the room) is usually recommended, with the fumigating time in an airtight chamber not exceeding 20 minutes.

Any sanitizer should be administered as soon as possible after the eggs are laid. Following laying, the number of bacteria on the shell increases tremendously, and some penetrate the shell.
4. **Hold minimum time.** Hatchery eggs should be held for as short a period of time as possible, for hatchability decreases as the time of holding is increased. A rule of thumb is that for every day eggs are held or stored after 4 days, hatching time is delayed 30 minutes and hatchability is reduced 4%. At the most, they should not be held longer than 10 days. Commercial hatcheries usually set twice a week.
5. **Holding temperature.** When the egg is laid and its temperature drops, embryological development ceases. As soon as possible, hatching eggs should be cooled to a temperature of 65°F. When it is necessary to hold hatching eggs longer than 7 days, it is recommended that they be warmed to 100°F (37.8°C) for 1 to 5 hours early in the holding period.
6. **Humidity.** High humidity in a hatchery tends to (a) prevent evaporation and an enlargement of

the egg's air cell, and (b) improve hatchability. A relative humidity of 75 to 80% is recommended.

7. **Position.** Hatching eggs are best held with the small ends up from the standpoint of increased hatch. However, most hatchery operations do not consider this holding position practical, because eggs are incubated with the large ends up; hence, eggs that are stored with the small ends up must be inverted.

8. **Turning.** If it is necessary to hold hatching eggs for more than 7 days, they should be turned. This prevents the yolk from sticking to the shell. Turning can be done by tipping the egg cases sharply. It is recommended that the cases be turned in this manner twice daily.

Incubator Operation

There are seven factors of major importance in incubating eggs artificially: temperature, humidity, ventilation, position of eggs and turning, testing (candling), setters and hatchers, and incubation time. Of these, temperature is the most critical. In natural conditions, heat comes from the body of the setting hen. This temperature is usually slightly lower than that of the nonbroody hen's average temperature of 106°F (41°C).

1. **Temperature.** The fertile egg will resume development when it is placed in the incubator. But maintenance of the proper temperature is of prime importance for good hatchability. Depending on the type of incubation, optimum temperatures range from 99 to 103°F (37.2–39.4°C). In the usual forced-air machine, the temperature should be maintained at about 99.5°F.

   Overheating is much more critical than underheating; it will speed up rate of development, cause abnormal embryos in the early stages, and lower the percentage of hatchability. Figure 3.3 shows the effect of temperature on the percentage of fertile eggs.

2. **Humidity.** Humidity is of great importance for normal development of the chicken embryo. During embryonic development (incubation), moisture is lost from the egg contents through the shell. This increases the size of the air cell, which after 19 days of incubation occupies about one-third of the egg. Although a variation of 5 to 10% may be acceptable, the relative humidity of the air within an incubator for the first 18 days should be about 60%. During the last 3 days, or the hatching period, it should be 70%. Best results are obtained with turkey eggs when the humidity is 2 to 3% higher than these figures. Lower humidity than recommended causes excess evaporation of water, while high

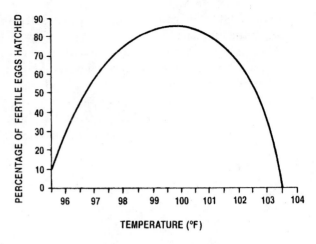

Figure 3.3 The effect of incubation temperature on percentages of fertile eggs hatched; with relative humidity 60%, oxygen 21%, carbon dioxide below 0.5%. *(Adapted from the University of Connecticut Extension Service publication)*

humidity prevents the evaporation of sufficient amounts of water from the egg. In both cases, hatchability is reduced.

3. **Ventilation.** As the embryo develops, it uses oxygen and gives off carbon dioxide. Thus, sufficient ventilation within the incubator is required to assure an adequate supply of oxygen and the proper removal of carbon dioxide. The best hatching results are obtained with 21% oxygen in the air—the normal oxygen level in the atmosphere. The embryo will tolerate a carbon dioxide level of 0.5%, but it will die if this level reaches 5%. Since the normal oxygen and CO_2 concentrations present in air seem to represent an optimum gaseous environment for incubating eggs, no special provision to control these gases is necessary other than to maintain adequate circulation of fresh air at the proper temperature and humidity.

4. **Position of eggs and turning.** The embryo head must occupy a position in the large end of the egg for proper hatching. Thus, the egg must be incubated large end up as gravity orients the embryo with its head uppermost. Somewhere between the fifteenth and sixteenth day, the head of the embryo is near the air cell.

   Eggs should be turned from 3 to 5 times a day between the second and the eighteenth day. The purpose of this turning is to prevent the germ spot from migrating through the albumen and becoming fastened to the shell membrane. That is, turning the eggs prevents an adhesion between the chorion and the shell membrane. Proper turning consists of rotating the eggs back and forth, not in one direction (a 30 to 45° angle is best).

   A hen sitting on a nest of eggs turns them frequently, using (a) her body as she settles on the eggs and (b) her beak as she reaches under her body. Modern incubators are equipped with turning devices that are able to rotate egg trays through a 90° angle. Also, they are equipped with timing mechanisms, and usually the eggs are turned every hour, which is probably more often than necessary.

5. **Testing (candling).** Under some circumstances, it may be advisable to check incubating eggs for fertility or embryo mortality. This is done by candling the eggs, using a 75-watt blue bulb. With suitable equipment, infertile eggs may be detected after 15 to 18 hours of incubation. The second test may be made 14 to 16 days after incubation, at which time the dead embryos may be removed.

6. **Setters and hatchers.** In commercial hatcheries two separate incubators are used during the incubation process. The bulk of the incubation (usually through the nineteenth day) is done in setters while the end of the process is in hatchers. The main reasons for separate setters and hatchers are:

   a. The machines are kept in two separate rooms, thereby (1) isolating the down, egg debris, and microorganisms that accompany hatching from the eggs in the hatchery, and (2) permitting the hatchers to be cleaned, disinfected, and fumigated without disturbing the eggs in the setters.

   b. The temperature of the hatchers is lowered to 98°F because there is some evidence that the hatch may be slightly improved by the lower temperature.

   c. The hatchers are equipped with special chick-holding trays, not needed in setters.

7. **Incubation time.** The incubation period varies for different species of birds. Generally speaking, the larger the egg, the longer the incubation period. Also, the incubation period may vary with the temperature and humidity of the incubator. The normal incubation periods for several species of birds follow:

Common Name	Incubation Period (in days)
Ostrich	42
Muscovy duck	33–35
Goose	28–32
Turkey	28
Duck	28
Peafowl	28
Guinea	27
Pheasant	24
Bob white quail	23–24
Chicken	21
Parakeet	19
Pigeon	18
Japanese quail	17
Canary	13

Embryonic Development

Structural development begins soon after fertilization (before laying) and continues during incubation. Very early, the embryo becomes differentiated into three distinct cell layers out of which the various organs and systems of the body develop.

Soon after fertilization, cell division proceeds. The blastoderm spreads out over the yolk and becomes differentiated into the following two layers of cells by a process called gastrulation:

1. The ectoderm, the first layer, forms the skin, feathers, beak, claws, nervous system, lens and retina of the eye, and the lining of the mouth and vent.

2. The endoderm, the second layer, produces the linings of the digestive tract and the respiratory organs.

Soon after incubation, a third germ layer, the mesoderm, originates. It gives rise to the bones, muscles, blood, and reproductive and excretory organs. Most of the development of the chick embryo occurs before day 7, thereafter predominantly growth takes place. But the producer continues to be very much interested because the objective is the hatching of a vigorous live chick; hence, it is important to know more about the development that occurs throughout the entire period of incubation.

Figures 3.4 and 3.5 show the development of the three cell layers, which give rise to all the organs, and

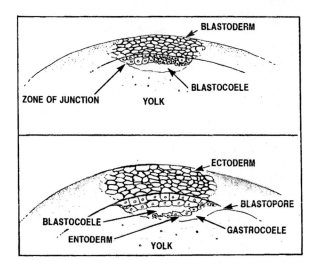

Figure 3.4 Gastrulation in form with telolecithal egg containing a large amount of yolk—a bird egg. These schematic diagrams show the effect of yolk on gastrulation. In the chick, the great amount of yolk effectively prevents the formation of an open blastopore. *(Original diagram from E. Ensminger)*

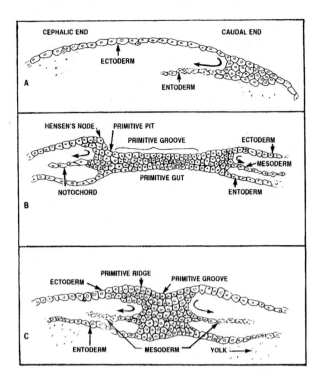

Figure 3.5 Transverse sections of early development of chick embryo, showing differentiation into three distinct cell layers out of which the various organs and systems of the body develop. *(Original diagram from E. Ensminger)*

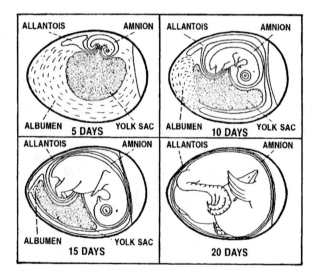

Figure 3.6 Successive changes in the position of the chick embryo and its embryonic membranes. *(Original diagram from E. Ensminger)*

systems of the body. Figure 3.6 shows successive changes in the position of the chick embryo and its embryonic membranes beginning with the fifth day of incubation and extending to just before hatching. Table 3.2 summarizes the important embryonic developments.

Embryo Communication

In some species of birds, eggs in a clutch tend to hatch at about the same time despite the fact that some eggs have been laid earlier and incubated longer than others. This is due to signals transmitted from egg to egg. This communication synchronizes hatching time, a feature that might be of considerable survival value in the wild. Slow clicking accelerates development, while fast clicking retards it. In quail, with an incubation period of about 16 to 18 days, embryo development can be accelerated by a full day and the young will still hatch fully mature.

Acceleration or retardation of hatching can be caused by use of artificial clicks. In chickens, clicking has been shown to advance the time of hatch, but not to retard it. Thus, it has been demonstrated that there is communication between embryos—that they "talk" to each other, and that nervous control of development does take place.

Yolk Makes for a Good Start in Life

Newly hatched chicks can be shipped long distances without food or water. The yolk is largely unused by the embryo and is drawn into the body of the chick on the nineteenth day, just before it hatches. It is highly nourishing and provides proteins, fats, vitamins, minerals, and water for the first several hours of the chick's life. The yolk is gradually used up during the first 10 days of life of the chick.

Hatcheries

Perhaps it is fair to say that hatcheries have exerted more influence on improving the general level of the poultry industry than any other segment of the business. This has been achieved through their breeding programs or through the breeding programs of the poultry breeders with whom they have a franchise agreement.

Changes in Hatcheries

NUMBER OF CHICKEN HATCHERIES

Year	Number of Hatcheries	Total Egg Capacity	Average Egg Capacity per Hatchery
1934	11,405	276,287,000	24,000
1953	8,233	616,976,000	80,000
1965	2,365	471,318,000	199,000
1975	797	416,000,000	521,000
1981	538	466,096,000	466,000
1989	376	524,615,000	1,395,000
2002	322	874,682,000	2,716,000

TABLE 3.2   Important Events in Embryonic Development

Stage or Period	Events—What Takes Place
Before egg laying	Fertilization, division, and growth of living cells, segregation of cells into groups of special function.
Between laying and incubation	No growth; stage of inactive embryonic life.
During incubation:	
First day:	
16 hours	First sign of resemblance to a chick embryo.
18 hours	Appearance of alimentary tract.
20 hours	Appearance of vertebral column.
21 hours	Beginning of formation of nervous system.
22 hours	Beginning of formation of head.
23 hours	Appearance of blood islands—vitelline circulation.
24 hours	Beginning of formation of eye.
Second day:	
25 hours	Beginning of formation of heart.
35 hours	Beginning of formation of ear.
42 hours	Heart begins to beat.
Third day:	
50 hours	Beginning of formation of amnion.
60 hours	Beginning of formation of nasal structure.
62 hours	Beginning of formation of legs.
64 hours	Beginning of formation of wings.
70 hours	Beginning of formation of allantois.
Fourth day	Beginning of formation of tongue.
Fifth day	Beginning of formation of reproductive organs and differentiation of sex.
Sixth day	Beginning of formation of beak and egg tooth.
Eighth day	Beginning of formation of feathers.
Tenth day	Beginning of hardening of beak.
Thirteenth day	Appearance of scales and claws.
Fourteenth day	Embryo turns its head toward the blunt end of egg.
Sixteenth day	Scales, claws, and beak becoming firm and horny.
Seventeenth day	Beak turns toward air cell.
Nineteenth day	Yolk sac begins to enter body cavity.
Twentieth day	Yolk sac completely drawn into body cavity; embryo occupies practically all the space within the egg except the air cell.
Twenty-first day	Hatching of chick.

NUMBER OF TURKEY HATCHERIES

Year	Number of Hatcheries	Total Egg Capacity	Average Egg Capacity per Hatchery
2002	65	47,879,000	736,000

The major changes in the hatching industry in recent years follow:

- **Fewer and larger hatcheries.** In the 1930s, there were > 10,000 chicken hatcheries in the United States; today only about 3% remain. But, in this same period of time, the average size of the hatchery increased tremendously. The increased size of hatcheries becomes obvious when it is realized that today chick hatcheries with incubators capable of holding 500,000 or more eggs account for > 60% of the United States' capacity. A similar trend has occurred in turkey poult hatchery capacity.

- **Year-round business.** Formerly, most hatcheries operated in the spring of the year only. In 1937, slightly under 2.0 hatches of chicks are taken off per year. Today > 11 chicken hatches and > 6 turkey hatches are obtained per year. (This is known as turnover; it is obtained by dividing hatch by capacity.) The primary reasons for extending the hatching season to a more nearly year-round business are (a) less seasonality in the production of eggs and turkeys, and (b) the development of a year-round commercial broiler industry.

- **Geographical shift.** Hatcheries have followed production, geographically speaking. Thus, with the shift of the broiler industry to the South and South Atlantic states, the total output of hatcheries has also increased in this area.
- **Franchise agreements.** Before 1950, franchising was practically unknown in the hatchery business. Today, many hatcheries have franchise agreements with breeders who use advanced methods to produce superior strains of birds. Under these agreements, the breeders furnish the eggs and the hatcheries agree to sell breeders' strain of birds in specified areas.
- **More integration.** Many hatcheries have become a part of broader businesses that involve poultry growing and processing, or sale of supplies. This may involve the selling of poultry feeds, medicines, equipment, farm supplies, and farm products.
- **Few started chicks sold.** With the growth of larger and more specialized poultry farms and their improved brooding technology, fewer and fewer started chicks (chicks 2 to 4 weeks of age) are being sold.
- **Business methods.** Today, hatchery business is big business, and it must be treated as such. A complete set of records must be kept in every successful hatchery. Such records should show all financial transactions. Additionally, there should be a record of inquiries received, the name and address of customers, the name and grade of chicks sold to each customer, the date of each sale, and a record of all hatching eggs purchased.
- **Marketing chicks.** The hatchery has three common channels through which chicks (see Figure 3.7) are sold: (a) contract and local sales at the hatchery, (b) mail orders direct to customers, and (c) wholesale.

Hatchery Manager

The manager can make or break any hatchery. Good managers are scarce. Traits of a good manager are presented in Chapter 8.

Hatchery Building

Certain general requirements of hatchery buildings should always be considered. It is with these that the ensuing discussion will deal. Once buildings are constructed, there is a practical limit to the changes that can be made in remodeling. Consequently, it is most important that very careful consideration be given to the following requisites:

1. **Egg-chick flow through hatchery.** Hatcheries should be designed so that the hatching eggs may be taken in at one end of the building and the chicks removed from the other. A knowledgeable architect should be employed to draw the details and write the specifications.
2. **Separate rooms.** Every commercial hatchery should have separate rooms for the office, fumigation, egg grading and traying, egg holding (see Figure 3.8), incubators, display chicks, display equipment and feed, chick sorting and boxing, and for chick boxes and other supplies.
3. **Construction.** The building housing the hatchery should be constructed of adequately durable material that will withstand outside temperature fluctuations reasonably well. A well-drained concrete floor is conducive to sanitation in the hatchery. Also, the building should be kept in good repair, be well painted, and present an attractive appearance.
4. **Lighting.** The hatchery should be well lighted. Good lighting makes for visibility and conven-

Figure 3.7 Newly hatched poult. *(ISU photo by Bob Elbert)*

Figure 3.8 Eggs from breeder farms being placed in racks at the hatchery, with a code identifying their farm of origin. This rack holds 6,480 eggs. *(Courtesy, Delmarva Poultry Industry, Inc.)*

ience of the caretakers, and makes a good impression on customers.

5. **Ventilation.** Ventilation refers to the changing of air—the replacement of foul air with fresh air. A well-ventilated hatchery makes possible proper incubator ventilation. Forced-air circulation is necessary in larger hatcheries to remove foul air and harmful humidity, without excess heat loss or creation of drafts.

6. **Heating and cooling.** It is common practice to install heating and cooling units in the same ventilating system. During cold weather, the heating unit operates; during hot weather, the cooling unit functions.

7. **Humidity.** Generally, additional humidity should be supplied. The recommended humidity for the various rooms follows: egg-holding room, 75 to 80%; incubator and hatching rooms, 50%; and chick-holding room, 60%.

8. **Battery brooders.** Most hatcheries use battery brooders for holding surplus chicks that accumulate as a result of weather conditions, order cancellations, or oversettings.

Hatchery Equipment

Many different kinds of equipment are necessary in a hatchery, with the size and number of these pieces of equipment determined by many factors. Generally, most modern hatcheries are equipped with the following:

1. Water softeners and heaters.
2. Egg handling equipment, including carts and conveyors.
3. Egg grading and washing equipment, including vacuum egg lifts, automatic egg graders, and egg washers.
4. A standby electric plant for use in case of power failure.
5. Modern incubators featuring forced-draft air circulation, small amount of floor space in relation to egg capacity, durable cabinet material, automatic temperature control, automatic humidifiers, carbon dioxide recorders, elimination of egg traying, mechanical egg turners, separate hatchers, efficient cooling, shortened egg transfer from setting trays to hatching trays, and various emergency devices.
6. Other hatchery equipment, including such items as egg candlers, test thermometers, chick box racks, chick graders, sexing equipment, tray washers, dubbing shears, beak trimming equipment, detoers, automatic syringes, Marek's vaccination equipment, *Mycoplasma gallisepticum* test equipment, and a waste disposal system.

Hatchery Sanitation

In order to reduce the possibility of disease, as well as make a good impression upon customers and potential customers, it is important that certain sanitation practices be rigidly adhered to in a hatchery; among them, isolation and disease-free chicks, clean eggs, sanitizing, separate rooms, and proper waste disposal (also see Chapter 11).

1. **Isolation and disease-free chicks.** The hatchery should take every precaution to avoid bringing in infection. Generally speaking, this calls for banning or controlling visitors, showers and clothing changes for employees, and obtaining eggs from disease-free stock. In particular, every effort should be made to avoid MG (*Mycoplasma gallisepticum*) and MS (*Mycoplasma synovial*).
2. **Clean eggs.** Only clean eggs should be used for hatching, and they should be collected and stored in clean equipment and containers. Although hatching eggs may be washed, there is always the hazard of introducing microorganisms into them.
3. **Sanitizing eggs.** All hatching eggs should be sanitized as soon as possible after collection. (See earlier section on Handling Hatching Eggs).

   Fumigation of eggs and incubators is an essential part of a hatchery sanitation program. When improperly done, fumigation can be a hazard; hence, it should always be done by, or under the supervision of, an experienced person. Routine preincubation fumigation of hatching eggs on the farm is highly recommended for eliminating *Salmonella* infection from poultry flocks. Each egg entering the hatchery should have been subjected to preincubation fumigation as soon as possible after its collection from the nest.

   High levels of formaldehyde gas will destroy *Salmonella* organisms on shell surfaces if used as soon as possible after the eggs are laid. An inexpensive cabinet for enclosing the eggs is required. Fans circulate the gas during the fumigation process and then exhaust it from the cabinet. Eggs for fumigation should be placed on racks so that the gas can reach the entire surface. Plastic trays used for washing market eggs are ideal for this purpose. A high level of formaldehyde gas is provided by mixing 1.2 cubic centimeters (cc) of formalin (37% formaldehyde) with 0.6 g of potassium permanganate ($KMnO_4$) for each cu ft (0.028 m^2) of space in the cabinet. An earthenware, galvanized, or enamelware container having a capacity at least 10 times the volume of the total ingredients should be used for mixing the chemicals. The gas should be circulated within the enclosure for 20 minutes, then expelled to the outside.

Humidity for this method of preincubation fumigation is not critical, but temperature should be maintained at approximately 70°F. Extra humidity may be provided in dry weather. Eggs should be set as soon as possible after fumigation and extra care taken to ensure that they are not exposed to new sources of contamination.

Eggs should be routinely refumigated after transfer to the hatchery to destroy organisms that may have been introduced as a result of handling. Recommendations for loaded-incubator fumigation vary widely, depending upon the make of the machine. Therefore the method, concentration, and duration of fumigation should be in accordance with the manufacturer's instructions. Empty hatchers should be thoroughly disinfected and fumigated prior to each transfer of eggs from setters.

4. **Separate rooms.** Hatcheries should be so designed that there are separate rooms for egg receiving, incubation, hatching, chick-holding, and waste disposal.

5. **Proper waste disposal.** Waste disposal may be a problem and a major source of infection in the hatchery unless it is properly handled. Incineration is an effective means of disposal of hatchery waste from the standpoint of sanitation and disease control, but it is an unprofitable method. The larger hatcheries now process poultry waste into livestock feeds, known as hatchery by-product. It consists of infertile eggs, eggs with dead embryos, and unsalable sexed male chicks. Hatchery by-product is high in protein and calcium.

Sexing Chicks

Chick and poult sexing is routinely offered. Males of laying lines are not grown for meat because of their low growth rate and poor feed efficiency. Hence, they are usually destroyed and incorporated in the hatchery by-product. Chicks are sexed by the following methods:

1. **Sex-linked genes.** Sex-linked genes for (a) rate of feathering or (b) color pattern are used by breeders to sex day-old chicks (also see Chapter 4). Each method examines either the rate of feathering or down coloring. These two methods of sexing visually are:

   a. **Feather sexing.** Day-old chicks may be sex separated accurately by examining the relative length of the primary and covert feathers of the wing, with the females carrying genes for fast feathering and the males carrying genes for slow feathering. It is noteworthy, however, that slow-feathering males usually do not feather well in the brooder house, particularly during hot weather, often resulting in slower growth and increased cannibalism.

   b. **Color sexing.** Usually, gold and silver genes are used for color sexing at day-old; gold, buff, or red chicks are females, and white or light yellow chicks are males. Chicks can be sexed rapidly and accurately in this manner. Where this method of sexing is used for broiler production, it is noteworthy that the color differences are also evident in the mature broiler; females are gold or brown, males are white. But the pinfeathers are usually white; hence, there is no processing problem.

2. **Vent sexing.** Examination of the cloacal wall is widely used by hatcheries to sex newly hatched chicks (see Figure 3.9).

3. **Examination of the rudimentary male organs.** Although it takes considerable skill to be accurate, the rudimentary copulatory organ, or male process, can be identified at hatching and used to identify sexes. This method can be referred to as the Japanese method. By using a special instrument, it is possible to see the testes of day-old chicks through the intestinal wall.

Beak Trimming and Dubbing

Most large hatcheries now offer beak trimming and dubbing services. Beak trimming of day-old broiler chicks or replacement pullets by hatcheries is now a

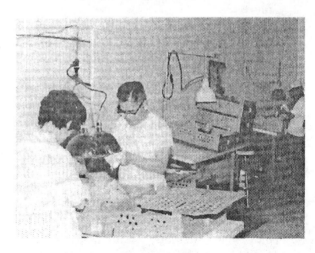

Figure 3.9 Sexing turkey poults. *(Courtesy Gretta Irwin, Iowa Turkey Federation)*

rather common practice for preventing cannibalism. When beaks are properly cut at hatching, they will not usually regrow to the point where picking is a serious problem during the broiler-growing period. However, replacement pullets, if beak-trimmed at hatching, will usually need to be beak-trimmed again.

Dubbing is the term commonly used to designate the removal of a part of the comb of day-old chicks. The large combs of Single-Comb White Leghorn hens are sometimes injured, particularly when the birds are housed in cages. Such injuries can be prevented by dubbing. This procedure is relatively harmless when performed at an early age, and the comb that develops will be smaller and less likely to be injured when the hen is an adult.

Vaccination

The immunity passed to chicks from the egg protects them for the first few days of life and often makes vaccination ineffective. For this reason, most vaccines are administered after chicks have left the hatchery. There is one exception: Today, many hatcheries vaccinate day-old chicks or *in ovo* for Marek's disease.

Delivering Chicks

Provided they are not overheated or chilled during shipment, chicks may be transported for up to 3 days' time without feed or water (except what comes from the yolk), and still be expected to arrive in good condition. With broiler chicks, however, most hatcheries attempt to deliver the chicks as expeditiously as possible in order to get them on feed and water immediately after they hatch. This practice reduces early mortality and improves growth.

Figure 3.10 Brooding of turkey poults. *(Courtesy, P. R. Ferket, North Carolina State University)*

BROODING

Brooding refers to the care of young poultry beginning from the time of hatching or from the time received from the hatchery. Wherever chicks are raised—whether for broilers or replacement layers—artificial brooding of some kind is necessary. No phase of the poultry business is as important as brooding—it is the part that makes for a proper start in life. This is covered in Chapter 8. Good brooding and rearing practices bring out the good qualities inherited by chicks or poult, whereas, poor brooding practices can ruin chicks or poult of the best breeding (see Figure 3.10). Success in breeding pullets for layers is indicated by a uniform flock of fast-growing, well-feathered, healthy pullets. Success in brooding broilers results in rapid and uniform growth, good efficiency of feed utilization, fast feathering, and a low rate of mortality.

FURTHER READING

Bell, D. D., and W. D. Weaver. *Commercial Chicken Meat and Egg Production*, 5th ed. Norwell, MA: Kluwer Academic Publishers, 2002.

Physiology and Biochemistry of the Domestic Fowl. Vol. 5. Edited by B. M. Freeman. New York: Academic Press, 1984.

Reproduction in Farm Animals. Edited by B. Hafez and E. S. E. Hafez. Philadelphia: Lippincott, Williams and Wilkins, 2000.

Sturkie's Avian Physiology, 5th ed. Edited by G. C. Whittow. San Diego: Academic Press, 2000.

4

Poultry Genetics and Breeding

Objectives

After studying this chapter, you should be able to:

1. Understand generally the importance of genetics and breeding to poultry.
2. Be able to define DNA, a gene, a chromosome, and the genome.
3. Give the differences when breeding chickens for meat or eggs or when breeding turkeys.
4. List the major poultry primary breeders.
5. Understand the goals for poultry breeders.
6. List the major breeds and varieties of chickens and turkeys.
7. Be able to define heterosis, transgenics, recessive, dominant, and sex-linked genes.

More has been accomplished by the application of genetics to poultry than with other livestock. Breeding of poultry differs from the breeding of other livestock animals in view of the following:

1. The breeding program is based on pedigree and/or grandparent or great-grandparent lines. These are subjected to intensive selection.
2. There is a much larger number of generations that have been subjected to the intensive selection due to very rapid reproduction.
3. Breeding/genetics is predominantly concentrated in a few large international companies.

The poultry industry takes advantage of hybrid vigor, or heterosis, to increase growth rate. For broiler chickens, closed grandparent or great-grandparent lines are subject to intense selection. These lines are crossed to give parent lines that are in turn crossed to produce the broiler chicks. A similar situation exists for turkeys and laying hens (for table eggs, see Figure 4.1).

In developing high-performing genetic stocks, breeders have applied quantitative population ge-

netics together with inbreeding, hybridization, genomics, and other techniques. Commercial breeding requires excellent beginning stock (germplasm), scrupulous record keeping, and the application of population genetics, statistics, and computing to evaluate genetic stocks and progress from selection.

Commercial breeding of poultry has three objectives: (1) increased product output per bird or unit space, (2) increased efficiency of production (per unit feed), and to a lesser extent (3) improved quality of the product and disease resistance. The major indices are improvements in growth rate and feed conversion (in broiler chickens, turkeys, and ducks, etc.) and in egg production and feed conversion (in laying chickens). In addition, improvements in the following are desirable: fertility, hatchability, body conformation, meat yield and quality, egg size and quality, and livability.

GENETICS

The monk, Gregor Mendel, working with peas, discovered the basis of heredity; that inheritance is by units (called genes). Genes determine the develop-

Figure 4.1 Adult white laying hen. *(Courtesy, Hy-Line International)*

ment, functioning, and characteristics of poultry (e.g., growth rate or feather color). Genes, composed of deoxyribonucleic acid (DNA), are predominantly present in paired chromosomes (and therefore paired genes). In the body cells of an animal, each of the chromosomes is in duplicate. This is known as the diploid situation. In addition, there are some genes in the energy-producing organelles of the cell, the mitochondrion (pl., mitochondria). This is mitochondrial DNA.

One of each pair of genes and chromosomes comes from the sex or germ cell, or gametes (spermatozoa and ovum), from each parent. In the formation of the gametes, the ovum (egg) and the spermatozoa (sperm), a reduction division occurs and half the number of chromosomes and genes goes into each gamete. Spermatozoa and ova contain only one copy of each gene/chromosome and are referred to as haploid. Thus each gene maintains its identity generation after generation. When mating and fertilization occur, the single chromosomes from the gamete from each parent unite to form new pairs, and the genes are again present in duplicate (diploid) in the embryo's cells.

Animals are made up of millions of microscopic cells. Each cell has a nucleus containing the chromosomes and multiple mitochondria, and together these contain the full genetic makeup, or genome, of the animal. Poultry genetics/breeding will be covered briefly.

Deoxyribonucleic Acid (DNA)

In recent years, our knowledge of the functioning of DNA has expanded tremendously with the development of the fields of molecular biology and genomics. Deoxyribonucleic acid (DNA) encodes the genetic information in animals, plants, bacteria, and many viruses. DNA consists of long sequences of four nucleotides, these being:

- Adenine (A)
- Guanine (G)
- Cytosine (C)
- Thymine (T)

The genetic code is based on sequences of these four nucleotide bases. The sequences are either:

1. **Expressed or transcribed as ribonucleic acid.** This RNA is processed to form messenger RNA (mRNA). In turn, the mRNA is used as a template for the synthesis of proteins with specific amino acids. This is known as translation.

Transcription	Processing	Translation
DNA $\Rightarrow$ RNA	$\Rightarrow$ mRNA	$\Rightarrow$ Protein
Expression		

2. **Other DNA.** This is not expressed and is known as untranslated or up- and downstream elements. These control expression of RNA from the DNA.

In the sequences of DNA that get expressed as RNA, the genetic code consists of sequences of three base pairs known as codons. Each codon represents an individual amino acid or "start" or "stop." The sequence of DNA acts as a code for all the proteins in the body. The DNA is duplicated and copied during cell division so that daughter cells get the full genetic code.

DNA gets expressed and translated to make proteins that do the following:

1. Control the functioning of a cell
2. Determine the basic morphology of the cell, organ, and body
3. Control development (cell division and differentiation) and thereby dictate the entire process by which a group of cells become an organ and how a fertilized ovum becomes a chicken, a cow, or a human

Genome and Chromosomes

In avian species there are two distinct sizes of chromosomes: macrochromosomes and microchromosomes,

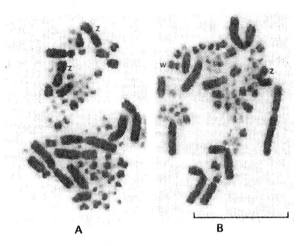

Figure 4.2 Chromosomes of the chicken showing the nine largest pairs and the ZW sex chromosomes. The chicken has 78 chromosomes.
(A) Metaphase of male chicken showing ZZ chromosomes;
(B) metaphase of female chicken showing ZW chromosomes; horizontal bar 10 μm. *(Reprinted from Mizuno, S. and MacGregor, H. (1998). The ZW lampbrush chromosome of birds: a unique opportunity to look at the cytogenetics of sex chromosomes, Cytogenetics and Cell Genetics, 80, 149–157 with permission from S. Karger - A.G., Basel)*

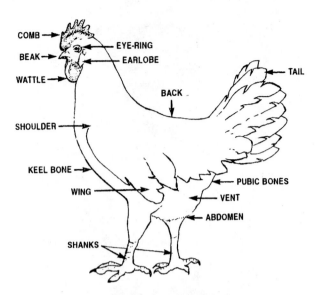

Figure 4.3 External anatomy of the chicken.
(Original diagram from E. Ensminger)

both containing many genes. Each chromosome is typed with special DNA called the telomere.

1. The *chicken genome* comprises 39 pairs of chromosomes (see Figure 4.2):
   - 8 macro- or large chromosomes
   - 30 micro- or small chromosomes
   - The sex chromosome—ZZ in males and ZW in females
2. There are 41 pairs of chromosomes in the *turkey genome.*

Size of Genome

There has been tremendous progress mapping the chicken genome recently. It is a relatively small but gene-rich genome with over 20,000 genes. There are 1.2×10^9 (1.2 billion) base pairs per haploid genome. There is also a well-developed genomic map for the chicken with about 2,000 genetic loci mapped.

Genetic Variation and Selection

Much of the genetic variation between individuals is due to point mutations, called single nucleotide polymorphisms (SNPs). These variations also influence the nucleotide sequence in RNA and the structure of proteins or control expression.

Poultry geneticists/breeders use the genetic variation and the techniques of population genetics to improve chickens, turkeys, and ducks (for ex-

ternal structure of the chicken see Figure 4.3). Much of the early research on poultry genetics used color as the end point. Multiple genes control the feather color in chickens and turkeys (for details see following). Most traits of economic interest are multiple genes (>30 for growth). Selection may be assisted by quantitative trait loci (QTL) or candidate genes. Genes may be introgressed from other populations of chickens, turkeys, and so on.

Transgenics

Transgenics is the transfer of a gene or series of genes from one organism (animal, plant, bacteria) or a synthetic gene(s) to another organism often of a different species. The recent development of transgenic chickens by the techniques of genetic engineering opens the door to intensifying research in new directions for commercial applications in commercial poultry production and for biomedical purposes. It will allow the introduction or deletion of genes. A single or multiple gene(s) for, say, resistance to a specific disease could be introduced. Poultry could be used to produce human antibodies and these could be harvested from the eggs. Opponents of transgenics or genetically modified organisms (GMOs) raise the issues of (1) food and environmental safety (DNA spreading from transgenic animals or plants to wild species with unexpected consequences) and (2) the moral responsibility of tampering with nature (removing nature's evolutionary barrier between species that do not mate). Nevertheless, poultry geneticists should keep abreast of new developments in this exciting new field.

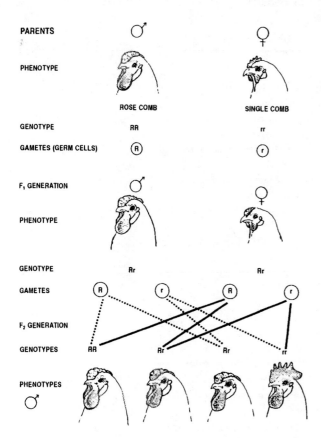

Figure 4.4 Inheritance of a pair of characters. *(Original diagram from E. Ensminger)*

Simple Gene Inheritance (Qualitative Traits)

In the simplest type of inheritance, only one pair of genes is involved. Thus, a pair of genes is responsible for comb type, plumage color, and skin color. Figure 4.4 shows the inheritance of a pair of characters; the result of mating a rose comb White Wyandotte male and a single-comb White Plymouth hen (comb types commonly found in chickens). An uppercase (capital) *R* is used as the symbol for the rose comb gene because it is dominant to the single-comb gene; and a lowercase *r* is used as a symbol for the recessive single comb.

Multiple Gene Inheritance (Quantitative Traits)

Relatively few characters of economic importance in poultry are inherited in as simple a manner as comb type. Commercially important characters such as egg production, growth rate, and feed efficiency are due to multiple genes; thus, they are called multiple-gene characters. Because these show a gradation from high to low, they are viewed as quantitative traits. Quantitative traits are of particular interest to poultry breeders to improve the average performance of a flock in several quantitative characters at the same time.

Growth rate is highly heritable and has been markedly improved by selection. It is estimated that poultry geneticists have been responsible for over three-quarters of the improvement in the efficient production of broiler meat. Egg production and hatchability have relatively low heritabilities. However, intensive selection has enabled considerable advances in these characteristics.

Dominant and Recessive Genes

Since animals are diploid, there are two copies of genes, or alleles, but with allelic variation. Genes may be:

- Recessive (not expressed) ($aa = 0$)
- Additive (both expressed equally ($a + A$) or incomplete dominance
- Dominant (expressed) ($AA = Aa$)
- Overdominant (more expression if with recessive genes) ($Aa > AA$)

Some genes have the ability to prevent or mask the expression of others, with the result that the genetic makeup of such animals cannot be recognized with perfect accuracy. This ability to cover up or mask the presence of one member of a set of genes is called dominance. The gene that masks the other one is the dominant gene; the one that is masked is the recessive gene.

Both rose comb and pea comb are dominant to single comb. When rose- and pea-combed birds are crossed, the comb is somewhat intermediate and given the name walnut comb. Color is based on multiple genes. White in the White Leghorn is dominant to colored plumage (as discussed following). Thus, if a White Leghorn is mated to a colored bird, the next (F1) generation will be all white. The dominance of white in the Leghorn is due to the presence of a gene that inhibits color (pigment formation). White is not always dominant, it depends on the genes causing it. An example of the inheritance of two pairs of characters may be obtained by crossing a Black Wyandotte with a White Plymouth Rock. In this case, rose comb (R) is dominant to single (r), and black (B) is dominant to white (b). All of the first-generation crosses from such a mating will be rose combed and black. However, the F2 generation will consist of nine rose comb/black (RB), three rose comb/white (Rb), three single comb/black (rB), and one single comb/white (rb); in a 9:3:3:1 ratio.

A dominant gene will cover up a recessive gene. Hence, a bird's breeding performance cannot be recognized by its phenotype (how it looks). This is of great significance in practical breeding. Dominance makes the task of identifying and discarding

TABLE 4.1   Examples of Dominant and Recessive Characters in Chickens

Character	Dominant or Recessive	Sex-Linked
Barred plumage	In Plymouth Rocks, dominant to nonbarring	Yes
Black plumage	Dominant to recessive white	
Broodiness	Dominant to nonbroodiness	Yes
Buff plumage	Dominant to recessive white	
Close feathering	Dominant to loose feathering	
Early sexual maturity	Dominant to late sexual maturity	Yes
Feathered shanks	Dominant to nonfeathered shanks	
Growth hormone receptor/ sex-linked dwarfism	Dominant to mutant receptor	Yes
Rose comb	Dominant to single comb	
Side sprigs	Dominant to normal comb	
Silver plumage	Dominant to gold plumage	Yes
Slow feathering	Dominant to rapid feathering	Yes
White plumage	In White Leghorns, dominant to color	
White skin and shank color	Dominant to yellow skin and shank color	

Adapted and revised from Winter A., and Funk, E. M. (1960). Poultry Science and Practice. *5th Edition, JB. Lippincott*

all animals carrying an undesirable recessive factor a difficult one. Recessive genes can be passed on from generation to generation, appearing only when two animals, both of which carry a recessive factor, happen to mate. Even then, only one out of four offspring produced will, on the average, be homozygous for the recessive factor and show it.

The recessive gene may be economically advantageous. The recessive gene can be masked by the presence of the dominant gene. Alternatively, there may be cases of overdominance where the maximal effect is observed in the heterozygous state. Some dominant and recessive characters in chickens are listed in Table 4.1. Many alleles exhibit additivity (incomplete or partial dominance) so that it is possible to distinguish the heterozygous individuals as well as both types of homozygote.

There are varying degrees of dominance. In the vast majority of cases, dominance is incomplete or partial. Also, it is now known that dominance is not the result of single-factor pairs, but that the degree of dominance depends upon the animal's entire genetic makeup together with the environment to which it is exposed, and the various interactions between the genotype and the environment.

Inheritance of Economically Important Traits in Poultry

Phenotype is controlled by both genotype and environment. For example, a hen may have genes that allow her to lay 300 eggs a year, but if she is not fed properly, given good housing, or protected from disease, these genes affecting egg production may not be expressed.

The inheritance of some economically important characters in poultry follows:

- **Plumage color.** White feathers are an important factor in the breeding of broiler chickens because they are easier to pluck clean. Similarly, white turkeys and white ducks are preferred.
- **Skin and shank color.** The different shank colors found in fowl result from various combinations of pigments in the upper and lower layers of skin. Yellow shanks are due to carotenoid pigments in the epidermis and the absence of melanin pigment. Black shanks are due to the presence of melanin in the epidermis. White shanks are due to the absence of both pigments.
- **Rate of feather development.** Early feathering is essential for minimizing pinfeathers on the dressed carcass of broilers. Modern broilers carry the early feathering gene. This is a recessive sex-linked gene. In order to identify the sexes at hatching, alleviating the necessity of vent sexing, the sire must have had early wing feathering and the dam late wing feathering. At hatching, their sons are late like their mother and the daughters are early like their father. To utilize sex-linked inheritance, it is important that the sire have the recessive trait, in this case, early, and the mother the dominant trait. This type of sex identification is important in broiler chicks in order to separate the more rapidly growing broiler males from their slower growing sisters (see Figure 4.5).
- **Color sexing.** The sex of baby chicks can be identified by silver (white) versus gold (red) down colors (see Box 4.1). The gold color is recessive and used as the male parent (s^+s^+) mated with the female,

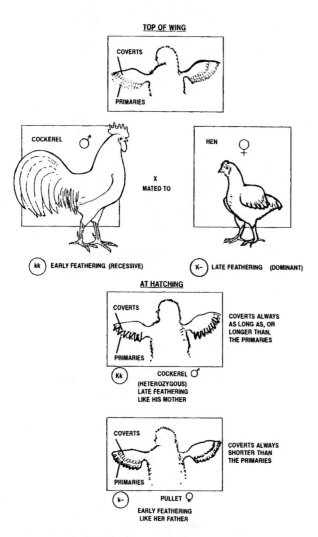

Figure 4.5 Sex-linked feather (wing) sexing by the primary and covert feathers. *(Original diagram from E. Ensminger)*

which was white (silver) (S⁻) at hatching. This mating produces white male (Ss⁺) and gold female (s⁺⁻) baby chicks. This approach is used in egg-layer strains.

- **Blood groups.** Blood-group information can be useful in detecting pedigree errors. Genetic fingerprints are, however, now preferable.
- **Egg production.** Greater efficiency of egg production can be achieved through selecting layers for the economically important characteristics such as:
    - Feed efficiency.
    - Eggs produced (rate of laying times versus length of the laying period).
    - Early sexual maturity (heritability ~30%), lowering the cost per egg.
    - Egg size (heritability ~50%).
    - Egg quality (exterior traits including shell color, texture, strength [resistance to breakage], and shape); and interior traits (lack of blood and meat spots, amount of thick white, and proportion of white to yolk).

- Shell color is heritable. Some breeds produce white eggs, others brown eggs. Varying shades of color may be expected among brown eggs. Tinted shells should be avoided for white eggs because consumers do not like them.
- Intensity of lay (heritability ~10%). Intensity or rate of lay is to the number of eggs laid by a hen during peak production. It is most important from a profit standpoint. Thus, if a flock peaks at 85% production, the average hen in the flock must be laying an egg, on the average, more than 8 out of every 10 days. Such a flock has a high intensity of lay.
- Hatchability (heritability ~12%). Hatchability is the percentage of fertile eggs that hatch under artificial incubation. It is strongly influenced by feeding and management. Poultry breeders select against lethal genes that either reduce or prevent hatching and against small egg size and poor shell quality, indirectly impairing hatchability.
- Broodiness. When the birds are brooding, they are not laying. Hence, a minimum of broodiness is desirable. Broodiness has been greatly reduced in laying chicken breeds but is still present in turkeys and bantum chickens, together with broiler chickens to a limited extent.
- **Broiler chickens.**
    - **Growth rate.** There are wide differences in body size between breeds. Large body size is of importance to breeders of broiler chickens and turkeys since mature body size is correlated with rate of growth and efficiency of feed utilization. The weight of broilers is highly heritable (~60%), and rapid progress has been made through the selection of the largest individuals at broiler age. Growth rate (~35% heritability) and feed efficiency are highly correlated. Hence, growth rate is of great importance in the breeding of meat chickens and turkeys. Improved growth leads to savings in feed consumption, labor, facilities, and so forth.
- **Feed efficiency**
    - **Body conformation.** Conformation is especially important in turkeys because they are marketed at higher weights. With broiler producers, however, conformation is of secondary importance. Body conformation is of less importance in layers since after they have finished producing eggs, their carcasses are usually processed into chicken soup and so on.

BOX 4.1   Feather Colors

It is important for broiler chickens and turkeys to have white feathers and little skin or fat coloration. Moreover, if the down of newly hatched chicks has different colors, males and females can be easily separated. Pigmentation is controlled genetically. Geneticists now understand this. It should be remembered that genes are linked on chromosomes. In the chicken, white coloration can be achieved by this combination:

II, C⁺C⁺, EE, co⁺ co⁺,   S, B, Id

Autosomal   Sex-Linked on Z chromosome

Where I is dominant white

    C is color

    E is extended black (the gene is the melanocortin I receptor gene on chromosome 30)

    co⁺ is Columbia

    S is silver (with s gold)

    B is barred

    Id is dermal melanin inhibitor

In turkeys, white feather coloration can be achieved by:

bb, CC, RR, SpSp, SlSl,   N⁻, I⁻

Autosomal   on Z chromosome

Where B is black, b is white

    C is colored

    R is red

    Sp is spotted

    Sl is slate

    N is Narragansett

    E is brown

• **Sex-linked dwarfism.** Dwarfism is a condition in which an animal (or plant) is below normal in size due to genetics. It is found in all species, including chickens. Both egg producers and broiler producers have been interested in sex-linked dwarfism. Hens that carry the sex-linked dwarf gene are about one-third smaller than normal body size, but their egg size is only about 8% smaller. Using these "mini" layers lowers body maintenance requirements and thus increases the feed efficiency. Another application of sex-linked dwarfism is in broiler production, to reduce the cost of hatching eggs. Broiler breeder hens are relatively poor layers and are large-bodied to ensure that their broiler progeny have high growth rate. As a consequence, feed costs for hatching eggs are high. However, when the dwarf gene is introduced into the female parent line, body size is reduced about 30%, hens require less feed, and the cost of hatching eggs and broiler chicks is correspondingly reduced. When dwarf female breeders are mated to normal nondwarf male line, the progeny are normal because the dwarf gene (dw) is recessive. Broiler progeny from dwarf hens show a slight reduction (2–3%) in growth to market age due to hatching from smaller eggs.

• **Livability.** Livability (viability) is greatly influenced by feeding and management. Chickens show genetic and family differences in resistance or susceptibility to diseases. Hence, poultry geneticists have concentrated on developing strains of higher livability. The possibility of developing genetically resistant strains has been demonstrated; resistant lines showed 2–3% mortality from leukosis, while susceptible strains had a 25% mortality.

Mutations

Gene changes are known as mutations. A mutation may be defined as a sudden variation that is later passed on through inheritance and that results from changes in a gene or genes. Mutations are not only rare but also they are predominantly harmful. Mutations provide the raw material for breeders to work with.

Lethal Genes and Abnormalities of Development

Lethal genes are congenital abnormalities that result in the death of an animal, most frequently during embryonic development. Other defects occur that are not sufficiently severe to cause death but that do impair the usefulness of the affected animals. Many such abnormalities are hereditary, being caused by certain "bad" genes. All lethal genes are recessive and may remain hidden for many generations. The prevention of such genetic abnormalities requires that the germplasm be purged of the "bad" genes. This means that where recessive lethals are involved, the breeder must be aware that both parents carry the gene.

Inbreeding

As practiced by the poultry breeder, inbreeding is the mating of closely related individuals, such as brother to sister, for successive generations. The procedure of developing inbred lines and forming hybrids was borrowed from hybrid corn breeders. Inbreeding is used in poultry to develop pedigree lines which, when crossed, will give progeny with high egg production, large eggs, good meat type, and good livability.

The steps in developing inbred hybrids are the following:

1. The formation of inbred lines by intensive inbreeding and selection.
2. The screening of crosses (hybrids) of inbred lines.

Heterosis

Heterosis (hybrid vigor) is a name given to the biological phenomenon that causes crossbreds to outproduce the average of their parents. For numerous traits, the performance of the cross is superior to the average of the parental breeds. Heterosis is used in many breeding programs. The first major example is hybrid seed for corn/maize (developed inbred lines are crossing). This approach is used extensively in commercial poultry production. The basis for heterosis is still not fully understood. It may involve overdominance and/or masking the effects of unfavorable recessive genes.

Heterosis is measured by the amount that the crossbred offspring exceeds the average of the two parent breeds or inbred lines for a particular trait, using the following formula for any one trait:

$$\frac{\text{Crossbred average} - \text{Purebred average} \times 100}{\text{Purebred average}} = \text{Percent Hybrid Vigor}$$

The level of hybrid vigor for all traits depends on the breeds crossed. The greater the genetic difference between two breeds, usually the greater the hybrid vigor expected.

Breeding Turkeys

Body conformation is especially important in turkeys because they are marketed at heavier weights than broilers. Since conformation, size, and color of turkeys are highly heritable, they have responded well to breeding and selection. In turkey breeding programs selections are largely based on physical appearance, growth rate, carcass quality (phenotype), and artificial insemination.

BREEDING OF CHICKENS

Types and Classes of Chickens

Type refers to the general shape and form, without regard to breed. Chickens are of three types: (1) the meat type, bred for meat production; (2) the egg type, bred for egg production; and (3) exhibition chickens, bred for competition in exhibitions. The term *class* designates groups of breeds developed in certain regions; hence, the class names: American, English, Asiatic, and so on.

Broiler Production Chickens

Broiler poultry meat comes from broiler chickens that are slaughtered at 6 to 7 weeks old at about 5 lb (2.3 kg). Depending on markets or further processing, chickens may be processed at light or heavier weights. In order to grow rapidly to market weight, the parents of broiler chickens have become very large. The broiler breeder hens are relatively poor egg layers (lower peak production and duration of lay than layer hens producing shell eggs). They have about a 9-month laying period as contrasted to almost 12 months for egg-type chickens.

Broiler chickens are white or reddish white in color. Broad-breasted males are crossed with females of a different origin. Crossbreeding is used to improve egg production of the mothers as well as the growth rate of the broiler chick. The pedigree lines (grandparent or great-grandparent stock) are selected for their ability to grow rapidly. Some breeders produce both maternal and paternal lines, others only maternal or paternal lines. For this reason, there is a tendency to merchandise broilers under the trademark names of both breeders. When a breeder produces both parents, only one trade name is used.

Egg Production Chickens

Chickens bred for egg production must be small bodied in order to consume the least feed while producing a large number of eggs. The most efficient hens weigh about 3 to 3.5 lb (1.4 to 1.6 kg) for White Leghorn–type white egg producers and about 4 lb (1.8 kg) for brown egg producers. Excellence in egg-type chickens is judged by the average egg yield per bird in large flocks, and egg size and quality, together with feed efficiency.

Chickens bred for egg laying are either (1) white and lay white eggs, tracing their inheritance to the White Leghorn, or (2) colored and produce brown eggs. Today's superior egg production is obtained by crossing synthetic strains. These chickens are sold under a breeder's trademark name, often followed by a number.

Exhibition Chickens

Exhibition chickens are produced both in the normal body sizes and in miniature (bantams). Breeding is restricted to members of the same breed and variety, and hence called purebreds. The primary purpose for these is competition in exhibitions/poultry shows. The specific standards of excellence of each variety are detailed in a book titled *Standard of Perfection*, which is published by the American Poultry Association (APA). Anyone contemplating breeding of exhibition poultry (chickens, ducks, turkeys, or geese) should obtain the latest edition, which has pictures of ideal specimens.

Since there are so few of some of the rarer varieties, problems of fertility, poor egg production, and hatchability can arise from inbreeding. Mating of close relatives can be risky, but it is used to emphasize a particular color or shape characteristic. Lack of vigor may also be a characteristic of inbred poultry.

Breeds and Varieties of Chickens

The term *breed* refers to an established group of fowls related by breeding and possessing a distinctive shape and the same general weight. A *variety* is a subdivision of a breed, distinguished either by color, color and pattern, or comb. Hence, a breed

may embrace a number of varieties, distinguished by different colors (e.g., white, black, buff, etc.); or by color/markings (e.g., Light Brahma, Dark Brahma, etc.); or by different combs (single comb, rose comb, etc.).

Standard of Perfection lists nearly 200 varieties of chickens (see Table 4.2). Examples of breeds of chicken are shown in Figures 4.6 to 4.11 and color plate 5. Two breeds (White Leghorn and White Plymouth Rock) are of major importance today, with three other breeds (Rhode Island Red, Barred Plymouth Rock, and New Hampshire) also of some importance. A total of 342 breeds and varieties of domesticated land fowl and waterfowl are also listed in *Standard of Perfection*. The American Poultry Association was organized in 1873 in the United States and Canada. Its primary objective was to standardize the breeds and varieties of domestic fowl shown in exhibitions. In the poultry industry, there are no breed registry associations like those for other livestock. There is a general lack of interest in purebreds by commercial *poultry breeders*.

Figure 4.6   Buff-Laced Polish. *(Courtesy, Watt Publishing)*

Figure 4.7   Single-Comb White Leghorns. *(Courtesy, Watt Publishing)*

TABLE 4.2 Some Breeds and Varieties of Chickens and Their Characteristics

Breed and Variety	Plumage Color	Standard Weight		Type of Comb	Color of Ear Lobe	Color of Skin	Color of Shank	Shanks/ Feathers?	Color of Egg
		Rooster	Hen						
		lb (kg)	lb (kg)						
American:									
Jersey Black Giant	Black	13 (5.9)	10 (4.5)	Single	Red	Yellow	Black	No	Brown
New Hampshire	Red	8 (3.6)	6.5 (3.0)	Single	Red	Yellow	Yellow	No	Brown
Rhode Island Red	Red	8.5 (3.9)	6.5 (3.0)	Single and rose	Red	Yellow	Yellow	No	Brown
White Plymouth Rock	White	9.5 (4.3)	7.5 (3.4)	Single	Red	Yellow	Yellow	No	Brown
White Wyandotte	White	8.5 (3.9)	6.5 (3.0)	Rose	Red	Yellow	Yellow	No	Brown
Asiatic:									
Black Langshan	Greenish black	9.5 (4.3)	7.5 (3.4)	Single	Red	White	Bluish-black	Yes	Brown
Brahma (light)	Columbian pattern	12 (5.4)	9.5 (4.3)	Pea	Red	Yellow	Yellow	Yes	Brown
Cochin (buff)	Buff	11 (5.0)	8.5 (3.9)	Single	Red	Yellow	Yellow	Yes	Brown
English:									
Australorp	Black	8.5 (3.9)	6.5 (3.0)	Single	Red	White	Dark Slate	No	Brown
Buff Orpington	Buff	10 (4.5)	8 (3.6)	Single	Red	White	White	No	Brown
Silver-Gray Dorking	Black body with silver white markings	9 (4.1)	7 (3.2)	Single	Red	White	White	No	White
Sussex	Speckled/Red/Light	9 (4.1)	7 (3.2)	Single	Red	White	White	No	Brown
White Cornish	White	10.5 (4.8)	8 (3.6)	Pea	Red	Yellow	Yellow	No	Brown
Mediterranean:									
Andalusian (blue)	Bluish slate (laced)	7 (3.2)	5.5 (2.5)	Single	White	White	Slaty-blue	No	White
Anocona	Black, may be white tipped	6 (2.7)	4.5 (2.0)	Single and rose	White	Yellow	Yellow	No	White
Minorca (S.C. Black)	Black	9 (4.1)	7.5 (3.4)	Single	White	White	Dark slate	No	White
White Leghorn	White	6 (2.7)	4.5 (2.0)	Single and rose	White	Yellow	Yellow	No	White

Figure 4.8 Barred Plymouth Rocks. *(Courtesy, Watt Publishing)*

Figure 4.11 White Cornish. *(Courtesy, Watt Publishing)*

Figure 4.9 White Plymouth Rocks. *(Courtesy, Watt Publishing)*

Figure 4.10 New Hampshires. *(Courtesy, Watt Publishing)*

Business Organization of Poultry Breeding

Commercial chicken producers are either egg producers or broiler producers. For either, it is net return that counts. The commercial producer is interested in increasing product output per bird,

the efficiency of production, and the product quality. Improvements in fertility, hatchability, growth rate, body conformation, egg yield, meat yield, feed conversion, egg quality, meat quality, and viability (e.g., pleuropneumonia-like organism-free [PPLO-free]) contribute to profitability. No industry has done a better job than the poultry industry in incorporating science and technology. Hand in hand with the transformation of the poultry industry, there has come a very high degree of specialization. From a breeding standpoint, today's poultry breeding is centered in two types of business enterprises: (1) the primary breeder and (2) the broiler breeder or multiplier together with hatcheries.

Primary Breeders

Modern primary breeders are large, well-financed and managed international companies. These are incorporated with various departments (e.g., breeding research and development, sales, advertising, etc.). Major primary breeders of chickens and turkeys are listed in alphabetical order as follows:

- Aviagen Group (a holding company set up by BC Partners)(global brands: Arbor Acres—broiler chickens), Nicholas Turkey Breeding Farms, producing turkey poults and Ross Breeders, producing broiler stock.
- Cobb-Vantness (a wholly owned subsidiary of Tyson Food) produces broiler chicks.
- Hendrix Poultry Breeding is owned by the Hendrick Group and Nutreco and operated under the Nutreco umbrella. Hendrix Poultry Breeding produces under the following trademarks: *Hisex* (chicks for both white and brown layers), *Bovans* (chicks for white and brown layers)(produced by Centurian Poultry in Georgia for the North American market), and *DeKalb* (chicks for

laying hens). Hybrid Turkeys (a subsidiary of the parent company, Nutreco) is a primary breeder of turkeys and is located in Canada.

- Hubbard-ISA (a subsidiary of Merial Animal Health) is associated with the following business groups: British United Turkeys (BUT) and British United Turkeys of America (BUTA), producing turkey poults; Hubbard-ISA, producing chicks for broiler production (brand names: *Hubbard*, *ISA*, and *Shaver*) and egg laying (brand names: *Babcock, Hubbard-ISA, Shaver lines*).

- Hy-Line International and Hy-Vac are U.S. subsidiaries of the Lohmann-Wesjohann Group. Hy-Line International produces chicks for egg-layer hens and Hy-Vac produces eggs for vaccines. Other parts of the group, Lohmann-Tierzucht and H&N International, are breeders of egg layers and specific pathogen-free (SPF) eggs.

Figure 4.12 DNA laboratory.

There are also primary breeders of ducks and geese. Cherry Valley Farms (part of the Nickerson Group) is the world's largest primary breeder of Pekin ducks. Sepalm produces customized lines of geese and ducks, particularly for "foie gras."

The primary breeders of chickens and turkeys are international businesses operating in various forms in North and South America, Europe, Asia, and Africa. They have recognized the importance of copyrighted trade and brand names. They have well-trained geneticists; computer specialists; veterinarians, including pathologists; and frequently nutritionists on their staffs (see Figures 4.12 and 4.13). In the case of broiler chickens, the primary breeder sells parent line broiler breeders derived from grandparent stock or uses the parent line in-house to produce the chicks destined to become broilers. With hens laying table eggs, either the parent line or the commercial line is sold as chicks.

In some cases, primary breeders will contract out the maintenance of pedigree and/or grandparent flocks. A business model may become contracting out even the ownership of the pedigree flocks, but retaining the control of the breeding program and improving the genetics. It is possible to envision breeding companies as information technology companies.

Figure 4.13 DNA testing.

Hatcheries/Multipliers (Including Broiler Breeders)

The multipliers multiply the stock supplied by the primary breeders. The hatcheries hatch the egg from the primary breeders or multipliers (see

Chapters 14, 15, and 16 for details on laying hens, broiler chickens, and turkeys, respectively). Chicks or poults are sold to producers, who then grow them out as commercial egg layers, broilers, or market turkeys (see Figure 4.14). Primary breeders and multipliers along with hatcheries operate together through a contract or in a single vertically integrated company. A contract, signed by the parties, specifies that the breeder will provide the hatchery with the breeding stock for the hatchery supply flock or parent flocks. These arrangements give the primary breeder control over the stock sold by the hatcheries/multipliers. The contract agreement is desirable to the multipliers because they can concentrate on producing and to the hatcheries because they can concentrate on hatching eggs and selling and providing chicks or poults.

Figure 4.14 Baby chicks. *(Courtesy, Hy-Line International)*

Methods of Mating

For chickens, the most common methods of mating are flock and pen mating, although artificial insemination can sometimes be used.

Flock or mass mating means that a number of males are allowed to run with the entire flock of hens. Better fertility may be obtained from flock mating than from pen mating. The number of hens per male will vary with the size and age of the birds. With the light breeds (e.g., Leghorns), it is customary to use one male for 15 to 20 hens. With the heavy breeds, one male is placed with each 8 to 12 hens. Age is also a factor; young males are more active than older ones, with males over 3 years old not satisfactory.

In pen mating, a pen of hens is mated with one male. If the birds are trapnested and each hen's leg-band number recorded on its egg, this system makes it possible to know the parents of every chick hatched from a pen mating. About the same number of hens are mated with one male in pen mating as in flock mating. However, fertility may not be as good.

Artificial insemination can be credited as a primary factor that has led to the rapid improvement of turkeys. It has served a twofold purpose:

- A single ejaculate can be collected and used to inseminate several females, whereas natural mating only allows the ejaculate to be used for a single female.
- Artificial insemination has permitted the selection of heavier, meatier birds see Figures 4.15 A–C. Male turkeys are so large that natural mating is difficult.

Turkey hens should be inseminated with fresh diluted semen (equivalent to whole 0.025 ml) per insemination. One- to two-week intervals between inseminations maintain maximum fertility.

Selecting and Culling Chickens

The terms *selection* and *culling* carry opposite connotations. *Selection* aims at progress; it deals with retaining the best in the flock—seldom more than the top 20 to 25% and generally not more than 10 to 15%—for carrying forward. However, *culling* refers to the removal of the least productive part of the flock. It is aimed at prevention of retrogression rather than making progress.

Methods of Selection

Except for the poultry fancier, who is interested in breeds and varieties from the standpoint of *Standard of Perfection*, only exhibition chickens are selected on the basis of show winnings.

The following methods of selection are used in poultry breeding:

Individual or mass selection. Selection is based on physical appearance. Many geneticists maintain that this system has little value. Yet, some poultry improvement may have resulted from removal of undesirable individuals that are slow gainers and poor layers. Where a flock is above average in productivity, and for those characteristics of economic importance influenced by

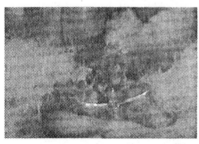

A

B

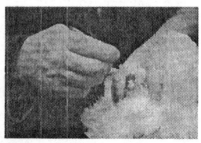

C

Figure 4.15 Artificial insemination. (A) Collecting semen ("milking" a male), (B) Poultry semen, (C) Inseminating female *(ISU photos by Bob Elbert))*

multiple genes, individual or mass selection is ineffective.

Pedigree selection. Pedigree selection is of special importance when production data are not available, or when selection is being made between two males that are comparable in all other respects. For example, pedigree selection is of value where one is selecting breeding males whose dams have very different production records (albeit unlikely), such as between 250 eggs and 175 eggs. In making use of pedigree selections, however, it must be remembered that ancestors closest in the pedigree are much more important than those many generations removed.

Family selection. Family selection refers to the performance or appearance of the rest of the members of the family, particularly the bird's sibs (sisters and brothers). Sons of a high egg-producing hen whose sisters each lay high numbers of eggs are more likely to transmit desirable genes for egg production to their daughters than are the sons of a very high egg-producing hen whose sisters had moderate egg production. Family testing has been important to genetic progress in producing high-laying strains of chickens.

Indeed, individuality, pedigree, and family are important, and all should be used as tools in selection, but the only really sure basis concerning the ability of an individual to transmit genes for the desired characteristics to most of its progeny is based on a breeding test or progeny test.

Measurement to Improve Egg-Layer Genetics

With layers, annual egg production is the single most important selection criterion to invoke. Yet, it is not the only trait of importance to the breeder. He or she must select for feed efficiency, egg quality (size, shape, color, shell texture, and interior quality), body size and general appearance, livability, and so on. It must be remembered, however, that selection for two or more characteristics automatically cuts down the effectiveness of selection for any one of them. Also, annual egg production is of low heritability, probably about 20%. This means that improvement in egg production through selection is relatively slow.

Measurement to Improve Broiler Genetics

Growth rate of broiler chickens to market weight has an estimated heritability of 35 to 40%. This trait is of major importance in the breeding of meat chickens as well as turkeys because rapid growth means a savings of labor, feed consumption, and overhead in production costs. Also, feed efficiency is highly correlated with growth rate.

Because of the high heritability of growth rate and the strong breeding programs, genetic improvement for market weight continues to make excellent progress. Carcass quality and uniformity have been improved through crossbreeding.

Random Sampling Performance Test/Multiple Unit Poultry Test

Random Sample Performance Tests were used from the 1940s to 1980s to compare the performance of commercial poultry (broiler chickens, laying hens for shell eggs, and turkeys) under uniform testing conditions.

National Poultry Improvement Plan (USA)

The National Poultry Improvement Plan for chickens was established in 1935 under the USDA. The National Turkey Improvement Plan started in 1943. In 1970, the two were consolidated into the National Poultry Improvement Plan. The plan's objective is to provide new technology through a cooperative state-federal program to effectively improve poultry health. The plan, developed by industry representatives and government officials, established standards for the evaluation of poultry breeding stock and hatcheries, for instance, freedom from hatchery disseminated diseases. Initially, the program aided in identifying the bacterial disease transmitted from the hen to the chick through the egg caused by the bacterium *Salmonella pullorum*. Today, breeders screen multiple diseases including pullorum, Arizona infection, fowl typhoid, infectious bronchitis, Newcastle, *Mycoplasma gallisepticum*, and *Salmonella* serotype Enteritidis.

Information on the National Poultry Improvement Plan is available on the USDA Animal and Plant Health Inspection Service Website (http://www.aphis.usda.gov/vs/npip/). In 2002, the plan was amended to be consistent with developments in the poultry industry and with technological advances. The present National Poultry Improvement Laboratories are the following:

- Animal Disease Diagnostic Laboratory
- Purdue University
- Missouri Department of Agriculture Veterinary Diagnostic Laboratory
- California Animal Health and Food Safety Laboratory System
- Nebraska Veterinary Diagnostic Laboratory
- Colorado State University Veterinary Diagnostic Laboratory

- North Carolina Veterinary Diagnostic Laboratories
- Cornell University Diagnostic Laboratory
- Ohio Animal Disease Diagnostic Laboratory
- Florida Veterinary Diagnostic Laboratories
- Oregon State University Veterinary Diagnostic Laboratory
- Georgia Poultry Laboratory Network
- Penn State University Veterinary Diagnostic Laboratory
- Iowa State Veterinary Diagnostic Laboratory
- Texas Veterinary Medical Diagnostic Laboratory
- Kansas State University Veterinary Diagnostic Laboratory
- University of Connecticut Veterinary Diagnostic Laboratory
- Michigan State University Animal Health Diagnostic Laboratory
- University of Illinois Veterinary Diagnostic Laboratory
- Minnesota Veterinary Diagnostic Laboratory
- University of Missouri Veterinary Diagnostic Laboratory
- Mississippi Veterinary Diagnostic Laboratory
- Washington State University Veterinary Diagnostic Laboratory

Acceptance of the plan is optional with states and the industry. There is strong support for the plan by industry and government. The plan has been highly effective in controlling poultry diseases. By 1980, pullorum disease had effectively ceased to exist in the United States. Of the 35.4 million chickens tested, only 0.000002% were reactors; and of the 3.2 million turkeys, there were no reactors. This is no small achievement since in 1920 when pullorum testing started, there were 11% pullorum reactors in chicken flocks.

Culling

The attitude toward culling has changed as laying flocks have increased in size. Culling is not practiced in most high egg-producing commercial flocks. With improved poultry genetics, management, nutrition, and health, only superior birds are placed in laying houses. Hence, flocks contain few poor layers.

For small-scale producers, culling can increase efficiency of egg production. Culling is particularly useful when hens are kept for a second year of egg production. Non-laying and low-producing birds are identified and removed. This keeps the egg production rate high and saves the feed costs for unproductive birds. Culling should take place throughout the year. Obviously weak or diseased birds should be culled, as should poor layers (see Appendix I).

BREEDS AND BREEDING TURKEYS

Only one breed of turkeys is recognized by *Standard of Perfection*; hence, we should refer to varieties rather than breeds. The varieties of turkeys of historical importance are listed in Table 4.3 (also see Chapter 16). Today, only the Broad-Breasted White is commercially important. Broad-Breasted White refers to strains developed since 1950 by first crossing Bronze and White Holland varieties and then backcrossing the second-generation white progeny to Bronze males. This procedure is repeated for several generations so that the resulting Broad-Breasted White is essentially a Bronze turkey with white feathering and broad-breasted in conformation. These birds were developed in response to processors' objections to the dark pins of the Broad-Breasted Bronze. The USDA developed a small white turkey variety. This was released as the Beltsville Small White in 1941. However, turkey growers did not adopt it because meat produced per bird, per unit facility, or per unit labor was low. The Beltsville Small White is little used today.

Most turkeys are bred as standardbreds; that is, bred pure rather than hybridized. Also, individual selections and mass inseminations are the usual practices. Color, size, and conformation are highly heritable, with the result that they have responded well to these standard breeding methods. A few breeders have trapnested their birds and selected for egg production and hatchability.

In the case of turkeys, it appears that the breeding accomplishments in developing fast-growing, broad-breasted birds have left the breeder with a good market bird but with breeding populations seriously lacking in reproductive qualities. Part of this is attributed to the fact that heavily fleshed males are clumsy in mating. For this reason, many breeders are using artificial insemination.

 USEFUL WEBSITES

Chicken genome

http://www.ri.bbsrc.ac.uk/chickmap/ChickMapHomePage.html
or

http://www.genome.iastate.edu/chickmap/

Chickens-Broilers

Arbor Acres: http://www.aa-na.aviagen.com/home.asp
Ross Breeders: http://www.rossbreeders.com/

TABLE 4.3 Varieties of Turkeys and Their Characteristics

Variety	Standard Weights		Plumage Color	Beak	Color of Throat Wattle	Beard	Shank and Toes	Comments
	Adult Tom	Adult Hen						
	lb (kg)	lb (kg)						
Black	33 (15)	18 (8.2)	Metallic black	Slaty black	Red, changeable to bluish white	Black	Pink in adults	Black color evolved from selecting darker birds in the population. Black turkeys were popular in Spain, France, and Italy.
Bourbon Red	33 (15)	18 (8.2)	Brownish red, with white wing and tail markings	Light horn at tip, dark at base	Red, changeable to bluish white	Black	Reddish pink in adults	Developed in Bourbon County, Kentucky. Very attractive.
Broad-Breasted White	33 (15)	18 (8.2)	Pure white	Light pinkish horn	Red, changeable to pinkish white	Deep black	Pinkish white	Developed from crossing Broad-Breasted Bronze with white feathered variety. Very similar to Bronze; only white, and slightly higher fertility.
Bronze	36 (16)	20 (9.1)	Black; with an iridescent sheen of red, green, and bronze	Light horn at tip, dark at base	Red, changeable to bluish white	Black	Dull black in young; smoky pink in mature birds	The Broad-Breasted Bronze is a sub-variety. Of all meat animals, the Broad-Breasted Bronze most uniformly produces a well-fleshed carcass.
Narragansett	33 (15)	18 (8.2)	Dull black, with white markings	Horn	Red, changeable to bluish white	Black	In adults, deep salmon	Developed in Narragansett Bay area of Rhode Island, by crossing domestic stock on wild turkeys.
White Holland	33 (15)	18 (8.2)	Pure white	Light pinkish horn	Red, changeable to pinkish white	Black	Pinkish white	White Holland is thought to be a sport (mutation) from bronze turkeys

Cobb-Vantress: http://www.cobb-vantress.com/

Hubbard-ISA: http://www.hubbard-isa.com/

Chickens-Layers

Hubbard-ISA: http://www.hubbard-isa.com/

Hendrix Poultry: http://www.hendrix-poultry.nl

Lohmann–Hy-Line International: http://www.hyline.com

Turkeys

British United Turkeys: http://www.but.co.uk

British United Turkeys of America: http://www.ansci.umn.edu/poultry/resources/buta-pubs.htm

Nicholas Turkey Farms: http://www.nicholas-turkey.com

Hybrid turkeys: http://www.hybridturkeys.com

Ducks

Cherry Valley Farms: http://www.cherryvalley.co.uk/

FURTHER READING

Poultry Breeding and Genetics. Edited by R. D. Crawford. Amsterdam, The Netherlands: Elsevier Science Publishers, 1990.

5

Fundamentals of Poultry Nutrition

Objectives

After studying this chapter, you should be able to:

1. List the basic nutrient categories and give functions of each.
2. List the major vitamins and give a deficiency sign for each.
3. List the macrominerals and give a deficiency sign for each.
4. List the microminerals and give a deficiency sign for each.
5. Define commonly used energy terms.
6. List and discuss factors that affect an animal's requirement for various nutrients.

The primary purpose of raising poultry is to transform feeds into meat and eggs. In order to do this efficiently and economically the principles of nutrition need to be understood (for healthy chickens feeding see Figure 5.1). Applying sound nutritional principles to genetically superior birds under good management has resulted in the rapid increase in poultry worldwide in recent decades.

A thorough knowledge of the anatomy and physiology of the avian digestive system is important to fully understanding the fundamentals of poultry nutrition. Refer to Chapter 2 for this topic.

NUTRIENT COMPOSITION OF POULTRY AND EGGS

Nutrition encompasses the various chemical and physiological reactions that change feed elements into body elements. It follows that knowledge of body and egg composition is useful in understanding poultry response to nutrition. Table 5.1 gives the body composition of poultry compared to that of some of their mammalian counterparts used in meat production.

Within each type of animal there is a wide range in body composition, depending on age and nutritional status. These ranges are evident in poultry, as shown in Table 5.1. However, the following conclusions relative to body composition may be drawn:

1. **Water.** On a percentage basis, the water content shows a marked decrease with advancing age, maturity, and fatness.
2. **Fat.** The percentage of fat normally increases with growth and fattening. It is recognized, of course, that the amount of fat is materially affected by the feed intake.
3. **Fat and water.** As the percentage of fat increases, the percentage of water decreases.
4. **Protein.** The percentage of protein remains rather constant during growth, but decreases as the animal fattens. On the average, there are 3 to 4 units (lb or kg) of water per 1 unit (lb or kg) of protein in the body.
5. **Ash.** The percentage of ash shows the least change. However, it decreases as animals fatten because fat tissue contains less mineral than lean tissue.
6. **Composition of gain.** The data presented in Table 5.1 clearly indicate that gain in weight may not provide an accurate measure of the actual gain in energy of the animal, because it tells nothing about the composition of gain. This is important because efficiency of feed utilization

TABLE 5.1   Body Composition of Animals at Different Weights and Ages, Ingesta-Free (Empty) Basis, and Egg

Species	Age or Status	Weight		Water	Fat	Protein	Ash
		(lb)	(kg)	(%)	(%)	(%)	(%)
Poultry:							
Chick	Newly hatched	0.09	0.041	78.8	4.0	15.3	1.9
Broiler	Market	3.5	1.6	64.0	14.2	18.8	3.7
Hen	Layer	4.5	2.04	62.4	15.6	18.7	4.0
Turkey	8 Weeks	5.1	2.3	70.7	4.5	20.7	4.1
Turkey	Market	18.0	8.2	60.2	18.7	19.4	3.1
Egg	Newly laid	0.136	0.062	66.0	10.0	13.0	11.0
Cattle:							
Steer	Choice grade	1,050	477.3	53.5	26	17	3.5
Sheep:							
Lamb	Market	100	45.4	53.2	29	15	2.8
Pig:							
Pig	Market	220	100	50	34.4	13	2.6

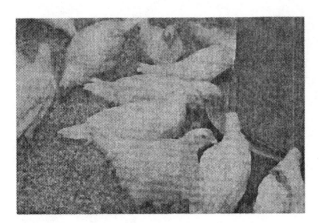

Figure 5.1   Healthy chickens feeding. *(ISU photo by Bob Elbert)*

(pounds feed per pound body gain) is greatly influenced by the amount of fat produced.

7. **Species comparison.** The following species' differences are noteworthy:
   a. The bodies of poultry contain less fat, but more water, than cattle, sheep, and pigs in comparable condition.
   b. Because of their smaller skeletons, the bodies of pigs contain less ash than those of cattle, or poultry.
   c. At normal market stage, broilers are higher in protein than four-footed animals.

The chemical composition of the body varies widely between organs and tissues and is more or less localized according to function. Thus, water is essential to every part of the body, but the percentage composition varies greatly in different body parts; blood plasma contains 90 to 92% water, muscle 72 to 78%, and bone 45%. Proteins are the principal constituents, other than water, of muscles, tendons, and connective tissues. Most of the fat is lo-

calized under the skin, near the kidneys, and around the intestines. But it is also present in the muscles, bones, and elsewhere.

Although carbohydrates are very important in nutrition, they account for less than 1% of the body composition, being concentrated primarily in the liver, muscles, and blood. Instead, they provide the basic building blocks for fat (energy reserves) and carbon skeletons for amino acids and are metabolized rapidly for the production of energy.

CLASSIFICATION OF NUTRIENTS

Birds do not utilize feeds as such but portions of feeds called nutrients that are released by digestion and then absorbed into the body to be used by the various organs and tissues.

There are more than 40 nutrients required by poultry. These are classified into six categories. They include carbohydrates, fats, proteins, minerals, vitamins, and water. Knowledge of the basic functions of the nutrients in the body and of the interrelationships between various nutrients and other metabolites within the cells of the bird is necessary before one can make practical scientific use of the principles of nutrition.

Energy is frequently listed with nutrients since it is an important consideration in diet formulation. It results from the metabolism of carbohydrates, proteins, and fats in the body and should be considered a property of feedstuffs and diets.

FUNCTIONS OF NUTRIENTS

Of the feed consumed, a portion is digested and absorbed for use by the bird. The remaining undigested portion is excreted and constitutes the major portion of the feces. Nutrients from the digested feed are used for a number of different body processes, the exact usage varying with the species, class, age, and productivity of the bird. All birds use a portion of their absorbed nutrients to carry on essential functions, such as body metabolism and maintaining body temperature and the replacement and repair of body cells and tissues. These uses of nutrients are referred to as maintenance. That portion of digested feed used for growth or the production of eggs is known as production requirements.

Maintenance

A bird's cells and organs are never idle. Birds use nutrients to keep their bodies functioning every hour of every day, even when they are not being used for production. Maintenance requirements may be defined as the combination of nutrients that are needed by the bird to keep its body functioning without any gain or loss in body weight or any productive activity. Although these requirements are relatively simple, they are essential for life itself. A bird must have the following:

- Heat to maintain body temperature
- Sufficient energy to keep vital body processes functional
- Energy for minimal movement
- The necessary nutrients to repair damaged cells and tissues and to replace those that have become nonfunctional

Thus, energy is the primary nutritive need for maintenance. Even though the quantity of other nutrients required for maintenance is relatively small, it is necessary to have a balance of the essential proteins, minerals, and vitamins.

No matter how quiet a bird may be, it requires a certain amount of fuel and other nutrients. The least amount on which it can exist is called its basal maintenance requirement There are only a few times in the normal life of a bird when only the maintenance requirement needs to be met. Such a status is closely approached by mature males not in service and by mature, nonproducing females. Nevertheless, maintenance is the standard benchmark or reference point for evaluating nutritional needs.

Even though **maintenance requirements** might be considered as an expression of the nonproduction needs of a bird, there are many factors that affect the amount of nutrients necessary for this vital function; among them are the following:

- Exercise
- Weather
- Stress
- Health
- Body size
- Temperament
- Level of production
- Individual variation

The first four are external factors. They are subject to control to some degree through management and facilities. The others are internal factors. They are part of the bird itself. Both external and internal factors influence requirements according to their intensity. For example, the colder or hotter it gets from the most comfortable (optimum) temperature, the greater will be the maintenance requirements.

Growth

Growth may be defined as the increase in size of bones, muscles, internal organs, and other parts of the body. It is the normal process before hatching and after hatching until the bird reaches its full, mature size. Growth is influenced primarily by nutrient intake. The nutritive requirements have increased as we have improved the genetic capabilities of poultry.

Growth is the very foundation of animal production. Poultry will not make the most economical finishing gains unless they have been raised to be vigorous. Nor can one expect the most satisfactory production of eggs from laying hens unless they are well developed during their growing period.

Generally speaking, organs vital for the maintenance of life (e.g., the brain, which coordinates body activities, and the gut, upon which the rest of the postnatal growth depends) are early developing; and the commercially more valuable parts, such as muscle, develop later. The nutritive needs for growth vary with age, breed, sex, rate of growth, and disease.

Age

In comparison with older birds, young birds generally have the following characteristics:

- Consume more feed per unit of body weight
- Utilize feed more efficiently, in pounds of feed eaten per pound of body gain
- Have a higher requirement for protein, energy, vitamins, and minerals per unit of body weight
- Require a more concentrated and more easily digested diet

The digestive systems of newly hatched birds are somewhat immature. Many digestive enzymes increase rapidly over the first 2 weeks of life. For this reason, the type of diet is particularly important during this period of functional and physical development of the digestive tract and should include a balance of all available nutrients in a readily available

TABLE 5.2   Days Needed to Double Birth Weight and Months Needed to Reach 50% of Mature Body Weight for Poultry and Humans

Species	Days to Double Birth Weight	Months to Reach 50% Mature Weight
	(days)	(months)
Chicken	5	2
Duck	4	1½
Turkey	5	4
Human	150	115–145

form. Young birds (chicks, poults, ducklings, and goslings) should be fed a highly nutritious diet.

After an initial adjustment period (which may be prolonged if the feed, environment, or disease level are unfavorable), the rate of gain of birds is very rapid when measured as a percentage of body weight. Table 5.2 shows the growth rate of chickens, ducks, and turkeys as measured by the days needed to double birth weight and the number of months needed to reach 50% of mature body weight. As shown, the juvenile period of humans is extremely long by comparison.

Breed

Larger breeds of birds grow more rapidly than smaller breeds and have a higher nutrient requirement.

Sex

Growth studies involving young birds of both sexes, and of all species, reveal the following:

- Males gain more rapidly than females and have a higher feed requirement.
- Males use feed more efficiently for body weight gain than females, because of the higher water and protein content and the lower fat content of the increased body weight.
- Mature average size is larger in males than in females.
- Females reach maturity faster than males.

Rate of Growth

In recent years, the accent in broiler production has been on faster growth with improved feed efficiency and marketing at an early age. Achieving this goal has involved improved nutrition. Today, broilers generally reach market weight in 6 to 7 weeks, with efficiencies in feed utilization commonly being less than 1.8 unit (lb or kg) of feed per unit (lb or kg) of bird.

Rapid gains call for more nutrients. In turn, this necessitates high-energy, well-balanced rations. For the most part, fast gains are efficient gains; when birds grow at maximum rates, they require fewer nutrients and fewer pounds of feed per pound of gain.

Health

Ill health—diseases and parasites—results in poor growth and development in young birds. When the causative factor is severe, growth may be stunted.

Reproduction

Nutrition plays an important role in reproduction. Egg production involves feeding for number of eggs, egg quality, hatchability, and control of molt and broodiness. The nutritive needs for commercial egg production include those for maintenance of the birds, growth of pullet layers, and the formation of eggs. The nutritive requirements are greater for birds with an inherited capacity for high egg production than for those that lay only a few eggs. A large egg contains about 75 calories of metabolizable energy, 7.5 g of crude protein, and 2 g of calcium.

With poultry, the development and hatching of a fertile egg constitute reproduction. As with mammals, the nutrient requirements of poultry breeders (including chickens, turkeys, ducks, geese, etc.) tend to be more rigorous than those for commercial laying hens. For optimal hatchability and embryonic development, breeders require greater amounts of vitamins A, D, E, B_{12}, riboflavin, pantothenic acid, niacin, and of the mineral, manganese. Birds intended as breeders should be started on breeder diets at least a month before hatching eggs are to be saved.

NUTRIENTS

Nutrients are utilized in one of two metabolic processes: for anabolism or for catabolism.

Anabolism is the process by which nutrient molecules are used as building blocks for the synthesis of complex molecules (e.g., proteins in muscles). Anabolic reactions require the input of energy into the system. Catabolism is the breakdown of complex molecules (e.g., proteins in muscles). This ultimately leads to the oxidation of nutrients, liberating energy that is used to fulfill the body's immediate needs.

Energy (Carbohydrates and Fats)

The energy requirement may be defined as the amount of available energy that will provide for growth or egg production at a high enough level to permit maximal economic return for the production unit. Energy is required for practically all life processes, for the action of the heart, maintenance of blood pressure and muscle tone, transmission of nerve impulses, ion transport across membranes, reabsorption in the kidneys, protein and fat synthesis, and the production of eggs.

A deficiency of energy is manifested by slow or stunted growth, body tissue losses, and/or lowered production of meat and eggs rather than by specific signs, such as those that characterize many mineral and vitamin deficiencies. For this reason, energy deficiencies may go undetected for extended periods of time.

Normal diets contain carbohydrates, fats, and proteins. Although each of these has specific functions in maintaining a normal body, all of them can be used to provide energy for maintenance, for growth, and for egg production. From the standpoint of supplying the normal energy needs of birds, however, the carbohydrates are by far the most important, with more of them being consumed than any other compound, whereas the fats are next in importance for energy purposes. Carbohydrates are usually more abundant and cheaper, and most of them are very easily digested, absorbed, and transformed into body fat. Also, carbohydrate feeds may be more easily stored in warm weather and for longer periods of time than fats. Feeds high in fat content are likely to become rancid, and rancid feed is unpalatable, if not actually injurious in some instances. Feed intake is largely governed by the energy concentration of the feed. Although the bulkiness of feed can alter feed intake, the bird, for the most part, will eat to satisfy its energy needs. Because of this, special attention must be given to nutrient ratios, especially the ratio of energy to various nutrients such as amino acids and minerals. For example, if the producer uses a high-energy feed, the protein content of the feed must be high if the bird is to ingest adequate amounts of protein. Conversely, if the energy content of a feed is low, the protein content should be low also; otherwise, the bird will consume excessive amounts of expensive protein.

The metabolic rate of poultry is an indication of the energy needs of the bird. Several factors can affect this rate, including the following:

1. **Breed and strain.** The development of breeds and strains for specific purposes has resulted in genetic differences in the efficiency of energy utilization. For example, researchers at the University of Guelph found that Leghorn chickens obtained 3% more metabolizable energy from their diet than broiler strain chickens. In turkeys, separate lines have been developed for males and females.
2. **Activity.** Birds that have access to large areas have a higher metabolic rate than their coun-

terparts that are confined to small cages, which restrict movement.
3. **Diurnal rhythm.** Within individual birds, the metabolic rate will be lower at night.
4. **Environmental temperature.** Poultry are extremely sensitive to environmental temperature, especially hot weather. Since there is little heat dissipated through sweating, much of the body heat must be lost through panting.
5. **Diet.** The metabolic rate of birds is affected by the type of diet. A low-energy, high-protein diet will necessitate different digestive processes than a high-energy, low-protein diet.
6. **Level of production.** A bird that is in heavy production will have different energy demands than one that is not in production.
7. **Other factors.** Other factors affecting dietary requirements for energy, in addition to those already listed, are stress, body size, and feather coverage.

Carbohydrates

Carbohydrates constitute about 75% of the dry weight of plants and grain and make up a large part of the poultry ration. They are the primary source of energy in poultry diets and can also be converted into fat and stored. Carbohydrates are made of units termed monosaccharides. These are simple sugars of which the most common ones are glucose, fructose, galactose, and mannose. By far the most common monosaccharide is glucose. Monosaccharides are combined principally into disaccharides and polysaccharides. The principal disaccharides are milk sugar, lactose, and common table sugar, sucrose. Another important disaccharide is maltose. It is a breakdown product of starch. The polysaccharide, glycogen, is the storage form of carbohydrate in the liver and muscles of poultry and other birds and mammals.

The most common polysaccharides in poultry diets are the following:

- Starch (storage form of carbohydrate in plant seeds)
- Pentosans
- Cellulose and hemicellulose (the structural carbohydrates in plants)

The monosaccharides, disaccharides, starch, and pentosans are readily digestible and sometimes referred to as nitrogen-free extract (NFE). Poultry do not produce the enzymes necessary to break down cellulose and hemicellulose; digestion of these is very limited. The term often used for these compounds of limited digestibility is *fiber*. Fiber is largely insoluble and represents structural components of plants.

Carbohydrate storage in the animal body is as glycogen and is limited. It may represent 1% of muscle and up to 8% of the liver. Blood glucose is about 200–300mg% for chickens while mammals range from 50–100mg%. This small quantity of glucose in the blood, which is constantly replenished by changing the glycogen of the liver back to glucose, serves as the chief source of fuel with which to maintain the body temperature and to furnish the energy needed for all body processes.

Fats

Lipids (fats), like carbohydrates, contain three elements—carbon, hydrogen, and oxygen. Fats are soluble in such organic solvents as ether, chloroform, and benzene.

As feeds, fats function much like carbohydrates in that they serve as a source of heat and energy and for the formation of fat. Because of the larger proportion of carbon and hydrogen, however, fats liberate more heat than carbohydrates when metabolized, furnishing approximately 2.25 times as much heat or energy per unit on oxidation as do the carbohydrates. A smaller quantity of fat is required, therefore, to serve the same function. Although fats are used primarily to supply energy in poultry diets, they also improve the physical consistency of rations and the dispersion of microingredients in feed mixtures, and serve as carriers for (and aid in the absorption of) fat-soluble vitamins. Fats used for feeding poultry are derived from three sources: animal or poultry fats obtained from the rendering industry, restaurant greases, acidulated soapstocks from the vegetable oil industry, and/or mixtures thereof. The nutritional value of fats for poultry feed is determined by moisture, impurities, unsaponifiables, free fatty acids, total fatty acids, and fatty acid composition.

The polyunsaturated linoleic and arachidonic acids are considered to be essential fatty acids. They have specific functions in the body that are not related to energy production. Birds exhibit poor growth, fatty livers, reduced egg size, and poor hatchability without these essential fatty acids. Fats for poultry feed should be stabilized against oxidation. The metabolizable energy (ME) contribution of fats may be influenced by their fatty acid composition, free fatty acid content, level of fat inclusion in the ration, ingredient composition of the ration, and age of poultry. Fats often increase the utilization of dietary energy by poultry in excess of the increase expected when the ME of the fat is added to the ME values of the other ration constituents. Supplemental fats may increase energy utilization in adult chickens due to the following:

- A decreased rate of food passage through the gastrointestinal tract
- The heat increment of fat being less than that of carbohydrates

Fats constitute about 17% of the dry weight of market broilers and about 40% of the dry weight of whole eggs. Food fats affect body fats. Thus, poultry consuming soft fat, such as vegetable oils, may accumulate fat that is more soft and oily than normal.

Measuring and Expressing Energy Value of Feedstuffs

One nutrient cannot be considered more important than another, because all nutrients must be present in adequate amounts if efficient production is to be maintained. Yet, historically, feedstuffs have been compared or evaluated primarily on their ability to supply energy to animals. This is understandable because of the following:

- Energy is required in larger amounts than any other nutrient.
- Energy may be the limiting factor in poultry production.
- Energy is the major cost associated with feeding poultry.

Our understanding of energy metabolism has increased through the years. With this added knowledge, changes have come in both the methods and terms used to express the energy value of feeds.

ENERGY DEFINITIONS AND CONVERSIONS. Some pertinent definitions and conversions of energy terms follow:

Calorie (cal). The amount of energy as heat required to raise the temperature of 1 g of water 1°C (precisely from 14.5° to 15.5°C). It is equivalent to 4.184 joules. Although not preferred, it is also called a small calorie and so designated by being spelled with a lowercase *c*. This is a very small unit. In poultry nutrition, the common unit used is the **kilocalorie**. This is equal to 1,000 calories. In popular writings, especially those concerned with human caloric requirements, the term Calorie (with a capital *C*) is frequently used for kilocalorie.

Kilocalorie (kcal). The amount of energy as heat required to raise the temperature of 1 kg of water 1°C (from 14.5° to 15.5°C). It is equivalent to 1,000 calories. In human nutrition, it is referred to as a kilogram calorie or as a large Calorie and is so designated by being spelled with a capital *C* to distinguish it from the small calorie.

Megacalorie (Mcal). Equivalent to 1,000 kcal or 1,000,000 calories. Also referred to as a therm, but the term *megacalorie* is preferred.

British Thermal Unit (BTU). The amount of energy as heat required to raise 1 lb of water 1°F; equivalent to 252 calories. This term is seldom used in animal nutrition.

Joule. An international unit (4.184 J = 1 calorie) for expressing mechanical, chemical, or electrical energy, as well as the concept of heat. In the future, energy requirements and feed values will likely be expressed by this unit.

CALORIE SYSTEM OF ENERGY EVALUATION. Calories are used to express the energy value of feedstuffs. To measure this heat, an instrument known as the bomb calorimeter is used, in which the feed (or other substance) tested is placed and burned in the presence of oxygen. The value determined in this analysis is termed **gross energy.**

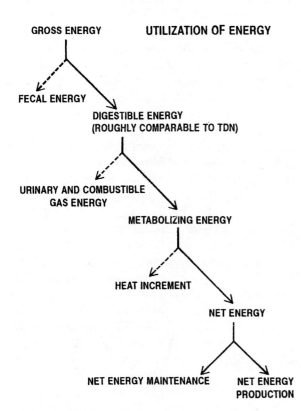

Figure 5.2 Utilization of energy. *(Original diagram from E. Ensminger)*

Through various digestive and metabolic processes, numerous losses of the energy in feed occur as the feed passes through the bird's digestive system. These losses are illustrated in Figure 5.2. Measures that are used to express energy requirements and the energy content of feeds differ primarily in the digestive and metabolic losses that are included in their determination. The following terms are used to express the energy value of feeds:

Gross energy (GE). Gross energy represents the total combustible energy in a feedstuff. It does not differ greatly between feeds, except for those high in fat.

Digestible energy (DE). Digestible energy is that portion of the GE in a feed that is not excreted in the feces.

Metabolizable energy (ME). Metabolizable energy represents that portion of the GE that is not lost in the feces, urine, and gas. Although ME more accurately describes the useful energy in the feed than does GE or DE, it does not take into account the energy lost as heat.

Net energy (NE). Net energy represents the energy fraction in a feed that is available for maintenance and productive purposes. It is what is left after the fecal, urinary, gas, and heat losses are deducted from the GE. Net energy, because of its greater accuracy, is being used increasingly in diet formulations, especially in computerized formulations for large animals and large operations.

Of the various systems of expressing energy values of feeds and nutrient requirements— gross energy, digestible energy, metabolizable energy, and net energy—the poultry industry has found metabolizable energy, commonly labeled ME, to be the most reliable expression of energy needs. In general, metabolizable energy represents 25 to 90% of gross energy.

Other nutrient requirements are always considered in reference to the energy requirement. Using the energy level for reference will enable the nutritionist to form a ratio of the amount of the nutrient per unit of energy, thereby keeping the nutrients in balance with available energy. In eating to satisfy its energy need, a bird will eat less of a high-energy diet and more of a low-energy diet. Having the nutrients in relation to dietary energy will ensure proper intake on a daily basis.

True metabolizable energy (TME). TME is the gross energy of the feed minus the gross energy of the excreta of food origin. When a chicken is fasted there will still be losses of energy in the excreta (feces and urine). These energy losses are thus not from the feed but contain metabolic and endogenous fractions from the body. Correcting for these energy losses gives a value termed true metabolizable energy. The term *apparent metabolizable energy* (AME) is often used for the traditionally determined energy values that include metabolic and endogenous fractions.

In simple formulas, AME and TME are the following:

AME = feed energy − (fecal + urinary + gaseous energy)

TME = AME + (metabolic + endogenous energy)

In poultry, the gaseous losses are negligible and usually ignored.

TME values are easier and much less expensive to determine than the traditional ME values. In growing animals, nitrogen is deposited in tissues primarily as protein. A correction is often made for the energy that would have been lost in association with nitrogen had it been excreted; the objective being to provide an energy measure that is independent of the growth of an animal and therefore more useful in formulating diets.

So ME and TME values when corrected for nitrogen excretion are expressed as ME_N and TME_N.

Following the nitrogen correction, the values obtained by use of TME assay are virtually identical to those obtained by using the more traditional apparent metabolizable energy; hence, nitrogen corrected TME (TME_N) and apparent metabolizable energy (AME) values are interchangeable.

Proteins

Proteins are complex organic compounds made up chiefly of amino acids, which are present in characteristic proportions for each specific protein. This nutrient always contains carbon, hydrogen, oxygen, and nitrogen; and in addition, it usually contains sulfur and frequently phosphorus. Proteins are essential in all plant and animal life as components of the active protoplasm of each living cell.

Crude protein refers to all the nitrogenous compounds in a feed. It is determined by finding the nitrogen content and multiplying the result by 6.25. The nitrogen content of protein averages about 16% (100/16 = 6.25).

In plants, the protein is largely concentrated in the actively growing portions, especially the leaves and seeds. Plants also have the ability to synthesize their own proteins from such relatively simple soil and air compounds as carbon dioxide, water, nitrates, and sulfates, using energy from the sun. Thus, plants, together with those bacteria that are able to synthesize these products, are the original sources of all proteins.

Proteins are much more widely distributed in animals than in plants. Thus, the proteins of the animal body are primary constituents of many structural and protective tissues such as bones, ligaments, feathers, skin, and the soft tissues, which include the organs and muscles.

Birds of all ages require adequate amounts of protein of suitable quality for maintenance, growth, reproduction, and egg production. Of course, the protein requirements for growth are the greatest and most critical. Typical broiler starter rations contain from 21 to 24% protein, and typical laying rations from 16 to 17% protein. Grain and millfeeds supply approximately one-half of the protein needs for most poultry rations. Additional protein is supplied from high-protein concentrates of either animal or vegetable origin. From the standpoint of poultry nutrition, the amino acids that make up proteins are really the essential nutrients, rather than the protein molecule itself. Hence, protein content as a measure of the nutritional value of a feed is becoming less important, and each amino acid is being considered individually. The requirements of poultry for the sulfur amino acids and lysine are given in Chapter 7.

The energy content of the diet must be considered in formulating to meet the desired intake of all essential nutrients other than the energy itself, including the intake of the essential amino acids. For example, if the producer uses a high-energy feed, the protein content of the feed must be high if the bird is to ingest adequate amounts of protein. Conversely, if the energy content of the feed is low, the protein content should be low also; otherwise, the bird will consume excessive amounts of expensive protein. Currently stated amino acid requirements have no reference to environmental conditions. Percentage requirements should probably be raised in warmer environments and lowered in colder environments in accordance with expected differences in feed or energy intakes. The amino acid concentrations presented in the tables in Chapter 7 are intended to promote maximum growth and normally optimal economic productivity. Maximum economic returns may not, however, always be assured by maximum growth, particularly when protein prices are high. So, at times the dietary concentrations may be somewhat reduced, lowering growth rate to some degree, but maintaining economic returns. Least-cost formulation can optimize economic returns.

It has been determined that the chick requires dietary sources of protein to furnish 13 different amino acids. These amino acids are referred to as essential, since the chicken cannot produce them in sufficient amounts for maximum growth or egg production, and because a dietary deficiency of any one of them interferes with body protein formation and affects growth or egg production. The primary object of protein feeding, therefore, is to furnish the bird with protein which, upon digestion, will yield sufficient quantities of the 13 essential amino acids needed for top performance. According to our present knowledge, the following division of amino acids as essential, semiessential, and nonessential for the chick seems proper:

Essential	Semiessential	Nonessential
(indispensable or required)	(semi-dispensable)	(synthesized by the body)
Arginine	Cystine	Alanine
Histidine	Glycine	Aspartic acid
Isoleucine	Tyrosine	Glutamic acid
Leucine		Hydroxyproline
Lysine		Proline
Methionine		Serine
Phenylalanine		
Threonine		
Tryptophan		
Valine		

When formulating poultry rations, they must be so designed as to supply all the essential amino acids in ample amounts. Special attention needs to be given to supplying the amino acids lysine, methionine, cystine, and tryptophan, which are sometimes referred to as the critical amino acids in poultry nutrition. Additionally, there must be sufficient total nitrogen for the chicken to synthesize the other amino acids needed. The amino acid requirements of growing chickens and turkeys together with those of laying hens are met by proteins from plant and animal sources together with added lysine and methionine. Protein supplements that most nearly supply the essential amino acids of the bird are known as high-quality supplements. Usually it is necessary to choose more than one source of dietary protein, then combine them in such a way that the amino acid composition of the mixture meets the requirements of the bird.

Some high-protein feedstuffs may contain toxic compounds. For example, cottonseed meal may contain gossypol, which discolors the yolks in stored eggs. Certain strains of rapeseed meal contain high enough levels of goitrogenic compounds to be toxic. Even soybean meal, the most used protein supplement, contains harmful substances such as trypsin inhibitor, but these are destroyed by proper heating.

Any excess protein consumed by the bird can be burned in the body to yield energy in somewhat the same manner as carbohydrates and fats. However, in practical feeding of poultry, it is seldom wise to use excessive protein because carbohydrates and fats are generally more economical sources of energy. In addition to dietary energy concentration and ambient temperature, the following factors affect the amino acid requirements:

1. The rate of growth or intensity of egg production. The more rapid the growth and the higher the egg production, the higher the amino acid requirements.

2. The strain/genetics. Even within birds of like body size, growth rate, or egg production, there may be differences in requirements among strains.

3. The protein level. The amino acid requirements tend to increase with dietary protein.

4. The amino acid relationships, specifically:

    a. *Methionine-cystine.* The requirement for methionine can be met only by methionine, while the requirement for cystine may be met by cystine or methionine. This is because methionine is readily converted to cystine metabolically, while the reverse is not possible.

    b. *Phenylalanine-tyrosine.* The requirement for phenylalanine may be met only by phenylalanine, while the requirement for tyrosine may be met by tyrosine or phenylalanine.

    c. *Glycine-serine.* Glycine and serine can be used interchangeably in poultry diets.

5. Antagonisms. There are specific antagonisms among amino acids that may be structurally related (e.g., valine-leucine-isoleucine and arginine-lysine). Increasing one or two of such a group may raise the need for another of the same group.

6. Imbalances. In supplementing diets with limited amino acids (lysine and methionine), it is important to supplement first with the most limiting one, followed by the second-most limiting one. Oversupplementation with only the second-most limiting one may create an imbalance and accentuate the primary deficiency.

7. Conversion of certain amino acids to vitamins. High levels of methionine may partly compensate for a deficiency of choline or vitamin B_{12} by providing needed methyl groups, and high levels of tryptophan may alleviate a niacin deficiency through metabolic conversion to niacin.

8. Amino acid availability. The usual assumption that amino acids are 80 to 90% available is not necessarily valid. For example, feathers or blood are either indigestible in native form or made indigestible by overheating in processing.

The consequences of a protein or amino acid deficiency vary with the degree of the deficiency. A borderline deficiency is characterized by poor growth and feathering, reduced egg size, poor egg production (but hatchability is not affected), tendency toward greater deposition of carcass and liver fat, poor feed conversion into eggs or meat, and lack of melanin pigment in black or reddish feathers with low lysine. A severe protein deficiency is marked by stopping of feed intake, stopping of egg production, loss of body weight, resorption of ova, a tongue deformity with leucine, isoleucine, and phenylalanine deficiency, stasis of the digestive tract, and death.

True Digestibility of Amino Acids

There is much interest in determining amino acid availability, using some of the same techniques that were originally developed for determination of true metabolizable energy. This interest stemmed from the fact that, frequently, amino acids cannot be completely utilized due to inherent characteristics of the feedstuff. Details follow.

(1) The available amino acids are the amino acids in the diet that are not combined with compounds interfering with their digestion, absorption, or utilization by the bird; they are actually supplied to the sites of protein synthesis.

(2) The digestible amino acids are those that are absorbed from the gut lumen; they are calculated as the difference between the amount of amino acids in the feed and the excreta.

Differentiation must be made between apparent and true digestibility. In addition to nonabsorbed feed, the excreta also contain materials originating in the tissue of the bird (e.g., cells sloughed from the gut wall, mucus, bile, unabsorbed digestive juices, etc.). Apparent digestibility makes no correction for these endogenous components. In simple formulas, apparent digestibility coefficient and true digestibility coefficient follow:

$$\text{Apparent Digestibility Coefficient} = \frac{\substack{\text{amino acid} \\ \text{consumed}} - \substack{\text{amino acid} \\ \text{excreted}}}{\text{amino acid consumed}} \times 100$$

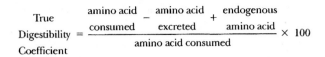

$$\text{True Digestibility Coefficient} = \frac{\substack{\text{amino acid} \\ \text{consumed}} - \substack{\text{amino acid} \\ \text{excreted}} + \substack{\text{endogenous} \\ \text{amino acid}}}{\text{amino acid consumed}} \times 100$$

The availability of amino acids in feedstuffs is estimated by three main methods:

1. *In vitro* tests
2. Growth tests
3. Digestibility tests

Most of the recent research has focused on digestibility determinations.

Digestibility trials measure the digestibility of all amino acids in a feedstuff in one run. Various experimental techniques may be applied to determine the digestibility of amino acids. Dr. I. R. Sibbald of the Animal Research Center, Ottawa, Canada, developed the rooster precision feeding technique, which estimates the true digestibility of amino acids. It consists of force-feeding a precise quantity of feed (1 to 2% of body weight) to a starved adult rooster, followed by a 48-hour collection period of the excreta. Other birds are fasted throughout the experiment to allow for collection of the endogenous excreta. It appears that true amino acid digestibility established in the chicken is valid for other bird species also.

Diet formulation on the basis of digestible amino acids has the following advantages:

1. It improves performance prediction under practical conditions.
2. It allows more by-product and alternate protein sources to be included in poultry feeds, along with the possibility of reducing costs while minimizing the risks of deterioration in performance.
3. It reduces the variability of poultry performance.
4. It evaluates feedstuff and rations more precisely.
5. It makes possible a more accurate supply of amino acids required for optimal performance.

The disadvantages or reservations relative (1) to the rooster bioassay technique for determining the true digestibility of amino acids in feedstuffs, and (2) to the use of such values in feed formulations are:

1. It does not consider the effects of microbes on dietary protein in the lower gut. Microbes both utilize and synthesize amino acids. Research has indicated that, under certain conditions, 20 to 25% of the amino acids found in the excreta are of microbial origin.

2. When conventional feed ingredients are used in poultry rations, consideration of amino acid digestibility is not particularly advantageous.

Crystalline amino acids, not being protein bound or enclosed in a feedstuff impairing their digestion, are 100% digestible and 100% available.

Minerals

Minerals are inorganic elements, frequently found as salts with either inorganic elements or organic compounds. Minerals are required for the formation of the skeleton, as parts of hormones or as activators of enzymes, and for the proper maintenance of necessary osmotic relationships within the body of the bird.

The minerals that have been shown to be essential for chickens and turkeys are calcium, phosphorus, magnesium, manganese, zinc, iron, copper, iodine, sodium, chlorine, potassium, sulfur, molybdenum, and selenium. Of these, calcium, phosphorus, manganese, sodium, chlorine, iodine, selenium, and zinc are considered to be of most practical importance since outside sources of them must be added to practical feed formulations for chickens and turkeys. Pertinent information relative to poultry minerals is summarized in Table 5.3.

Some minerals are required by poultry in relatively large amounts. They are referred to as major or **macrominerals.** Others are needed in very small amounts; they are referred to as trace or **microminerals.**

Major or Macrominerals

The major or macrominerals of importance in poultry are salt (sodium chloride), calcium, phosphorus, magnesium, and potassium.

SALT (SODIUM CHLORIDE/NaCl). Sodium and chlorine are essential for all animals, including poultry. Dietary proportions of sodium, potassium, and chlorine are important determinants of acid-base balance. The proper dietary balance of sodium, potassium, and chlorine is necessary for growth, bone development, eggshell quality, and amino acid utilization. It is generally recommended that 0.2 to 0.5% of the diet consist of salt.

As in most nutrient deficiencies, a lack of salt will reduce reproductive performance and retard growth. Also, cannibalism occurs in flocks when diets are deficient in salt.

CALCIUM (CA) AND PHOSPHORUS (P). The common calcium supplements used in poultry feeding are ground limestone and oystershells for broiler breeders. The most important phosphorus sources are dicalcium phosphate, defluorinated rock phosphate, and steamed bonemeal together with phytate phosphate in corn and soy meal following enzymic treatment with phytase. Bonemeal, dicalcium phosphate, defluorinated rock phosphate, colloidal phosphate, and raw rock phosphate are used where both calcium and phosphorus are needed. The biological availability of calcium is high for most commonly used supplements, but the biological availability of phosphorus in supplements varies.

The calcium allowance of the laying hen is difficult to define. A high producing hen is usually fed a diet containing 3.4% calcium. This is believed to represent the mean dietary concentration for the quantities of feed likely to be consumed (110 g per hen per day) over a considerable range of environmental temperature. Most of the calcium in the diet of the mature laying fowl is used for eggshell formation, whereas most of the calcium in the diet of the growing bird is used for bone formation. Other functions of calcium include roles in blood clotting and neuromuscular function. High concentrations of calcium carbonate (limestone) and calcium phosphate make the ration unpalatable. Also, excess dietary calcium interferes with the availability of other minerals such as magnesium, manganese, and zinc.

In addition to its function in bone formation, phosphorus is required in the metabolism of carbohydrates and fats and is a component of all living cells. It is important that the minimum level of inorganic phosphorus be provided. The stated requirement is based on the generally greater availability of inorganic phosphorus than that of phytate phosphorus. Only about 30% of the phosphorus in plant products is available to the young chick, poult, or duckling. For growing chickens and turkeys, a calcium to phosphorus ratio of 1.2:1 is considered ideal. However, ratios of 1:1 to 1.5:1 are well tolerated.

Leg problems and thin eggshells occur when calcium and/or phosphorus are deficient (see Figure 5.3). It is also essential (especially in poultry housed in confinement) that adequate amounts of vitamin D_3 be supplied in the ration to ensure proper absorption of these elements (see Figure 5.4).

MAGNESIUM (MG). Most poultry feeds contain adequate amounts of magnesium, and supplementation generally is not necessary. A dietary level of 600 ppm of magnesium is required by broilers. Excessive magnesium levels can interfere with the absorption of other elements—notably calcium and phosphorus. If it is known that a diet contains a high level of magnesium, levels of other elements should be increased.

Chicks that are fed magnesium-deficient rations exhibit depressed growth rates and lethargy.

TABLE 5.3 Poultry Minerals

Mineral That May Be Deficient Under Normal Conditions	Conditions Usually Prevailing Where Deficiencies Are Reported	Function of Mineral	Some Deficiency Symptoms	Types of Poultry Rations Usually Requiring Supplementation			
				Starting	Growing	Laying	Breeding
Major or macrominerals: Salt (sodium and chlorine—NaCl)	Omitted at the mill.	Improves appetite, promotes growth, helps regulate body pH, and is essential for hydrochloric acid formation in the stomach.	Chloride-deficient chicks show poor growth, high mortality, nervous symptoms, and reduced blood chloride level. Sodium deficiency in layers results in decreased egg production, poor growth, and cannibalism.	Yes	Yes	Yes	Yes
Calcium (Ca)	Imbalance of Ca:P ratio. Presence of interfering elements.	Bone formation, eggshell formation, blood clotting, and neuromuscular function.	Anorexia, thin eggshells, rickets, or osteoporosis, tetany, and abnormal walk.	Yes	Yes	Yes	Yes
Phosphorus (P)		Bone formation, metabolism of carbohydrates and fats, a component of all living cells, maintenance of the acid-base balance of the body, and calcium transport in egg formation.	Anorexia, weakness, rickets. Cage-layer fatigue, characterized by birds being paralyzed and unable to rise from a recumbent position. But there is evidence that this condition is not due to this factor alone.	Yes	Yes	Yes	Yes
Magnesium (Mg)	Diets containing high levels of Ca and P.	Essential for normal skeletal development, as a constituent of bone, enzyme activator, primarily in glycolytic system.	Decreased egg production, depressed growth and lethargy, and convulsions.	No	No	No	No
Potassium (K)	Low plant protein level/high animal protein level.	Major cation of intracellular fluid where it is involved in osmotic pressure and acid-base balance. Muscle activity. Required in enzyme reaction involving phosphorylation of creatine. Influences carbohydrate metabolism.	*Chicks:* Retarded growth and high mortality.	No	No	No	No
Trace or microminerals: Copper (Cu)		Essential element in a number of enzyme systems and necessary for synthesizing hemoglobin and preventing nutritional anemia.	Anemia, depigmentation of feathers, digestive disorders. *Poults:* Marked cardiac hypertrophy.	Yes	Yes	Yes	Yes

Mineral	Function	Deficiency Source	Deficiency Symptoms				
Iodine (I)	Needed by the thyroid gland for making thyroxine, an iodine-containing hormone that controls the rate of body metabolism or heat production.	Feeds produced on iodine-deficient soils.	Enlarged thyroid. Eggs produced from deficient breeder hens have a lowered hatchability, prolonged hatching time, and a subsequent retardation of yolk-sac absorption.	Yes	Yes	Yes	Yes
Iron (Fe)	Necessary for formation of hemoglobin, an iron-containing compound that enables the blood to carry oxygen. Iron is also important to certain enzyme systems.		*Chicks and poults:* Microcytic, hypochromic anemia. In red-feathered chickens, complete depigmentation of the feathers occurs.	Yes	Yes	Yes	Yes
Manganese (Mn)	Necessary for growth, bone structure, and reproduction.		*Chicks and poults:* Perosis, shortened leg bones, skull deformation, parrot beak. Poor egg production, shell quality, and hatchability.	Yes	Yes	Yes	Yes
Selenium (Se)	Involved in the destruction of peroxides within the cell as a constituent of glutathione peroxidase. Useful in preventing exudative diathesis.	Feeds that are grown in selenium-deficient areas.	Exudative diathesis, pancreatic fibrosis, steatitis, and muscular dystrophy. With a severe selenium deficiency, growth rate is reduced and mortality is increased. *Turkeys.* Myopathies of the gizzard and heart.	Yes	Yes	Yes	Yes
Zinc (Zn)	Zinc is a component of several enzyme systems, including peptidases and carbonic anhydrase. Also, zinc is required for normal protein synthesis and metabolism and is a component of insulin.		Bone problems, poor feathering (feather fraying occurs near the ends of the feathers), retarded growth, and loss of appetite. A zinc deficiency in breeder diets reduces egg production and hatchability.	Yes	Yes	Yes	Yes

As used herein, the distinction between *mineral requirements* and *recommended allowances* is as follows: In *mineral requirements*, no margins of safety are included intentionally; whereas in *recommended allowances*, margins of safety are provided in order to compensate for variations in feed composition, environment, and possible losses during storage or processing.

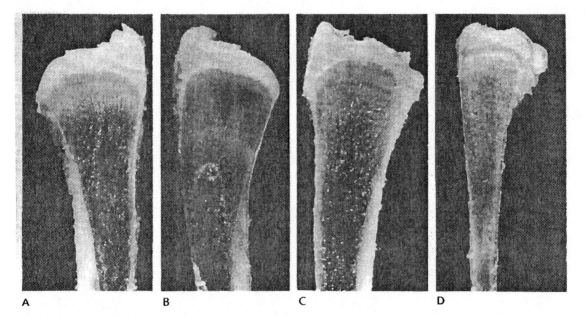

A   B   C   D

Figure 5.3 Deficiencies of mineral (phosphorus and/or calcium) and amino-acids influence bone growth (tibiotarsal bone).
Plate A Chicks on control diet;
Plate B Chicks on phosphorus deficient diet (Note: wide zone of hypertrophy);
Plate C Chicks on calcium and phosphorus deficient diet (Note: wider zone of proliferation);
Plate D Chicks on lysine deficient diet (Note the hypoplasia). *(Reprinted from* Diseases of Poultry, *10th edition, edited by B. W. Calnek, with permission from Iowa State Press)*

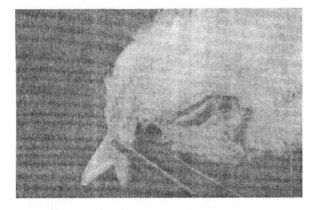

Figure 5.4 "Rubbery beak" of a chicken with rickets caused by feeding a diet deficient in vitamin D. *(Reprinted from Diseases of Poultry, 10th edition, edited by B. W. Calnek, with permission from Iowa State Press)*

TABLE 5.4 Suggested Trace Mineral Supplements to (Chemically Defined) Diets[1]

Element	Amount of Diet	
	(mg/lb)	(mg/kg)
Boron (B)	0.91	2
Chromium (Cr)	1.36	3
Fluorine (F)	9.08	20
Inorganic sulfate	[2]	[2]
Molybdenum (Mo)	0.45	1
Nickel (Ni)	0.05	0.1
Silicon (Si)	113.50	250
Tin (Sn)	1.36	3
Vanadium (V)	0.09	0.2

[1]Adapted from *Nutrient Requirements of Poultry*, 8th rev. ed., NRC, National Academy Press, Washington, DC, 1984, p. 7, Table 1.
[2]Sulfur is supplied to the diet by methionine and cystine. There may be a response to inorganic sulfate if the diet is low in cystine.

When disturbed, magnesium-deficient chicks become hyperirritable and go into convulsions. If the condition is not corrected, chicks become comatose and eventually die.

POTASSIUM (K). Potassium is found in large amounts in most plant feedstuffs, and generally supplementation is not required. It has an essential role in cellular homeostasis, along with sodium and chlorine. In cases of potassium deficiency, generalized muscle weakness is observed.

TRACE OR MICROMINERALS

Trace minerals are minerals that are required in small amounts. The trace minerals of special concern in poultry (see Table 5.4) are the following:

- Copper
- Iron
- Iodine
- Manganese
- Selenium
- Zinc

As soils become leached, their content of trace minerals and the feedstuffs grown on them become borderline or deficient, thereby necessitating that more of certain minerals be added to poultry rations. Interactions between various trace minerals—such as copper and molybdenum, selenium and mercury, calcium and zinc, or calcium and manganese—are also important in poultry nutrition. Selenium is also metabolically involved with vitamin E and arsenic.

COPPER (CU) AND IRON (FE). Both of these elements are necessary in the prevention of anemia. If a deficiency of either element exists, there is a reduction in the size of the red blood cells as well as a decreased oxygen-carrying capacity of the cells. Poultry rations should contain about 80 ppm of iron and 6 to 8 ppm of copper.

Iron salts may be used in poultry rations containing cottonseed meal to tie up gossypol—a compound that causes discoloration of yolks. However, iron salts do not prevent the pink discoloration of egg whites caused by cyclopropenoid fatty acids in cottonseed meal.

IODINE (I). In certain areas of the United States (Northwest and Great Lakes regions), soils are deficient in iodine. In these deficient areas, it may be advisable to add iodine to the ration to prevent goiter. Feeds for laying hens should contain 300 ppb of iodine. For growing chicks, feeds should contain 350 ppb of iodine.

MANGANESE (MN). This element is routinely added to poultry diets in such forms as manganous chloride, manganous carbonate, manganese dioxide, and manganese sulfate. Dietary levels of 30 ppm for layers and 60 ppm for broilers should be sufficient. Chicks that are deficient in manganese characteristically display slipped tendons (perosis). A deficiency of either choline or biotin will produce similar symptoms. Hens suffering from a manganese deficiency will exhibit a reduction in egg production and hatchability. Many of the eggs that are laid are either thin-shelled or without shells.

SELENIUM (SE). Chicks deficient in selenium develop exudative diathesis, steatitis, and pancreatic fibrosis. Selenium-deficient turkeys develop a condition commonly referred to as white gizzard disease—a form of muscular dystrophy. Care should be taken in the incorporation of selenium in poultry rations since it is toxic at a level of 10 ppm.

ZINC (ZN). Chicks that are deficient in zinc exhibit bone problems, poor feathering, anorexia, and retarded growth. Zinc and calcium absorption are interrelated; hence, as the calcium level is increased, the zinc level should be increased also.

Vitamins

Vitamins are defined as complex organic compounds that are required in minute amounts by one or more animal species for normal growth, production, reproduction, and/or health. The vitamins required by poultry, along with their deficiency symptoms and dietary sources, are shown in Table 5.5.

The column of Table 5.5 headed Types of Poultry Rations Usually Requiring Supplementation indicates the types of poultry rations in which special attention must be paid to the inclusion of dietary sources of the vitamins. As shown, vitamin A, vitamin D, riboflavin, and vitamin B_{12} are commonly low in most poultry rations. It is also to be emphasized that vitamin D_3, the animal form of vitamin D (made by the irradiation of 7-dehydrocholesterol), is more active for poultry, and should, therefore, be used instead of vitamin D_2, the plant form of the vitamin.

The fat-soluble vitamins (A, D, E, and K) can be stored and accumulated in the liver and other parts of the body, while only very limited amounts of the water-soluble vitamins (biotin, choline, folacin, niacin, pantothenic acid, riboflavin, thiamin, vitamin B_6, and vitamin B_{12}) are stored. For this reason, it is important that the water-soluble vitamins be fed regularly in the ration in adequate amounts. Vitamin C is synthesized by poultry; hence, it is not considered a required dietary nutrient. There is some evidence, nevertheless, of a favorable response to vitamin C by birds under stress.

Requirements for some of the vitamins may be met by the amounts occurring in natural feedstuffs. However, formulators of poultry feeds should be alert to the need for dietary supplementation with vitamins usually assumed to be supplied by the feedstuffs.

Fat-Soluble Vitamins

Vitamin A and vitamin D have long been routinely added to poultry feed. Vitamin E and vitamin K are also included in vitamin supplements for poultry feeds.

VITAMIN A. The requirement of vitamin A in poultry is dependent upon the following:

1. **Individual bird variations.** Metabolic rates and subsequent nutrient requirements vary from bird to bird, even though they may be of the same strain.

TABLE 5.5 Poultry Vitamins

Vitamins That May Be Deficient Under Normal Conditions	Function of Vitamin	Conditions Usually Prevailing Where Deficiencies Are Reported	Some Deficiency Symptoms	Types of Poultry Rations Usually Requiring Supplementation			
				Starting	Growing	Laying	Breeding
Fat-soluble vitamins:							
Vitamin A	Essential for normal maintenance and functioning of the epithelial tissues, particularly of the eye and the respiratory, digestive, reproductive, nerve, and urinary systems.	Old vitamin premix.	*Chicks:* Depressed growth, weakness, loss of coordination, xerophthalmia, anorexia, lowered resistance to infection, and alterations in mucous membranes. *Adults:* Depressed production, low hatchability, discharge from nose and eyes, lowered resistance to infection, and alterations in mucous membranes.	Yes	Yes	Yes	Yes
Vitamin D$_3$	Aids in assimilation and utilization of calcium and phosphorus, and necessary in normal bone development.	Birds that are in confinement.	*Chicks:* Rickets, poor feathering, reduced growth. *Adults:* Weak bones, poor eggshell formation, reduced production and hatchability.	Yes	Yes	Yes	Yes
Vitamin E	Antioxidant. Muscle structure. Reproduction.	Destruction by oxidation of the diet.	*Chicks:* Encephalomalacia, exudative diathesis, muscular dystrophy. *Adults:* Poor reproductive performance; prolonged vitamin E deficiency results in permanent sterility in the male and reproductive failure in the female. *Poults:* Myopathy of the gizzard.	Yes	Yes	Yes	Yes
Vitamin K	Essential for prothrombin formation and blood clotting.	Coccidiosis. When high levels of antibodies or sulfa drugs are fed. Newly hatched chicks from deficient females.	Hemorrhaging. Increased clotting time.	Yes	Yes	Yes	Yes
Water-soluble vitamins:							
Biotin	Involved in carbohydrate, lipid, and protein metabolism.	Broilers fed a milo, wheat, or wheat–barley–based diet. Feeding avidin, a protein in uncooked egg white, which binds biotin and renders it unavailable nutritionally.	*Chicks:* Cracking and degeneration of skin on feet, around beak, and perosis (slipped tendon). *Adults:* Reduced hatchability. *Poults:* Broken flight feathers, bending of the metatarsus, and dermatitis of the footpads and toes, base of beak, eye ring, and vent.	Yes	No	No	Yes
Choline	Involved in nerve pulses. A component of phospholipids. Donor of methyl groups.		*Chicks, poults, ducklings:* Retarded growth and perosis (slipped tendon). *Adults:* Increased mortality, lowered egg production, and increased abortion of egg yolks from ovaries.	Yes	Yes	No	No

Vitamin	Metabolic function	Conditions usually prerequisite to a deficiency	Deficiency symptoms				
Folacin (folic acid)	Related to vitamin B_{12} metabolism. Metabolic reactions involving incorporation of single carbon units into larger molecules.		*Chicks:* Poor growth, poor feathering, perosis, and anemia. *Adults:* Reduced hatchability and egg production. *Turkey poults:* Nervousness, droopy wings, and a stiff extended neck. *Turkey breeder hens:* Normal egg production, but reduced hatchability.	Yes	Yes	Yes	Yes
Niacin (nicotinic acid, nicotinamide)	Required by all living cells, and an essential component of important metabolic enzyme systems involved in glycolysis and tissue respiration.	A predominantly corn-soybean ration.	*Chicks:* Enlargement of hock joints and perosis, retarded growth, and inflammation of mouth and tongue ("black tongue"). *Adults:* No symptoms observed in hen except on protein-deficient diet. *Turkey poults:* A hock disorder similar to perosis.	Yes	Yes	Yes	Yes
Pantothenic Acid (vitamin B_3)	Part of coenzyme A, a necessary factor for intermediate metabolism.	Use of artificially dried corn (heating destroys the pantothenic acid) and the omission of milk by-products from the diet.	*Chicks:* Poor growth, ragged feather development, degeneration of skin around beak, eyes, and vent, and liver damage. *Adults:* Reduced hatchability. Mortality is high in newly hatched chicks from pantothenic acid-deficient hens.	Yes	Yes	Yes	Yes
Riboflavin (vitamin B_2)	A component of enzyme systems essential to normal metabolic processes.		*Chicks:* Curled toe paralysis, reduced growth, and diarrhea. *Adults:* Poor hatchability with many dying during second week of incubation.	Yes	Yes	Yes	Yes
Thiamin (vitamin B_1)	As a coenzyme in energy metabolism. Promotes appetite and growth, and required for normal carbohydrate metabolism.		*Chicks:* Anorexia, loss of coordination, poor feathering, polyneuritis. *Adults:* Blue comb, paralysis.	No	No	No	No
Vitamin B_6 (pyridoxine, pyridoxal, pyridoxamine)	As a coenzyme in protein and nitrogen metabolism. Involved in red blood cell formation. Important in endocrine systems.		*Chicks:* Poor growth, lack of coordination, and convulsions. *Adults:* Reduced body weight, egg production, and hatchability.	No	No	No	No
Vitamin B_{12} (Cobalamins)	Numerous metabolic functions, and essential for normal growth and reproduction in poultry.		*Chicks:* Poor growth, perosis, mortality. *Adults:* Reduced hatchability, fatty heart, liver, and kidneys.	Yes	Yes	Yes	Yes

As used herein, the distinction between *vitamin requirements* and *recommended allowances* is as follows: In *vitamin requirements,* no margins of safety are included intentionally; whereas in *recommended allowances,* margins of safety are provided in order to compensate for variations in feed compositions, environment, and possible losses during storage or processing.

2. Type of bird. There are genetic differences between different strains of birds with regards to their respective abilities to utilize vitamin A.

3. Production and stress. Vitamin A requirements vary according to the type and stage of production. The layer needs more vitamin A than the growing chicken, since vitamin A is passed into the egg. Likewise, birds that are subjected to environmental stress probably have different vitamin A requirements than birds under minimal stress.

4. Destruction of the vitamin. Vitamin A in feed can be destroyed when fats in the feed become rancid. Likewise, certain processing methods can reduce the activity of the vitamin in the feed. Parasites and bacteria in the gut can also destroy vitamin A before it is absorbed.

5. Absorbability. Since vitamin A absorption is dependent on a lipoprotein in the blood, deficiencies of protein and/or fat can reduce absorption. It is noteworthy that poultry are just as efficient as the rat in the conversion of carotene to vitamin A (see Table 5.6)

Young chicks are more susceptible to vitamin A deficiencies than adults because it takes a relatively long period for adult birds to deplete their body stores. Deficiency symptoms of vitamin A in chicks are characterized by retarded growth, emaciation, general weakness, staggered gait, ruffled plumage, lowered resistance to infection, xerophthalmia, and disruption of mucosal membranes. In adults, eye problems are prevalent along with decreased egg production and hatchability.

Excessive amounts of vitamin A can be toxic, but these levels must be on the order of 500 times the recommended allowances. Symptoms of excess vitamin A (hypervitaminosis) are anorexia, emaciation, inflammation of epithelial tissues, abnormalities of bone, swelling of the eyelids, and mortality.

VITAMIN D. Cholecalciferol (D_3) is the form of vitamin D that has the highest activity in poultry feed. Today, most birds are reared in confinement where exposure to sunlight is insufficient for the conversion of 7-dehydrocholesterol—the precursor of vitamin D in the skin—to vitamin D_3. Thus, vitamin D_3 is routinely added to poultry feed. The cost of supplementation is small, especially when one considers the consequences of such deficiencies. The dietary requirements of vitamin D depend on four factors: (1) exposure to sunlight, (2) Ca:P ratio, (3) levels of calcium and phosphorus in the feed, and (4) intensity of production.

Leg problems, poor growth, and poor feathering are the common deficiency symptoms in growing birds (see Figures 5.4 and 5.5). Egg production,

TABLE 5.6 Conversion of Beta-Carotene to Vitamin A for Different Species

Species	Conversion of mg of Beta-Carotene to IU of Vitamin A		IU of Vitamin A Activity (Calculated from Carotene)
	(mg)	(IU)	(%)
Standard (rat)	1	1,667	100.0
Poultry	1	1,667	100.0
Beef cattle	1	400	24.0
Dairy cattle	1	400	24.0
Sheep	1	400–500	24.0–30.0
Swine	1	500	30.0
Horses:			
Growth	1	555	33.3
Pregnancy	1	333	20.0
Mink	Carotene not utilized		—
Human	1	556	33.3

Source: Adapted from the Atlas of Nutritional Data on United States and Canadian Feeds, *NRC, National Academy of Sciences, Washington, DC, 1971, p. XVI, Table 6.*

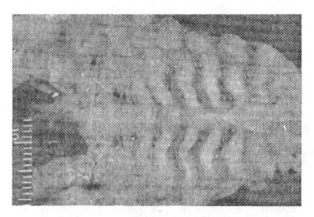

Figure 5.5 Abnormalities of the bones–ribs in chickens with rickets caused by feeding a diet deficient in vitamin D. Note the beaded appearance and the bent shape. *(Reprinted from* Diseases of Poultry, *10th edition, edited by B. W. Calnek, with permission from Iowa State Press)*

hatchability, and eggshell quality will be decreased in deficient hens.

VITAMIN E. As with vitamins A and D, vitamin E is extremely susceptible to destruction from the oxidation of fats in the feed. To prevent this, antioxidants are commonly added to poultry feeds. Also, vitamin E is often added to feed in an esterified form to protect it from destruction.

The three classical symptoms of vitamin E deficiency in chicks are encephalomalacia, exudative diathesis, and nutritional muscular dystrophy. Encephalomalacia is a condition whereby there is a necrosis in the brain. Chicks exhibit an outstretching of the legs with toes curled; and the head is often in a retracted position. Prior to these symptoms of acute toxicity, chicks display a generalized lack of coordination. Exudative diathesis is a condition in which the walls of capillaries become highly permeable. Nutritional muscular dystrophy in chicks is analogous to stiff lamb disease in sheep and white muscle disease in calves.

Reproduction is impaired in vitamin E-deficient adult birds. Degeneration of the testes is observed in deficient males—a condition that can lead to permanent sterility if not corrected in time. Layers suffering from a vitamin E deficiency do not show a dramatic drop in egg production, but hatchability is severely reduced.

In vitamin E-deficient turkey poults, a myopathy of the gizzard can be observed. Selenium and vitamin E are closely related in physiological functions, but they cannot replace each other in the diet.

VITAMIN K. Vitamin K occurs in a number of naturally occurring and synthetic compounds with varying solubilities in fat and water. Menadione is a fat-soluble, synthetic compound that can be considered as the reference standard for vitamin K activity. Two naturally occurring forms are K_1 or phylloquinone and K_2 or minaquinone. Water-soluble forms include menadione sodium bisulfate (MSB), menadione sodium bisulfite complex (MSBC), and menadione dimethylpyrimidol (MPB). The theoretical activity of these compounds can be calculated on the basis of the proportion of menadione present in the molecule.

Vitamin K should be added to starter rations and rations that incorporate drugs, such as sulfaquinoxaline, which reduce the microbial population of the gut. When heavy parasitic infections occur, such as in coccidiosis, the vitamin K requirement is increased. Birds that are deficient in vitamin K have a greatly increased susceptibility to hemorrhaging and an increased clotting time. Newly hatched chicks from vitamin K-deficient females will have low storage of vitamin K and are generally susceptible to injury.

Water-Soluble Vitamins

Of the water-soluble vitamins, biotin, niacin, pantothenic acid, riboflavin, and vitamin B_{12} may be low in poultry feeds. Young chicks are most susceptible to vitamin deficiencies.

The requirements for the water-soluble vitamins are interrelated in some instances. They are also dependent upon the nature of the diet. The types of carbohydrate, protein concentration, and amino acid balance are major factors determining the dietary requirements for several vitamins.

BIOTIN. The availability of biotin in many of the cereal grains is very low. Hence, biotin levels and availability should be carefully monitored in poultry rations. Eggs are high in biotin. Therefore, careful consideration should be given to biotin in rations for layers. The symptoms of biotin deficiency in chicks are cracking and degeneration of skin on the feet and around the beak, and perosis. Reduced hatchability is observed in biotin-deficient hens.

CHOLINE. Since choline levels can be marginal in starter and grower rations, choline is generally added to these types of feeds. Much like the deficiency symptoms of several of the other water-soluble vitamins, perosis (slipped tendon) can be observed in choline-deficient chicks. Since older birds can synthesize choline on the cellular level, it is difficult to produce a choline deficiency in laying hens.

FOLACIN (FOLIC ACID). Because feeds rich in folacin—for example, alfalfa meal and liver meal—are not used today in practical poultry feeds, it is necessary to supplement poultry feeds with folacin. Deficiency symptoms of folacin in chicks are poor growth, abnormal coloration of feathers, perosis, and anemia. Hens that are deficient in folacin are observed to have reduced egg production and hatchability. Poults fed folacin-deficient rations may develop a paralysis of the neck.

NIACIN (NICOTINIC ACID/NICOTINAMIDE). Synthetic niacin is routinely added to starter and breeder rations. Some niacin can be synthesized in the body from tryptophan; but since corn, which is notably low in tryptophan, is the most widely used feedstuff in poultry feed, this means of fulfilling the niacin requirement should be viewed with skepticism.

Chicks deficient in niacin show an enlargement of the hock joints, perosis, dermatitis, retarded growth, and an inflammation of the mouth and tongue. In addition to these symptoms, poor feathering and hyperirritability may be observed.

PANTOTHENIC ACID (VITAMIN B₃). Pantothenic acid is widely distributed in nature. However, it is routinely added to starter and breeder rations—often in the form of calcium pantothenate—to ensure against any possibility of deficiency. Deficiency symptoms in chicks are rather nonspecific—poor growth, poor feathering, liver damage, and lesions around the beak, eyes, and vent (see Figure 5.6). Lowered hatchability and a high mortality rate of newly hatched chicks can result from feeding hens a diet deficient in pantothenic acid.

RIBOFLAVIN (VITAMIN B₂). Poultry feeds should be supplemented with riboflavin. Chicks suffering

Figure 5.6 Pantothenic acid deficiency. Note the curled feathers indicating poor feathering.

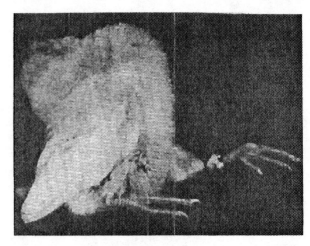

Figure 5.8 Thiamin deficiency with chick showing "classical" symptoms of "star gazing". *(Reprinted from* Diseases of Poultry, *10th edition, edited by B. W. Calnek, with permission from Iowa State Press)*

Figure 5.7 Riboflavin deficiency or curled-toe paralysis in a young chick. Note the curled feathers.

from a riboflavin deficiency display a characteristic curled-toe paralysis as well as depressed growth and diarrhea (see Figure 5.7). In curled-toe paralysis, the brachial and sciatic nerves become greatly enlarged. Poor egg production and hatchability are observed in riboflavin-deficient hens.

THIAMIN (VITAMIN B_1). Although the requirement for thiamin is high in poultry, it is rarely added as a supplement to the ration because most feed ingredients contain high levels. Polyneuritis, a type of paralysis, is common in thiamin-deficient birds. Prior to this acute deficiency condition, anorexia, emaciation, ruffled feathers, and incoordination are observed in thiamin-deficient chicks (see Figure 5.8). When polyneuritis sets in, a progressive paralysis is observed, beginning first in the toes and ultimately reaching the head, whereupon the head is retracted so that it lies on the back. Deficient adults frequently have a blue comb.

VITAMIN B_6 (PYRIDOXINE, PYRIDOXAL, PYRIDOXAMINE). Deficiencies of vitamin B_6 are rare in poultry,

since most feeds used in poultry rations contain relatively high levels of the vitamin. Deficiency symptoms in chicks are depressed growth and neurological problems such as poor coordination and convulsions Mature birds deficient in vitamin B_6 exhibit anorexia, rapid loss of weight, and lowered egg production and hatchability.

VITAMIN B_{12} (COBALAMINS). Vitamin B_{12} is found only in animal products and bacterial fermentation products. Consequently, poultry feeds that consist primarily of plant products are supplemented with vitamin B_{12}.

Liver stores of vitamin B_{12} may be high enough to sustain adults for several months, but high-protein diets can accelerate this depletion process. Birds deficient in vitamin B_{12} exhibit poor feed conversions, depressed growth, reduced hatchability, and in some cases perosis (slipped tendon). Fatty livers, kidneys, and hearts can be observed in some deficient birds. Vitamin B_{12} deficiency depresses growth.

Water

Poultry should have free access to clean, fresh water at all times. It is needed as a solvent, a lubricant, and a temperature control device. A general rule is that chickens drink approximately twice as much water by weight as the feed they consume.

The amount of water required by poultry varies considerably, as evidenced in Table 5.7. The factors that affect the amount of water birds will consume are the following:

- Age
- Body weight

TABLE 5.7 Daily Water Consumption by Chickens and Turkeys of Different Ages[1]

Age	Leghorn-Type Pullets		Chicken Broilers[2]		Turkeys[2]	
(week)	(U.S. gal)	(liter)	(U.S. gal)	(liter)	(U.S. gal)	(liter)
1	5	19	5	19	10	38
2	10	38	13	50	20	76
3	12	45	24	90	30	114
4	17	64	37	140	40	151
5	22	83	53	200	50	189
6	25	95	69	260	60	227
7	28	106	85	320	75	284
8	30	114	100	380	95	360
9	35	132			115	435
10	38	144			125	473
12	40	151			150	568
15	42	158			160	606
20	45	170			200	757

Laying or Breeding

35	50	189			M 240	908
					F 130	492

[1]Will vary considerably depending on temperature and ration composition.
[2]Mixed sexes.
Source: Adapted from Nutrient Requirements of Poultry, *9th rev. ed., NRC, National Academy Press, Washington, DC, 1994, p. 16, Table 1-1.*

BOX 5.1 Some New Developments in Nutrition

- In ovo feeding. Research has found that injection of amino acids and starch into the amnion results in more rapid growth posthatch. With adaptation, it may be possible to use the technology for in ovo vaccination to do this in an automated manner. The amino acids and starch in the amniotic fluid are swallowed by the chick embryo and the nutrients digested and absorbed from the intestine (also see color plate 16).
- Rendering. Rendering is the process by which mortalities and nonused parts of animal carcasses are processed to yield a high protein animal feed. The spread of the prion diseases of bovine spongiform encephalitis (BSE) from rendered cattle led to the ban on cattle by-products in cattle feed in the United States and of animal by-products in animal feed in Europe. It is not clear whether rendering will continue in North America and elsewhere. It is possible that rendering will be improved, since prions may be vulnerable to enzyme treatment. Governmental regulation restricts feeding rendered products back to the same species and may eventually eliminate all rendered products.
  - Restrictions on use of rendered products. Some countries have restricted use of rendered animal by-products. With the reduction in animal fat in the diet, the carcass fat has less saturated fat. This has the following consequences:
    1. Improved healthfulness of the product
    2. Reduced quality for use in cooking
    Unsaturated fatty acids can be oxidized and become rancid (tasting bad!). This is particularly a problem where cooking oil for fried chicken is reused multiple times.
- Methionine. Diets can be supplemented with methionine as L-methionine (not done commercially), DL-methionine (dry, 99% pure), and DLM-hydroxymethionine analogue (liquid, 88% active). The latter two are used extensively by the poultry industry.
- Botanics/Botanicals. It may be possible to use botanicals partially to replace growth-promoting antibiotics. Botanical extracts of specific plants include garlic, rosemary, anise, horseradish, juniper, yarrow, cinnamon, and others. An example in poultry feed is Apex R3010 (from Braes Feed Ingredients). The company has reported improved performance in broiler chickens with their product, which meets FDA GRAS (generally recognized as safe) criteria. Botanicals can be used in the European Union and with producers of natural and organic poultry in North America.
- Probiotics and Prebiotics
  1. Probiotics are live beneficial microorganisms (e.g., lactic and bacteria) that aid digestion and/or reduce pathogenic bacteria (disease causing in poultry or food-borne pathogens for people) in the gastrointestinal tract.
  2. Prebiotics are feed ingredients that modify the population of microorganisms in the gastrointestinal tract. Prebiotics include nutrients used preferentially by benign bacteria and dead bacteria (e.g., Aspergillus), which enhance the gastrointestinal tract's immune responses.

- Production
- Environment/weather (heat and humidity)
- Type of ration

The intensity of production dramatically affects the water requirement.

Water is the largest single constituent of poultry tissue; it constitutes 85% of the body of a baby chick, 58% of the body of an adult bird, and 66% of an egg. Yet, water is often neglected. Birds can lose 98% of their body fat or 50% of their body protein and still survive. However, a 10% loss in body water causes serious physiological disorders, and 20% loss in body water will cause death.

In addition to being readily available, water quality is important. Water should be tested to determine that salts, pesticides, and microorganisms are at acceptable levels and that the water is palatable to poultry. Water that adversely affects growth, reproduction, or productivity should not be used.

FURTHER READING

Austic, R. E., and M. C. Nesheim. *Poultry Production*, 13th ed. Philadelphia, PA: Lea & Febiger, 1990.

Bell, D. D., and W. D. Weaver. *Commercial Chicken Meat and Egg Production*, 5th ed. Norwell, MA: Kluwer Academic Publishers, 2002.

Jurgens, M. H. *Animal Feeding and Nutrition*, 9th ed. Dubuque, IA: Kendall/Hunt Publishing, 2002.

Leeson, S., and J. D. Summers. *Commercial Poultry Nutrition*. Guelph, Ontario: University Books, 1997.

Leeson, S., and J. D. Summers. *Broiler Breeder Production*. Guelph, Ontario: University Books, 2000.

Nutrient Requirements of Poultry, 9th rev. ed. Washington, DC: National Research Council, National Academy of Sciences, 1994.

6

Feeds and Additives

Objectives

After studying this chapter, you should be able to:

1. List the major categories of feedstuffs.
2. List the major feedstuffs providing protein for poultry diets and advantages and disadvantages of each.
3. List the sources of minerals for poultry diets and advantages and disadvantages of each.
4. List the major by-products used in poultry diets and advantages and disadvantages of each.
5. List the major additives used in poultry diets and the purpose of each.
6. Discuss the major methods of evaluating poultry feedstuffs and diets.

The economic importance of poultry feeding becomes apparent when it is realized that 55 to 75% of the total production cost of poultry is from feed, with the production of eggs toward the lower side of this range and production of broilers and turkeys toward the upper side. For this reason, the efficient use of feed is extremely important to poultry producers.

The major objective of poultry feeding is the conversion of feedstuffs into human food. In this respect, the domestic fowl is quite efficient.

FEEDS FOR POULTRY

A wide variety of feedstuffs can be used in poultry feeds. Broadly speaking, these may be classed as energy feedstuffs, protein supplements, mineral supplements, vitamin supplements, and nonnutritive additives (see Figure 6.1 for photograph of feed mill).

Energy Feedstuffs

The major energy sources of poultry feeds are the cereal grains and their by-products and fats.

Grains

Corn is the most important grain used by poultry, supplying about one-third of the total feed (see Figures 6.2 to 6.4). **Oats, barley, and the sorghum** grains are also used extensively in poultry rations. Oats are lower in energy than corn and are generally too expensive for broiler and layer rations. But oats can be used very effectively in feeds for replacement birds. Barley is less palatable than corn and is lower in vitamin A and energy. The sorghum grains can be readily substituted in place of corn as an energy feed, and they are being used extensively for this purpose in the southern states. **Wheat** may replace corn when available and the price is right. It is slightly lower in energy than corn, but higher in protein. Because wheat is gelatinous and has a tendency to "paste" on the beaks, it should be coarsely ground and pelleted when fed at high levels. Surplus **rice** along with broken and low-grade rice can be used effectively in poultry rations. Milled rice is quite comparable to corn in feeding value, except that it is lacking in vitamin A activity and pigmenting qualities.

Figure 6.3 A field of corn/maize. *(Courtesy, Iowa Agricultural and Home Economics Experiment Station)*

Figure 6.1 This feed mill combines (1) a grain receiving operation, (2) an ingredient storage area, (3) a computer-controlled mixing system, and (4) a pellet-making operation. A feed mill similar to this provides the nutritionally balanced diets to Delmarva's chickens. *(Courtesy, Delmarva Poultry Industry, Inc.)*

Figure 6.4 Harvesting corn/maize. *(Courtesy, Iowa Agricultural and Home Economics Experiment Station)*

Figure 6.2 Corn, the leading grain used for poultry. *(Courtesy, USDA)*

Triticale, an intergeneric cross between wheat and rye, may be used as a replacement for corn or wheat with no adverse effect on growth or feed efficiency. Although **millet** is seldom used in poultry rations, it can be freely substituted for corn. **Rye and buckwheat** are seldom used in poultry feeds; rye because of being unpalatable and producing

sticky, pasty droppings when fed at moderate to high levels, and buckwheat because of scarcity and unpalatability. When used, rye should not replace more than one-fourth of the cereal grains in the ration of young chicks or more than one-third of the cereal grains of older birds; and buckwheat should be limited to 15% of the cereal grains of the ration.

CEREAL MILLING BY-PRODUCT FEEDS. Numerous by-products result from the milling and processing of grain. Many contain large amounts of protein as well as energy. Table 6.1 gives the cereal sources and lists the milling by-products that are commonly used in poultry feeds.

BREWERS' AND DISTILLERS' GRAINS. Considerable quantities of grains are used in the brewing of beer, in the distilling of liquors, and for fuel-ethanol production. After processing, the remaining by-product can be readily adapted to many feeding programs. In addition to those feeds, solubles and yeast products from these industries are used in livestock feeds, although to a much lesser extent.

TABLE 6.1 Cereal Milling By-Product Feeds

Cereal Source	Milling By-Product[1]	Description	Comments
Corn, rice, wheat	Bran	Outer, coarse coat (pericarp) of grain separated during processing.	Laxative in action. Rich bran must have at least 14% crude protein and less than 14% crude fiber.
Corn, sorghum, wheat	Flour	Soft, finely ground and bolted meal from the milling of cereal grains and other seeds. Consists primarily of gluten and starch from endosperm.	Sorghum grain flour must have less than 1% crude fiber. Wheat flour must have less than 1.5% crude fiber.
Corn, sorghum, wheat	Germ meal	Embryo of the seed.	Wheat germ meal must contain at least 25% crude protein and 7% crude fat.
Corn, sorghum	Gluten (feed and meal)	Tough, viscid, nitrogenous substance remaining when flour is washed to remove the starch.	As currently manufactured, corn gluten meal contains approximately 60% protein.
All grains	Grain screenings	Small imperfect grains, weed seeds, and other foreign material of value as a feed that is separated through the cleaning of grain with a screen.	Quality varies according to percentage of weed seeds and other foreign material. Should be finely ground in order to kill noxious weed seeds.
Corn	Hominy feed	This is the by-product from producing pearl hominy.	Good hominy feed should contain a minimum of 1,350 kcal ME/lb (2,970 kcal ME/kg).
Oats, rice	Groats	Grain from which the hulls have been removed.	A high-grade feed, but usually expensive. Sometimes used in chick mashes.
Corn, oats	Meal	Feed ingredient in which the particle size is larger than flour.	Must contain at least 4% fat. Oatmeal must contain less than 4% fiber.
Rye, wheat	Middlings	A by-product of the flour milling industry comprising several grades of granular particles containing different proportions of endosperm, bran, and germ, each of which contains different percentages of crude fiber.	Deficient in calcium, carotene, and vitamin D. Wheat middlings cannot contain more than 9.5% crude fiber. Rye middlings cannot contain more than 8.5% crude fiber.
Barley, oats, sorghum, rice, rye, wheat	Mill-run (mill by-product or pollards)	State in which a grain product comes from the mill, usually ungraded and having no definite specifications.	Grain sorghum mill feed must contain more than 5% crude fat and less than 6% crude fiber. Oat mill by-product cannot contain more than 22% crude fiber. Rice mill by-product cannot contain more than 32% crude fiber. Rye mill-run cannot contain more than 9.5% crude fiber. Wheat mill-run cannot contain more than 9.5% crude fiber.
Rice	Polishings	By-product of rice, consisting of a fine residue that accumulates as rice kernels are polished (often hulls and bran have been removed).	High in fat; hence, rancidity can pose problems. Good source of thiamin.
Wheat	Red dog	By-product of milling spring wheat. Consists primarily of the aleurone with small amounts of flour and fine bran particles.	Cannot contain more than 4% crude fiber.
Wheat	Shorts	A by-product of flour milling consisting of a mixture of small particles of bran and germ, and the aleurone layer and coarse fiber.	Cannot contain more than 7% crude fiber.

[1]Cereal by-products recognized by the Association of American Feed Control Officials.

When available at competitive prices, the mash from the brewing and distilling industries, as well as some of the pharmaceutical fermentations, can be used in poultry rations. Usually they contain some of the B vitamins and other nutrients that are supplied by fermentation.

Fats

Animal and vegetable fats are now used extensively in poultry feed. They serve the following functions:

- To increase the caloric density of the ration
- To control dust
- To reduce the wear and tear on feed-mixing equipment
- To facilitate pelleting of feeds
- To increase palatability
- To help homogenize and stabilize certain feed additives, especially those of a very fine particle size.

However, the use of fats in poultry feeds requires good mixing equipment. Also, it is necessary that the fat be properly stabilized in order to prevent rancidity. Chickens are capable of tolerating high levels of fats, but costs usually permit only a maximum of 4 to 5% fat added to the ration.

Molasses

Molasses can be used effectively as an energy feed in poultry rations provided the level of usage is closely monitored. Excessive amounts cause wet droppings. Molasses levels are generally restricted to 2 to 5%. In addition to its use as an energy feed, molasses is used in the following ways in poultry feeds: (1) to reduce the dustiness of a ration, and (2) as a binder for pelleting.

Cane and beet molasses are by-products of the manufacture of sugar from sugarcane and sugar beets, respectively. Citrus molasses is produced from the juice of citrus waste. Wood molasses is a by-product of the manufacture of paper, fiberboard, and pure cellulose from wood; it is an extract from the more soluble carbohydrates and minerals of the wood material. Starch molasses is a by-product of the manufacture of dextrose from starch derived from corn or grain sorghums in which the starch is hydrolyzed by use of enzymes and/or acid. Cane or blackstrap is, by far, the most extensively used type of molasses. The different types of molasses are available in both liquid and dehydrated forms.

Protein Supplements

The usefulness of a protein feedstuff for poultry depends upon its ability to furnish the essential amino acids required by the bird, the digestibility of the protein, and the presence or absence of toxic substances. As a general rule, several different sources of protein produce better results than single protein sources. Both vegetable and animal protein supplements are used for poultry. Most of the protein supplements of animal origin contribute minerals and vitamins, which significantly affect their value in poultry rations, but they are generally more variable in composition than the vegetable protein supplements.

Plant Proteins

Even though they are not especially high in protein by comparison with other feedstuffs, the vegetative portions of many plants supply an extremely large portion of the protein in the total ration of poultry, simply because these portions of feeds are consumed in large quantities. Needed protein not provided in these feeds is commonly obtained from one or more of the oilseed by-products. The protein content and feeding value of the oilseed meals vary according to the seed from which they are produced, the geographic area in which they are grown, the amount of hull and/or seed coat included, and the method of oil extraction used.

Additional plant proteins are obtained as by-products from grain milling, brewing and distilling, and starch production. Most of these industries use the starch in grains and seeds, then dispose of the residue, which contains a large portion of the protein of the original plant seed.

OILSEED MEALS. Several rich oil-bearing seeds are produced for vegetable oils for human food (margarine, shortening, and salad oil), and for paints and other industrial purposes. In processing these seeds, protein-rich products of great value as poultry feeds are obtained. Among such high-protein feeds are soybean meal (see Figures 6.5 and 6.6), coconut meal, cottonseed meal, linseed meal, peanut meal, rapeseed/canola meal, safflower meal, sesame meal, and sunflower seed meal.

Oil is extracted from these seeds by one of the following basic processes or modifications thereof: solvent extraction, hydraulic extraction, or expeller extraction.

Soybean Meal. Soybean meal has the highest nutritive value of any plant protein source and is the most widely used protein supplement in poultry rations.

In the past, oil was extracted by hydraulic or expeller/extruder processes (see Figure 6.5). Today, almost all soybeans are solvent extracted. Soybean meal normally contains 41, 44, 48, or 50% protein, depending on the amount of hull removed. Because of its well-balanced amino acid profile, the protein of soybean meal is of better quality than

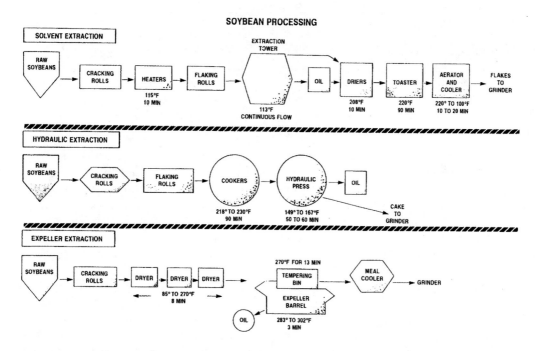

Figure 6.5 Processing soybeans. *(Original diagram from E. Ensminger)*

Figure 6.6 A field of soybeans. *(Courtesy, American Soybean Association)*

other protein-rich supplements of plant origin. However, it is low in calcium, phosphorus, carotene, and vitamin D.

Because raw soybeans contain several antinutritional factors, they should never be fed to poultry. However, heat-processed soybean products are very acceptable poultry feeds.

Coconut Meal (Copra Meal). This is the by-product from the production of oil from the dried meats of coconuts. The oil is generally extracted by either (1) the hydraulic process or (2) the expeller process. Coconut meal averages about 21% protein content. The quality of the protein is not high; hence, its use should be restricted in poultry ra-

tions. If copra meal is fed to poultry, it should be supplemented with the amino acids, lysine and methionine. When properly supplemented with amino acids, copra meal can compose up to 20% of poultry rations. Light-colored meal is more digestible than the dark meal.

Cottonseed Meal. The protein content of cottonseed meal varies from about 22% in meal made from undecorticated seed to 95% in flour made from seed from which the hulls have been removed completely. Thus, by screening out the residual hulls, which are low in protein and high in fiber, the processor is able to make a cottonseed meal of the protein content desired—usually 36, 41, 44, or 48%.

Cottonseed meal is low in lysine and tryptophan and deficient in vitamin D, carotene (vitamin A value), and calcium. Also, it contains toxic substances known as gossypol and cyclopropenoic fatty acids, varying in amounts with the seed and the processing. Gossypol can cause discoloration of egg yolks, and cyclopropenoic fatty acids impart pink color to egg whites. But, cottonseed meal is rich in phosphorus. It can be fed to growing chickens provided free gossypol levels are below 0.03% and supplemental lysine is provided.

Today, glandless cottonseed, free of gossypol, is being grown and improved. Someday, meal made from it may replace conventional cottonseed meal in poultry rations and alleviate (1) many of the restrictions as to levels of meal, and (2) the need to add iron to tie up free gossypol.

Linseed Meal. This is the finely ground residue (known as cake, chips, or flakes) remaining after the oil extraction of flax seed. It averages about 35% protein content (33 to 37%). Linseed meal is lacking in carotene and vitamin D, and is only fair in calcium and the B vitamins. It is laxative, and can depress growth in poultry. Hence, linseed meal levels should be no higher than 5% of the poultry diet.

Peanut Meal. Peanut meal, a by-product of the peanut industry, is ground peanut cake, the product that remains after the extraction of part of the oil from peanuts by pressure or solvents. It is a palatable, high-quality vegetable protein supplement used extensively in poultry feeds. Peanut meal ranges from 41 to 50% protein and from 4.5 to 8% fat. It is low in methionine, lysine, and tryptophan, and low in calcium, carotene, and vitamin D.

Since peanut meal tends to become rancid when held too long—especially in warm, moist climates—it should not be stored longer than 6 weeks in the summer or 2 to 3 months in the winter.

Rapeseed/Canola Meal. Rape is grown extensively throughout the world. However, meal made from unimproved rape is rather unpalatable. Also, it contains several goitrogenic compounds that can pose potential hazards to poultry. Canola was developed by selection of rapeseed by Canadian plant scientists in the 1970s. The new canola is low in glucosinolates in the meal, and low in erucic acid (a long-chain fatty acid) in the oil. Canola is grown mainly in Canada and some states in the United States (e.g., North Dakota).

Canola meal averages about 36% crude protein and its amino acids compare favorably with soybean meal. When the price is favorable, canola meal may be used as a protein supplement for all classes of livestock and poultry, but the amount in the total ration should be limited to about 10% for layers and breeding ducks, and 15% for breeding turkeys.

Safflower Meal. A large proportion of the safflower seed is composed of hull—about 40%. Once the oil is removed from the seeds, the resulting product contains about 60% hulls and 18 to 22% protein. Various means have been tried to reduce this high-hull content. Most meals contain seeds with part of the hull removed, thereby yielding a product of about 15% fiber and 40% protein. Safflower meal is deficient in lysine and methionine. Decorticated meal may be used successfully in layer rations at levels up to 15%.

Sesame Meal. Little sesame is grown in the United States, despite the fact that it is one of the oldest cultivated oilseeds. The oil meal is produced from the entire seed. Solvent extraction yields higher protein (45%) but lower fat levels (1%) than either the screw press or hydraulic methods, which produce meals containing about 38% protein and 5 to 11% oil. Sesame meal is extremely deficient in lysine. It is generally recommended that some soybean meal, along with added lysine, be fed with sesame meal to achieve optimum utilization.

Sunflower Seed Meal. Sunflower meal—a newcomer to the oilseed industry in the United States—is rapidly gaining acceptance as a high-quality source of plant protein in both North America and Europe.

Sunflower oil meal varies considerably, depending on the extraction process and whether the seeds are dehulled. Meal from prepressed solvent extraction of dehulled seeds contains about 44% protein, as opposed to 28% for whole seeds. Screw-pressed sunflower meal ranges from 28 to 45% protein. When sunflower meal is used in poultry feeds, it should be combined with high-lysine supplements, such as meat scrap or fish meal, and/or lysine should be added. When sunflower meal constitutes all (or most) of the protein supplement, the ration should be pelleted or crumbled to prevent stickiness and necrosis of the beak.

Animal Proteins

Protein supplements of animal origin are derived from (1) meatpacking and rendering operations, (2) poultry and poultry processing, (3) milk and milk processing, and (4) fish and fish processing. Before the discovery of vitamin B_{12}, it was generally considered necessary to include one or more of these in the rations of chickens. With the discovery and increased availability of synthetic vitamin B_{12}, high-protein feeds of animal origin have become less essential, although they are still included to some extent in rations for most monogastric animals.

With further improvement in the protein quality of plants, such as the development of high-lysine corn, the use of animal proteins may decline in the future. The cost of such proteins will need to be more competitive than they are at the present time if they are to be included in any major quantity in animal rations. Blending several proteins with complementary balances of amino acids and supplying more concentrated sources of individual amino acids may also be a factor affecting the future role of animal proteins in poultry rations.

Many protein supplements of animal origin are difficult to process and store without some spoilage and nutrient loss. If they cannot be dried, they must usually be refrigerated. If not heated to destroy disease-producing (pathogenic) bacteria, they may be a source of infection. However, protein availability will be reduced and some nutrients lost if the feed is heated excessively.

MEAT PACKING BY-PRODUCTS. Although the meat or flesh of animals is the primary object of slaughtering, modern meatpacking plants process numerous and valuable by-products, including protein-rich poultry feeds.

Tankage and Meat Meal. About 2.5 million tons of tankage and meat meal are fed to livestock. Tankage and meat meal are made from the trimmings that originate on the killing floor, inedible parts and organs, cleaned entrails, fetuses, residues from the production of fats, and certain condemned carcasses and parts of carcasses.

The end products and the methods of processing are:

1. Tankage (or digester tankage, meat meal tankage, feeding tankage) is produced by the older wet-rendering method, in which all of the material is cooked by steam under pressure in large closed tanks; hence, the derivation of the name—tankage. Tankage may also be made by the dry-rendering method (in which case it is known as meat meal), or by mixing products containing both wet-rendered and dry-rendered materials. The level of protein of tankage (generally 60%) is often standardized during manufacturing by the addition of enough blood to raise the total protein to the desired level.

2. Meat meal (or meat scrap) is produced by the newer and more efficient dry-rendering method, in which all of the material is cooked in its own grease by dry heat in open steam-jacketed drums until the moisture has evaporated.

The level of protein of meat meal (generally 50 to 55%) was originally established because the normal proportions of raw materials available for rendering resulted in a product that, after being pressed and ground, contained approximately 50% protein. The protein content of meat meal is adjusted up or down by raising or lowering the quantity of bone and fat in the raw material.

Protein content alone is not an infallible criterion of the feeding value of tankage or meat meal, for it varies considerably according to the kind of raw material from which it is produced. For this reason, the poultry producer should (1) purchase tankage or meat meal from a reputable source, and (2) mix these products with other protein supplements, especially when the birds are confined.

Meat and Bone Meal Tankage, and Meat and Bone Meal. When because of added bone, tankage or meat meal contains more than 4.4% phosphorus (P), the word *bone* must be inserted in the name; and it must be designated, according to the method of processing, as either (1) meat and bone meal tankage or (2) meat and bone meal. Thus, when such high-phosphorus products are prepared by the older wet-rendering method, they are known as meat and bone meal tankage. Likewise, when products in excess of 4.4% phosphorus are prepared by the newer dry-rendering method, they are designated as meat and bone meal.

Meat and bone meal tankage and meat and bone meal, or other similar bone-containing products, contain less protein than tankage or meat meal, usually 45 to 50%.

Liver Meal, and Liver and Glandular Meal. These products are produced in limited quantities by U.S. meatpacking plants. Most of the supply comes from South America and Australia. Liver meal is made entirely by drying and grinding livers; whereas liver and glandular meal is made by drying and grinding liver and other glandular tissue, but >50% of the dry weight of the product must be derived from liver. These products contain proteins of high quality and are rich sources of both the fat-soluble and water-soluble vitamins.

Blood Meal. Blood meal is ground, dried blood. When prepared by a special process and reduced to a fine powder, the end product is called blood flour. It contains 80 to 82% protein, more than any other packinghouse product, and it is an excellent source of the amino acid lysine, of which about two-thirds is available to the bird. However, due to the high temperature of processing, the protein is less digestible and of lower quality than that in high-grade tankage or meat meal, being especially low in the essential amino acid isoleucine. Also, blood meal and blood flour differ from tankage and meat meal in that they are low in calcium and phosphorus.

In recent years, new methods of drying blood, such as flash drying, have shown considerable promise in producing a uniform high-lysine product.

POULTRY WASTES. By-product feedstuffs are derived from all segments of the poultry industry—from hatching all the way through processing for market; and they come from the broiler and turkey segments of the industry as well as from egg production. Centralization of these industries into large units with enough volume of wastes to make it feasible to process the potential feeds has opened new markets for what was previously a disposal problem. Certain precautions have had to be included to make the products useful and safe, but considerable amounts of poultry products are currently being used in poultry rations.

The three most extensively used high-protein by-products of the poultry industry are hatchery by-products, poultry by-products, and poultry feathers. Cull birds, unsalable eggs, eggshells, and slaughter wastes are also used in poultry feeds.

Hatchery By-Products. Of the various by-products from poultry, the most valuable is hatchery by-products consisting of infertile eggs, eggs with dead embryos, and unsalable sexed male chicks. However, these products deteriorate quite rapidly if not cooled promptly—a factor that is true of most poultry, fish, and meat products.

Poultry By-Product Meal. Poultry by-products consist of nonrendered clean parts of carcasses of slaughtered poultry—such as heads, feet, and viscera—free from fecal content and foreign matter, except for such trace amounts as are unavoidable in good factory practice. The meal is the dried, rendered, and ground poultry by-product. Because of the heads and feet, poultry by-products are lower in nutritional value than the flesh of animals, including poultry. The biological value of the proteins is lower than the other animal proteins and the better plant proteins. They may be successfully used in animal rations, however, provided they are not the sole source of proteins.

Hydrolyzed poultry by-product aggregate is the product resulting from treatment under pressure of all by-products of slaughtered poultry, clean and undecomposed, including such parts as heads, feet, undeveloped eggs, intestines, feathers, and blood.

Feather Meal. Feather meal is a by-product that is nearly all protein. It can be used in rations after the feathers are hydrolyzed with heat and pressure to make the proteins available. Hydrolyzed feather meal is defined as the product resulting from the treatment under pressure of clean, undecomposed feathers from slaughtered poultry, free of additives and/or accelerators. Not less than 75% of its crude protein content must be digestible by the pepsin digestibility method. Although hydrolyzed feather meal is high in protein (from 85 to 90%), it is rather low in nutritional value, being low in the amino acids histidine, lysine, methionine, and tryptophan. The amino acids present are readily available.

Because of the deficiencies of several amino acids, care must be used when incorporating feather meal into poultry feed. The addition of fish meal or meat meal tends to complement feather meal and facilitates its use. In practice, feather meal rarely exceeds 5% of poultry rations.

DAIRY PRODUCTS. Along about 1910, processes were developed for drying buttermilk. Soon thereafter,

special plants were built for dehydrating buttermilk, and the process was extended to skimmed milk and whey. Beginning about 1915, dried milk by-products were incorporated in commercial poultry feeds. The superior nutritive values of milk by-products are due to their high-quality proteins, vitamins, and good mineral balance, and the beneficial effect of the milk sugar, lactose. In addition, these products are palatable and highly digestible. They are an ideal supplement for balancing out the deficiencies of the cereal grains. The chief limitation to their wider use is price.

Dried Skimmed Milk and Dried Buttermilk. As the names indicate, these products are dehydrated skimmed milk and buttermilk, respectively. They contain less than 8% moisture and average 32 to 35% protein. One pound of dried skimmed milk or dried buttermilk has about the same composition and feeding value as 10 lb of their respective liquid forms. Although dried skimmed milk and dried buttermilk are excellent poultry feeds, they are generally too high priced to be economical for this purpose.

Prior to World War II, dried skimmed milk was the most widely used dried milk product included in feeds. During and since the war, however, much of it has been marketed as a human food.

Whey. Whey is a by-product of cheese making. Practically all of the casein and most of the fat go into the cheese, leaving the whey, which is high in lactose (milk sugar) and ash but low in protein (0.6 to 0.9%) and fat (0.1 to 0.3%). However, its protein is of high quality.

Numerous whey products are commercially available. Those used in poultry feeds when the price is right are:

1. **Dried hydrolyzed whey.** This product is the residue obtained by drying lactose-hydrolyzed whey. It contains at least 30% glucose and galactose.
2. **Dried whey.** This product is derived from drying whey from the manufacture of cheese. It is high in lactose (milk sugar), containing at least 65%, and rich in riboflavin, pantothenic acid, and some of the important unidentified factors.
3. **Dried whey product.** When a portion of the lactose (milk sugar) that normally occurs in whey is removed, the resulting dried residue is called dried whey product. According to the Association of American Feed Control Officials, the minimum percentage of lactose must be prominently declared on the label of the dried whey product. Dried whey product is a rich source of the water-soluble vitamins.

4. **Dried whey solubles.** This product is obtained by drying the whey residue from the manufacture of lactose following the removal of milk albumin and the partial removal of lactose. The minimum percentage of lactose must be prominently declared on the label.

Milk Protein Products. The four milk protein products currently available are:

1. **Dried milk protein.** This feed is obtained by drying the coagulated protein residue resulting from the controlled coprecipitation of casein, lactalbumin, and minor milk proteins from defatted milk.
2. **Dried milk albumen.** This product is produced by drying the coagulated protein residue separated from whey. It contains at least 75% crude protein on a moisture-free basis.
3. **Casein.** Casein is the solid residue that remains after the acid or rennet coagulation of defatted milk. It contains at least 80% crude protein.
4. **Dried hydrolyzed casein.** This is the residue obtained by drying the water-soluble product resulting from the enzymatic digestion of casein. It contains at least 74% crude protein.

Although the quality and quantity of protein from milk protein products are excellent, these sources of protein are generally too expensive to be used routinely as poultry feeds.

MARINE BY-PRODUCTS. Marine by-products are generally considered by poultry producers to be excellent sources of nutrients. Proteins, vitamins, and minerals are all readily available in most fish products.

Fish Meal. Fish meal is a by-product of the fisheries industry. It consists of dried, ground whole fish or fish cuttings—either or both—with or without extraction of part of the oil. If it contains more than 3% salt, the salt content must be a part of the brand name. In no case shall the salt content exceed 7%.

The feeding value of fish meal varies somewhat, according to:

1. **The method of drying.** It may be either vacuum, steam, or flame dried. The older flame-drying method exposes the product to a higher temperature. This makes the proteins less digestible and destroys some of the vitamins.
2. **The type of raw material used.** It may be made from the offal produced in fish packing or canning factories, or from the whole fish with or without extraction of part of the oil. Fish meal made from offal containing a large proportion of heads is less desirable because of the lower quality and digestibility of the proteins. Although few feeding comparisons have been made between the different kinds of fish meals, it is apparent that all of them are satisfactory when properly processed raw materials of good quality and moderate fat content are used.
3. **The amount of oxidation.** Much of the variation in the efficiencies of fish meal is due to the oxidation. Today, ethoxyquin is being added to many fish meals to prevent oxidation.

It is of interest to poultry producers to know the sources of the commonly used fish meals. These are:

1. **Menhaden fish meal.** This is the most common kind of fish meal used in the eastern United States. It is made from menhaden herring (a very fat fish not suited for human food), caught primarily for their body oil (see Figure 6.7). The meal is the dried residue after most of the oil has been extracted.
2. **Sardine meal or pilchard meal.** This is made from sardine canning waste and from the whole fish.
3. **Herring meal.** This is a high-grade product produced in the Pacific Northwest and Alaska.
4. **Salmon meal.** This is a by-product of the salmon canning industry (e.g., in the Pacific Northwest and in Alaska).
5. **White fish meal.** This is a by-product from fisheries making cod and haddock products for human food. Its proteins are of very high quality.

Figure 6.7 Menhaden herring are a fat fish considered to be unsuitable for human consumption. They are caught primarily for their body oil, but the dried residue after oil extraction provides an excellent high-quality protein feed. *(Courtesy, National Fisheries Institute)*

Fish meal should be purchased from a reputable company on the basis of protein content. It varies in protein content from 57 to 77%, depending on the kind of fish from which it is made. When of comparable quality, fish meal is superior to tankage or meat meal as a protein supplement for poultry. The protein of a good-quality fish meal is 92 to 95% digestible. If fish meal is poorly processed or improperly stored, the digestibility of protein decreases dramatically. Since fish meals are cooked, there is danger that certain amino acids—notably lysine, cystine, tryptophan, and histidine—will be denatured, but these losses are minimized when proper processing techniques are used.

Fish meals containing high levels of fat are considered to be low quality. If they are incorporated into poultry feeds, they tend to impart a fishy flavor to poultry products. Also, problems of rancidity are greater in high-fat fish meals.

Fish meal is an excellent source of minerals. Calcium and phosphorus are especially abundant, being present in the amounts of 3 to 6% and 1.5 to 3.0%, respectively. Many of the trace minerals required by poultry, especially iodine, can be supplied in part by fish meal. Fish meal is not a particularly good source of vitamins. Most of the fat-soluble vitamins are lost during the extraction of oil, but a fair amount of the B vitamins remain. Fish meal is one of the richest sources of vitamin B_{12}.

Fish Residue Meal. This is the dried residue from the manufacture of glue from nonoily fish. If the product contains more than 3% salt, the amount of salt must be included in the product name. However, salt content must not exceed 7%.

Condensed Fish Solubles. This is a semisolid by-product obtained by evaporating the liquid remaining from the steam rendering of fish, chiefly sardines, menhaden, and redfish. The water that comes from processing contains about 5% total solids. After this liquid is evaporated or condensed, it contains about 50% total solids. Condensed fish solubles, containing approximately 30% protein, are a rich source of the B vitamins and unknown factors. They are particularly rich in pantothenic acid, niacin, and vitamin B_{12}.

Dried Fish Solubles. Dried fish solubles are obtained by dehydrating the glue water of fish processing. Dried fish solubles contain at least 60% crude protein.

Fish Liver and Glandular Meal. This marine by-product is obtained by drying the entire viscera of fish. The Association of American Feed Control Officials (AAFCO) specifies that it must contain at least 18 mg of riboflavin per pound, and that at least 50% of dry weight must consist of livers.

Fish Protein Concentrate. This by-product is prepared through the solvent extraction processing of clean, undecomposed whole fish or fish cuttings. The product cannot contain more than 10% moisture and must contain at least 70% protein, according to AAFCO. The solvent residues must conform to the rules as set forth by the Food Additive Regulations.

Shrimp Meal and Crab Meal. These are high-quality protein feeds for poultry. Shrimp meal is the ground, dried waste of shrimp processing. It may consist of the head, hull (or shell), and/or whole shrimp. The provisions relative to salt content are the same as those for fish meal. It is either steam dried or sun dried, with the former method being preferred. On the average, it contains 32% protein and 18% mineral. Crab meal, the by-product of the crab industry, is composed of the shell, viscera, and flesh. Mineral content is exceedingly high, about 40%, and protein content is above 25% (generally about 30%). A rule of thumb for feeding crab meal is that 1.6 units (lb or kg) of crab meal can replace 1 unit (lb or kg) of fish meal.

Whale Meal. This is the clean, dried, undecomposed flesh of the whale after part of the oil has been extracted. If it contains more than 3% sodium chloride, the amount of salt must constitute a part of the product name. But the amount of salt cannot exceed 7%. Harvesting of whales is presently prohibited by international agreement.

Yeasts

Yeasts are listed as protein supplements, but they may also be classed as vitamin supplements. Three forms of dried yeast are used in poultry feeding: (1) dried yeast, which is a by-product of the brewing industry; (2) torula yeast, resulting from the fermentation of wood residue and other cellulose sources; and (3) irradiated yeast, which is used because of its vitamin D_2 content. Of these, only brewer's dried yeast, which must contain not less than 35% crude protein, is used to any extent in poultry feeds.

A wide variety of materials can be used as substrates for the growth of yeasts. Current research deals with the use of industrial by-products, which otherwise would have little or no economic value. By-products from the chemical, wood and paper, and food industries have shown considerable promise as sources of nutrients for single-cell organisms; among them are (1) crude and refined petroleum products, (2) methane, (3) alcohols, (4) sulfite waste liquor, (5) starch, (6) molasses, and (7) cellulose.

Yeast proteins are generally of good quality, although they are deficient in methionine. They are good sources of lysine and excellent sources of the B complex vitamins.

Mineral Supplements

Mineral supplements are required by poultry for skeletal development of growing birds, for eggshell formation of laying hens, and for certain other regulatory processes in the body.

Of the macrominerals required by poultry, salt (sodium chloride), calcium, magnesium, potassium, and phosphorus are routinely considered for poultry rations. The producer should add trace minerals that are believed to be in short supply. This is usually accomplished through the use of a commercial trace mineral mix.

Salt is added to most poultry rations at a 0.2 to 0.5% level. Too much salt will result in increased water consumption and wet droppings (also see Chapter 5).

Guidelines For Mineral Supplementation

No single plan can be proposed as being the best for mineral supplementation. Rather, the producers must tailor their supplement regimens to encompass the following considerations:

1. **Needs of the birds.** Age, sex, weight, and production parameters must all be considered.
2. **Types of feed.** A high-energy ration will require a different mineral supplement than a low-energy ration.
3. **Region from which the feeds were obtained.** The mineral content of the feed will reflect the mineral composition of the soil on which it was grown.

Calcium and Phosphorus Supplements

The common calcium supplements used in poultry feeding are ground limestone, crushed oystershells or oyster flour, bone meal, calcite, chalk, and marble.

Most of the phosphorus in plant products is in organic form and not well utilized by young chicks or turkey poults. Hence, for poultry, emphasis is placed upon inorganic phosphorus sources in feed formulation. The most important phosphorus sources in poultry feeding are dicalcium phosphate, defluorinated rock phosphate, and steamed bone meal. All of these are calcium phosphates, which can supply both calcium and phosphorus. Bone meal, dicalcium phosphate, defluorinated phosphate, colloidal phosphate, and raw rock phosphate are used where both calcium and phosphorus are needed in the ration.

PRECAUTIONS RELATIVE TO CALCIUM AND PHOSPHORUS SUPPLEMENTS. Phytate phosphorus is unavailable to poultry and other simple-stomached animals, because of general lack of the enzyme phytase in the gastrointestinal tract. In plants, approximately one-third of the phosphorus has been found to be present as nonphytin phosphorus and is available to chickens. In calculating the available phosphorus content of feed, the phosphorus from inorganic supplements and animal feedstuffs is considered to be 100% available while that from plants is assumed to be 30% available.

Raw, unprocessed rock phosphate usually contains from 3.25 to 4.0% fluorine, whereas steamed bone meal normally contains only 0.05 to 0.10%. Fortunately, through heating at high temperatures under conditions suitable for elimination of fluorine, the excess fluorine of raw rock phosphate can be removed. Such a product is known as defluorinated rock phosphate.

The Association of American Feed Control Officials has established maximum fluorine content for mineral substances and total ration (see Table 6.2).

Vitamin Supplements

As with mineral supplements, careful consideration must be given to the vitamin supplementation of poultry feeds. While the requirements of vitamins are extremely small in comparison with energy and protein, the omission of a single vitamin from the diet of a species that requires it will produce specific deficiency symptoms, thereby reducing production. Moreover, the cost for vitamin supplementation constitutes a very small fraction of the total feed bill.

It is to be emphasized that subacute deficiencies can exist, although the actual deficiency symptoms do not appear. Such borderline deficiencies are the most costly and the most difficult with which to cope, going unnoticed and unrectified; yet they may result in poor and expensive production. Also, under practical conditions one will usually not find a vitamin deficiency that involves only a single vitamin. Instead, deficiencies usually represent a combination of factors, and usually the deficiency symptoms will not be clear-cut.

TABLE 6.2   Maximum Fluorine Content for Mineral Substances and Total Ration

Class of Animals	Maximum Fluorine Content of Any Mineral or Mineral Mixture That Is to Be Used Directly for the Feeding of Animals Shall Not Exceed—	Fluorine Content of Rock Phosphate (or other Ingredients) Shall Be Such That the Maximum Fluorine Content of the Total Ration Shall Not Exceed—
	(%)	(%)
Poultry	0.60	0.035
Cattle	0.30	0.009
Sheep	0.35	0.01
Swine	0.45	0.014

Source: Feed Control, official publication, Association of American Feed Control Officials, Washington, D.C. 1975, p. 46.

Formerly, a wide variety of feed ingredients were added to poultry rations for their vitamin content. But it was found that the vitamin concentration of feedstuffs varied tremendously, being affected by plant species and part (leaf, stalk, or seed), harvesting, storing, and processing. Generally speaking, vitamins are easily destroyed by heat, sunlight, oxidation, and mold growth. So, today, nutritionists rely on vitamin supplements, which in many cases are chemically pure sources that need to be used only in very minute amounts. In modern feed formulation, premixes are typically used to add trace ingredients in a manner that ensures thorough mixing. These trace ingredients include vitamins, minerals, and other non-nutritive additives.

Alfalfa Products

Alfalfa hay or leaf meal is included in some poultry rations to supply vitamin A, riboflavin, vitamin E, vitamin K, unidentified factors, minerals, and protein. Dehydrated alfalfa meal is superior to sun-cured meal because of its higher vitamin content. Leaf meal is better than hay because of its lower fiber and greater nutrient content.

Poultry nutritionists interested in high-energy rations are reducing or eliminating alfalfa and substituting synthetic forms of vitamins, because alfalfa meal is fibrous and unpalatable, and its vitamins are not stable. A good dehydrated alfalfa meal should contain 100,000 IU of vitamin A activity per pound. Alfalfa processors are improving alfalfa products by the addition of antioxidants to preserve the fat-soluble vitamins and carotenoid pigment, and pelleting alfalfa meal to prevent dustiness and help preserve nutritional values.

Nonnutritive Additives

Modern poultry feeds commonly contain one or more nonnutritive additives. These additives are used for a variety of reasons. They are not nutrients, but some of them improve production under certain circumstances. Others prevent rancidity in the feed. There is no evidence of a nutritional deficiency when they are omitted from a ration. When considering the use of a feed additive, the following questions should be asked:

1. **What are the specific needs of the birds?** As intensity of production increases, the need for production stimulants increases.
2. **Does the additive have a withdrawal period or a product discard period?** Quite often a drug will have a required withdrawal period before the birds can be slaughtered or before the eggs can be marketed. If a withdrawal period is required, the producer must have separate holding facilities for unmedicated and medicated feeds. Additionally, feed-mixing facilities must be thoroughly cleaned after mixing medicated feed. If residues are found in the market products, the producer is subject to prosecution. The ultimate guide to using any drug properly is the label or tag. These should be read and followed carefully.
3. **Can the additive be used in combination with other additives?** The FDA has strict regulations as to which additives can be used in combination. If there is any doubt, the producer should contact the county agent or a reputable feed company.
4. **What is the best form of the product to be used?** Likewise, the active ingredient of one product may be more stable or readily available than that of a competing brand.
5. **Must methods of mixing and storing be considered?** The properties of various additives are such that methods of mixing and storing should be considered.

Production-promoting additives (antibiotics, etc.) are approved by FDA for incorporation in poultry feeds. But this list is being constantly revised since each product is reviewed on the basis of (1) its safety to the birds, and (2) its safety to humans who consume the poultry products. Thus, the status of each of these products is subject to change. Hence, the user is urged to keep abreast of current recommendations. The label of each additive contains information concerning the restrictions of use in combination with other drugs, level of incorporation, and withdrawal periods for meat-type birds. Additional information pertaining to additives appears in the *Feed Additive Compendium* and in the *Federal Register*. These instructions must be followed carefully because misuse of feed additives can lead to costly condemnations and fines.

Antioxidants

All feeds are susceptible to spoilage, but those high in fat content are especially prone to autoxidation and subsequent rancidity. To curb the oxidation of feeds, antioxidants are routinely added to many poultry feeds.

Antioxidants are compounds that prevent oxidative rancidity of polyunsaturated fats. It is important that rancidity of feeds be prevented because it may cause destruction of vitamins A, D, and E, and several of the B complex vitamins. Also, the breakdown products of rancidity may react with the epsilon amino groups of lysine and thereby decrease the protein and energy value of the diet.

The antioxidants that are presently approved for addition to fat in poultry feeds are butylated hydroxyanisole (BHA), butylated hydroxytoluene (BHT), and ethoxyquin. These antioxidants may be used to prolong the induction period in fats and to prevent oxidation in mixed feeds. They are used at a level of ¼ lb per ton. Antioxidants are capable of temporarily inhibiting the destructive effects of oxygen on sensitive feeding ingredients—the unsaturated fats, fat-soluble vitamins, and other constituents. They are normally incorporated in the vitamin–trace mineral premix to prevent vitamins A and E from oxidative destruction. Some are added to feed fats to stabilize them against rancidity.

Flavoring Agents

Flavoring agents are feed additives that are supposed to increase palatability and feed intake. Chickens have limited ability to differentiate between sucrose solutions, for which they show preference, and saccharine solutions, which they avoid. Other studies have shown that chickens possess a sense of taste, but very limited ability to smell. Numerous chemical agents are so objectionable to chickens that they cause a decrease in normal feed consumption, but no flavoring agent has yet been found that will increase feed consumption above that normally obtained from well-balanced poultry rations composed largely of good-quality corn and soybean meal. However, there is evidence that certain natural feedstuffs are relatively unpalatable to chickens; among them, barley, rye, and buckwheat. Whether chickens avoid certain feedstuffs on the basis of taste, lack of eye appeal, or because of adverse effects upon metabolism or sense of well-being is unknown.

There is need for flavoring agents that will help to keep up feed intake (1) when highly unpalatable medicants are being administered, (2) during attacks of diseases, (3) when birds are under stress, or (4) when a less palatable feedstuff is being incorporated in the ration.

Pellet Binders and Additives that Alter Feed Texture

Pellet binders are products that enhance the firmness of pellets. Several feed additives are known to produce a marked increase in the firmness of pellets. These include:

- Sodium bentonite (clay)
- Liquid or solid by-products of the wood pulp industry, consisting mainly of hemicelluloses or combinations of hemicelluloses and lignins
- Guar meal, an annual legume produced in Asia

Although bentonites have no nutritive value, several reports indicate that at the level of common usage (2 to 2.5% of the ration) they may even improve the growth and/or feed utilization of animals. Hemicellulose preparations at levels up to 2.5% may serve as good energy sources, but lignin has practically no nutritive value. In addition to its binding properties, guar meal is a satisfactory source of protein and energy when limited to 2.5 to 5.0% of well-balanced diets. Molasses or fat are sometimes added to feed as an aid in pelleting, as well as a concentrated source of energy.

Additives that Enhance the Color or Quality of Poultry Products

Feeds that contain large amounts of xanthophylls and carotenoids produce a deep yellow color in the beak, skin, and shanks of yellow-skinned breeds of chickens. The consumer associates this pigmentation with quality and in many cases is willing to pay a premium price for a bird of this type. Also, processors of egg yolks are frequently interested in producing dark-colored yolks to maximize coloration of egg noodles and other food products. The latter can be accomplished by adding about 60 mg of xanthophyll per kilogram of diet. In recognition of these consumer preferences, many producers add ingredients that contain xanthophylls to poultry rations. It is not necessary to incorporate high levels of xanthophyll in starter and grower rations. Low levels can be maintained through these periods of feeding, but finishing rations should contain high levels of these pigment-producing compounds.

Only a few natural products contain significant quantities of xanthophyll. The xanthophyll content of some feedstuffs follows:

Feedstuff	Xanthophyll Content (mg/kg)
Marigold petal meal	7,000
Algae, common, dried	2,000
Alfalfa juice protein, 40% protein	800
Alfalfa meal, 20% protein	240
Corn gluten meal, 41% protein	132
Corn, yellow	22

Although the xanthophylls in corn and alfalfa are utilized efficiently, both ingredients lose xanthophyll, or lutein, rather quickly when stored.

Today, certain synthetic xanthophylls are produced for use to supplement the natural xanthophylls of feedstuffs. Their use should be in keeping with Food and Drug Administration approval and manufacturers' directions.

Grit

Since poultry do not have teeth to facilitate grinding of feed, most grinding takes place in the thick-muscled gizzard. The more thoroughly feed is ground, the more surface area is created for digestion and subsequent absorption. Hence, when hard, coarse, or fibrous feeds are fed to poultry, grit is sometimes added to supply additional surface for grinding within the gizzard. Additionally, grit serves to break down ingested feathers and litter, which can sometimes lead to gizzard impaction. When mash or finely ground feeds are used, the value of grit is greatly diminished.

Oyster-, clam-, and coquina shells and limestone are sometimes used for grit. Being relatively soft and calcareous, they provide a source of calcium since they, too, are ground in the process. Gravel and pebbles have been used successfully as long-lasting sources of grit. Since modern diets are composed of ground ingredients, grit is no longer needed in poultry diets.

Enzymes

Enzymes are complex protein compounds produced in living cells, which cause changes in other substances without being changed themselves. They are organic catalysts.

Normally, the enzymatic output of the digestive system of birds is adequate for maximum digestion of the starches, fats, and proteins. However, repeated experiments in many laboratories with western barley have shown that the metabolizable energy value of barley for poultry produced in the West under semiarid conditions is improved either by soaking in water before being fed or by the addition of enzyme preparations derived from fungal fermentations. Yet, little or no improvement in metabolizable energy has been shown from the use of the same fungal enzymes with other common poultry feedstuffs, or even with barley produced in the eastern part of the United States.

Antifungal Additives (Mold Inhibitors)

Antifungals are agents that destroy fungi. Feeds provide an excellent environment for the growth of fungi, such as *Aspergillus flavus*, *Fusarium*, and *Candida albicans*, which are detrimental to the health of poultry. *C. albicans* is the causative agent of a condition in poultry called thrush or moniliasis.

Nutritionists have given considerable attention to the effects of fungal infestations of feeds. It has been speculated that perhaps many nutritional problems of the past (for example, suspected nutrient deficiencies) were, in fact, caused by feeds contaminated with fungi. Fungi can affect feed intake and subsequent production through contamination at one or more of four stages in the feeding chain: (1) in the field (preharvest), (2) during storage, (3) at mixing, and (4) in the bird itself. Fungal contamination can pose problems through the production of toxins, alterations of the chemical composition of the feed, or alterations of the metabolic functioning of the bird ingesting or harboring the fungus.

Fungi, for example, *Aspergillus flavus*, produce toxins. The toxin produced by *A. flavus* is known as aflatoxin, called mycotoxins. Aflatoxin, which has clearly been shown to be a carcinogen (tumor producing), causes much trouble in animals. But it is not the only mycotoxin to be feared. Mycotoxins affect all species, especially the young. Generally, ruminants appear to tolerate higher levels of mycotoxins and longer periods of intake than simple-stomached animals. Growing chickens are markedly less susceptible to aflatoxins than ducklings, goslings, pheasants, or turkey poults. Fish are one of the most susceptible animal species to aflatoxin poisoning.

The primary condition conducive to the growth of *A. flavus*, and therefore the production of aflatoxin, is moisture; hence, proper harvesting, drying, and storage will minimize mold growth and subsequent toxin production. All feedstuffs should be dried below the critical moisture content that permits the growth of molds—approximately 12%. Additionally, mold inhibitors should be added to high-moisture feeds that are exposed to air during storage. Propionic acid, acetic acid, and sodium propionate are used in high-moisture feeds to inhibit mold growth. Many feed manufacturers add such antifungals as nystatin and copper sulfate to concentrate feeds to prevent further growth by molds. It should be noted, however, that these compounds do not improve the value of moldy feeds. The toxicity of aflatoxin-contaminated feed can be reduced by irradiation with ultraviolet light or exposure to anhydrous ammonia under pressure.

Drugs

Poultry rations frequently contain drugs designed to promote production or prevent a specific disease. For example, a wide variety of chemical substances, sold under various trade names, are available for use in the prevention of coccidiosis. These drugs are known as coccidiostats. Turkey rations are frequently formulated with drugs for the prevention of blackhead. This class of drugs, known as histomonostats, also contains a wide variety of chemical substances sold under various trade names.

When any drug is incorporated into poultry feed, it is mandatory that the producer follow all restrictions concerning (1) the length of usage, (2) specific types of birds the drug may be used for, and (3) withdrawal periods prior to marketing.

PRODUCTION-PROMOTING DRUGS. The list of drugs that exert production-promoting effects is long and diverse. They can be classified as antibiotic, arsenical, or a nitrofuran.

Antibiotics. Antibiotics are chemicals produced by living organisms (molds, bacteria, or green plants) and that have bacteriostatic or bactericidal properties.

In 1949, quite by accident, while conducting nutrition studies with poultry, Jukes of Lederle Laboratory and McGinnis of Washington State University obtained startling growth responses from feeding a residue from Aureomycin production. Later experiments revealed that the supplement used by Jukes and McGinnis—the residue from Aureomycin production—supplied the antibiotic chlortetracycline. Such was the birth of feeding antibiotics to livestock.

The primary reason for using antibiotics in poultry feeds is for their growth-stimulating effect, for which purpose they are generally used in both broiler and market turkey rations. The reasons for the beneficial effects of antibiotics still remain obscure, but the best explanation for their growth-stimulating activity is the disease level theory, based on the fact that antibiotics have failed to show any measurable effect on birds maintained under germ-free conditions. In addition to their use as growth stimulators, antibiotics are used to increase egg production, hatchability, and shell quality in poultry. They are also added to feed in substantially higher quantities to remedy pathological conditions.

Antibiotics are generally fed to poultry at levels of 5 to 50 g per ton of feed, depending upon the particular antibiotic used. Higher levels of antibiotics (100 to 400 g per ton of feed) are used for disease-control purposes. The antibiotics most commonly used in poultry rations are bacitracin, virginiamycin, bambermycin, and lincomycin. High levels of calcium in a laying mash will inhibit assimilation of certain tetracycline-type antibiotics to the bloodstream and reduce their effectiveness.

In all probability, antibiotics will always be used as feed additives to control and treat health problems in poultry. But the status of subtherapeutically used antibiotics as production stimulators is, at the present time, tenuous. Pressure from consumer groups and medical people may result in banning many of the antibiotics that are primarily used for medicinal purposes in humans from the list of approved production promoters. However, in the future, an increasing number of antibiotics will likely be developed specifically for the purpose of improving poultry performance. One example is that of bambermycin. This antibiotic was developed solely for use as a production promoter, serving to increase rate of gain and feed efficiency in chickens and swine. It has no medicinal applications, and, therefore, poses no health hazard with regards to bacteria becoming resistant to it.

Arsenicals and Nitrofurans. The use of arsenic and its compounds dates back to antiquity. Hippocrates (460–377 B.C.) used realgar (arsenic sulfide) in the treatment of ulcers. In the Middle Ages, its poisonous properties became known and widely used, particularly by women to dispose of their adversaries or unwanted suitors.

Arsenicals and nitrofurans exert many of the same effects as the antibiotics; hence, they are often added to poultry feeds to improve performance. It would appear that the action of arsenicals and that of antibiotics are very similar, since the effects of the two are not considered to be additive. For broilers, arsenilic acid is used at 45 to 90 g per ton, and 3 nitro (Roxarsone) is used at 22.5 to 45.0 g per ton of ration; but, in keeping with the recommendation made relative to the use of any drug, the manufacturer's directions should be followed.

ANTIPARASITIC DRUGS. To combat parasitic infestations, drugs in the form of feed additives are available. These can be separated into two categories—anticoccidials and anthelmintics. See Chapter 11 for details on specific parasites.

Anticoccidials (Coccidiostats). Anticoccidial drugs are used to control coccidial infection—a parasitic disease caused by microscopic protozoan organisms known as Coccidia, which live in the cells of the intestinal lining of animals. Each class of domestic livestock harbors its own species of Coccidia, thus there is no cross-infection between species. However, within species, coccidiosis can spread very rapidly among animals.

This disease affects poultry, cattle, sheep, goats, and rabbits. However, poultry are most affected by coccidiosis. Hence, anticoccidials are routinely added to poultry feed. Each coccidiostat should be used at the level designated by the manufacturer.

Anthelmintics. Helminths are many-celled worm parasites varying greatly in size, shape, structure, and physiology. They may be classified as (1) flukes (or trematodes), (2) tapeworms (or cestodes), (3) roundworms (or nematodes), or (4) thorny-headed worms (or acanthocephala). With a few exceptions, the eggs or larvae must leave the host animal in which they originate to undergo further development on the ground, elsewhere in the open, or in intermediate hosts. Anthelmintics (wormers; vermifuges) generally require more than one administration. The first administration kills those worms that are present in the body, and subsequent wormings kill those worms that hatched from eggs after the initial dose. Some anthelmintics can be fed continuously. The prevention and control of parasites is one of the quickest, cheapest, and most dependable methods for increasing production with no extra birds, no additional feed, and little more labor. This is important, for, after all, the producer bears the brunt of this reduced production and wasted feed.

From time to time, new vermifuges, or wormers, are approved and old ones are banned or dropped. Where parasitism is encountered, therefore, it is suggested that the producer obtain from local authorities the current recommendation relative to the choice and concentration of the vermifuge (wormer) to use. This information can be obtained from the county agent, entomologist, veterinarian, or agricultural consultant.

EVALUATING POULTRY FEEDS

Poultry producers may evaluate the different feeds that they use. It is also important that they have a working knowledge of the value of different poultry feeds from the standpoint of purchasing and utilizing them. The nutritive requirements for a specific substance are determined by finding the minimum amount of that particular nutrient or substance that will permit maximum development of the physiological function or economic characteristics of concern. In general, the economic characteristics of importance in poultry are growth, efficiency of feed utilization, egg production, and hatchability.

For example, the requirement of a specific nutrient for growth can be determined. Groups of birds are fed on an experimental diet containing different levels of the nutrient. The level of nutrient is increased until the quantity does not result in further increases in growth. If the test ration is complete in all other respects, then the nutrient requirement will be equal to the minimum supplemental level found to give maximum growth. Some feeds are more valuable than others; hence, measures of their relative usefulness are important. Among such methods of evaluating the usefulness of poultry feeds are: (1) physical evaluation, (2) chemical analysis, (3) biological tests, and (4) cost factor.

Physical Evaluation of Feedstuffs

In order to produce or buy superior feeds, poultry producers need to know what constitutes feed quality, and how to recognize it. They need to be familiar with those recognizable characteristics of feeds that indicate high palatability and nutrient content. If in doubt, observation of the birds consuming the feed will tell them, for birds prefer and thrive on high-quality feed.

The easily recognizable characteristics of good grains and other concentrates are:

1. Seeds are not split or cracked.
2. Seeds are of low-moisture content—generally containing about 88% dry matter.
3. Seeds have a good color.
4. Concentrates and seeds are free from mold.
5. Concentrates and seeds are free from rodent and insect damage.
6. Concentrates and seeds are free from foreign material, such as iron filings.
7. Concentrates and seeds are free from rancid odor.

Figure 6.8 The technician is analyzing the protein content of feed ingredients to ensure a proper feed formulation for poultry. *(Courtesy, Delmarva Poultry Industry, Inc.)*

Chemical Analysis

While the biological response of animals (feeding trials) is the ultimate indicator of nutritive adequacy in a ration, tests of this type are difficult to perform, require extended periods of time, and are usually expensive. Thus, certain chemical analyses have been developed that are rough indicators of the value of a feedstuff or ration with regard to specific nutrient substances. The usual chemical analysis of feeds includes crude protein, ether extract or crude fat, crude fiber, ash or mineral, and moisture (see Figure 6.8). It is recognized, however, that such proximate analysis of poultry feeds leaves much to be desired because in many cases the protein and nitrogen-free extract indicated may not be available to poultry.

In addition to the so-called proximate analysis, specific chemical and microbiological determinations can be made from many of the vitamins and individual mineral elements. Feed composition tables serve as a basis for ration formulation and for feed purchasing and merchandising. Commercially prepared feeds are required by state law to be labeled with a list of ingredients and a guaranteed analysis. Although state laws vary slightly, most of them require that the feed label (tag) show in percent the minimum crude protein and fat; and maximum crude fiber and ash. Some feed labels also include maximum salt, and/or minimum calcium and phosphorus. These figures are the buyer's assurance that the feed contains the minimal amounts of the higher cost items—protein and fat—and not more than the stipulated amounts of the lower cost, and less valuable, items—the crude fiber and ash.

Bomb Calorimetry

When compounds are burned completely in the presence of oxygen, the resulting heat is referred to as gross energy or the heat of combustion. The bomb calorimeter is used to determine the gross energy of feed, waste products from feed (for example, feces and urine), and tissues. The calorie is defined as the amount of heat required to raise the temperature of 1 g of water 1°C. With this fact in mind, we can readily see how the bomb calorimeter works.

The bomb calorimeter procedure is as follows:

1. An electric wire is attached to the material being tested so that it can be ignited by remote control.
2. 2,000 g of water are poured around the bomb.
3. 25 to 30 atmospheres of oxygen are added to the bomb.
4. The material is ignited.
5. The heat given off from the burned material warms the water.
6. The thermometer registers the change in temperature of the water. For example, if 1 g of feed is burned and the temperature of the water increases 1°C, 2,000 calories/gram are given off.

The determination of the heat of combustion with a bomb calorimeter is not as difficult or time consuming as the chemical analyses used in arriving at TDN values.

Biological Tests

Most chemical and microbiological tests for nutrient substances give information about the total amount of nutrient present in a particular feedstuff or diet. However, these tests do not tell anything about the digestibility and utilization of the feedstuff or ration in the digestive tract of the animal. Hence, biological tests directly involving the bird are required to establish the true usefulness of feed supplying the nutrient needs of the bird. These biological tests are particularly important in evaluating protein and energy-yielding nutrients like carbohydrates and fats.

1. Biological measure of protein utilization. The amount of protein or nitrogen digested by the bird can be determined by a balance experiment in which a measured intake of protein is compared to the measured undigested protein in the feces of the bird. The biological value of a protein source is defined as the amount of protein retained in the body expressed as a percentage of the digestible protein available. Thus, this expression is a reflection of the kinds and amounts of amino acids available to the bird after digestion. If the amino acids available to the bird closely match those needed for body protein formation, the biological value of the protein is high. If, however, there are excesses of certain amino acids and deficiencies of other amino acids as a result of digestion, the biological value of the protein is low because of the increased number of amino acids which must be excreted via the kidney.

2. Biological measure of energy utilization. The total energy content of a feed can be measured by completely burning the feed in an apparatus known as a bomb calorimeter. Birds, like other animals, are not able to extract all of the energy present in feeds. Hence, the term *digestible energy* is used to describe the total energy of the feed minus that which remains undigested. Metabolizable energy is the total energy in the feed minus both fecal and urinary energy; it represents all the available energy for any use in the animal. The net energy value of a feed is the metabolizable energy content minus the energy employed in utilizing it; thus, net energy may be used for body storage or the production of heat and muscular activity. Metabolizable energy values are used to describe the energy content of poultry feedstuffs and rations. Metabolizable energy values are relatively easy to measure in poultry where the feces and urine are voided together and are little affected by various physiological conditions.

Cost Factor

From the standpoint of a poultry producer the most important measurement of a feed's usefulness is in terms of net returns. Cost per pound or per ton of feed, and pounds of feed required to produce a pound of broiler or a dozen eggs are important only as they reflect or affect the cost per unit of poultry products produced. For example, if the cost of a broiler ration is $0.09 a pound and 1.9 lb of the feed is required to produce 1 lb of body weight, then the feed cost per pound of body weight can be arrived at by multiplying the above figures (0.09 × 1.9). This gives a feed cost of $0.171 per pound. Obviously when rations are compared, the ration that produces a unit of poultry product at the lowest total feed cost is the most desirable from an economic point of view.

BUYING FEEDS

Poultry producers who have familiarized themselves with the various types of feeds available are in a position to make reasonable and responsible decisions about what feeds to include in their rations. In order to maximize production and profit, they must choose the feeds that are most economical for the particular demands of the birds to be fed. A high-energy, high-protein feed that is fed to low-producing layers is unnecessarily expensive. Conversely, a low-cost but low-energy feed that is fed to birds at a high production level will depress potential for production and should be considered an expensive feed. It is important, therefore, that poultry producers know and follow good feed buying practices. Likewise, they should consider the possibility of using commercial feeds, all or in part, and be knowledgeable about feed laws.

What Producers/Feed Buyers Should Know

Buying feeds is an integral part of modern poultry production. Moreover, the trend to purchase feeds, rather than grow them on location, will continue. In a broad sense, modern sophisticated buying involves knowledgeable buyers, futures trading, consideration of feed substitutions, volume buying, storage, capital outlay, and how to determine the best buy in feeds.

Poultry producers (integrators, growers, etc.) have a great incentive to purchase wisely and well. In the past, they may have thought only in terms of cost per unit weight. Today in addition to considering the feed analysis or the guarantee based on averages, producers are becoming increasingly aware of differences in the nutritional value of corn, soybeans, and others. Factors such as plant variety, soil, and weather affect quality of the feed as well. More and more poultry producers are purchasing formula feeds, rather than separate ingredients. By so doing, they are acquiring a service—diet formulation—as well as a feed. Of course, price of feed is very important. For this reason, see the section in this chapter on Best Buy in Feeds. Additionally, a number of complex and interrelated factors should be considered by the producer/buyer when purchasing feeds. Successful feed buying necessitates knowledge of all the factors that affect net returns, from the time a deal is made to buy the feed until the end product is marketed. Today, sophisticated poultry producers/feed buyers need to know the following:

1. The nutritive requirements of their birds.
2. Feed terms and feed processing methods.
3. Production and economic trends.
4. Business aspects, such as sources of credit, interest rates, contracts, futures, and possible tax savings to accrue from purchasing feeds before the end of the year.
5. Different feed grade and quality classifications.
6. Restrictive use of certain feedstuffs, such as cottonseed meal.
7. Associative, or additive, effects of certain feedstuffs, especially of protein feeds.

8. The origin and composition of the feed ingredients (soils in different areas produce different levels of minerals in plants).
9. The local potential to grow certain feeds.
10. The long-term availability of feed.
11. The moisture content of the feed ingredients.
12. Transportation costs.
13. The storage capabilities of the feed.
14. Feed shrinkage.
15. The risks, such as where a perishable product is involved.
16. What processing will be involved.
17. Certain feedstuffs affect the product produced, such as alfalfa meal, which can impart favorable color characteristics to poultry products.
18. Toxic residues.
19. Government regulations, especially as to levels, drug combinations, and withdrawals.
20. Impact of foreign feed purchases and feed requirements for meat and eggs for export.

Feed Requisites

In addition to the considerations already noted, it is important that all feeds—both bought and home-grown—meet the following requisites:

1. **Palatability.** "If they don't eat it, they won't produce"; and if they don't eat enough, feed efficiency will be poor. The relationship of feed consumption to feed efficiency becomes clear when it is realized that the maintenance requirement of a bird producing at a low rate represents a much greater percentage of the total feed required than for a bird producing at a more rapid rate.

Palatability is the result of the following factors sensed by the bird in locating and consuming feed: appearance, odor, taste, texture, and temperature. These factors are affected by the physical and chemical nature of the feed.

2. **Variety.** Some variety in the feed may be desirable, particularly from the standpoint of assuring balance of nutrients—for example, all the essential amino acids.

3. **Digestive disturbances.** The choice of feeds can give a big assist in minimizing such disturbances.

4. **Cost.** Cost is important, but net return is even more important; hence, it may well be said that it is net return rather than cost per ton, or per bag, that counts.

Best Buy in Feeds

Feed prices vary widely. For profitable production, therefore, feeds with similar nutritive properties should be interchanged as price relationships warrant.

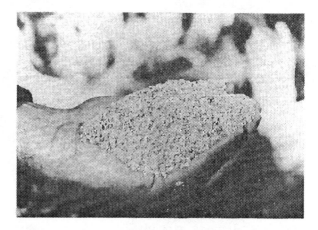

Figure 6.9 Chicken feed. *(Courtesy, Hy-Line International)*

In buying feeds, the poultry producer should check prices against value received. Most poultry establishments and feed companies now use computers to formulate least-cost rations.

Buying Commercial Feeds

Numerous types of commercial feeds, ranging from additives to complete rations, are on the market, with most of them designed for a specific species, age, or need. Among them are complete rations, concentrates, protein supplements (with or without reinforcements of minerals and/or vitamins), mineral and/or vitamin supplements, additives (antibiotics, etc.), and medicated feeds.

So, it may be said that there exist two good alternative sources of most feeds and rations—home-mixed or commercial—and the able manager will choose wisely between them (see Figure 6.9).

Farm-Mixed versus Complete Commercial Feeds

The ultimate criterion for choosing between farm-mixed and complete commercial feeds is which program will make for maximum returns to the producer for labor, management, and capital.

The poultry producer has the following options from which to choose for home-mixing feeds:

1. Purchase of a commercially prepared protein supplement (likely reinforced with minerals and vitamins), which may be blended with local or homegrown grain.
2. Purchase of a commercially prepared mineral-vitamin premix, which may be mixed with a protein supplement, then blended with local or homegrown grain.
3. Purchase of individual ingredients (including minerals and vitamins) and mixing the feed from the "ground up."

How to Select Commercial Feeds

There is a difference in commercial feeds! That is, there is a difference from the standpoint of what poultry producers can purchase with their feed dollars. Smart operators will know how to determine what constitutes the best in commercial feeds for their specific needs. They will not rely solely on the appearance or aroma of the feed, nor on the salesperson.

REPUTATION OF THE MANUFACTURER. The reputation of the manufacturer can be determined by (1) conferring with other producers who have used the particular products, and (2) checking on whether or not the commercial feed under consideration has a good record for meeting its guarantees. The latter can be determined by reading the bulletins and reports prepared by the respective state department in charge of monitoring feed quality and enforcing feed laws.

SPECIFIC NEEDS OF THE BIRDS. Feed requirements vary according to (1) the class, age, and productivity; and (2) whether the birds are fed primarily for maintenance, growth, commercial egg production, or hatching. The wise producer will buy different formula feeds for different needs.

Feeding poultry has become a sophisticated and complicated process. Feed manufacturers have extensive resources with which to formulate and test rations for different needs. As a result, most manufacturers have a large selection of feeds—one should be applicable to the needs of the producer. It is essential that the producer make clear to the feed company the kind of birds to be fed.

Feed Laws

The U.S. Food and Drug Act was passed in 1906, giving the federal government authority to regulate and inspect feeds shipped in interstate commerce. Additional controls were authorized in the Food, Drug, and Cosmetic Act of 1938. In addition to the federal laws, nearly all states have laws regulating the sale of commercial feeds. These benefit both feeders and reputable feed manufacturers. In most states the laws require that every brand of commercial feed sold in the state be licensed, and that the chemical composition be guaranteed.

Since most commercial feeds are closed formulas, it is necessary to sample and analyze the feed periodically to ensure that it is fulfilling its guarantees. Generally, samples of each commercial feed are taken periodically and analyzed chemically in the state's laboratory. Additionally, skilled microscopists examine the sample to ascertain that the ingredients present are the same as those guaranteed. Flagrant violations on the latter point may be prosecuted since they represent willful mislabeling.

FEED SUBSTITUTIONS

Successful poultry producers are keen students of values. They know that feed conversion rate alone does not determine success. The cost of producing each pound of broiler, turkey, or each dozen eggs is the determinant of success. For this reason, producers want to obtain a balanced ration that will make for the highest net returns. They recognize that feeds of similar nutritive properties can and should be interchanged in the ration as price relationships warrant, thus making it possible at all times to obtain a balanced ration at the lowest cost. Thus, (1) the cereal grains may consist of corn, barley, wheat, oats, and/or sorghum; and (2) the protein supplements may consist of soybean, cottonseed, peanut, rapeseed, sunflower, and/or linseed meal.

Table 6.3 summarizes the comparative values of the most common U.S. poultry feeds. In arriving at these values, two primary factors besides chemical composition and feeding value have been considered—namely, palatability and product quality. It is emphasized that the comparative values of feeds shown in the feed substitution table are not absolute. Rather, they are reasonably accurate approximations based on average-quality feeds, together with experiences and experiments.

FEED PREPARATION

Grains for poultry are prepared in the following three forms:

1. **Mash.** This is a ground feed and the usual end product resulting from mixing poultry feedstuffs.
2. **Pellets.** Composed of mash feeds that are pelleted. Birds usually consume more of a pelleted ration than the same ration in mash form. Usually there is more cannibalism with pellets than with mash or crumbles.
3. **Crumbles.** Produced by rolling pellets.

Pellets or crumbles cost slightly more than the same ration in mash form. Yet, they are used for broilers and turkeys because of improved feed efficiency—less feed to produce a unit (lb or kg) of gain. When alfalfa is to be included in poultry feeds, it should be ground.

TABLE 6.3   Feed Substitution Table for Poultry (As-Fed)

Feedstuff	Relative Feeding Value (lb for lb) in Comparison with the Designated (Underlined) Base Feed Which = 100	Maximum Percentage of Base Feed (or Comparable Feed or Feeds) It Can Replace for Best Results	Remarks
Grains, By-Product Feeds:			
Corn, No. 2	100	100	Corn is the most widely used grain in poultry feed. It is a good source of vitamin A, xanthophyll pigments, and linoleic acid.
Bakery wastes	75	50	Bakery wastes are very similar to cereal grains in composition.
Barley	80–85	50	Barley is very low in vitamin A; less palatable than corn; western grown barley may be improved through enzyme treatment.
Beans (cull)	90	5	Unpalatable.
Buckwheat	90	15	Extremely low in protein.
Cassava	85	20	A good source of linoleic acid.
Hominy	95	50	Best when used as a 50:50 mix with barley, corn, or oats; can be used as a scratch feed.
Millet	95–100	65	High levels of molasses will produce wet droppings.
Molasses	70	5	Usually too bulky for broilers and too expensive for layers, but used extensively in replacement rations.
Oats	70–80	50	
Peas, dried	90–100	5–10	Better source of protein and lysine than grain.
Rice, rough	80–85	20–50	Deficient in vitamin A.
Rice, bran	50	5–10	Rice bran is high in fat and susceptible to oxidation.
Rice polishings	85–90	5–10	High in fat; good source of thiamin.
Rye	90	25–30	Unpalatable; will cause sticky, pasty droppings if excessive amounts are fed.
Sorghum	100	100	May replace corn, but lacks xanthophyll pigments and is lower in linoleic acid.
Triticale	80–90	30	Triticale is somewhat lower in feeding value than either wheat or corn.
Wheat	90–95	100	Wheat is the most variable of the cereal grains in protein; very low in xanthophyll; should be processed to increase palatability.
Wheat bran	75	10–15	Higher in protein than most grains, good source of lysine.
Protein Supplements:			
Soybean meal (48%)	100	100	Well balanced in amino acids. Best quality of all plant protein supplements. Very palatable.
Babassu oil meal	50	20	Similar to copra meal.
Blood meal, flash- or ring-dried	120	5–20	Excellent source of lysine.
Copra meal (coconut meal)	50	25	Copra meal should be supplemented with lysine and methionine.
Corn gluten meal	50–75	25	Used extensively to supply xanthophyll pigments in broiler finishing rations. Low in lysine and tryptophan.

(continued)

TABLE 6.3 Feed Substitution Table for Poultry (As-Fed) *(continued)*

Feedstuff	Relative Feeding Value (lb for lb) in Comparison with the Designated (Underlined) Base Feed Which = 100	Maximum Percentage of Base Feed (or Comparable Feed or Feeds) It Can Replace for Best Results	Remarks
Cottonseed meal	85	80[1]	Gossypol, a compound found in cottonseed meal, can cause discoloration of egg yolks. The addition of iron in the ration may minimize the dangers of gossypol poisoning. Cyclopropenoic fatty acids in cottonseed meal can cause discoloration of egg whites. Glandless cottonseed meal is recommended.
Feather meal	50	5	Feather protein is deficient in methionine, lysine, histidine, and tryptophan. Poor quality protein.
Fish meal	115	50–65	Excellent balance of amino acids and good source of calcium and phosphorus. Most poultry rations incorporate some fish meal at levels of about 2–5% of the ration. Fish meal can impart fishy flavors if fed at levels in excess of 10%.
Linseed meal	80	10	Linseed meal depresses growth and can cause diarrhea. Levels should be restricted to a maximum of 5% of the ration.
Meat and bone meal	100	20–50	Variable protein quality; good source of calcium and phosphorus; deficient in methionine and cystine.
Meat meal (50–55% protein)	100	50–65	Meat meal is high in phosphorus and low in methionine and cystine. It is recommended that the maximum level of usage not exceed 10% of the ration.
Peanut meal (41% protein)	95	75–100	If peanut meal is to be substituted for soybean meal, lysine must be added.
Poultry by-product meal	100	20–50	Excellent protein if properly prepared.
Rapeseed meal (canola meal)	80	30	Not recommended for poultry starter ration. It may be used at levels of up to 10% of the total ration for fattening birds or layers. Meal from new low glucosinolate rapeseed may constitute 20% of layer rations and 10% of starter feeds.
Safflower seed meal (decorticated)	95	50–100	Safflower seed meal can be incorporated at levels up to 15% of the layer ration. Deficient in lysine and methionine.
Sesame meal	95–100	100	Extremely deficient in lysine, but good plant source of methionine.
Sunflower seed meal	95–100	100	Low in lysine and methionine.
Yeast, torula	100	60	Yeast is a good source of the B complex vitamins.

[1]When cottonseed meal is substituted for soybean meal at this level, it must be degossypolized.

PRESENCE OF SUBSTANCES AFFECTING PRODUCT QUALITY

The composition of the feed can affect the product. The color of the skin or shanks of a broiler or of the yolk of an egg is primarily due to the carotenoid pigments consumed in the feed. Corn, alfalfa meal, and corn gluten meal are the main feeds used to contribute these pigments.

Screw-processed cottonseed meal, which is high in gossypol, may, when fed to laying hens, cause egg yolk discoloration in stored eggs. Some fish products may impart off-flavors to poultry meat or eggs. Thus, certain feedstuffs may be undesirable because of their effect on the end product.

TOXIC LEVELS OF INORGANIC ELEMENTS FOR POULTRY

Poultry are susceptible to a number of toxins, any one of which may prove disastrous in a flock. Among them are a number of inorganic elements. It is important, therefore, that the poultry producer guard against toxic levels of inorganic elements. Toxicity is influenced by (1) the form of the element (for example, methyl mercury is considerably more toxic than inorganic mercury); and (2) the composition of the diet (particularly with respect to the content of other mineral elements).

BOX 6.1 Some New Developments in Feeds and Additives

- Designer feeds. The approaches of traditional plant breeding and agricultural biotechnology are allowing the development of corn, soybeans, and other grains that more nearly meet the needs of the various stages of growth of the different poultry species and pigs. Grains that are presently available or likely to be available very shortly include the following:
  - High-oil corn
  - Low-phytate corn and soybeans
  - High-protein corn and soybeans

  Grains under development include (note yield and specific attributes are needed):
  - Elevated lysine corn
  - Elevated essential/limiting amino acids (e.g., methionine, lysine, threonine, and tryptophan) in corn and soybeans
  - Reduced indigestible complex sugars/carbohydrates (oligosaccharides) in soybeans
- Alternative feeds. While corn-soybean diets for poultry predominate in the United States, there are alternatives:
  - Replacement of soybean meal with canola (rape) or sunflower meal.
  - Replacing corn with wheat, barley, and so on.
  - A very high soy diet may create litter and health problems because soy is high in potassium and the complex sugars (oligosaccharides) are poorly digested in young birds.
- Enzymes. Enzymes can be useful to improve the digestibility of nutrients.
  - Phytase. Phytic acid or phytate is a complex sugar with 6 phosphate molecules—The phytate binds calcium and other key nutrients (e.g., zinc and copper). There is marked differences in both of the following:
    1. Phytate content of grain (both corn and soybeans contain phytate)
    2. Phytate digestion in poultry (with age, genetic variation, and gut microorganisms)

    Addition of the enzyme phytase to the diet can improve phosphate digestibility. Note that calcium inhibits phytase.
  - Keratinase Keratinase is a proteolytic enzyme that can break down feathers and connective tissue in meat and bone meal. It has the potential for improving the digestibility of feed and for food safety because the enzyme has been reported to destroy prions. These are the agents that cause scrapie in sheep and bovine spongiform encephalitis (BSE, or mad cow disease) in cattle.
- Probiotics and Prebiotics
  1. *Probiotics* are live beneficial microorganisms (e.g., lactic and bacteria) that aid digestion and/or reduce pathogenic bacteria (disease causing in poultry or food-borne pathogens for people) in the gastrointestinal tract.
  2. *Prebiotics* are feed ingredients that modify the population of microorganisms in the gastrointestinal tract. Prebiotics include nutrients used preferentially by benign bacteria and dead bacteria (e.g., Aspergillus), which enhance the gastrointestinal tract's immune responses.
- Antibiotics. It is possible that the use of antibiotics for growth promotion and improved health of poultry will be greatly restricted or ultimately banned in the United States, as is the case in the European Union.

FURTHER READING

Austic, R. E., and M. C. Nesheim. *Poultry Production,* 13th ed. Philadelphia, PA: Lea & Febiger, 1990.

Bell, D. D., and W. D. Weaver. *Commercial Chicken Meat and Egg Production,* 5th ed. Norwell, MA: Kluwer Academic Publishers, 2002.

Leeson. S., and J. D. Summers. *Commercial Poultry Nutrition.* Guelph, Ontario: University Books, 1997.

Nutrient Requirements of Poultry, 8th rev. ed. Washington, DC: National Research Council, National Academy of Sciences, 1984.

Nutrient Requirements of Poultry, 9th rev. ed. Washington, DC: National Research Council, National Academy of Sciences, 1994.

7

Poultry Feeding Standards—Diet Formulation and Feeding Programs

Objectives

After studying this chapter, you should be able to:

1. List the factors involved in feed formulation.
2. List the factors affecting nutrient requirements for poultry.
3. Define phase feeding and list factors influencing change of diet.
4. List reasons for slowing the development of broiler breeders and turkey breeders.
5. List ways to limit feed intake during breeder development.

Production of high quality poultry feed like that being consumed by the turkeys in Figure 7.1 will be discussed in this chapter. Although there is arguably more information available on the nutritional requirements of poultry than other farm animals, this is limited primarily to chickens and turkeys. Even here there are many gaps in our knowledge. Poultry feeding has changed more than the feeding of any other species—it has outpaced the entire livestock field. Today, the vast majority of commercial poultry is produced in large units wherein the maximum of science and technology exists. Confinement production is the industry standard, and well-balanced diets containing adequate sources of all known nutrient materials are fed for maximum production. The current trend in poultry production is toward controlled environment, which usually results in lowered feed consumption. Under these conditions, the daily feed consumption is often taken into consideration and the nutrient content of feed (energy, amino acids, vitamins, and minerals) varied so as to compensate for the reduced feed intake and meet the requirements.

The feed required to produce a unit of eggs, broilers, and turkeys has declined since 1940 (see Table 7.1). The most dramatic change has occurred in broilers. In 1940, it required 4.7 lb of feed to produce 1 lb of weight gain in broilers; in 2000, it took only 1.7 lb. The nutrient composition of broiler and egg shown in Table 7.2 is indicative of the relative importance of these nutrients as broiler and egg constituents.

This chapter details two aspects of poultry feeding: (1) feeding standards and diets, and (2) feeding programs. Poultry producers have the following alternatives for diets:

- Purchase of a commercially prepared, complete feed.
- Purchase of a commercially prepared protein supplement, reinforced with vitamins and minerals, which may be blended with local or homegrown grain.
- Purchase of a commercially prepared vitamin-mineral premix that may be mixed with an oil meal, and then blended with local or homegrown grain.
- Purchase of individual ingredients (including vitamins and minerals) and mixing the feed from the ground up.

Figure 7.1 Young turkey at a typical feed pan of an automatic feeder line. *(Courtesy P. R. Ferket, North Carolina State University)*

Figure 7.2 Poultry feed formulated with the use of a computer. *(ISU photo by Bob Elbert)*

TABLE 7.1 Feed Units Required per Unit of Eggs and Poultry Produced, Selected Years, 1940–2000

		Per Pound Liveweight	
Year Ending October	Per Dozen Eggs (feed units)	Turkey (feed units)	Broiler (feed units)
1940	7.4	4.5	4.7
1960	6.4	3.4	3.0
1980	4.1	2.9	2.1
2000	3.7	2.7	1.7

TABLE 7.2 Nutrient Composition of Broiler Meat and Eggs

Nutrient	Broiler	Egg
	(%)	(%)
Water	64.0	66
Protein	18.8	13
Fat	14.2	10
Minerals	3.7	11

Today, few large poultry producers purchase a commercially prepared, complete feed. Most of them own their own feed-mixing plant or contract directly with one, thus using the fourth option. Several days' feed supply is contained in large bulk tanks as seen in Figure 7.2.

FACTORS INVOLVED IN FORMULATING POULTRY DIETS

Before anyone can formulate a poultry diet intelligently, it is necessary to know the following:

1. The nutrient requirements of the particular birds to be fed, which calls for feeding standards

2. The availability, nutrient content, and cost of feedstuffs (feed ingredients)
3. The acceptability and physical condition of feedstuffs
4. The average daily consumption of the birds to be fed
5. The presence of substances harmful to product quality

NUTRIENT REQUIREMENT DETERMINATION

The nutritive requirements for a specific substance are determined by finding the minimum amount of that particular nutrient or substance that will permit maximum development of the physiological function or economic characteristic of concern. In general, the economic characteristics of importance in poultry are growth, efficiency of feed utilization, egg production, and hatchability. If the test diet is complete in all other respects, then the nutrient requirement will be equal to the minimum supplemental level found to give maximum growth.

FEEDING STANDARDS

Feeding standards are tables listing the amounts of one or more nutrients required by different species of animals for specific productive functions, such as growth, fattening, and reproduction. Most feeding standards are expressed in either (1) quantities of nutrients required per day, or (2) concentration in the diet. The first type is used where animals are provided a given amount of a feed during a 24-hour period, and the second is used where animals are provided a diet without limitation on the time in which it is consumed.

Today, the most up-to-date feeding standards in the United States are those published by the National Research Council (NRC) of the National Academy of Sciences. Periodically, a specific committee, composed of outstanding researchers who have worked extensively with the class of animal whose requirements are being reviewed, revises the nutrient requirements of each species for different functions. Thus, the nutritive needs of each type of livestock are dealt with separately and in depth. Feeding standards have been established for the various types of poultry, and through the use of these standards, producers can formulate diets tailored to meet the nutrient requirements of their birds.

Although feeding standards are excellent and necessary guides, there are still many situations where nutrient needs cannot be specified with great accuracy. Also, in practical feeding operations, economy must be considered; for example, producers are interested in obtaining that level of egg production that will make for the largest net returns in light of current feed costs and the market price of eggs. Moreover, feeding standards tell nothing about the palatability, physical nature, or possible digestive disturbances of a diet (see Figure 7.3 for chickens feeding). Neither do they give consideration to individual bird differences, management differences, and the effects of such stresses as weather, disease, parasitism, and surgery (beak trimming, caponizing, etc.). Thus, there are many variables that alter the nutrient needs and utilization of birds—variables that are difficult to include quantitatively in feeding standards, even when feed quality is well known (see Figure 7.4 for storage of poultry feed).

National Research Council (NRC) Requirements for Poultry

The requirements for most of the nutrients needed by poultry have been established. These differ ac-

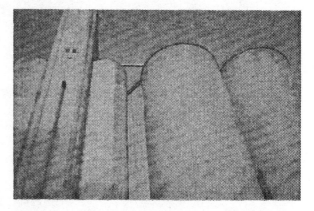

Figure 7.4 Grain elevators with large bins. *(Courtesy, Jerry Sell)*

cording to the kind and age of bird and the purpose for which it is being fed. A deficiency of a nutrient can be, and often is, a limiting factor in egg production or growth. NRC standards do not provide for margins of safety. Rather, the values reported represent adequacy, using as criteria growth, health, reproduction, feed efficiency, and quality of products produced.

It is recognized that a number of forces may affect the nutritive requirements of poultry; among them, the following:

1. **Temperature and humidity.** When temperature and humidity conditions deviate much from 60° to 75°F and 40 to 60%, respectively, adjustments in nutrients levels should be made to compensate for changes in feed intake.
2. **Genetic differences.** Genetic differences among strains affect nutrient requirements; hence, in this chapter consideration has been given to differences in the requirements between broiler-type (as seen in Figure 7.3) and egg-type strains of chickens.

Also, the nutrient composition of individual feedstuffs is variable. In order to compensate for these conditions, the nutritionist usually adds a margin of safety to the stated requirements in arriving at nutritive allowances to be used in diet formulation.

HOW TO BALANCE DIETS

The increasing complexity of poultry diets, along with larger and larger enterprises, makes it imperative that the producers who choose to mix feed be absolutely sure that they will have a nutritionally balanced and adequate diet.

- When fed free-choice, birds tend to eat to satisfy their energy requirements. Consequently, it is possible, within limits, to regulate the intake of

Figure 7.3 Chickens feeding at a metal feeder. *(Courtesy, USDA)*

Figure 7.5 Poultry feed is augered into poultry houses. *(ISU photo by Bob Elbert)*

all nutrients, except water, by including them in the diet in specific ratios to available energy. Thus, the energy content of the diet must be considered in formulating to meet a desired intake of all essential nutrients other than energy itself.

- The larger commercial feed companies, and the larger poultry producers who do their own mixing or formulating, generally rely on the services of a nutritionist and the use of a computer in formulating their diets.

- Even though it is more time-consuming, and fewer factors can be considered simultaneously, a good job can be done in formulating diets by the hand method. Figure 7.5 shows an auger system for moving feed from a storage bin into a house.

- Good poultry producers should know how to balance diets. They should be able to select and buy feeds with informed appraisal; to check on how well their manufacturer, dealer, or consultant is meeting their needs; and to evaluate the results.

- Diet formulation consists of combining feeds in the proper ratio in order to supply the daily nutrient requirements of the bird. This may be accomplished by the methods presented later in this chapter.

1. In computing poultry diets, more than simple arithmetic should be considered, for no set of figures can substitute for experience and intuition. Formulating diets is both an art and a science—the art comes from bird know-how, experience, and keen observation; the science is largely founded on mathematics, chemistry, physiology, and bacteriology. Both are essential for success.

2. Before attempting to balance a diet, the following major points should be considered:

   a. **Availability and cost of the different feed ingredients.** The first step in diet formulation is to determine what feeds are available and their cost. Cost of ingredients should be based on delivery after processing—because delivery and processing costs are quite variable.

   b. **Moisture content.** When considering costs and balancing diets, feeds should be placed on a comparable moisture basis; usually, either as-fed or moisture-free.

   c. **Composition of the feeds under consideration.** Feed composition tables or average analysis should be considered only as guides, because of wide variations in the composition of feeds. For example, the protein and moisture contents of sorghum can be quite variable. Similarly, soybeans have marked differences in protein content. Whenever possible, especially with large operations, it is best to take a representative sample of each major feed ingredient and have a chemical analysis made of it for the more common constituents—protein, fat, fiber, nitrogen-free extract, and moisture; and often calcium, phosphorus, and carotene. Such ingredients as oil meals and prepared supplements, which must meet specific standards, need not be analyzed as often, except as quality-control measures.

   d. **Quality of feed.** Numerous factors determine the quality of feed, including:

      1. **Stage of harvesting.** For example, early harvested grains are higher in moisture than those that are mature.

      2. **Freedom from contamination.** Contamination from foreign substances such as dirt, sticks, and rocks can reduce feed quality, as can mycotoxins, pesticide residues, and a variety of chemicals.

      3. **Uniformity.** Does the feed come from one particular area or does it represent a conglomerate of several sources?

      4. **Length of storage.** When feed is stored for extended periods, some of its quality is lost due to its exposure to the elements.

   e. **Degree of processing of the feed.** Often, the value of feed can be either increased or decreased by processing. For example, grinding some types of grains makes them more readily digestible to poultry and increases their feeding value. Extension can improve digestibility of proteins.

   f. **Soil analysis.** If the origin of a given feed ingredient is known, a soil analysis or knowledge of the soils of the area can be very helpful; for example,

1. the phosphorus content of soils affects plant composition,
2. soils high in molybdenum and selenium affect the composition of the feeds produced,
3. iodine- and cobalt-deficient areas are important in animal nutrition, and
4. other similar soil-plant-animal relationships exist.

g. **The nutrient requirements and allowances.** These should be known for the particular class of poultry for which a diet is being tailored. Also, it must be recognized that nutrient requirements and allowances must be changed from time to time, as a result of new experimental findings.

3. In addition to providing a proper quantity of feed and to meeting the nutritive requirements, a well-balanced and satisfactory diet should be:
   a. **Palatable and digestible.**
   b. **Economical.** Generally speaking, this calls for the maximum use of feeds available in the area.
   c. **One that will enhance.** Rather than impair, the quality of the product (meat or eggs) produced should be enhanced.
4. In addition to considering changes in availability of feeds and feed prices, diet formulation should be altered at stages to correspond to changes in weight and productivity of birds.

Computer Methods

Poultry and feed companies use computers in diet formulation. Also, many of the state universities, through their Federal-State Extension Services, are offering diet balancing computer services to farmers within their respective states on a charge basis. Consulting nutritionists are available throughout the United States and provide computer services, as well as other services. With the recent advent of the low-cost personal computer, this powerful technology is available to almost everyone.

Despite their sophistication, there is nothing magical or mysterious about the use of computers in diet balancing. Their primary advantages are accuracy and speed of computation. In addition, computer programs (software) used in diet balancing provide a means of organizing needed information in a logical and systematic manner. The computer should be viewed as an extension of the knowledge and skills of the formulator. At this time, there is no push-button system of feed formulation available. The degree of success realized is very dependent on the management of data put into the computer, and on the evaluation of the resulting formulations that the computer generates. In the hands of experienced users, the computer enables the producer and nutritionist to be more precise in carrying out diet formulation.

Two basic approaches to diet formulation are practiced with computers:

1. Trial-and-error formulation
2. Linear programming (LP)

Trial-and-Error Formulation with the Computer

Many diet balancing software programs written for the computer allow for trial-and-error diet balancing. Feed mill nutritionists frequently use this technique to enter into the computer diets that are given to them by other nutritionists or by a producer. The objective in this case is to confirm the nutrient values for the diet based on the specific ingredients used by the feed manufacturer. In many cases, these diets are not to be altered without permission. In other cases, the number of ingredients for a specific diet may be limited so that the trial-and-error technique is just as fast as using linear programming to arrive at the desired nutrient levels in the diet.

It does not necessarily take specialized computer software to use the trial-and-error method. Spreadsheet (or Financial Spreadsheet) programs, for instance, organize data into rows and columns. Information, such as nutrient values for a feedstuff, may be entered into data cells. Simple and complex arithmetic operations can be controlled by the user to the extent that rather large trial-and-error method diets can be programmed and run.

Linear Programming (LP)

The most common technique for computer formulation of diets is the linear programming (LP) technique. At times, this is referred to as least-cost diet formulation. This designation results from the fact that most LP techniques for diet formulation have as their objective minimization of cost. A few LP programs are in use that solve for maximization of income over feed costs. Regardless, the poultry producer and nutritionist should always keep in mind that maximizing net profit is the only true objective of most diet formulations. A skilled user of the LP system will control diet quality by writing specifications that lead to diets that will maximize profit.

Briefly described, the LP program is a mathematical technique in which a large number of simultaneous equations are solved in such a way as to meet the minimum and maximum levels of nutrients and levels of feedstuffs specified by the user at the lowest possible cost. It is not necessary to understand the inner workings of the computer program to use LP, though it does take experience to use it to good advantage and to avoid certain pitfalls. The most common pitfalls are incorrectly entered or missing data and the specification of minimums and maximums that cannot be met with the feedstuffs available. The latter is called an infeasible solution. When an infeasible solution is encountered, the user must determine (1) if this is due to incorrect or missing data, or (2) if the specifications must be relaxed.

FEEDING PROGRAMS

The nutritive requirements of poultry vary according to species (between chickens, turkeys, ducks, and geese), according to age, and according to the type of production—whether the birds are kept for meat production, layers, or breeders. For this reason, many different diets are required.

To be successful, diets must meet the nutritive requirements of the birds to which they are fed, as shown in the National Research Council's nutritive requirement tables. The reference for these tables will be found at the end of this chapter. In using these tables to formulate practical diets, it must be remembered that they are minimum requirements, which means that they do not provide for any margins of safety. Further, the protein-energy relationships shown therein should be retained.

Birds eat primarily to satisfy their energy needs. Also, the temperature of the environment has an important influence on feed intake—the warmer the environment, the less the feed intake. Therefore, the requirement of all nutrients, expressed as a percentage of the diet, is dependent upon the environmental temperature. Other factors affecting feed intake are health, genetics, form of feed, nutritional balance, stress, body size, and rate of egg production or growth.

It is believed that feed intake is, in part, controlled by the amount of glucose in the blood. It has been observed that the addition of fat to the diet results in overconsumption on the part of the bird. As a result, some variation in the protein:energy ratio may be tolerated. In general, when free-choice feeding dietary protein levels that are low in relation to energy, fat deposition is markedly increased; with higher levels of protein, less fat is deposited. Increasing the protein level above that required for maximum growth rate reduces fat deposition still further.

Figure 7.6 Layers at feeder pan of an automatic feeder line. (*Courtesy, Hy-Line International*)

Feeding Chickens

Today, broiler chickens and laying hens are provided complete diets with all the needed nutrients available in the quantities necessary, and with the feed and water available on an *ad libitum* or free-choice basis. Broiler breeders are feed restricted. Formulations are varied according to the type of production—whether the birds are bred and kept for egg production (layers), hatchery production (breeders), or meat production (broilers). Also, consideration is given to age, sex, stage and level of production, temperature, disease level (and other stresses), management, and other factors.

Feeding Layers

Laying hens (Leghorn-type chickens kept for the production of eggs) have small body size and are prolific layers. They are generally fed *ad libitum* during the laying period. (Also see Figure 7.6 for laying feeding system). Occasionally, layers will consume excess feed during the latter phases of egg production (following peak production) with resultant obesity, reduced feed efficiency, and higher incidence of fatty liver syndrome. When this situation is detected, limiting feed intake to 85 to 95% of full-feed consumption is desirable. Data on feed consumption in a particular flock, together with information on body weight, ambient temperature, and rate of egg production, may be used to determine the degree of feed restriction. Feed consumption is easily monitored by placing bulk feed tanks on scales as seen in Figure 7.4. The effect of rate of egg production on nutrient requirements is shown in Table 7.3.

The following additional information is pertinent to feeding layers:

1. The largest item of cost in the production of eggs is feed.

TABLE 7.3   Metabolizable Energy Required Daily by Chickens in Relation to Body Weight and Egg Production[1,2]

Body Weight		*Rate of Egg Production (%)*					
		0	50	60	70	80	90
(lb)	*(kg)*	*Metabolizable Energy/Hen Daily (kcal)[3]*					
2.2	1.0	130	192	205	217	229	242
3.3	1.5	177	239	251	264	276	289
4.4	2.0	218	280	292	305	317	330
5.5	2.5	259	321	333	346	358	371
6.6	3.0	296	358	370	383	395	408
7.7	3.5	333	395	408	420	432	445

[1]Adapted from *Nutrient Requirements of Poultry.*
[2]A number of formulas have been suggested for prediction of the daily energy requirements of chickens. The formula used here was derived from that in *Effect of Environment on Nutrient Requirements of Domestic Animals* (NRC, 1981).

$$ME/hen\ daily = W^{0.75}(173\text{-}1.95T) + 5.5\ \Delta W + 2.07EE$$

where: W = body weight (kg), T = ambient temperature (°C), ΔW = change in body weight in g/day, and EE = daily egg mass (g).
[3]Temperature of 22°C, egg weight of 60 g, and no change in body weight were used in calculations.

2. In the final analysis, the objective of feeding laying hens is to produce a dozen eggs of good quality at the lowest possible feed cost. Thus, the actual cost of the feed that a layer eats in producing a dozen eggs—not the price per pound of feed—determines the economy of the diet.
3. Feed consumption per bird varies primarily with egg production and body size. It is also influenced by the health of the birds and the environment, especially the temperature.
4. Normally, a mature Leghorn, or other lightweight bird, eats about 82.5 lb (37.5 kg) of feed per year and produces about 22 doz eggs in that same period of time. Hence, it requires about 3.75 lb (1.7 kg) of feed to produce 1 doz eggs. A bird of the heavier breeds eats >100 lb (.45 kg) of feed per year; hence, the heavier breeds are not as efficient egg producers. With lightweight layers, the producer should aim for a feed efficiency of 3.5 to 4.0 lb of feed per dozen eggs.
5. Feed may affect egg quality. Deficiencies of calcium, phosphorus, manganese, and vitamin D_3 lead to poor shell quality. Yolk color is almost entirely dependent on the bird's diet. Low vitamin A levels may increase the incidence of blood spots.
6. Table 7.4 summarizes nutrient requirements for first cycle pullets.

Phase Feeding

This refers to changes in the layer diet adjusting for the following:

1. Age and level of egg laying
2. Temperature and climatic changes
3. Differences in body weight and nutrient requirements of different strains of birds
4. Changes in nutrients as feeds are changed for economic reasons

Feeding Replacement Pullets

Pullets generally perform well during their laying year when their nutrient requirements have been met during the growing period. Leghorn-type pullets are seldom restricted-fed during the growing period because varying the lighting during growth from 6 to 20 weeks of age can be used to control feed consumption and sexual development. Also, some producers restrict the feed intake of light breeds because, even among these, restricted feed intake results in slightly higher mortality during the rearing period, but in lower mortality and higher egg production during the laying year. Recommended nutrient requirements are shown in Table 7.5.

The following additional information is pertinent to feeding replacement chicks (pullets):

1. Feed accounts for approximately 60% of the cost of raising replacement pullets.
2. Replacement chicks are usually fed a diet lower in energy than broiler chicks. Also, feed and daily light periods may be restricted, so as to permit the pullets to reach larger body size before they start to lay than would be the case were they full-fed, and fully lighted.
3. Always use complete starter feeds for chicks, and give chicks starter feeds without grain supplement until they are 5 weeks old.
4. When chicks are 5 weeks old, change to the growing diet.

TABLE 7.4   Recommendations for First-Cycle Egg Production[1]

Component	18–32 Weeks	33–44 Weeks	45–58 Weeks	59+ Weeks
Metabolizable energy, kcal/kg	2,900	2,900	2,850	2,800
Protein, %	16	15.50	15.25	15.00
Methionine and cystine, %	0.70	0.66	0.62	0.58
Lysine, %	0.80	0.78	0.73	0.74
Calcium, %	3.60	3.75	3.85	3.95
Available phosphorus, %	0.40	0.38	0.32	0.30

[1]Assumes 100 g feed intake per bird per day.

TABLE 7.5   Recommendations for Started Egg-Type Pullets

Component	Starter (0–6 weeks)	Grower 1 (7–8 weeks)	Grower 2 (8–15 weeks)	Preproduction (16–18 weeks)
Metabolizable energy, kcal/kg	3,000	3,050	3,050	3,050
Protein, %	20	18	16	15.5
Methionine and cystine, %	0.80	0.75	0.65	0.60
Lysine, %	1.10	0.90	0.75	0.75
Calcium, %	1.00	1.00	1.00	2.75
Available phosphorus, %	0.45	0.45	0.40	0.40

TABLE 7.6   Recommendations for Growing Broilers

Component	Starter (0–3 weeks)	Grower (3–5 weeks)	Finisher (5–7 weeks)
Metabolizable energy, kcal/kg	3,100	3,200	3,200
Protein, %	23	20	18.5
Methionine and cystine, %	0.95	0.80	0.70
Lysine, %	1.20	1.05	0.95
Calcium, %	1.00	0.90	0.80
Available phosphorus, %	0.45	0.40	0.35

Feeding Broilers

The following information pertaining to feeding broilers, roasters, and capons is pertinent:

1. Feed is the largest item of cost in broiler production, representing > 60% of the total cost.
2. Producers aim for broilers with an average weight of over 5.0 lb (2.3 kg) at 6 to 7 weeks of age, feed conversion of < 2.0, and mortality < 3.0%.
3. Most producers are using multiple stages in their feeding programs, with at least three stages: starter, grower, and finisher.
4. Table 7.6 summarizes nutrient requirements for growing broilers.

Feeding Broiler Breeders

The following information is pertinent to feeding breeders:

1. The nutritive requirements for breeding flocks are more rigorous than those for commercial laying flocks. Breeders require greater amounts of vitamins A, D, E, and B₁, and of riboflavin, pantothenic acid, niacin, and manganese than do laying flocks. Diets with these added ingredients in the right proportions give high hatchability and good development of chicks. Such diets cost more than normal layer diets.
2. Broiler breeder replacement pullets should receive low-energy diets, in the range of 1,090 to 1,135 kcal/lb (~2,450 kcal/kg), and/or the feed intake should be restricted to avoid excess fat accumulation at the time they reach sexual maturity.
3. Broiler breeder hens, which are heavy and have a high energy requirement for maintenance, require approximately 400 to 450 kcal ME per hen per day for maximum egg production. Since these hens tend to become overfat when fed high-energy diets, the energy content of their diets is limited to about 1,200 to 1,250 kcal/lb (2,640 kcal/kg) or to restrict their feed intake (< 420 kcal ME per hen per day).
4. Male breeders require slightly less energy than females during growth. The lower fat deposi-

tion in the male compared with the female is offset by the energy needs for more rapid growth. Being larger, the adult male cock requires considerably more energy than the hen for maintenance, but this is largely offset by the hen's need for energy for egg production.

Feeding Growing Broiler Breeders

Heavy breeds accumulate excessive amounts of body fat, so it is common practice to restrict the feed intake of these birds to produce pullets with leaner bodies at the time of sexual maturity. This is beneficial because (1) it produces healthier birds, and (2) it reduces feed costs during the rearing period. The methods commonly employed follow:

1. Skip-a-day method. This involves feeding pullets on alternate days only, from 9 weeks to sexual maturity. When fed every other day, pullets consume more feed on the days that feed is available than they would normally consume on a daily basis. They are unable, however, to consume enough feed in 1 day to satisfy their total energy requirements for 2 days. Thus, growth and body fat content are reduced.

2. Daily restriction of feed. Under this system, the producer determines the amount of feed that would normally be consumed by the pullets each day, then provides them with a fraction of that amount on a daily basis. Often this fraction is 75 to 85% of the amount of feed that would be consumed on a free-choice basis.

3. Bulky, low-energy or low-protein and/or amino acid imbalanced diets. Another form of restriction involves (a) the feeding of bulky, low-energy diets, or (b) the feeding of low-protein and/or amino acid imbalanced diets on a free-choice basis during the period 12 to 20 weeks of age. The diets are formulated to be adequate in all nutrients, but are sufficiently low in energy or low in protein and/or imbalanced in amino acids that pullets cannot consume enough to satisfy their energy or protein needs for maximum growth. Under such a program, it is possible to restrict the growth of young pullets by 10 to 15%, an amount comparable to the growth depression with the skip-a-day and daily restriction methods.

FEEDING TURKEYS

Feeding turkeys involves two distinct areas of emphasis: feeding market turkey and feeding turkey breeders.

Feeding Market Turkeys

Most market turkeys are of the large type. The formulations of the diets fed to market turkeys should

Figure 7.7 Young turkeys grown in spacious confinement houses. *(Courtesy P. R. Ferket, North Carolina State University)*

be changed as the birds grow. Thus, the nutrient requirements provided by the NRC provide for such changes at interval. Nutritional adjustments are often made for expected ambient temperature variations in order to ensure that the birds consume the necessary amount of protein, minerals, and vitamins, regardless of changes in feed consumption. See Figure 7.7 for an example of confinement turkey production.

Feeding Turkey Breeders

The feeding programs for breeder stock are usually divided into prebreeding (holding) and breeding periods. The prebreeding, or holding, diets may be fed from the time the breeders are selected, at about 16 weeks of age. Holding diets are usually formulated at medium energy levels in order to stabilize development and weight gains after market age. Hens are fed the holding diet until the time of light stimulation, at about 30 weeks of age; thereafter, breeder diets are fed. Toms may be fed a nutritionally balanced holding diet from the time of breeder selection throughout the breeding season. In some programs, the body weight of toms is controlled by limited feeding. The light stimulation of toms is normally initiated at about 26 weeks of age.

It is not necessary to feed low-energy diets or to restrict feed intake of turkey hens in the prebreeding, or holding, period. Corn–soybean meal-type diets may be fed *ad libitum*. Growth restriction does not result in any consistent improvement in reproductive performance. Nevertheless, the use of holding diets for turkey breeders is common practice. These diets usually contain medium energy concentrations to stabilize development and weight gains after mature body weight is attained. Care should be taken to provide turkey breeder hens adequate intake of minerals and vitamins during the holding period so that they are not depleted of these nutrients prior to the onset of lay. The following information is pertinent to feeding turkeys:

1. Prevent poult "starve out." Upon arrival, poults should be encouraged to consume feed and water as soon as possible. It may be necessary to dip the beaks of some of them in feed and water to start them eating and drinking.
2. Turkeys grow faster than chickens; hence, they have relatively higher feed and protein requirements.
3. Young turkeys use feed efficiently. Large White turkeys require about 2.5 to 3.5 lb of feed to produce 1 lb of liveweight, when grown to a market weight of about 35 lb and 18 weeks of age.
4. A high-fiber, low-energy holding diet retards sexual maturity and may result in some desirable effects upon later reproductive performance. The holding diet limits energy intake, but should not limit protein, vitamins, and minerals. When a holding diet is used, the birds should be switched to the breeder diet 2 weeks prior to egg production.
5. Good range provides green feed and tends to reduce feed costs. However, it may make for higher losses from blackhead and other diseases, and predators; and range turkey operations may make the neighbors unhappy because of dust, odors, and noise.
6. As they approach maturity, turkeys fed for market purposes should be fed diets that are quite different from those that are fed for turkey breeders.
7. Table 7.7 summarizes nutrient requirements for growing turkeys.

FEEDING DUCKS

The NRC publishes requirements for ducks. Typically, ducks are provided with two or three feeds during the growing period. When only two feeds are provided during the growing period, a 22% protein starter diet is usually fed the first 2 weeks, followed by a grower-finisher diet. When three feeds are provided during the growing period, they consist of a 22% protein starter diet, an 18% and a 16% diet. The following information is pertinent to feeding ducks:

1. Ducks should be fed pellets rather than mash. Use 1/8-in. pellets for starter diets, and 3/16-in. pellets for older ducks. Pellets will save 15 to 20% in the feed required to produce a market duck.
2. Ducks are very susceptible to aflatoxicosis, so monitoring feeds for aflatoxin is important.
3. Ducks are nearly as good foragers as geese.
4. When used, holding diets are designed to maintain breeding ducks from about 8 weeks of age until the breeding season commences, without their getting too fat. It is recommended that birds fed holding diets be limited to about 1/2 lb per bird per day.
5. When a holding diet is used, the breeder diet should be substituted for it about 4 weeks before eggs are desired for hatching purposes.
6. When feeding ducks, pellet quality, proper feather development, and limiting carcass fat disposition are concerns, in addition to proper growth and satisfactory feed conversion.
7. Commercial ducks grow as rapidly and efficiently as commercial broilers.

FEEDING GEESE

The NRC publishes requirements for geese. Geese are raised under the following variety of feeding programs:

1. The production of farm geese, with the goslings given a starter feed for about 2 weeks, followed by foraging the farm for a variety of pasture and grain feedstuffs; then, marketed at about 18 weeks of age after liberal grain feeding for the last 2 or 3 weeks.
2. The goslings are limit-fed prepared feed throughout the growing period, but are allowed considerable foraging in addition; then, marketed at about 14 weeks of age following liberal feeding of a high-energy finishing diet.

TABLE 7.7 Recommendations for Growing Turkeys, Males and Females

Tom, weeks	0–3	3–6	6–9	9–12	12–15	15–18	18 to end
Hen, weeks	0–3	3–6	6–8	8–10	10–12	12–14	14 to end
Metabolizable energy, kcal/kg	2,900	3,000	3,150	3,280	3,400	3,475	3,525
Protein, %	27	26	24	22	18	15	14
Methionine and cystine, %	1.15	1.05	1.00	0.90	0.75	0.65	0.60
Lysine, %	1.80	1.65	1.55	1.40	1.15	0.90	0.85
Calcium, %	1.35	1.30	1.20	1.15	1.10	1.00	0.90
Available phosphorus, %	0.78	0.75	0.70	0.65	0.62	0.55	0.50

3. The goslings are full-fed in confinement and marketed as junior green geese at about 10 weeks of age.
4. The production in some European countries of goose liver for *pate de foie gras.* In this program, the geese are grown to about 12 weeks of age, following which they are force-fed a high-grain diet for the production of livers of high-fat content.

For breeding purposes, geese are fed a prebreeding (holding) diet beginning 6 to 8 weeks before the breeding season, followed by a breeding diet formulated for the intensive production of fertile eggs. The following additional information is pertinent to feeding geese:

1. Rations for geese should be pelleted, with 3/32- or 3/16-in. pellets preferred. Mash and crumbles cause too much feed wastage and should not be used.
2. Although all diets may be home-mixed, a commercially prepared diet is recommended for young goslings and breeders during the laying season.
3. Succulent green feed should provide the bulk of the diet for young growing geese.

FEEDING OSTRICHES

Good nutrition is essential to the success of ostrich production. Because of the rapid growth of ostrich chicks (they reach full size at about 6 months of age), they must receive a diet that is high in protein, and that contains the essential amino acids, along with adequate energy, minerals, and vitamins. Also, the nutrition of breeders affects egg production and hatchability. Nutrient requirements for growing ostriches are given in Table 7.8.

FEEDING PIGEONS

Pigeons grow more rapidly than most birds during the first 20 days of life. They receive their first nourishment from pigeon milk regurgitated from the parent pigeon's crop. Pigeon milk is a thick, creamy, semidigested substance high in protein and fat, but low in carbohydrates. When 20 to 40 days of age, squabs may be fed a pigeon feed. Unlike other forms of poultry, pigeons will not eat mash, so pigeon feed either consists of whole or cracked grains or commercially prepared pellets.

Most pigeon producers feed (1) a complete pelleted diet, or (2) a complete pelleted feed plus whole or cracked grain. The most common grains are corn, wheat, sorghum, and peas. Grains can be offered in an open trough or cafeteria style where the self-feeder has individual compartments for each type of grain. If an open trough is used, it is recommended that pigeons be fed twice daily. Only enough feed should be offered at each feeding period as will be eaten in 1 hour. Commercial pigeon feeds are available.

FEEDING GAME BIRDS

The NRC publishes requirements for pheasants and quail. Commercial game bird feeds are available in most areas. Generally, these are of the following types:

- Starter diet containing 28% protein, for the first 6 weeks.
- Grower diet containing 20% protein, for 7 to 14 weeks of age.
- Depending on whether they are being marketed for game-release or for dressed game birds, finisher diet or flight conditioner diet containing 15% protein, from 15 weeks until market. Commercial game bird feeds may be in the form of mash, crumbles, or pellets.

Note: Game birds require a higher level of protein in early life than chickens. Grit should be available in a separate feeder, with the size of the grit determined by the size of the birds.

Specific information relative to feeding pheasants, Bobwhite quail, and Japanese quail follows.

TABLE 7.8 Recommendations for Growing Ostriches

	0–9 Weeks	*9–42 Weeks*	*42 Weeks → Market*
Metabolizable energy, kcal/kg	2,465	2,450	2,300
Protein, %	22	19	16
Methionine and cystine, %	0.70	0.68	0.60
Lysine, %	0.90	0.85	0.75
Calcium, %	1.50	1.20	1.20
Available phosphorus, %	0.75	0.60	0.66

Source: Modified from S. E. Scheideley and J. L. Sell. Nutrition Guideline for Ostriches and Emus, *Iowa State University Extension Publication Pm-1696.*

Of course, in a bad year the bigger the flock, the greater the potential losses.

Also, it should be recognized that the bigger and the more complicated the operation (integrated operations are more complicated than specialized operations), the more competent the management required. Although it is easier to achieve efficiency of equipment, labor, purchases, and marketing in big operations, size per se will not make for greater efficiency. Management is still the key to success.

MANAGEMENT

Four major ingredients are essential to success in the poultry business: (1) good management, (2) good birds, (3) good feed, and (4) good records/data. Management can make or break any poultry enterprise. The importance of good managers and supervisors cannot be overestimated.

Traits of a Good Manager

There are established bases for evaluating many articles of trade. They are graded according to well-defined standards. Additionally, they may be chemically analyzed. But no such standard or system of evaluation has evolved for managers of poultry enterprises. Table 8.1 provides a Poultry Manager Checklist that may have the following uses:

1. Students may use the checklist for guidance as they prepare themselves for managerial positions.
2. Employers may find the checklist useful when selecting or evaluating a manager.
3. Managers may apply the checklist to themselves for self-improvement purposes.

No attempt has been made to assign a percentage score to each trait. Rather, it is hoped that this checklist will serve as a useful guide to the traits of a good manager. A good manager is an investment, not a cost. He or she will make money for the producer.

Standard Operating Procedures, Organization Chart, and Job Description

It is important that all workers know to whom they are responsible and for what they are responsible. This should be written down in an organization chart and a job description.

An Incentive Basis

The poultry industry relies on hired labor and contract growers. Good employees—the kind that everyone wants—are hard to come by and keep. The agricultural labor situation may become more difficult in the years ahead. There is need,

TABLE 8.1 Poultry Manager Checklist

CHARACTER—(Integrity)
Ethics: Sincere, honest, and loyal.
INDUSTRY—(Work, work, work!)
Enthusiasm, initiative, and assertiveness.
ABILITY—
Poultry technical "know-how" and experience; business acumen—including the ability to systematically arrive at the financial aspects and convert this information into sound and timely management decisions; strong interpersonal skills (communication, team-building skills, motivation, problem solving); computer literate; common sense; organization skills; imagination; growth potential.
PLANS—("Plans work, and works plans")
Sets goals, prepares organization chart and job description, ensures that standard operating procedures (SOPs) are in place and up-to-date, engages in strategic planning and its implementation, understands SWOT analysis (strengths, weaknesses, opportunities, and threats) of the situation of the company, unit, or farm.
ANALYZES—
Identifies the problem, determines options with "pros and cons," then comes to a decision. Studies the latest industry information (e.g., poultry magazines, *Feedstuffs*), financial data (e.g., *Wall Street Journal*), publications (print and electonic media), and data books (e.g., *Blue Book*).
COURAGE—
To accept responsibility to make difficult decisions; to innovate, and to keep on going when times are tough.
PROMPTNESS AND DEPENDABILITY—(A self-starter)
Is a "doer" not a "talker." Has "get-up-and-go" or TNT, which means that the work is done "today, not tomorrow."
LEADERSHIP—
Identifies problems. Prioritizes addressing these challenges. Stimulates subordinates and delegates responsibility. A good manager is someone who *does the job right, a leader who does the right job and is ahead of "the curve."*
PERSONALITY—
Cheerful, not a "complainer or whiner"! Always learning. Aware of industry trends.

therefore, for the industry to accomplish the following:

1. Help recruit and hold "topflight" employees.
2. Cut costs and boost profits.

Incentive pay and/or profit sharing may be the answer. Many manufacturers have long had an incentive basis. Executives are frequently accorded stock option privileges, through which they prosper as the business prospers. Blue-collar workers may be paid on piecework (number of units or pounds produced) or receive bonuses based on quotas. Also, workers get overtime pay and have group insurance and a retirement plan. A few industries have a true profit-sharing arrangement based on net profit as such, a specified percentage of which is divided among employees. No two systems are alike. Yet, each is designed to pay more for labor, provided labor improves production and efficiency. In this way, both owners and workers benefit from better performance.

The incentive basis chosen should be tailored to fit the specific operation, with consideration given to kind and size of operation, extent of owner's supervision, present and projected productivity levels, mechanization, and other factors. Family-owned and family-operated poultry farms have a built-in incentive basis; there is pride of ownership, and all members of the family are fully cognizant that they prosper as the business prospers.

Indirect Incentives

Normally, we think of incentives as monetary in nature. However, there are other ways of encouraging employees to do a better job. These are indirect incentives and include, in addition to good wages, the following: good working conditions and labor relations, vacation time with pay, time off and sick leave, group health insurance, and opportunities for continuing education/training and promotion.

How Much Incentive to Pay?

After reaching a decision to pay incentives, it is necessary to arrive at how much to pay. There are some guidelines that may be helpful:

1. Pay the going base or guaranteed salary; then add the incentive pay above this.
2. Determine the total stipend (the base salary plus incentive) to which you are willing to go,
3. Before making any offers, always check the plan on paper to see (a) how it would have worked out in past years based on your records, and

(b) how it will work out as you achieve the future projected production.

IMPROVING EFFICIENCY

There are two ways in which to increase profits in any business, including the poultry business: (1) increase returns, and/or (2) cut costs.

An individual poultry producer can do very little to increase returns, because supply and demand determine prices. However, a producer can cut costs. Cutting on those items accounting for the largest percentage of cost of production offers the greatest possibility to lower costs. For example, cutting feed costs is extremely important because feed costs account for up to 75% of the total costs.

Major Production Costs of Layers

The items of highest cost for shell egg production are:

- Feed
- Pullets
- Housing/equipment (e.g., caging) (interest and depreciation)
- Labor

These account for 96% of the cost of producing eggs, with feed being 55 to 75%. Hen replacement costs (raising replacement pullets) are ranked in order: feed, chicks, and labor, with feed being twice the cost of the chicks.

Major Production Costs of Broilers

The major broiler production costs are:

- Feed
- Chicks
- Housing/equipment
- Labor

Feed accounts for about 70% of the total cost of production.

Major Production Costs for Turkeys (see Figure 8.1 for growing turkeys)

The chief turkey production costs are:

- Feed
- Poults
- Fixed costs (land, buildings and equipment, maintenance, taxes, and insurance)
- Labor

Before adding expensive equipment, owners or managers should determine if it will pay. They should compare the cost of mechanization with its

Figure 8.1 Young turkey poults on litter. *(Courtesy Gretta Irwin, Iowa Turkey Federation)*

savings in labor. The break-even point on how much you can afford to invest in equipment to replace employees can be arrived at by the following formula:

Equation 1

Annual savings in labor from new equipment = Amount you can afford to invest in the new equipment divided by a factor based on the depreciation on the equipment together with operating and service costs (.15 – 0.20 or even 0.25 depending on the financial or accountants' advice and tax consequences).

Example

If hired labor costs $20,000 per year (including health insurance, social security, etc.), this becomes

Equation 2

$20,000/the factor, say 0.2 = $100,000
This is the break-even point. As labor costs are going up, it may be good business to exceed this limitation under some circumstances. Nevertheless, the break-even point ($100,000 in this case) is probably the maximum expenditure that can be economically justified at the time.

MANAGEMENT OF POULTRY

Techniques such as brooding, forced molting, and artificial insemination are considered in Chapter 14 on the laying hens, Chapter 15 on broiler produc-

tion, and Chapter 16 on turkey production (see Figure 8.1). In view of its importance, Chapter 9 is dedicated to animal waste.

Molting is a natural process of shedding and renewing feathers. Hens usually molt in the following order: head, neck, body (including breast, back, and abdomen), wing, and tail.

- Birds inherit the tendency to shed their plumage annually. Under normal conditions, an early molter is a poor layer; and a late molter is a good layer.
- Hens seldom lay and shed feathers at the same time. A high-producing bird may, for a time, molt and lay simultaneously, but usually she is in declining production when molting begins.
- Egg type birds have been bred so intensely for egg production that some continue to lay rather than molt. Producers interested in obtaining a second or third cycle of egg production may force rest their birds and induce a molt. In recent years, there has been increased interest in forced resting.

Forced resting or induced molting can be achieved by a variety of techniques. However, all methods aim at shutting off the production of eggs as rapidly as possible, giving the hens a suitable rest period, and having the hens return to maximum production. In most cases, molting is induced at the end of a production season. However, it can also be used at the beginning or in the middle of a reproductive season to delay or stop production. For example, the procedure can be used in the middle of a season when hens show very poor production due to disease or some other stress. The procedure may also have application in situations where marketing of eggs has been restricted due to accidental dietary contamination. Molting is normally done to reduce replacement cost on a per dozen egg basis. In general, higher net replacement costs (initial cost less salvage) and lower egg prices tend to favor the use of forced molting (also see Chapter 14).

CAPONIZING

A capon is a male bird with the reproductive organs removed. It bears the same relation to a rooster (cock) as a steer does to a bull, a wether to a ram, or a gelding to a stallion. Caponizing used to be performed extensively to improve the quality of the poultry meat. This represents a disappearing part of the industry due to the gains in growth rate of broiler chickens. The art of caponizing is very old.

As early as 37 B.C., Cato and Varro referred to the altered males, called capons, in their book *Roman Farm Management.*

CANNIBALISM

Cannibalism is the eating of a species by others of the same species. It can be encountered in birds of all ages. Among baby chicks, the trouble is usually confined to toe and tail picking. With mature birds, the vent, tail, and comb are the regions most frequently picked.

The cause of cannibalism is not fully understood. It is known that it is more frequent under confined conditions. The stresses of overcrowding, too much light, and an unbalanced diet are among the factors that can lead to cannibalism in chickens and other domestic fowl. Also, poor ventilation, heat, and injury or sickness can cause birds to peck and eat at each other.

The best time to deal with the problem is before it starts. Producers should be sure that birds have enough space, especially at feed and watering facilities, and when nesting. Lighting should neither be too bright nor kept on too long; red lights are a temporary solution where cannibalism is already started (with such lighting, the birds are not able to see the blood or wounds). Low-intensity lights are the best permanent solution.

Good ventilation of the poultry house can lower heat and remove foul air, lessening the stress in the birds. Any diet in which there is too much high-energy feed and too little fiber can also lead to cannibalism. The birds should be watched closely and any injured, sick, or crippled individuals should be removed immediately.

Beak trimming should be a normal practice when the birds are confined. This is the practice of removing the pinching and tearing part of a bird's beak in order to lessen cannibalism and feather picking. It is a widely accepted practice in the poultry industry. Beak trimming is a precision operation and experience is a great asset in doing it properly. Moreover, there are advantages and disadvantages to beak trimming, but the advantages far outweigh the disadvantages.

Machines for beak trimming are in common use in most poultry areas. Most hatcheries and poultry equipment supply houses have them for sale or can get them. Hatcheries, veterinarians, and poultry service establishments will usually either rent the equipment or do the work at low cost. Figure 8.2 shows one kind of beak trimming equipment.

Most producers recommend beak trimming day-old chicks at the hatchery. They are more easily handled at this age, and hatchery beak trimming is convenient. Other producers feel that stress is min-

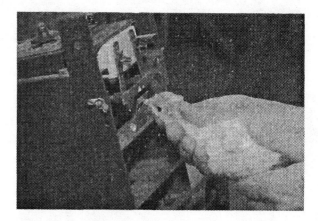

Figure 8.2 Beak trimming. *(ISU photo by Bob Elbert)*

Figure 8.3 Methods of beak trimming of chickens and turkeys. *(Original diagram from E. Ensminger)*

imized if beak trimming is delayed until chicks are 7 to 14 days of age. Beak trimming at the latter *stage is* usually more uniform and effective for a longer period. Figure 8.3 illustrates properly beak trimmed poultry.

TROUBLESHOOTING

Producers should know the signs of bird well-being. Likewise, they should be able to do the following:

1. Recognize trouble once it strikes.
2. Diagnose the cause.
3. Institute corrective measures; thereby holding mortality and economic losses to a minimum.

This means that they must know what to look for when troubleshooting. Poultry health is covered in detail in Chapter 11.

Chick Troubleshooting

Producers should look for the following when troubleshooting:

- Are the chicks spread out over the floor area and at the feeding and watering equipment? If so, that is good.
- Are they panting and gasping? If so, they may (a) be too hot or (b) have a disease problem.
- Are they grouped close to the heat unit or huddling and piling in small groups? If so, they are too cold (see Figures 8.4 and 8.5).
- Are feeders and waterers crowded? If so, there is not enough space.
- Is the litter packed and wet? If so, improve the ventilation and/or put down new litter.
- Are ammonia odors strong? If so, improve ventilation and/or change litter.

Layer Troubleshooting

When layer troubleshooting, the worker should look for the following signs that all is well:

- Alert birds
- Red heads
- Feed consumed
- Lack of overt disease or wounds
- Good egg production

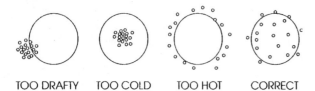

TOO DRAFTY   TOO COLD   TOO HOT   CORRECT

Figure 8.4 Chick behavior under different brooding conditions/temperatures.

Figure 8.5 Example of pen set-up for brooding chicks.

The following are signs that all is not well:

- New plumage and yellow shanks often characterize hens that have molted and are poor layers
- Signs of molting and shrunken head parts characterize hens that are out of condition
- Mortality and on-farm condemnations
- An unsteady gait and other signs of diseases and parasites

Check length of the laying period if there is concern relative to (a) egg size, (b) shell texture, or (c) profitable lay.

- Pullets that started laying at an early age or when undersized lay small eggs longer than pullets that were more mature and larger when they started laying.
- Shell texture of the hen's egg decreases in quality toward the end of the laying period due to fatigue of the shell gland.

Egg Troubleshooting

The most common egg troubles and their causes are:

1. **Weak shells**—which may be caused by:
   - Disease
   - Age of bird
   - Strain of chickens
   - Hot weather
   - Lack of calcium in the feed
   - Deficiency of vitamin D
2. **Too many cracks**—which may be caused by:
   - Weak shells
   - Poor egg collecting (e.g., rough handling of eggs and container)
   - Faulty washing of eggs and container
3. **Mottled shells**—which may be due to:
   - Eggs not gathered often enough
   - Egg-holding room too dry—below 75% relative humidity
   - Egg-packing material and cases too dry
4. **Weak whites**—which may result from:
   - Eggs not gathered often enough
   - Eggs not cooled quickly
   - Eggs not stored at cool temperature and high humidity
   - Eggs not packed daily as soon as cooled
   - Egg-packing material too warm or too dry
   - Disease
   - Age of birds
5. **Cooked whites**—which are generally caused by:
   - The egg wash water being too hot (use thermometer to check temperature)
   - Eggs left in water too long

6. **Yolk too colored**—due to:
   - Feed (yellow corn or green feed)
   - Heat

COMPUTERS

The poultry industry uses computers as a critically important management tool. A successful poultry manager must be computer literate. Feed mills and poultry companies use computers extensively in ration formulation. Computers are also being used by poultry businesses as ways to analyze a great mass of data and consider many alternatives.

COMMUNICATION SYSTEM

Business visitors to poultry farms are not interested in seeing the birds. They are there on business—to talk to the manager or owner. There needs to be a communication system, otherwise visitors may be tempted to walk around the poultry houses looking for the manager or one of the employees. This breaches *biosecurity* (see Chapter 11).

To prevent the spread of poultry diseases, only those who need to come into the unit should be allowed to enter. This necessitates fencing and locked gates, as well as a communication system so that potential visitors can be screened. (Necessary sanitary precautions must be taken first.)

The requirements for a workable communication system are:

- Fences
- A locked gate(s)
- A communication device
- A sign at the gate telling visitors how to reach the person they want to see

In the ideal layout, the establishment is designed so that the office is located at the main gate to allow office personnel to control all entry. This is common on larger operations, but it means that someone must be in the office at all times.

Five different types of communication systems are commonly employed. They are:

1. *Horns, buzzers, or bells* that are activated by a push button at the gate. Of all devices, these are the simplest and least costly. However, someone must go to the gate in order to see who is there.
2. *Two-way speaker system* ("squawk box"). It may consist of a multiple set, with units in various buildings and a central unit in the office. This arrangement is versatile. It can be set up so that managers can screen visitors without leaving their work.

3. *Belt-mounted beeper.* This device may signal the wearer to the phone, to the office, or to the front gate.
4. Hand-carried or vehicle-mounted *two-way radio or cell phone.*
5. A *combination communication device* and gate lock deactivator. Visitors can let the manager or office personnel know they are at the entrance. The gate lock can then be deactivated from the office or other location on the ranch.

The producer should decide what system best suits the particular operation. The important thing is that there be an effective communication system as a prerequisite of poultry disease prevention.

In addition to installing a communication device, the owner or manager should establish a certain time during the day to be in the office to receive telephone calls or use the cell phone. Salespeople should visit only by appointment. Business should be conducted in the office or by telephone.

MANAGEMENT OF LIGHT

It has long been known that light stimulates egg production in chickens and other birds. Records show that the ancient Chinese made their canaries sing more by placing a lighted candle by the cage at night. Much later, early in the 1900s, poultry farmers in the state of Washington found that they could increase winter egg production by placing a lighted lantern in the chicken house for a few hours each evening. At that time, however, it was thought that the role of light was primarily a matter of increasing the "workday" of the bird. Today, the action of light is considered physiological; light enters the eye of the bird and stimulates the pituitary gland. In turn, the pituitary gland releases luteinizing hormone (LH), which causes ovulation. Because of this phenomenon, the management of lighting systems is an integral part of modern poultry production.

Several lighting systems have been designed, differing primarily in windowless versus open houses, in light-to-dark ratio, and in continuous versus intermittent light. It is noteworthy that chickens do not respond to all wavelengths of light. Orange and red lights are most effective. Shorter wavelengths are not effective. Normal white incandescent bulbs give sufficient light at effective wavelengths. It is noteworthy, too, that light stimulus seems to have a threshold level of intensity beyond which further increases in brightness of light have no effect. A level of 0.5 to 1.0 foot-candle of light should be provided at the darkest points of exposure of the hens. Excessive light is both unnecessary and uneconomical. Automatic time switches should be installed in poultry houses for pullets or layers.

BROODING

Brooding is essential for the successful growth of day-old broiler chicks and turkey poults, together with chicks to be reared as replacement pullets. Critical issues are the following:

- Temperature
- Water availability
- Feed availability
- Adequate space and secure environment
- Ready access for personnel

Temperature Control

There are optimum temperatures for chicks of different ages. When chicks are a day old, the temperature should be from 90° to 95°F (32°–35°C) at the level of the chicks on the floor. The temperature is usually lowered about 5°F (2.75°C) each week until a temperature of 70° to 75°F (21°–24°C) is reached, or until the chicks are 6 to 8 weeks old and fully feathered. A reliable thermometer will enable the caretaker to provide comfortable temperatures. If in doubt, however, the chicks will tell you (see Figure 8.4).

If the chicks huddle or crowd together and cheep, they are too cold. If they move away from the source of heat or from under the brooder, or if they pant and hold their wings away from their bodies, they are too warm. Chilling chicks may result in their piling and smothering. Too cool temperatures also cause diarrhea and increase susceptibility to infectious disease, together with reducing growth rate. Too high temperatures result in reduced appetite and retarded growth.

Ventilation

Ventilation is required to provide fresh air for the chicks—to remove carbon dioxide and carbon monoxide, to remove ammonia and other fumes, and to aid in keeping the litter dry. Gas, oil, and other flame-type brooders that burn the oxygen out of the air require more ventilation than other types of brooders. A good ventilating system provides plenty of fresh air without drafts. Chicks huddling in certain areas or spots may indicate floor drafts. A strong odor of ammonia means that there is not enough air movement. If the floor litter is relatively dry and the air in the house has little or no ammonia or other odors, this is an indication that the ventilation is adequate.

Moisture Control

A fairly humid environment, around 50 to 60% relative humidity, is desirable and conducive to good feathering. A very dry atmosphere will cause poor feathering. Too much moisture in the brooder house may cause trouble. If the litter or walls and ceiling become wet, there is too much moisture and it should be reduced. The litter must be kept relatively dry. By increasing ventilation, the condition of the litter can be improved and kept dry. If moisture condenses on the ceiling and walls until it drips, better insulation and ventilation are needed. Wet litter may lead to an outbreak of diseases.

Brooder Houses and Equipment

In modern commercial production, chicks are reared in specially designed and constructed brooder houses, with 50,000 or more birds brooded as a unit and cared for by one person. Common features for brooding include the following (also see Figure 8.5):

- Heater or stove (pancake [jet] brooders, infrared [radiant] gas brooders, or forced-air furnaces)
- Supplemental waterers or minidrinkers in addition to the automatic water (nipple) line (with electrolytes added to the water for the first 2 to 3 days following hatching)
- Supplemental feeders and feed spread on the paper
- A brooder guard enclosing chicks or poults
- Litter covered with paper for secure footing (litter at a temperature of ~85°F [30°C])
- Good ventilation to prevent buildup of toxic gases including ammonia

Management Schedule

Successful brooding requires a well-programmed and well-executed management schedule. The following management schedule is divided into two parts: (1) before the chicks arrive, and (2) after the chicks arrive.

1. Before chicks arrive:
   - **Clean thoroughly.** Thoroughly clean the brooder house—floor, walls, and overhead.
   - **Disinfect.** Disinfect the brooder house, using one of the commercial disinfectant materials according to the manufacturer's directions. For the disinfectant to be most effective, the brooder house must first be thoroughly cleaned.

- **Set up and make a trial run with brooders.** Set up and start the brooders 2 or 3 days before the chicks arrive to make sure they are properly adjusted and working satisfactorily. This will reveal any missing or malfunctioning parts, permit temperature adjustment, and help remove moisture from the house and litter.
- **Clean all equipment.** All equipment, such as feeders and waterers, should be thoroughly cleaned and disinfected before the chicks arrive.
- **Consider providing clean, fresh litter.** Put clean, fresh litter 3 to 6 in. (~8 cm) deep on the floor before the brooders are set up.
- **Install brooder guards.** If brooder guards are used, one which is 14 to 18 in. (~45 cm) high will be sufficient. Allow about 36 in. (1 m) between the outer edge of the brooder and the brooder guard. This guard will keep the chicks confined to the brooder area and result in their eating and drinking faster. A brooder guard made of poultry netting is satisfactory for brooding during the summer months. The brooder guard should be removed by the end of the first week.

2. After chicks arrive:
   - **Brooder temperature.** Follow the manufacturer's recommendations.
   - **Vaccination.** An appropriate vaccination program should be followed.
   - **Feeding.** As appropriate for broiler chicks, broiler breeders, replacement pullets, turkeys, and so on.
   - **Water.** Plenty of clean, fresh water should be provided at all times. Water should be distributed so that the chicks or poults can drink conveniently.
   - **Light.** Some light should be on all night, except for 1 hour of darkness, during the first few days because this may be helpful in keeping birds from crowding or piling.

SAVE ENERGY, SAVE COSTS

Energy shortages and high costs are here to stay and may worsen. It is important, therefore, that poultry producers establish energy conservation practices to stretch fuel supplies and lower production costs.

Some estimate that the poultry industry could save 20% of the energy used. This is a matter of attention to details and modification of existing practices. Little expense, if any, is needed to institute energy-saving methods. Some ways in which to save energy follow:

Brooding and growing. Brooding uses the most energy of any phase of poultry production. Costs can be reduced by:

a. Partial house brooding.

b. Clustering three or four brooders together and encircling the cluster with a single solid chick guard.

c. Keeping the recommended number of chicks under each hover (undercapacity increases fuel cost per bird).

d. Adjusting brooding temperatures to conserve heat.

Housing. Good management of existing housing, along with modifying it if necessary (such as by adding insulation), saves energy. Regular checking and sealing air leaks around doors, air intakes, and fans will reduce heat loss. Adding insulation to existing houses is expensive. But if the current insulation is inadequate, it will likely be cheaper to buy and install more insulation than to buy feed or fuel. Always make sure that wall and ceiling insulation has the recommended R-value.

Ventilation. Energy can be saved by a well-managed ventilation system. One way is to reduce ventilation rates. Temperature within the house can be allowed to rise to 70° to 75°F (21°–23°C), provided air quality is kept acceptable and if ammonia does not exceed 50 ppm. This saves electricity and reduces feed usage, too. Other ways of saving energy include selection of the most efficient fans and keeping fans clean.

Lighting. Energy can be saved by careful management of the lighting program, including:

a. Installing reflectors (a 25-watt bulb with a reflector gives as much light as a 40-watt lamp without a reflector).

b. Reducing light intensity, which can cut electricity use by 25 to 50%. (**CAUTION:** Never decrease the intensity of light of laying flocks in production.)

c. Using intermittent lighting in windowless houses. (**CAUTION:** Birds already in egg production should not be switched from some other lighting program.)

d. Going through the house every 2 to 3 weeks and replacing lightbulbs that do not emit enough light.

Feeding, watering, and management. The energy requirements for watering and feeding are relatively small. Nevertheless, some savings can be effected in most operations. The number of times automatic feeders run each day should be reduced to the minimum needed for the desired level of feed consumption—3 to 4 runs a day may suffice.

Energy is conserved by keeping the watering system in good repair to prevent leakage and spilling. Also, consideration should be given to switching from continuous-flow troughs to a valve-controlled or discontinuous system to reduce pumping in the water used. If continuous-flow troughs are used, turn the water off at night and operate the system on an intermittent schedule in the daytime.

Among the management programs that may be instituted to save energy are keeping all moving equipment in good repair, cleaning manure-removal equipment after each use, and using only the recommended size of motors on feeders, waterers, egg-collection equipment, and other automation, as well as maintaining and adjusting such equipment regularly.

CONTROL PESTS

The necessity of getting the greatest production per bird and the most profitability requires the timely use of pesticides to control insects (flies, lice, and mites), rodents, birds, and other pests (also see Chapters 9, 11, and 12). At the same time, the producer must be increasingly concerned about timing, choice, and dosage of registered pesticides, along with the method of application, in order to stay within the residue tolerances or limits allowed in eggs and meat by the Food and Drug Administration. Government agencies are continually sampling eggs and fowl for pesticide residues, feed additives, and antibiotics. Misuse of these chemicals can result in financial loss through confiscation and reflect discredit on the poultry industry. In order that pesticides be used properly and efficiently, the following precautions and suggestions should be observed:

Good sanitation and regular inspection. The use of pesticides should be accompanied by good sanitation and regular inspection of buildings and birds.

Start using pesticides early. Pest control methods should be instituted before the problem builds up and reduces production efficiency.

Use only approved pesticides. Only registered, approved pesticides should be used on birds, in and around poultry houses, in egg rooms, or in storage areas. New products are developed, and sometimes old products are banned. Accordingly, it is recommended that the producer follow the current recommendations of the Cooperative Extension Service, or other recognized specialists, for the control of pests.

Read and follow label directions. When using any pesticides, the operator should always read and follow the label directions.

Mix and use pesticides with care. Pesticides should be mixed where birds cannot get to containers, equipment, or spillage. Likewise, pesticides should be kept away from eggs, feed, feeders, water, and watering equipment. Good ventilation should be provided in confined areas.

Store pesticides and application equipment properly. Pesticides and application equipment should be stored in a separate, marked, locked building or storage area away from children, poultry, feed, and water sources.

Dispose of empty containers properly. Empty paper and plastic containers should be burned in an approved manner, then the ashes should be buried.

RODENTS

Poor sanitation attracts rodents. Rodents have severe negative consequences for the poultry industry due to the following:

- Their consumption of feed (100 rats consume ~1 ton of feed per year!).
- Contamination of feed with feces and urine.
- Spreading diseases of poultry and zoonotic diseases that affect people. (Rats and mice are sources and reservoirs of disease-causing microorganisms (e.g., *Salmonella*). The pathogens may be transmitted via the feces or external parasites.
- Destruction of buildings (undermining foundations, gnawing electrical lines, destruction of insulation, chewing through wood, rubber, vinyl, etc.).
- For range poultry, rodents may attract predators such as foxes and coyotes.
- Rapid reproduction with a female mouse capable of producing 35 to 60 offspring in a year, which are in turn sexually mature in about 7 weeks (rats are marginally less fecund).
- Upset to neighbors.

The major rodents that infest poultry facilities are:

- The Norway, or brown, rat
- The roof, or black, rat
- The house mouse

Rats will tend to eat grain at one or two locations while mice are less restrictive. Rats also need a source of water. These rodents are predominantly nocturnal. Seeing them during the day is very suggestive of high populations.

The Norway Rat (*Rattus norvegicus*)

The Norway, or brown, rat is a relatively large rodent. Its coat is predominantly reddish to gray-brown with the underneath gray to yellow-white. An adult has a head and body 7 to 10 in. (~17–25 cm) long with a tail 5.5 to 7.5 in. (~16 cm) long. It weighs about 11 oz (~300 g).

The Roof Rat (*Rattus rattus*)

The roof rat is also known as the plague, house, or black rat. It is sleeker than the Norway rat and black

to gray in color with a lighter (gray to white) underside. The roof rat has a tail that is longer than the head plus body.

The House Mouse (*Mus musculus*)

The house mouse is gray-brown. An adult has a head and body about 3 in. (~8 cm) long with a tail 3 to 4 in. (~8 cm) long. It weighs about 0.5 oz (~15 g).

Observation of Rodents

Signs of rodents include the following:

- Droppings
- Gnawed holes in wood and electrical wiring
- Urine stains that fluoresce under black (UV) light
- Rodent holes/burrows (see Figure 8.6)
- Numbers caught as part of a rodent control program

Control of Rodents

The rapid rate of reproduction and the destructive power of rodents, together with concerns for poultry

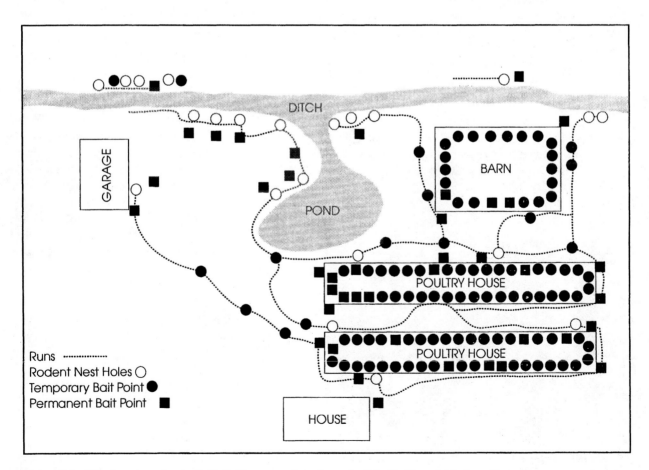

Runs ················
Rodent Nest Holes ○
Temporary Bait Point ●
Permanent Bait Point ■

Figure 8.6   Sites for rats and rodenticide/baiting around poultry units. *(Modified from Arbor Acres Manual)*

health, make rodent control essential. The success of these efforts should be monitored. Approaches used to control rodent populations include:

- Physical exclusion from facilities and feed (rodent proofing)
- Rodenticides
- Fumigation
- Trapping (snap trap plus glue boards, called "sticky traps")

An example of where to bait with rodenticides is shown in Figure 8.6. (See also Figure 8.7.)

Physical Exclusion from Facilities and Feed

Buildings should be rodent proofed with concrete construction plus the use of galvanized steel, brick, aluminum, and so forth. Also a curtain wall of wire mesh might be included in the constructed wall to prevent rats burrowing beneath the foundation walls. Seal holes (rats can pass through a 0.5 in. or ~1.2 cm hole and mice through a 0.25 in. or 0.6 cm hole) with steel wool, temporarily, followed by concrete.

Trapping

When using snap traps, bait with bacon or peanut butter or oats. Also leave traps open until the bait has been taken once since rodents can be "trap shy." Note that rats and mice frequently have runs along the inside or outside of walls. This is where to trap or bait.

Rodenticides

Rodenticides are chemicals that kill rodents. They are frequently referred to as rat or mouse poison. Rodenticides are incorporated into rodent-attractive foods or baits. Rodenticides can be categorized as:

- Anticoagulants
- Vitamin D or vitamin D-type
- Others

ANTICOAGULANTS. The anticoagulant rodenticides are related to coumarin or vitamin K. The first of these was warfarin. A second generation of anticoagulants has been developed that is effective even against rats and mice resistant to first-generation anticoagulants. An example is bromodiolone. A list of anticoagulant rodenticides follows:

- Brodifacoum (Havoc®, Talon®) use at 0.005%
- Bromodiolone (Maki®, Contrac®) use at 0.005%
- Chlorophacinone (RoZol®) use at 0.005%
- Difethialone (Generation®) use at 0.0025%
- Diphacinone (Ramik®) use at 0.005%
- Pindone (Pival®) use at 0.0025% (this can also be used to control wild bird populations)
- Other anticoagulant rodenticides include coumachlor, coumafuryl, coumatetralyl, difenacoum, and flocoumafen.

VITAMIN D. Vitamin D or vitamin D metabolite-containing rodenticides include: ergocalciferol and cholecalciferol (vitamin D_3).

OTHER RODENTICIDES. They include the following:

- Bromethalin (Assault®, Vengeance®)—a central nervous system toxin causing paralysis.
- Zinc phosphide (Ridall®, ZP®)—phosphide gas (smelling like garlic) is released and kills via paralysis of the heart plus damage to the liver and intestines. Phosphide is also available as aluminum and magnesium phosphides.
- Other poisons are or have been used—strychnine (a botanical), the inorganics such as thallium (thallium sulfate), arsenic (arsenous oxide, potassium arsenite, and sodium arsenite), sodium or potassium nitrate, yellow phosphorus, and hydrogen cyanide, the organochlorides (e.g., lindane), the organophosphorus rodenticides (e.g., phosacetim), and the unclassified rodenticides (e.g., chloralose, sodium fluoroacetate).

These can be hazardous to pets, poultry, and people.

LITTER

Litter is used primarily for the purposes of keeping the birds clean and comfortable. It absorbs mois-

Figure 8.7 Rodenticides that may be used around poultry units. *(ISU photo by Bob Elbert)*

ture from the droppings and then loses this moisture to the air brought in by ventilation. Excreta accumulate in the litter and compost. A good litter is highly absorbent and fairly coarse, to prevent packing. It should be free from mold and contain a minimum amount of dust. Availability and cost will determine the type of litter used. Table 8.2 lists some common litter materials and gives the average water absorptive capacity of each.

Naturally, the availability and price per ton of various litter materials vary with area and from year to year. In some regions, shavings and sawdust are available. In other regions, corn or wheat straw is more plentiful. Facts of importance relative to litter materials and litter uses follow:

- Wood products include sawdust, shavings (see Figure 8.8), tree bark, chips, and others. The suspicion that these will hurt the land is unfounded. They decompose slowly, but this can be speeded up by nitrogen.

TABLE 8.2 Water Absorption of Various Litter Materials

Material	Units of Water Absorbed/100 Units of Air-Dry Bedding
Barley straw	210
Cocoa shells	270
Corn stover (shredded)	250
Corn cobs (crushed or ground)	210
Cottonseed hulls	250
Flax straw	260
Hay (mature, chopped)	300
Leaves (broadleaf)	200
(pine needles)	100
Oat hulls	200
Oat straw (long)	280
(chopped)	375
Peanut hulls	250
Peat moss	1,000
Rye straw	210
Sand	25
Sawdust (top-quality pine)	250
(run-of-the-mill hardwood)	150
Sugarcane bagasse	220
Tree bark (dry, fine)	250
(from tanneries)	400
Vermiculite[1]	350
Wheat straw (long)	220
(chopped)	295
Wood chips (top-quality pine)	300
(run-of-the-mill hardwood)	150
Wood shavings (top-quality pine)	200
(run-of-the-mill hardwood)	150

[1]This is a micalike mineral mined chiefly in South Carolina and Montana.

Figure 8.8 Litter is very important for poultry health. Photo shows wood shavings litter. *(ISU photo by Bob Elbert)*

- Cut straw will absorb more liquid than long straw. But there are disadvantages to chopping; chopped straws may be dusty.
- The desirable amount of litter to use is the minimum amount necessary to absorb completely the moisture in the droppings for multiple grow out periods, and to allow composting without the production of ammonia.

Reducing Litter Needs

Litter materials are becoming more costly because of (1) other uses for some of the materials, and (2) the current trend toward more confinement rearing of all livestock, which requires more bedding. Poultry producers may reduce litter needs and costs as follows:

1. **Ventilate quarters properly.** Proper ventilation lowers the humidity and keeps the litter dry. The condition of the litter and the amount of the fumes (ammonia fumes) are good indicators of the adequacy of ventilation.
2. **Chop litter.** Chopped straw, waste hay, fodder, or cobs will go further and do a better job of keeping the birds dry than long materials. Chopped straw, for example, will soak up approximately 25% more moisture than long straw.
3. **Consider slotted floors or cages.** Slotted floors and cages, for which no litter is needed, are extensively used for layers.

If properly conserved, poultry manure is rich in nutrients. Since the manure of birds contains both the feces and the excretion from the kidneys, it is much richer in nitrogen than the manure of four-footed animals. As much as 80% of the urinary nitrogen is present as uric acid. So, unless the manure is properly handled, to prevent break down by bacteria, much of the uric acid will be changed to ammonia, with consequent loss of fertilizing value. See Chapter 9 for details on management and composition of poultry waste.

 **USEFUL WEBSITE**

Arbor Acres: http://www.aa-na.aviagen.com/home.asp

FURTHER READING

Bell, D. D., and W. D. Weaver. *Commercial Chicken Meat and Egg Production*, 5th ed. Norwell, MA: Kluwer Academic Publishers, 2002.

9

Animal Waste and Other Environmental Issues

Objectives

After studying this chapter, you should be able to:

1. Understand the issues related to poultry waste.
2. List the major regulations affecting poultry waste.
3. Understand the phosphorus cycle.
4. List environmental locations where nitrogen can end up.
5. Understand how litter can be treated.
6. Give the major reasons for using various methods of disposal of mortalities.
7. List the nuisances from poultry production.

ANIMAL (POULTRY) WASTE

Poultry produce considerable amounts of excreta and other waste (see Figure 9.1 and Table 9.1). Poultry producers have long recognized the need to dispose of this in a responsible, environmentally sound manner using the resources in the waste profitably.

Whether real or perceived, there is little doubt that the public's, the policy makers', and the media's concerns about pollution from animal waste or manure will persist. However, government regulations about animal waste need to be followed with concern for concomitant costs. Manure should be looked upon as a resource and a cost. Therefore, planned animal waste management is an important part of modern poultry management. The collection, transport, storage, and use of manure must meet sanitary and pollution control regulations. Modern poultry buildings and equipment should be designed to handle the manure produced by the birds that they serve; and this should be done efficiently, with a minimum of labor and pollution, so as to retrieve the maximum value of the manure and achieve the maximum animal sanitation and comfort.

Poultry operations are having an increasing number of complaints lodged against them because of manure, odor, and flies. Lawsuits, based on the nuisance law, have been filed against some poultry producers. Excreta/manure should be looked upon as a resource (Table 9.2). From the standpoint of using it as a fertilizer, along with implementing pollution controls, it is important to do the following:

- Exercise certain precautions when applying manure.
- Know how much manure can be applied to the land.

Precautions when using Manure as a Fertilizer

The following precautions should be observed when using manure as a fertilizer:

- Avoid applying waste closer than 100 ft to waterways, streams, lakes, wells, springs, or ponds.
- Do not apply where percolation of water down through the soil is not good, or where irrigation water is very salty or inadequate to move salts downward.
- Do not spread on frozen ground.

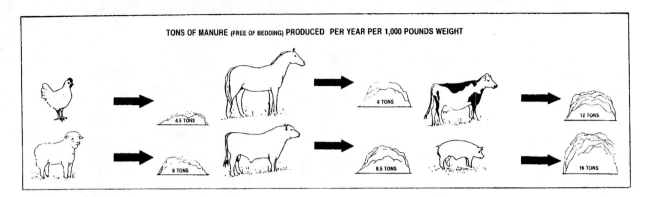

Figure 9.1 Manure production by class of livestock. Each class of confined animals produces per year per 1,000 lb (450 kg) liveweight the above tons of manure free of bedding. *(Original diagram from E. Ensminger)*

TABLE 9.1   Quantity of Pure Manure (Free of Litter) from Various Flocks

Birds	Type of Flock	Average Bird Weight (lb)	(kg)	Quantity Manure Produced (dry basis) (lb)	(kg)	Time Period
100	Laying hens	4.5	2.0	2,400	1,091	12 months
1,000	Broilers (chickens)	4.0	1.8	2,700	1,227	9 weeks
1,000	Turkeys	8.0	3.6	4,320	1,964	16 weeks

Source: University of Maryland Extension Service, Fact Sheet 39.

TABLE 9.2   Composition and Value of Chicken Manure, Fresh versus Different Degrees of Dryness

Kind of Manure	Water Content (%)	Solid Material (%)	Weight per Cubic Yard (lb)	(kg)	Nitrogen (N) (%)	Phosphorus (P) (%)	(P₂O₅) (%)	Potassium (K) (%)	(K₂O) (%)	Estimated Value of Manure/Ton[1]
Fresh	75	25	1,750	795	1.0	0.6	1.3	0.6	0.7	15.28
Partially dried	50	50	—	—	2.0	1.1	2.6	1.1	1.4	30.56
Dry manure	25	75	800[2]	364	3.0	1.7	3.9	1.7	2.1	45.84
Completely dry	0	100	—	—	4.0	2.3	5.2	2.3	2.8	61.12

Header note: *Approximate Percentage of Plant Food by Weight (Multiply by 20 to get pounds per tons)*

Source: Adapted from data reported by California Agricultural Extension Service.
[1]Calculated on the following retail price per pound basis: nitrogen (N), 29¢; P_2O_5, 30¢; and K_2O, 12¢.
[2]The volume-weight relationship may vary considerably at the lower moisture content due to handling, compaction, and biological decomposition.

- Distribute the waste as uniformly as possible on the area to be covered.
- Incorporate preferably by injecting (or possibly plowing or discing) manure into the soil as quickly as possible after application. This will maximize nutrient conservation, reduce odors, and minimize runoff pollution.
- Minimize odor problems by the following:
  - Adopt a "good-neighbor" policy. Discuss with neighbors about the timing of manure application, avoiding times when neighbors have family gatherings, weddings, or outdoor activities planned.

- Inject manure into the soil.
- Spread only on days when the wind is not blowing toward neighbors or populated areas.
- In irrigated areas, (a) irrigate thoroughly to leach excess salts below the root zone, and (b) allow about a month after irrigation before planting, to enable soil microorganisms to begin decomposition of manure.

Pollution Laws and Regulations

The Refuse Act of 1899 gave the Corps of Engineers control over runoff or seepage into any stream that flows into navigable waters. The U.S. Environmental Protection Agency (EPA) evoked that law to launch a program to control water pollution by requiring that all cattle feedlots that had 1,000 head or more the previous year must apply for a permit by July 1, 1971. The states followed suit, increasing legal pressures for clean water and air. Then came the Federal Water Pollution Control Act Amendments, enacted by Congress in 1971, charging the EPA with developing a broad national program to eliminate water pollution.

Owners and operators of animal-feeding facilities with more than 1,000 animal units must comply with regulations. Before constructing a poultry facility, the owner should become familiar with both state and federal regulations.

Nitrogen

Nitrogen from poultry waste [excreta, processing, and dead birds (mortality)] can be land-applied as a replacement source of fertilizer for crop production. However, nitrogen from poultry waste can represent a source of pollution as illustrated in Figure 9.2.

Losses in Water

Nitrogen as ammonia or nitrates can leach into surface or groundwater and hence to waterways and to the oceans, as, of course, can nitrogen from commercial fertilizer (e.g., anhydrous ammonia). There is evidence that excess concentrations of this nitrogen allow multiplication of microorganisms, leading to very low oxygen tension with water (hypoxia) such that animal life cannot survive. An example of this may be the "Dead Zone" in the Gulf of Mexico.

Air Loss

Nitrogen as ammonia may be lost as either a gas or absorbed onto dust and released into the air. The ammonia may be redeposited into the land or oxidized into oxides of nitrogen (NO_x), or be a noxious odor to neighbors. The NO_x and the ammonia are the "greenhouse gases" that are thought to contribute to global warning. When poultry manure is held in anaerobic lagoons, nitrogen will be lost as either ammonia or, following oxidation at the surface, as nitrogen gas. In addition, carbon in the waste can be processed by microorganisms in anaerobic lagoons to methane plus some carbon dioxide, both being greenhouse gases.

Potential Remedies

It is possible to reduce nitrogen in poultry excreta by providing the birds with diets where the amino acids are balanced to meet the nutritional needs of the bird at the specific phase of growth or egg production. This is referred to as the "ideal protein." The ideal protein is defined as "the balance of

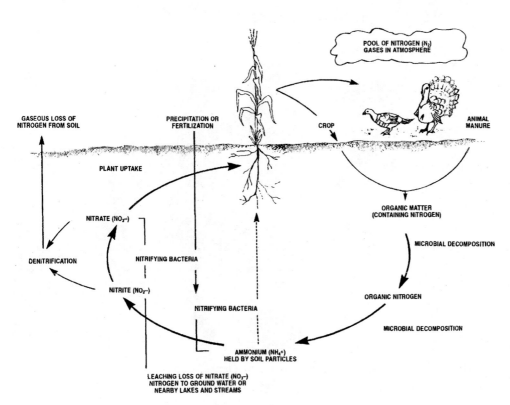

Figure 9.2 The nitrogen cycle for manure. *(Original diagram from E. Ensminger)*

indispensable amino acids that exactly meets the animal's requirements with no deficiencies or excesses." The ideal protein potentially provides an approach to reduce nitrogen excretion. It is estimated that idealizing poultry diets could reduce nitrogen in excreta by about 35%. This can have an impact on the amount of land on which the manure can be applied. Many states still use nitrogen as the legal standard for land application. It is likely that a phosphorus standard will be in place by 2005.

Phosphorus

Phosphorus is an essential nutrient that is obtained from plants, bones and minerals. Much of the phosphorus in poultry diets is in the form of phosphate bound to plant sugars—to phytate. This phytate phosphorus is not digestible and hence passes through the bird's gastrointestinal tract with the feces. This represents a dual problem in that inorganic (digestible) phosphate has to be added to the feed and that phytate phosphate is found in the excreta. The phytate is degraded by soil microorganisms, thereby releasing the phosphate.

Phosphate is a potential environmental hazard. If manure is applied to land at levels above those that can be taken up by growing crops, the phosphate builds up in the soil. There is considerable potential for phosphate getting into runoff water, surface water, watersheds, and then rivers. While some dissolved phosphate leaches into the surface and groundwater, the majority of phosphate enters waterways attached to soil and other particles. It can then leach off.

Phosphorus pollution of waterways can be of agricultural (manure or inorganic phosphorus fertilizer), industrial, or domestic (e.g., phosphate-containing detergents) origins. Irrespective of origin, phosphate in waterways is a critical problem. Eutrophic microorganisms require phosphate. In the presence of enough phosphate, the eutrophic microorganisms multiply rapidly. These use up the oxygen dissolved in the water such that fish and other animals die. There is evidence that phosphate from the Mississippi River is responsible for the "Dead Zone" in the Gulf of Mexico.

The amount of phosphorus in the excreta is also a problem since state and federal regulations are moving toward a "phosphorus standard" for land application of manure. The more phosphorus, the more land is required to apply manure. This creates logistic problems and increases the transportation costs.

Potential Remedies

If phytate phosphate is reduced in feed, the amount of phosphate in the excreta will decline in essen-

tially a "one to one" arrangement. Phosphate in phytate can be released by treatment of feed with the microbial enzyme, phytase. The phosphorus is then available to the bird. Therefore, the amount of phosphorus in the diet needs to be decreased. An alternative approach is to use low-phytate corn (maize), which has less phosphate that is much more available or digestible. This approach decreases phosphorus in the manure.

Processing of Poultry Waste

Litter

Broiler chicks and turkey poults are raised in litter. When this is removed, what is to be done with it? Most commonly, litter is land-applied providing nitrogen, potassium, and phosphate for growing crops. In addition, litter can be utilized for:

1. Electricity generation by incineration (directly or following gasification or pelletization)
2. Composting (Figure 9.3) followed by land application (Figure 9.4)

Figure 9.3 Composting turkey litter. *(Courtesy Gretta Irwin, Iowa Turkey Federation)*

Figure 9.4 Moving composted turkey litter. *(Courtesy Gretta Irwin, Iowa Turkey Federation)*

Poultry Mortalities

On-farm poultry mortalities and culled birds create a problem for the producer. Methods to dispose of mortalities include:

* Burial in open-bottom pit
* Composting with a carbon source (straw, sawdust, wood chips, chopped corn stalks, peat) followed by land application of the compost mixed with manure as plant nutrients
* Rendering following on-farm storage in freezers or following chopping fermentation
* Incineration

BURIAL IN OPEN-BOTTOM PITS. Burial in open-bottom pits is cheap and easy. However, there may be problems such as:

* Slow loss of poultry residue
* Seepage of nitrogen, phosphate, and pathogens into the groundwater

COMPOSTING. Mortalities can be decomposed by bacterial action in composting. This process requires the following:

* Balance of carbon to nitrogen (~25:1)
* Moisture about 50%
* Enough porosity to allow carbon dioxide and other gases generated by decomposition
* As heat is generated, a temperature in the range of 135° to 145°F (such that most pathogens are destroyed)
* A pH of 6.5 to 7.2 (with pH above 8.0, odor becomes a problem)

In many of the United States, composting is recommended and there may be a cost share. However, in other states, it may be restricted, requiring a permit. If done correctly, the process is not associated with odor, fleas, or vermin. However, there is

generation of carbon dioxide and other greenhouse gases.

The following are a set of guidelines for composting, based on the *Arbor Acres Breeders Manual*:

* Animals should be composted within 24 hours of death.
* Carcasses should be fully degraded before removal or land application.
* Runoff water should be avoided (e.g., by covering).
* Composting should not occur close to waterways or wells.
* Composting should be done in a weather-resistant structure.

Other Nuisances

Rodents

Rodents associated with poultry facilities can represent a problem to neighbors. People do not want rats or mice in or near their homes. Rodents are associated with disease and lack of sanitary conditions. Minimizing rodent populations is not only good business practice for the poultry producer but also essential for a good-neighbor policy.

Flies

Flies (e.g., the common housefly—*Musca domestica*) and also blowflies, soldier flies, and others are a nuisance associated with poultry production. Neighbors can easily be disturbed by large numbers of flies coming from poultry houses nearby. Concerns focus on the association of flies with disease, spoilage of food, and the unpleasant sounds made by flies around the house, at barbecues and picnics, and other outdoor activities.

Siting Poultry Facilities

Pollution control is the first and most important requirement in locating a new poultry establishment (Figure 9.7), or in continuing an old one. The location should be such as to avoid:

1. The neighbors complaining about odors and insects.
2. Pollution of surface and underground water.

Without knowledge of pollution control, no amount of capital, native intelligence, and sweat will make for a successful poultry enterprise. Beautification is also helpful (see Box 9.1).

How Environment Affects Poultry

No matter how good the genetics, a good environment is essential to obtain high production

BOX 9.1 Neighbors and Nuisances

Particularly in these days of lawsuits and contentiousness, a good-neighbor policy is the right policy. Odor represents a significant potential problem with neighbors (see the section on Animal Waste).

One approach to having a good-neighbor policy is to plant trees around poultry units (see Figures 9.5 and 9.6). This may provide the advantages of:

- Reducing odor due to removing odor-containing particles and feathers, and/or changing air current to dissipate odor.
- Reducing air and building temperatures.
- Providing a pleasant visual screen (essentially the old adage—"out of sight, out of mind").

In some states, the federal and state governments provide financial support for planting trees around poultry units.

Figure 9.5 Planting trees around a broiler house. *(Courtesy George W. Malone)*

Figure 9.6 Trees around a broiler house providing a visual screen and reducing odor problems by inducing turbulence (mixing of air) and up-draughts of air currents (moving air to above ground height). *(Courtesy George W. Malone)*

and bird comfort. Environmental factors affecting animals include:

1. Nutrition (Chapter 7)
2. Light (Chapter 14)
3. Temperature/weather (Chapter 7)
4. Air quality (Chapter 13)
5. Pathogens (Chapter 11)
6. Stress (Chapter 10)

 USEFUL WEBSITES

Arbor Acres Breeder Manual (2002)

http://www.aaf.com

Figure 9.7 Siting a turkey house. *(Courtesy Gretta Irwin, Iowa Turkey Federation)*

FURTHER READING

Bell, D. D., and W. D. Weaver. *Commercial Chicken Meat and Egg Production*, 5th ed. Norwell, MA: Kluwer Academic Publishers, 2002.

10

Poultry Behavior, Stress, and Welfare

This chapter will briefly cover the behavior of poultry and the issue(s) of poultry welfare. While there has been tremendous progress in the genetic selection of poultry, it is not clear the extent to which essential physiological, behavioral, and adaptive mechanisms (heat, respiration, etc.) have kept pace.

ANIMAL BEHAVIOR

Ethology is the study of animal behavior. We know that birds, including poultry, exhibit complex behaviors. The Austrian zoologist Konrad Lorenz pioneered studies on imprinting with geese. He found that if a newly hatched gosling was exposed immediately after hatching to some moving object, person, or animal, it would adopt that object as its parent-companion.

Welfare and comfort for poultry embraces far more than ventilated housing, along with ample feed and water. Poultry do more than eat, sleep, and produce eggs and/or meat. Selection may provide an answer to correcting behavioral problems; we need to breed poultry adapted to their environment.

Learning and Memory

Studies demonstrate that poultry have a capacity to learn; an obvious example is the location of feed and water. In chicks, the ability to learn improves with age following hatching.

Another example of learning is habituation—getting used to, or ignoring, certain stimuli. An example of habituation is the response of turkey poults to the warning call of the mother when there is impending danger. At first, they "scamper" for cover or to the corners of their pen, pile in clumps, and remain motionless for several minutes. However, they gradually get used to a repetition of this call, with each response becoming less intense.

Memory is the capacity to retain or recall what was learned or experienced (e.g., location of feed, water, and peck order). Animals learn to do some things, yet they inherit the ability to do others. This is often called instinct. Thus, ducks do not have to learn to swim, they instinctively take to water.

Senses

Vision and hearing are the most highly developed senses in the bird. They play a crucial role in social behavior, communication, and responses to predators. Chemical senses are also important.

Vision

Birds have a well-developed sense of vision. The eyes are large compared to the size of the head and brain. The number of optic fibers in the chicken optic nerve is 2.5 times that in man. The position of the eyes in the orbits gives chickens and turkeys a visual field of approximately 300 degrees. Because of the rather flat eyeball, there is little eye movement. Birds can follow moving objects by moving their head and neck. Chickens and turkeys appear to possess color vision.

Hearing

Birds have a very well developed sense of hearing despite having an inconspicuous external ear. They are as sensitive as mammals, including humans. This is not surprising given the importance of "calls" between both birds in the wild and poultry (e.g., birds in the wild establishing territories and "baby" chicks being attracted by maternal calls and repetitive tapping).

Chemical Senses

The ability of an animal to detect chemicals is divided into the following:

- Taste (gustation)
- Olfaction, or smell
- Trigeminal chemoreception, or chemesthesis

TASTE. Taste is the detection of chemicals in direct contact with the taste buds at fairly high levels. Birds have few taste buds. These are located on the base of the tongue and the floor of the pharynx. The total number of taste buds in the chicken is 24 compared with 9,000 in man. The taste buds can detect salt, acid (sour), and possibly sugars. Taste is involved in ensuring the well-being of both chickens and turkeys.

OLFACTION. Olfaction is the ability to detect very low levels of chemicals in the air. The olfactory receptors are in the nasal conchae. The ability of birds to detect odors seems to be comparable with that in mammals. In some bird species, olfaction is extremely important and very sensitive; for example, vultures finding dead animals by the odors.

TRIGEMINAL CHEMORECEPTORS. There is a third chemoreception system. Trigeminal chemorecep-

tors in the mouth detect chemically induced pain. An example of this in humans and mammals is the response to hot peppers (in fact, to the ingredient capsaicin). Thus, capsaicin can be a repellent for mammals. Birds do not avoid capsaicin, although they can be trained to avoid it. This demonstrates that they can detect capsaicin. There is a well-developed trigeminal system in birds. Some aromatic chemicals affect it, leading birds to be averse to these chemicals.

Communication Methods

Communication between animals involves giving by one individual some sign or signal that, on being received by another, influences its behavior. In poultry, these stimuli are auditory and/or visual.

Vocal

Vocal communication is of special interest because it forms the fundamental basis of human language. Sound is also an important means of communication among animals. Vocalizations in chickens are especially important in mother-young relationships. The cock's crow functions as a means of recognition, since a given cock crows much the same, and crows differ greatly between roosters. Also, the duration of crow differs between breeds and is highly heritable.

Among the vocalizations of turkeys are the following:

- The trill emitted during threat behavior and as the vocal component of the male strut.
- Yelps given by hens in attracting the poults.
- Various vocalizations given by the poults—a distress peep, a trilling contentment call, a high-pitched screamlike call when pecked, and the sleepy call just before going to sleep.
- The gobbling call of the male appears to serve as individual recognition within the flock. If one turkey gobbles, usually most of the males in the vicinity follow suit. At the height of the breeding season, a male turkey will respond with gobbling to a wide variety of stimuli.

Visual Displays

Among the important visual displays of poultry are:

- Raised hackles (feathers on the top of the neck) to denote aggressive intentions and make birds look larger and more formidable.
- Low crouch in a submissive response.
- The waltz plus wing-flutter in a displacement reaction. Birds are especially noted for their sexual behavior in the act of courtship.

In chickens, the male typically takes the initiative in sexual behavior. There are a variety of approaches by which a cock/rooster may evoke a sexual response in the hen. The most spectacular of these is the wing-flutter, waltz, or dance. These initial sexual reactions of cocks have been called courting. A hen may be indifferent to courting, or she may respond either negatively or positively. As a negative reaction, she may step aside, walk or run away, or struggle if captured. These types of avoidance of the male by the female may be accompanied by vocalizations varying in intensity from faint screams to loud squawks. A positive reaction to courting takes the form of a crouch, often with head low and wings spread. This behavior has been called a sex invitation, or crouching. The sexual crouch is a strong stimulus for the rooster to mount and tread, particularly when the rooster approaches from the rear. The male stands on the outstretched wings, grasps the comb or hackle of the hen, and moves the feet up and down in a treading manner. Subsequently, the male rears up, the hen moves the tail to one side, and each everts the cloaca as vents meet. The male usually steps off in a forward direction and the hen shakes herself vigorously as she gets to her feet. She may run in an arc and the cock may execute a waltz (also see Figure 10.1).

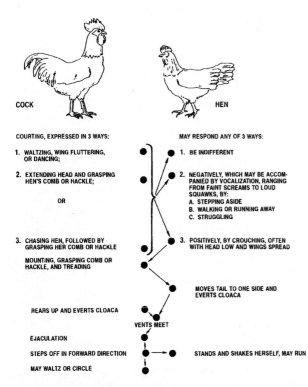

Figure 10.1 Mating behavior in chickens ("the waltz"), showing chain reaction between rooster (left) and hen (right). *(Original diagram from E. Ensminger)*

When attacked by a predator, young poults exhibit a frenzy of vocalization, dart about, freeze momentarily, and dash again in all directions.

Territorial Behavior

Territoriality often results in wide spacing of individuals in natural species, and elaborate behavioral patterns have evolved to maintain it. When individuals are crowded together, as is common with poultry, the behaviors associated with territory must be modified. Long-term modification of poultry can occur in display, threshold to fighting movements, or value of the birds. Short-term modification can occur through beak trimming.

How Poultry Behave

Poultry exhibit multiple behaviors, some of which are discussed here.

Behaviors Associated with Feeding

Genetically increasing feed intake is said to be the primary cause for rapid improvement in growth of broiler chickens. Food intake is regulated in poultry as it is other animals by specific areas in the hypothalamus within the brain. The amount of feed consumed is controlled by:

- Appetite (increasing feed intake)
- Satiety or fullness (decreasing feed intake)

There are multiple factors that influence feed intake including:

- Crop and gizzard fill or distension (reducing feed intake) (e.g., influenced by the hormones, ghrelin and cholecystokinin)
- Hormones from the intestine and adipose tissue (leptin) (reducing feed intake)
- Nutrients (e.g., sugars) acting on the brain or elsewhere (reducing feed intake)
- A complex series of peptides and other neurotransmitters in the hypothalamus

Free-range chickens may spend about half their time foraging and feeding (with ~ 14,000 pecks at food per day). Each species has its own particular method of ingesting feed. Chickens and turkeys ingest their feed by pecking, while ducks scoop their feed with their broad, soft bills. Chicks do not peck much until their second day after hatching, presumably due to digestion of yolk since the yolk sac is resorbed during and following hatching. Normal pecking experience requires some light. Initially, chicks peck and ingest both nutritive and nonnutritive substances. Chickens feed predominantly during the day (hours of light) with increased feeding prior to the beginning of the night.

The number of birds feeding at any given time is influenced by:

- Dominance relations
- Appetite
- Feeder space

After having been fed, a dominant bird may return when subordinates begin feeding, thereby increasing its consumption and potentially reducing the feed consumption of those in the lowest rank.

Chickens prefer crumbles. Low-fiber, high-energy pelleted rations make for maximum feed efficiency and growth of turkeys. But, because such rations are consumed rapidly and the birds have more idle time, they predispose the flock to feather picking (denuded backs) and so forth.

Sexual Behavior

Reproduction is the most important requirement of livestock production. Without young being produced, other economic traits are of little interest. Thus, it is important that those who breed animals have a working knowledge of sexual behavior. Sexual behavior involves courtship and mating and is under hormonal control. Each species has a specific sexual behavior to prevent interspecies mating.

> **Mating behavior in chickens.** Chickens are polygamous. Sexual behavior is referred to as mating behavior. Mating in chickens is preceded by various behavior patterns known as displays or courting, which synchronize sexual activities of males and females (see Figure 10.1). Preferential or nonrandom mating has been observed in both roosters and hens. Additional information on the sexual behavior of chickens follows:
>
> 1. Both early social experience and hormonal level influence the level of sexual behavior.
> 2. The earlier the separation of males from females, the less the sexual performance of males.
> 3. The relation of sex hormones and sexual behavior is established.
> 4. When several roosters are placed with a flock of hens, the dominant one (the one ranking highest in the peck order) is usually most successful in mating, fertilizing a large number of eggs, and siring the most chicks.

Figure 10.2 Courtship display of the male turkey ("strutting"). This display is characterized by slow and restricted movements and an elaborate feather display. *(Courtesy, Watt Publishing)*

> 5. Dominance among hens can interfere with mating.
> 6. There are differences in mating frequency of males belonging to different sire families.

Mating behavior in turkeys. The mating behavior in turkeys tends to follow a chain reaction similar to chickens. That is, the behavior of one sex partner elicits a specific response from the other, and, in turn, that response elicits a further response from the first partner. However, the following differences exist between the sexual behavior of chickens and turkeys:

1. The movements of chickens during courtship and mating are more rapid and the feather display of the male is less elaborate than in turkeys (see Figure 10.2).
2. Male chickens may force matings, but this behavior is not seen in turkeys.
3. Turkey females sometimes follow the male and crouch near him, whereas female chickens seldom approach the male before crouching.
4. Although males of both species move away if another male attempts to mount, roosters may tread each other, whereas male turkeys will not.

Parental Behavior (Maternal Behavior)

Parental behavior is largely confined to females among poultry, with the care-seeking behavior being normal for young birds.

Modern incubators have replaced the setting hen and precluded the need for broody hens. As a result, breeders have increased egg production by selecting against this trait. Few hens and chicks are allowed to run together. Nevertheless, when this happens, maternal behavior can be intense. For example, hens hover over their chicks by covering them with their spreading wings and nestling them close to their bodies during the night or at other periods when they rest or need protection. A hen with chicks exhibits a definite antagonistic behavior and will attack any enemy that bothers her young. To warn her chicks of danger, she emits a loud, shrill cry. The chicks react quickly and seek protection.

Parental care is usually absent in the male. Domestic hens adapt to laying in artificial nests, although some individuals must be trained to use them. The termination of laying and the start of incubation are closely related. Decreased levels of broodiness are desirable under domestication. Fortunately, selection against broodiness is very effective.

Aggressive or Agonistic Behavior

FIGHTING IN CHICKENS. This type of behavior includes fighting and threatening behaviors. Among all species, males are more likely to fight than females. Nevertheless, females may exhibit fighting behavior under certain conditions. Castrated males are usually quite passive, which indicates that the hormone testosterone is required for these behaviors. Fighting rarely results in death; it usually continues until one gives up. In chickens, agonistic behavior includes attack, escape, avoiding, and submissive behavior.

Cockfighting is illegal in most of North America and Western Europe. However, it was widespread through human history and still occurs in some countries. The ancient Greeks regarded the cock as the symbol of pugnacity. Legend has it that the sight of two fighting cocks had emboldened the Athenians to take up the struggle with the Persians. In commemoration, annual cockfights were held in Athens and other Greek cities. The Romans added iron spurs to the sharp claws of the cocks so that a fight became a lethal business.

Peck Order. When a number of strange hens are placed together in a pen, fights or threatening occurs to establish a dominance order or "peck order." The "winner" of each contest has the "right" to peck the loser, and the latter usually avoids the former. Some individuals give way without a fight, whereas others may challenge the winner again and again before dominance is settled. At subsequent meetings, one member of each pair pecks or threatens the other, and definite dominance-subordination patterns (the peck order) are established. Submission in hens is characterized by crouching. Fighting between hens is much less serious than between roosters.

Where roosters are run with a flock of hens, fighting may continue for several days with both combatants covered with blood. Eventually, one gives up and escapes. A mixed flock of males and females has two peck orders, one for each sex. Factors influencing the peck order of chickens are aggressiveness, appearance, level of gonadal hormones, experience, and breed differences.

Breed Differences. There are great breed differences in aggression in chickens, with meat breeds being quite placid.

AGGRESSIVE BEHAVIOR IN TURKEYS. Fighting often occurs when two strange turkeys meet for the first time, and the winner of the fight subsequently dominates the loser. The aggressiveness of male turkeys depends on their reproductive condition. Turkeys on short-day lengths are quiescent reproductively and show low aggressiveness. Turkeys on long-day lengths are reproductively active and can show tremendous aggressiveness. Groups of young turkeys raised together establish their social hierarchies at 3 to 5 months of age.

Turkeys exhibit a variety of threat displays, ranging from mild threat (head raised, looking at the opponent) to strutting (against other males). During the threat display, the birds vocalize in a distinctive trill of relatively high pitch. Actual fighting begins by both birds jumping at each other with their feet extended forward. If one bird lands a stroke on the other's back, the latter gives up. Following from 1 to 20 jumps, combat shifts to a tugging battle, in which the head is darted forward to grasp the caruncles, snood, wattle, or beak of the opponent. Simultaneously, they tug and push in an attempt to force the opponent's head downward. A fight is terminated by submission, with the loser retracting its snood, lowering its head, attempting to hide under the victor's breast, and fleeing. The winner may chase and peck the defeated bird.

Behavioral Norms

The poultry producer needs to be familiar with behavioral norms of birds in order to detect and treat abnormal situations, especially illness.

Some of the behavioral signs of good health are:

1. Alertness
2. Eating

Thermoregulatory Behaviors

The high deep-body temperature, the absence of sweat glands, and the very effective insulation provided by the plumage characterize avian thermoregulation. Panting is the main method of evaporative water and heat loss in birds. The most conspicuous thermoregulatory behavior of birds is movement to warmer or cooler areas.

Poultry also respond to thermal stress by the following:

- Movement from a heat source
- Increasing respiration rates (to assure evaporation cooling of the air sac system and oral mucosa surfaces)
- Decreasing feed consumption
- Changed body position—spreading their wings as they crouch on the ground (so that air can circulate past the less insulated undersurface of the wings, and so that squashing of the breast feathers facilitates heat loss to the soil), and opening the mouth and panting

In a cold environment, the chicken reduces surface area, and hence its heat loss, by hunching. An additional reduction in heat loss, amounting to 12%, may be achieved by tucking the head under the wing. A still further savings of 20 to 50% can be made if the chicken sits rather than stands, thereby reducing the heat loss from the unfeathered legs and feet.

It should be noted that poultry are exposed to daily changes in temperature and humidity even in the best environmentally controlled housing.

Dust Bathing

If litter or other loose materials are available, chickens and turkeys show "dust bathing" behavior. This removes excess oil from the feathers. Adult hens can "dust bathe" for about 30 minutes on alternate days. Ducks and geese groom by shaking movements to remove water from the feathers, distributing oil to the feathers from the uropygial gland above the tail, and wetting the feathers during bathing. These activities are necessary to keep the feathers in good condition, preserving waterproofing and thermal insulation.

Sleeping

Normal behavior in sleep should be recognized, especially since it differs widely between species. Chickens and turkeys wind their claws tightly around a pole, or roost, and snuggle closely together. They like to close their eyes and hide their heads in the feathers of their wings. Ducks and geese sleep on both land and water. On land they often drowse while standing on one leg; on water they paddle every now and then, in order not to drift ashore. The eyes frequently blink open, and sleeping birds can become fully alert instantly if disturbed.

Behaviors Associated with Oviposition/Egg Laying

Prior to egg laying, hens may show a variety of behaviors including:

- Nest building (rudimentary)
- Restlessness, including pacing
- Vocalization (prelaying calls)

Undesired Behavior

Abnormal behavior can provide a way in which to recognize diseases early. Sick birds usually eat less, may be dull and inactive, and may isolate themselves from the rest of the flock. Layers produce fewer eggs, and fertility and hatchability of eggs decline.

Feather Picking

Poultry can pull or peck at feathers of other individuals. This vice can lead to cannibalism for dead or dying birds. It can be controlled by:

- Beak trimming
- Reduced light intensity
- Reduced population density

Cannibalism

Cannibalism may be encountered among birds of all ages. Among "baby" chicks, the trouble is usually confined to toe and tail picking. With mature birds, the vent, tail, and comb are the regions most frequently picked.

The cause of cannibalism is not fully understood. It is more frequent under confined conditions. Without a doubt, it may be accentuated by deficiencies in management and nutrition. Also, it may be caused by too high light intensity. The best way to control cannibalism is by beak trimming.

Stereotypies

Stereotypies are repetitive behaviors. Chickens confined to small cages in laying batteries will develop stereotyped head movements.

Flightiness

Another example of abnormal behavior is flightiness. Flightiness occurs particularly in Leghorn chickens. In response to noise or sudden movements and high light intensity, birds may "jump" on top of each other resulting in injury and death.

Social Relationships

Social behavior may be defined as any behavior caused by or affecting another animal, usually of the same species.

Social organization may be defined as an aggregation of individuals into a fairly well-integrated and self-consistent group in which the unity is based upon the interdependence of the separate organisms and upon their responses to one another. The ancestral form for domestic chickens is the jungle fowl. The social unit is a rooster with 4 to 12 adult females and immature offspring. The females have a social hierarchy or pecking order.

When we restrict or confine birds and force them into spaces that bring them within the individual distance that has been established, we may create stress. Thereupon, the dominants have to pay more attention to maintaining their dominance and to protecting their own field of territory. They may be more aggressive. The subordinates become more nervous and this can spread throughout the group.

Social Interaction

If not too crowded, placing poultry together can accomplish two things: (1) greater feed consumption, due to the competition between birds (mutual facilitation), and (2) a quieting effect.

STRESS

Stress is physiological, physical, or psychological tension or strain. Stress affects birds. Social stresses are those changes in social behavior and population density that may influence growth and reproductive performance. Among the environmental factors that stress poultry are temperature, nutrition, disease, space per bird, social stress, gathering for market, and transportation. An example of social stress is mixing unfamiliar chickens together. This is not only associated with aggressive behavior but also increases in physiological indicators of stress, such as the adrenal hormone, corticosterone. Moreover, there are other changes in behavior such as decreased pecking at feed. Generally, stress is associated with physiological changes such as:

- Increased blood levels of the classic stress-related adrenal hormone, corticosterone
- Increased blood levels of epinephrine (adrenaline) and norepinephrine (noradrenaline)
- Increased heart rate

Moreover, there are behavioral changes such as immobility.

Stress activates the *hypothalamo-pituitary-adrenal axis:*

Hypothalamus
   ⇓ Corticotropin-releasing hormone (CRH)
Anterior Pituitary Gland
   ⇓ Adrenocorticotropic hormone (ACTH)
Adrenal Cortex
   ⇓ Corticosterone
Physiological Changes

The corticosterone orchestrates the physiological changes necessary to adapt to the stress. For instance, corticosterone increases blood glucose and liver glycogen.

WELFARE

Today, there is renewed interest in the study and application of animal behavior; we are trying to make it right with animals. For the time being, this calls for emulating the natural conditions of the species, including their space requirements, social organization, and training and experience. It calls for breeding and selecting animals better adapted to artificial environments. It is hoped that the principles and applications of animal behavior presented in this chapter will speed the process.

It should always be the case that farmers and producers are concerned for the welfare of their animals. Animal and poultry scientists have repeatedly demonstrated that a severely stressed animal will not show good performance. Hence, it is not in the economic interest of the producer to have stressed animals. Moreover, virtually all producers are humane. There are considerable changes in the standards for raising and keeping poultry and livestock. A growing, but still relatively small, percentage of the public are concerned about the welfare of food animals reflecting adverse publicity on "factory farms" and "battery" hens. Our knowledge of the science of behavior of animals is also impacting the discussion.

Welfare and Poultry Production in Europe

In Western Europe, there has been increasing concern for animal welfare by the public.

Welfare and Poultry Production in the United Kingdom

In 1979, the British government established an advisory body on the welfare of livestock and poultry. The Farm Animal Welfare Council includes producers/farmers, veterinarians, animal scientists, animal behaviorists, ethicists, and members of animal welfare groups. The council has stated that animals kept by man must be protected from unnecessary suffering. This encompasses both physical and mental well-being. The council has established *Five Freedoms* as the framework for establishment of standards of animal management/husbandry, environment (e.g., caging), handling, and humane slaughter. These Five Freedoms are:

1. Freedom from hunger and thirst
2. Freedom from discomfort
3. Freedom from pain, injury, or disease
4. Freedom to express normal behavior
5. Freedom from fear and distress

In the United Kingdom, the Royal Society for the Prevention of Cruelty to Animals (RSPCA) has supported these Five Freedoms by a farm monitoring coupled with marketing labels of foods that meet the specific standards. The labels are "freedom food."

Welfare and Poultry Production in the European Union

The European Union has established rigorous welfare standards for cages for laying hens. Conventional cages (also referred to as "barren" cages) will be banned after January 1, 2012, and no new conventional cages can be installed after January 1, 2003. The Directive allows producers to still use "enriched" cages. These have more space per hen together with a nest, a perch, and litter. Germany is going even further by banning enriched cages, and the United Kingdom is also considering this.

The European Union's welfare standards will also ban beak trimming, but because of concerns of feather pecking and cannibalism, this will not be implemented until January 1, 2011. However, some European governments are considering earlier implementation. For instance, the British government has a code of practice that encourages removal of the sharp beak tip ("beak tipping") rather than trimming the beak. See Figure 10.3 for mechanical harvesting of chickens with reduced stress.

Welfare and Poultry Production in the United States

In the United States, the poultry industry is addressing the issue of welfare. Traditionally, the industry has not liked the term *welfare*, preferring *bird comfort*. Obviously, for maximum productivity, the birds' physical needs are being met through the following:

- High-quality feed, supplying all known nutrients
- Clean water (checked regularly for impurities)
- Excellent management and care, with heat stress being reduced by tunnel ventilation (heat during transportation is still an issue, though)
- Unsurpassed poultry health programs

Figure 10.3 Mechanical harvester for broiler chickens. The harvester has rubber-"finger like" rotors to move birds into cages for transportation. The operator drives the harvester slowly into groups of birds in the grow-out house.
(Reprinted from Lacy, M. and Czarick, M. (1998). Mechanical harvesting of broilers. Poultry Science *77, 1794–1797 with permission from* Poultry Science*).*

Broiler chicks in the growing stages have the ability to exhibit behaviors such as eating, drinking, preening, and sleeping and are said to be free from the following:

1. Fear
2. Frustration
3. Pain

Some U.S. broiler companies have established the following to ensure welfare:

1. Company-wide poultry welfare officers
2. Poultry welfare councils with outside members
3. Standard operating procedures (SOPs) for brooders, growers, hatchers, and others.

In the United States, rigorous animal welfare requirements are increasingly being set by the food industry after pressure from animal rights activists. For instance, McDonald's has stated its commitment to animal welfare and the humane treatment of animals. A scientific advisory committee has also been established.

Goals for the humane treatment of laying hens have been established. Space requirements are 72 sq. in. per bird together with 4 in. of feeder. Feed and/or water withdrawal to force molt has been prohibited. Beak trimming is not supported. Inspections are carried out. Suppliers not in compliance with the company's requirements have 30 days to correct deficiencies or cease to be suppliers. Similarly in 2001, Burger King announced that it was requiring its suppliers to adhere to strict welfare requirements for the care, housing, transportation, and slaughter of poultry and other livestock. This is being "policed" by both announced and unannounced inspections or audits of facilities.

Food Industry Standards for the Welfare of Poultry

In 2002, the U.S. food industry (the Food Marketing Institute and National Council of Chain Restaurants, together with two major grocery chains) announced the introduction of standards for animal welfare. For laying hens, these are the United Egg Producers' (see Figure 10.4) practices and they include:

1. Increased cage space
2. Break trimming only to avoid cannibalism
3. Fresh feed and water requirements
4. Air ventilation requirements
5. Handling and transportation standards
6. More humane molting procedures
7. Daily inspection of birds to assure well-being

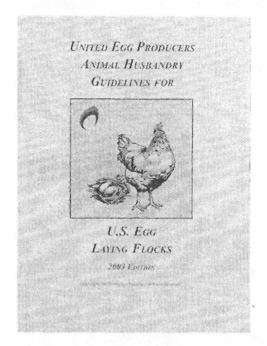

Figure 10.4 United Egg Producers' guidelines for egg layer welfare.

FURTHER READING

Honore, E. K., and P. H. Klopfer. *A Concise Survey of Animal Behavior.* San Diego, CA: Academic Press, Harcourt Brace Jovanovich, Publishers, 1990.

Sturkie's Avian Physiology, 5th ed. Edited by G. C. Whittow. San Diego, CA: Academic Press, 2000.

11

Poultry Health

Objectives

After studying this chapter, you should be able to:

1. Understand the importance of diseases (pathogenic, parasitic, and metabolic) to poultry.
2. List disease organisms and the diseases that are caused by them.
3. Know what coccidiostats are used for.
4. List the major components of the bird's immune system.
5. Understand how diseases are spread.
6. Give the reasons for biosecurity programs.
7. List the components of biosecurity programs.
8. Give the reasons for vaccination.

Healthy birds are a requirement for profit; unhealthy birds cause financial losses. This is well recognized by poultry producers. Deaths and condemnations result in economic losses. Animal health programs have reduced mortality and poor performance (growth, etc.) in diseased poultry. In the 1920s, the mortality of growing poultry was about 18%, but by the 1970s, this had fallen to about 5%.

Total mortality = mortality + number culled
("On-farm condemnations" as diseased + moribund)

Even greater losses come from decreased egg production and growth, lowered feed efficiency, and so on. Disease prevention is critical to the success of the poultry industry in a community, a region, or the nation.

NORMAL MORTALITY AND MORBIDITY LOSSES

In the broiler industry, mortalities run about 5% for the entire growing period. Increases above this indicate a problem. A rule of thumb regarding the evaluation of disease problems is if >1% of the birds are sick.

MONITORING HEALTH

It is important to know the normal to recognize the abnormal. Signs of good health or disease are the following:

- Amount of feed and water consumption
- Rate of growth/egg production
- Normal-appearing droppings
- Physiological indicators: temperature 106°F (range 105°–107°F) (41°C), pulse rate (200–400 beats per minute), and rate of breathing (15–36 breaths per minute)
- General sounds and activity of healthy birds compared to droopiness and ruffled feathers in diseased birds
- Watery excreta (diarrhea)

Symptoms of disease can be general (see above). Other symptoms are specific, seen only with certain diseases (e.g., discharge from the respiratory tract). There are also pathological changes in the structure (color, shape, etc.) of an organ. These are macrolesions (large changes obvious to the trained eye) or histological microlesions (small changes only observable using a microscope).

CAUSES OF AVIAN DISEASES

Avian diseases are due to the following causes:

1. Pathogenic Microorganisms (viruses, bacteria, and fungi) and parasites
2. Nonpathogen-caused diseases including:
   - Nutritional deficiencies (very rare in commercial poultry)
   - Genetic diseases (while very rare in commercial poultry, genetics can predispose commercial poultry to metabolic, pathogenic, and environmental diseases together with behavioral problems)
   - Metabolic diseases
   - Environmental diseases (predisposing poultry to pathogen invasion and/or metabolic diseases)
   - Behavioral problems (predisposing poultry to pathogen invasion)

This chapter will address pathogenic and nonpathogenic diseases, metabolic diseases, and environmental and behavioral problems in poultry.

PATHOGENIC OR INFECTIOUS DISEASES

Pathogenic microorganisms in poultry are viruses, bacteria, and fungi. In mammals, prions also cause disease. Stress (e.g., heat or cold, poor ventilation, overcrowding) predisposes poultry to disease. When microorganisms enter the cells of the body and multiply, they either disturb cell functioning directly or their toxins (poisons) affect the animal.

The ability of an organism to cause disease in a host is its virulence or pathogenicity. This can be altered (drastically reduced) to develop vaccines. Variation in pathogenicity also partially explains why the same disease may present different forms and degrees of severity.

Infectious diseases of poultry are usually contagious. A contagious disease is one transmitted from one animal to another within a flock or from one flock to another or from or to other animal species. A zoonotic disease is transmitted from animals to people. Infectious diseases are considered under viral, bacterial, and fungal diseases. Examples of poultry infected with various diseases are shown in Figures 11.1 to 11.9.

Viral Diseases

There are a large number of viruses that cause diseases in poultry. See Table 11.1 for a comprehensive list. Viruses may contain either DNA or RNA as their genetic material. Viruses are so small that they can-

Figure 11.1 Ducklings with *duck viral hepatitis (DVH)*. Symptoms: Ducklings lie on side with heads drawn back and paddle feet spasmodically. *(Copyright © Fort Dodge Animal Health and used with permission)*

Figure 11.2 Epidemic tremor (avian encephalomyelitis), a viral disease that can kill chicks and cause a lowering of egg output in laying flocks. Note that the chicks rest on their haunches or on their sides, and appear too weak to move about. The lack of muscle coordination prevents them from reaching feed and water. *(Copyright © Fort Dodge Animal Health and used with permission)*

Figure 11.3 Fowl cholera, an advanced, localized case. The swelling of the wattle is caused by accumulation of cheesy pus. If opened, these swellings have a foul odor. *(Copyright © Fort Dodge Animal Health and used with permission)*

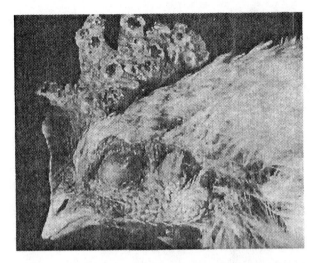

Figure 11.4 Chicken showing severe gross lesions of fowl pox. *(Reprinted from* Diseases of Poultry, *10th edition, edited by B. W. Calnek, with permission from Iowa State Press)*

Figure 11.5 Laryngotracheitis (gapes), a highly contagious viral disease. The gasping or cawing position is typical. Note the dried blood around the nostril and lower beak. *(Reprinted from* Diseases of Poultry, *10th edition, edited by B. W. Calnek, with permission from Iowa State Press)*

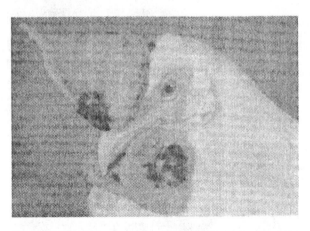

Figure 11.6 White Leghorn chicken with an experimental infection with avian influenza. Note the lesions on the comb and wattle. *(Reprinted from* Diseases of Poultry, *10th edition, edited by B. W. Calnek, with permission from Iowa State Press)*

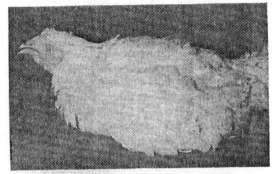

Figure 11.7 Chick with botulism. This is caused by the endotoxin from *Clostridium botulinum*. Note partial paralysis of the wing and lower eyelid, difficult breathing, and ruffled feathers. *(Reprinted from* Diseases of Poultry, *10th edition, edited by B. W. Calnek, with permission from Iowa State Press)*

Figure 11.8 Chick with aspergillosis (thrush). Note severe conjuntivitis. *(Reprinted from* Diseases of Poultry, *10th edition, edited by B. W. Calnek, with permission from Iowa State Press)*

not be seen through an ordinary microscope (they can be seen by using an electron microscope), they are capable of passing through the pores of filters that retain ordinary bacteria, and they propagate only in living cells.

Respiratory diseases affect the air passages, lungs, and air sacs and can be caused by viruses, for example, avian influenza, infectious bronchitis, laryngotracheitis (see Figure 11.5), and Newcastle

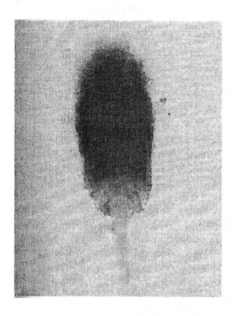

Figure 11.9 Organism causing aspergillosis in poultry—*Aspergillus fumigatus. (Reprinted from* Diseases of Poultry, *10th edition, edited by B. W. Calnek, with permission from Iowa State Press)*

disease. Other viral diseases produce lesions. Among them are some of the most devastating diseases of chickens and turkeys such as avian encephalomyelitis (epidemic tremor) (see Figure 11.2), avian pox, leukosis, and Marek's disease.

Avian Influenza

Avian influenza (previously known as fowl plague) is a significant problem for poultry and game birds. It is caused by many type A influenza viruses (*Orthomyxoviridae*), which are RNA viruses. There are two forms that affect poultry: highly pathogenic avian influenza and low-pathogenic avian influenza.

Highly pathogenic avian influenza is not endemic in the United States and responds to eradication programs including flock depopulation. However, low-pathogenic avian influenza is a chronic problem. In the United States, the federal government (USDA) indemnifies producers for up to half the losses after infected birds are euthanized. There is a waiting period of 7 days, which includes cleaning and disinfecting facilities, before repopulation with birds following depopulation.

Avian Leukosis

Avian leukosis includes Marek's disease.

MAREK'S DISEASE. Marek's disease was relatively common, resulting in >1% condemnations of broiler chickens in the 1970s. With vaccination, its impact has dropped 50-fold. The disease involves abnormal proliferation (cell division) of lymphocytes

and their infiltration of organs such as the skin, muscle, and intestines. The causative agent is the Marek's disease virus (MDV 1), a DNA herpes virus.

Avian Pneumovirus

The avian pneumovirus (an RNA virus related to the Newcastle disease virus) causes:

- Swellhead syndrome (chickens)
- Rhinotracheitis (turkeys)

Symptoms of avian pneumoviruses include upper respiratory tract fluid secretion due to inflammation, with the virus stimulating cytokine release. In addition, interferons are released, which impair virus replication.

Newcastle Disease Virus

Newcastle disease was first observed in England and Indonesia in 1926. It is a disease affecting the respiratory and gastrointestinal tracts and nervous system. It is widespread (endemic) in Africa, Asia, Europe, and South America with sporadic outbreaks in North America. Live vaccines are used with attenuated or avirulent viruses. Exotic Newcastle disease is not endemic in the United States.

The Newcastle disease virus (paramyxovirus, or PMV 1) comprises six genes encoded in RNA. When an avian cell is infected, the RNA is translated to form viral proteins. Later, an antigenomic template RNA is formed as an intermediate for viral RNA replication.

Infectious Bursar Disease

Infectious bursar disease is a highly contagious disease that can cause rapid death or marked reduction in growth in young chickens. This is caused by an RNA virus—infectious bursar disease virus (IBDV 1).

Emerging Diseases

New and emerging diseases include the following:

- Multicausal respiratory disease
- Viral enteritis (in turkeys and other poultry)
- Infectious stunting-runting-hypoglycemic-spiking mortality syndrome in chickens and turkeys
- Hydropericardium-hepatitis syndrome

Bacterial Diseases

Bacteria are the simplest forms of single-celled life. All bacteria are not detrimental to health. Some may be involved in digestion. Bacteria can be classified as pathogens (disease-producing organisms) or harmless or beneficial.

Successful control of bacterial diseases entails preventing bacterial multiplication in the bird and

TABLE 11.1 Nonnutritional Diseases of Poultry

Disease	Species Affected	Cause	Symptoms and Signs	Postmortem
Arizonosis (avian arizonosis)	Chiefly turkeys.	Gram negative bacterium (*Salmonella arizonae*).	Acute septicemia. Poults unthrifty, and may develop eye opacity/blindness. Mortality usually confined to first 3 to 4 weeks of age.	Yolk sacs slowly absorbed. Liver enlarged and mottled. Infected intestinal tract.
Aspergillosis (brooder pneumonia)	Almost all birds and animals, including humans.	Mold (*Aspergillus*)	Fever, difficult breathing, nervous symptoms.	Nodules in lungs and air sacs, pus in air sacs.
Avian encephalomyelitis (AE) (epidemic tremor)	All birds.	RNA Virus. (*Avian encephalomyelitis virus, AEV*)	Unsteadiness, sitting on hocks, inability to move, muscular tremors of head, neck, and limbs.	No gross lesions of the nervous system seen.
Bluecomb disease (coronaviral enteritis)	Turkeys.	RNA Virus. (*Coronviral enteritis virus*)	Poults appear cold and seek heat, stop eating and lose weight, and have frothy or watery droppings. In growing turkeys, the appearance of the disease is sudden with a concurrent drop in feed and water consumption. Sick birds show darkening of the head and skin.	Birds show few lesions. The contents of the duodenum, jejunum, and ceca are watery and gaseous.
Botulism (limberneck; food poisoning; Western duck sickness)	All birds except vultures.	Toxin produced by the anaerobic bacterium *Clostridium botulinum*.	Convulsions, paralysis, and sudden death.	Enteritis.
Chronic respiratory disease of chickens (Infectious sinusitis of turkeys) (also see mycoplasmosis)	Chickens and turkeys.	Bacterium *Mycoplasma gallisepticum*.	Coughing, gurgling, sneezing, nasal exudate, slow spread, loss of weight.	Mucus in trachea, air sacs thickened and containing yellow pus, thickened membrane over heart.
Coryza (roup) (Infectious coryza)	Primarily chickens.	Gram negative bacterium (*Hemophilus paragallinarum*)	Gasping, swollen eyes, nasal discharge, offensive odor.	White to yellow pus in eyes and sinuses.
Duck virus enteritis (duck plague)	Ducks, geese, and swans.	DNA Virus. (*duck enteritis virus—a herpes virus*)	Spreads rapidly. Can cause heavy mortality. Affected birds reluctant to walk.	Hemorrhages of internal organs.
Duck hepatitis	Ducks.	Virus. (*duck hepatitis virus type 1, 2, or 3*)	Acute disease. Especially affects ducklings up to 3 weeks of age. Mortality ranges from 5 to 90%.	Characteristic liver lesions.
Epidemic tremor (avian encephalomyelitis)	All birds, but primarily young chickens.	RNA Virus.	Tremors of head and neck, muscular incoordination. Temporary drop in egg production of layers.	None.
Erysipelas	Turkeys primarily, but other fowl affected. Also humans and swine.	Gram positive bacterium (*Erysipelothrix rhusiopathiae*).	Sudden losses, swollen snood, discoloration of parts of face, droopy.	Hemorrhages in muscles, mucus in mouth; reddened intestines.
Fowl cholera (avian cholera, avian pasteurellosis, avian hemorrhagic septicemia)	Chickens, turkeys, water fowl, and other birds.	Gram negative bacterium (*Pasteurella multocida*). (also known as *P. septica*)	Fever, purplish head, greenish-yellow droppings, sudden deaths.	Enlarged liver, hemorrhages in heart and in other organs.
Fowl pox (avian pox)	Chickens, turkeys, and other birds.	Avian pox viruses are DNA viruses of the *Avipoxvirus* genus	Small clear to yellow blisters on comb and wattles that soon scab over; decreased egg production.	May have lesions in throat and trachea.

(text continued on page 172)

Distribution and Losses Caused by	Treatment	Prevention	Remarks
Widespread.	Chemotherapy may reduce losses in acute outbreaks of avian arizonosis.	Elimination of infected breeder flocks. Hatchery fumigation and sanitation.	It is an egg-transmitted infection.
The incidence of the disease is not great.	No treatment. Remove source of infection.	Avoid musty and moldy feed and litter, provide good ventilation.	Young fowl most susceptible.
Worldwide.	Affected birds are usually destroyed.	Vaccination of breeder pullets at 10 to 15 weeks of age, repeated at molting if held second year.	Differentiated from vitamin E deficiency by history, signs, and histological study.
Heaviest losses are in condition and production, but death losses may be high in young turkey poults.	Antibiotics or nitrofurans according to directions may be helpful in reducing the mortality.	Depopulation and decontamination of turkey buildings and surrounding areas with a rest period before restocking.	Turkeys that recover from the disease are immune to challenge, but remain carriers for life.
Botulism occurs worldwide.	If isolated and provided with water and feed, many sick birds will recover. Move birds to clean facilities. Antitoxin may be used on valuable birds, but it is difficult to obtain and expensive.	Do not feed spoiled or decomposing feed. Promptly dispose of dead birds and rodents. Avoid wet spots in litter.	The toxin is very potent, being 17 times more deadly than cobra venom for the guinea pig.
Worldwide.	Antibiotics in feed or water according to directions. In severe outbreaks, inject birds with appropriate antibiotics.	Secure mycoplasma-free stock.	*Mycoplasma gallisepticum* (MG) infection is commonly designated as chronic respiratory disease (CRD) of chickens and infectious sinusitis of turkeys.
Worldwide.	Various sulfonamides or antibiotics used according to directions.	Keep age groups separate. Periodic complete depopulation.	Do not expose susceptible birds to recovered birds. Latter are lifetime carriers.
Worldwide.	No successful treatment.	Different kinds of vaccines give variable results.	This is a reportable disease.
Worldwide.	Administration of serum from immune ducks.	Vaccination of breeder ducks and vaccination of young ducklings. Strict isolation.	
Worldwide. Morbidity in affected flocks averages 5 to 10%	No satisfactory treatment is known.	Vaccination of breeders.	Vaccination of commercial laying flocks is of questionable value.
The disease is of economic concern to turkey growers throughout the world.	Use antibiotics according to recommendations.	Vaccinate.	Transmitted via wounds or skin abrasions.
Worldwide. At times it causes high mortality; at other times the losses are nominal.	Sulfonamides and antibiotics according to directions.	Sanitation, disposal of sick birds, isolation of new stock. Vaccination Commercially produced bacterins and live vaccines are available.	When there is an outbreak, work fast in treatment.
Worldwide. Mortality is not high. Economic loss is in reduced feed efficiency and production.	Treatment is of little value.	Vaccination.	Control mosquitoes.

(*continued*)

TABLE 11.1   Nonnutritional Diseases of Poultry (*continued*)

Disease	Species Affected	Cause	Symptoms and Signs	Postmortem
Fowl typhoid	Chickens, turkeys, ducks, pigeons, and pheasants.	Gram negative bacterium (*Salmonella gallinarum*).	Inactive, fever, greenish-yellow droppings.	Liver and spleen enlarged, bronze or greenish colored liver with some lesions.
Gumboro (infectious bursal disease)	Chickens.	RNA Virus. (*Infectious bursal disease virus*)	Sleepy. White, watery diarrhea.	Enlarged bursa, hemorrhages.
Hemorrhagic enteritis (enteritis)	Turkeys.	DNA virus (*adenovirus*)	Usually the only sign is one or more dead birds.	Severe hemorrhagic inflammation of intestinal lining from gizzard to ceca.
Infectious bronchitis	Chickens only.	RNA Virus. (*Infectious bronchitis virus*)	**Young birds:** Gasping, wheezing, nasal discharge. **Older birds:** Sharp and prolonged drop in egg production, and soft-shelled eggs.	Yellowish mucus or plugs in lower trachea and air passage of lungs.
Infectious synovitis (*Mycoplasma synovial* infection)	Chickens and turkeys.	*Mycoplasma synoviae.*	The disease occurs primarily in growing birds from 4 to 12 weeks of age. Enlarged hocks, foot pads, lame. Breast blisters.	Exudate at joints. Enlarged liver, spleen.
Laryngotracheitis (grapes, chicken flu)	Chickens.	DNA Virus. (*Laryngotracheitis virus*)	Gasping, coughing, loss of egg production, soft-shelled eggs, extending of neck outward on inhalation and slumping on exhaling, weeping of eyes.	Blood-stained mucus in trachea.
Leukosis (big liver disease; lymphoid leukosis)	Chickens, turkeys, and other fowl.	Retroviruses (RNA virus with DNA provirus during replication)	Loss of weight, diarrhea, thickened bones, gray eyes.	Enlarged liver and spleen, tumors in various parts of body. Benign or malignant neoplasms
Marek's disease (range paralysis, acute leukosis)	Chickens and other fowl.	DNA virus (*Herpes virus*; *Marek's disease virus*/ *MDV*)	Sudden death, loss of weight, diarrhea, paralysis of legs or wings. Skin lesions in young birds. May occur as early as 5 to 8 weeks of age.	Enlarged liver, spleen, kidney, ovary, and testicles; or nodular tumors on these organs. Enlarged nerves in wings or legs. Skin lesions in young birds.
Mycoplasmosis	All poultry.	Primarily the following three: *Mycoplasma gallisepticum*, *M. meleagridis*, *M. synovial.*	Infection of the air passages. Coughing, sneezing, nasal discharge. *M. synovial* may affect the joints, producing an exudative tendonitis and bursitis.	Thickened air sacs filled with exudates.
Mycosis of the digestive tract	Chickens, turkeys, and other fowl.	Pathogenic fungi.	Signs are not particularly characteristic. Affected birds show poor growth, stunted appearance, listlessness, and roughness of feathers.	Cheesy scum on crop lining.
Mycotoxicoses	All poultry.	Ingestion of toxic substances caused by molds growing on feeds and possibly litter. *Aspergillus flavis*, which produces aflatoxins, is of most concern. The B-1 toxin is the most toxic and of greatest concern to the poultry industry.	Reduced growth and egg production and high mortality.	Hemorrhages. Pale, fatty liver, kidneys.

Distribution and Losses Caused by	*Treatment*	*Prevention*	*Remarks*
Worldwide. Mortality of affected birds ranges from 1 to 40% if treatment not instituted promptly. The cost of fowl typhoid is primarily in testing under the National Poultry Improvement Plan.	Nitrofurans or sulfa drugs according to directions. But every effort should be made to eradicate the disease.	Get stock from disease-free sources.	Egg and mechanical transmission.
Disease occurs in most concentrated poultry-producing areas of the world. Heaviest losses in chicks up to 6 weeks.	None.	Live and dead vaccines.	The disease damages the birds' immune processes. Thus, vaccines for other diseases are less effective.
The disease has been reported in the U.S., Canada, Japan, Australia, India, and Israel.	Injection of convalescent antiserum, which is obtained from healthy flocks and usually collected at slaughter.	Vaccine.	
Worldwide. Economic loss is in lowered production and quality of eggs, and in mortality and lowered gains and feed efficiency of young chickens.	No specific treatment.	Strict isolation and repopulation with only day-old chicks following the cleaning and disinfecting of the poultry house. Inactivated and live vaccines.	
Probably worldwide. Mortality varies from 2 to 75% with 5 to 15% being the average.	Antibiotics.	Test breeders and purchase clean chicks and poults. Treatment of eggs with antibiotics; egg inoculation; or heat.	
Laryngotracheitis has been identified in most countries.	No drug treatment is effective.	Vaccination.	Farm eradication can be accomplished if security measures are superior.
With few exceptions, leukosis virus infection occurs in all chicken flocks; by sexual maturity most birds have been exposed.	None.	No vaccine. Buy birds from complement fixation avian leukosis (Cofal) flocks. Raise birds in isolation away from old or adult stock.	Now virtually eliminated from major breeding units.
Worldwide. Prior to development of a vaccine, losses ranged from 25 to 60%.	None.	Vaccination of day-old chicks at the hatchery.	Genetic resistance and isolation-rearing are important adjuncts to vaccination.
Worldwide.	Antibiotics.	Eradication from breeding flocks.	Mycoplasmosis infection is commonly designated as a chronic respiration disease of chickens and infectious sinusitis of turkeys.
Mycosis probably occurs frequently, but most cases are mild.	Fungicidal drugs.	Sanitation. Do not overcrowd. Nystatin: 50 gm/ton continuously.	
The disease first became prominent in 1960 when 100,000 turkeys died, later found to be caused by a toxin in moldy peanut meal.	Remove source of aflatoxin from the diet.	Avoid feed spoilage. Treat high moisture grains with propionic or acetic acids.	Is on increase. Young are more susceptible than mature birds.

(continued)

TABLE 11.1   Nonnutritional Diseases of Poultry (*continued*)

Disease	Species Affected	Cause	Symptoms and Signs	Postmortem
Newcastle disease	Chickens and turkeys. Other fowl.	RNA Virus. *Newcastle Disease Virus (NDV)* or *Paramyxovirus-1 (PMV-1)*	Gasping, wheezing, twisting of neck, paralysis, severe drop in egg production, soft-shelled eggs.	Often none. Sometimes mucus in trachea and thickened air sacs containing yellow exudate.
Paratyphoid	Chickens, turkeys, waterfowl, and other birds.	Gram negative bacteria; Salmonella species other than *S. pullorum* and *S. gallinarum.*	Seen mainly in poults.	Enteritis, nodules in wall of intestines.
Pullorum	Chickens, turkeys, other domestic and wild fowl.	Gram negative bacterium (*Salmonella pullorum*).	Sleepy, pasted up, inactive, high mortality in young birds.	Lesions on lungs, liver, and intestines. Unabsorbed egg yolks.
Tuberculosis	All poultry.	Bacterial (*Mycobacterium avium*). Serovar1 found in wild birds and patients with AIDS. Serovar2 found in chickens.	Unthriftiness, lowered egg production, and finally death.	Characteristic grayish-white or yellowish nodules of varying sizes in the liver, spleen, and intestines.
Turkey coryza (bordetellosis)	Turkeys.	Gram negative bacterium (*Bordetella avium*).	Snicking, rales, and discharge of excessive nasal mucus.	Lesions and excessive mucus in the upper respiratory system.
Ulcerative enteritis (quail disease)	Chickens, turkeys, and game birds.	Gram positive bacterium (*Clostridium colinum*).	Sleepy, loss of appetite.	Ulcers on intestines, enteritis.

Distribution and Losses Caused by	Treatment	Prevention	Remarks
Worldwide. Mortality of affected chickens varies from 0 to nearly 100%.	There is no effective treatment.	Vaccination.	It was first recognized in England in 1926; and it is named after the town of Newcastle. Newcastle disease was first reported in the U.S. in 1944. Symptoms in turkeys are mild; reduction of egg production of turkey breeders is main economic loss. This is a notifiable disease.
Worldwide. Mortality in turkey poults usually 1 to 20%. Outbreaks in ducks (keel disease) often run very high.	Sulfonamides, antibiotics, and nitrofurans may be employed to reduce mortality.	Hatchery and flock sanitation are the most important factors in paratyphoid prevention.	Egg and mechanical transmission; and through "blow-up" of infected eggs during incubation, and through some feeds.
Presently, the main economic loss from pullorum in the U.S. is due to the necessity of testing breeding flocks of chickens and turkeys to ensure freedom from infection.	Sulfonamides, nitrofurans, and antibiotics may be used to check mortality, but there is no substitute for a sound eradication program.	Eggs from disease-free breeders, hatched in disease-free incubators. Eradication from breeding stock. Buy chicks and poults only from pullorum typhoid-clean breeding stock.	Primarily egg-transmitted, but transmission may be by other means.
Worldwide, but occurs most frequently in the North Temperate Zone. In 1972, tuberculosis was the cause for condemnation of 0.04% of the 186.9 million mature chickens slaughtered under federal inspection in the U.S.	None.	Sanitation; put disease-free birds in a clean house or on clean ground.	Avian tuberculosis is transmissible to swine, so keep swine and chickens separated. M. avian infection is found in people with AIDS.
Worldwide.	No treatment is entirely effective, although use of antibiotics in the early stages of the disease may be helpful.	Commercial vaccines are available.	Morbidity is generally 100% while mortality ranges from 5 to 75%.
Death losses may be high in replacement pullets and quail.	Antibiotics.	Sanitation. Raising birds on wire is an effective preventive measure.	Increasing problem in chickens.

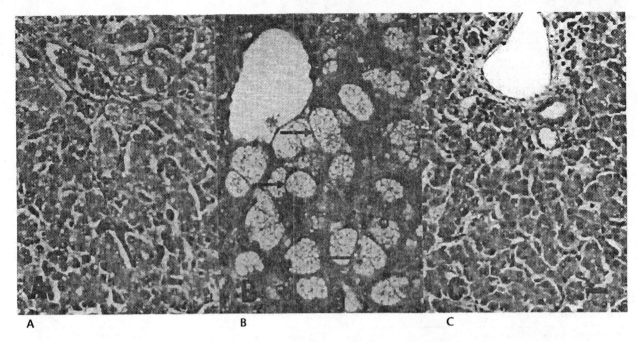

Figure 11.10 The products of molds or fungi include aflatoxins. When broiler chicks eat a diet containing aflatoxins there are profound lesions in the liver as can be seen from the micrographs of sections through the liver of control (A) and aflatoxin treated broiler chicks (B and C). *(Reprinted from Ledoux, D. R., Rottinghaus, G .E., Bermudez, A. J., and Alonso-Debolt, M. (1999). "Efficacy of a hydrated sodium calcium aluminosilicate to ameliorate the toxic effects of aflatoxin in broiler chicks." Poultry Science 78, 204–210 with permission from Poultry Science)*

the spread between birds. Bacterial respiratory diseases include chronic respiratory disease, coryza (roup), and infectious sinusitis of turkeys, mycoplasmosis.

Fungal or Fungal Product Diseases

Aspergillosis

Respiratory diseases can be caused by molds. A major example is aspergillosis, or brooder pneumonia in young chickens. The causative agent is normally *Aspergillus fumigatus*, but *A. glaucus* and *A. niger* may be involved.

Mycotoxins

Mycotoxins are toxins produced by fungi that can contaminate feed. In 1999 a significant proportion (~25%) of the world's feedstuffs were estimated to be contaminated. There are more than 300 mycotoxins. The most problematic are aflatoxin (from *Aspergillus* species (see Figure 11.10); deoxynivalenol (vomitoxin) (from *Fusarium* species); zearalenone (from *Fusarium* species); fumonisin (from *Fusarium* species); T_2 toxin (from *Fusarium* species); and ochratoxins (*Aspergillus*). Mycotoxins depress growth in poultry. Growth in chickens is reduced by aflatoxin, cyclopriazonic acid, deoxynivalenol (vomitoxin), fumonisin B_1,

and ochratoxin A. Sorgum ergot (produced by *Claviceps* species) does not affect the growth of broiler chickens.

Prions

While prions (protein disease-causing agents) are thought to cause diseases in cattle, sheep, and people, there is no evidence that poultry have prion-induced diseases.

PARASITES

External and internal parasites affect poultry. Control of parasites is a very effective way of increasing egg and meat production. Table 11.2 summarizes the internal and external parasites of poultry.

Internal Parasites

Table 11.2 lists the common internal parasites of poultry together with their treatment and prevention.

Protozoan Diseases

Protozoa are simple single-celled animals. Although many microscopic protozoan organisms are harmless, others produce severe disease. Among the more common and serious poultry diseases are the following:

- Coccidiosis. (causative agent: *Eimeria tenella* and other species of *Eimeria*) (organs affected: intestine).
- Histomoniasas or blackhead. This is found particularly in turkeys (economic damage more than $2 million in the United States). Outbreaks in chickens have caused significant mortality though. (causative agent: *Histomonas meleagridis*) (organs affected: ceca and liver) (also see Figure 11.11).
- Cryptosporidiosis. (causative agent: in chickens, *Cryptosporidium bailyi*, in turkeys, *C. meleagridis*).
- Trichomoniasis is found particularly in pigeons where it is called canker. It is transmitted with the crop milk. It can also affect chickens and turkeys. (causative agent: *Trichomonas gallinae*) (organs affected: upper digestive tract).
- Hexamitiasis. This once significant disease is now rarely found in turkeys in the United States. It also affects game birds. (causative agent: *Hexamita meleagridis*)
- Leucocytozoonosis. This is seen particularly in wild and domestic ducks and geese. (causative agent: *Leucocytozoon simondi* and other species of the genus *Leucocytozoon*) (organs affected: blood and internal organs).

COCCIDIOSIS. Coccidiosis is still one of the most common diseases of poultry. It is spread by the

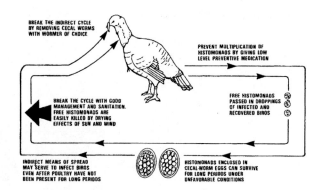

Figure 11.11 Blackhead (Histomoniasis), to which turkeys are especially susceptible. Preventive measures are: (1) good sanitation; (2) control of cecal worms; (3) turkeys kept on wire, away from chickens; and (4) a histomonastat administered continuously in feeds to turkeys over 6 weeks old. *(Original diagram from E. Ensminger)*

oocysts that can live in the excreta and are found in poultry houses. The oocysts are the infective form. They have thick walls and can remain in this quiescent form in the environment until eaten by poultry. The oocysts are easily transferred between poultry houses by personnel and equipment. The life cycle of *Eimeria tenella* is shown in Figure 11.12.

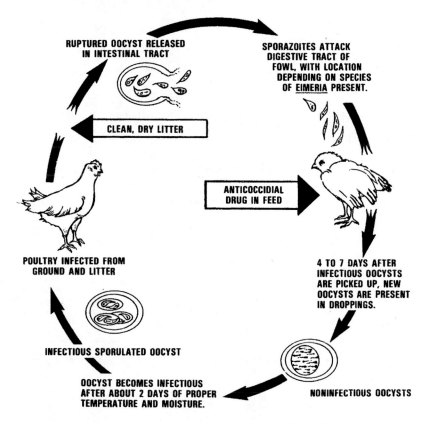

Figure 11.12 Life cycle of *E. tenella*, one of the genus containing many species causing coccidiosis in poultry, showing (1) the typical 7- to 9-day cycle, and (2) the stage of development at which anticoccidial drugs kill the organism. *(Original diagram from E. Ensminger)*

TABLE 11.2   Parasites of Poultry

Parasite	Species Affected	Cause	Symptoms and Signs	Postmortem
Blackhead (histomoniasis)	Chickens, but more resistant than turkeys. Turkeys.	Protozoa (*Histomonas meleagridis*).	Droopiness, loss of weight, sulfur-colored diarrhea, and darkened heads.	Lesions in liver and ceca, ceca enlarged, liver enlarged and spotted with dark red or yellow circular areas.
Coccidiosis	Chickens, turkeys, ducks, geese, and game birds.	Protozoa (*Eimeria* species). Six species of *Eimeria* cause the disease in chickens and three in turkeys. Other species cause the disease in ducks, geese, and game birds.	Bloody droppings (usually cecal-type only), sleepy, pale, ruffled feathers, unthrifty.	Bloody or cheesy plugs in ceca (cecal-type). Intestinal wall thickened with small white or reddish areas (intestinal-type).
Hexamitiasis(infectious catarrhal enteritis)	Turkeys.	The one-celled parasite *Hexamita meleagridis*.	Listlessness, foamy and watery diarrhea, and convulsions.	Dehydration, emaciation, thin and watery intestinal contents, and bulbous areas in intestines.
Large roundworms	Chickens and turkeys.	Ascarid infestation (*Ascaridia galli*).	Droopiness, emaciation, and diarrhea.	Roundworms, 1½ to 3 in. long, in intestines.
Leucocytozoonosis	Turkeys and ducks.	Protozoa (*Leucocytozoon*).	Symptoms are not apparent in older birds. Loss of appetite, droopiness, weakness, increased thirst and rapid, labored breathing in young birds.	
Lice	All poultry.	More than 40 species of lice.	Frequent picking, pale head and legs, loss of weight.	
Mite infestation (mites on body only during night)	Chickens are the commonest hosts, but these mites may occur on all poultry.	Common red, or roost, mites (*Dermanyssus gallinae*).	Reduced egg production, retarded growth, lowered vitality, damaged plumage, and even death.	
Mite infestation (mites always on body)	All poultry.	Northern fowl mites, the most common of which is *Ornithonyssus sylviarum*.	Droopy, pale condition and listlessness.	
Tapeworms	Chickens, turkeys, and other birds.	Several species of tapeworms.	Pale head and legs, poor flesh.	Tapeworms in intestines.
Trichomoniasis	Chickens, turkeys, pigeons, and quail.	Protozoa (*Trichomonas gallinae*).	Loss of appetite, droopiness, loss of weight, and darkened head.	Lesions—necrotic ulcerations—in the upper digestive tract, affecting the crop in particular.

Distribution and Losses Caused by	Treatment	Prevention	Remarks
Average annual losses in turkeys estimated to exceed $2 million.	A number of drugs are on the market, from which a selection may be made and used according to manufacturer's directions.	Sanitation; frequent range rotation. Do not crowd. Preventive medication in feed and water according to directions. Do not keep chickens and turkeys on the same premises.	Transmitted by droppings from infected birds. Control cecal worms.
Coccidia are found wherever poultry are raised. Estimates of annual losses in the U.S. range up to $200 million. More than $80 million spent on preventive medication, annually.	Anticoccidials in feed or soluble form of drugs in water according to directions.	Preventive medication in feed or water according to directions. Vaccination is useful in certain types of operations, but seek expert advice before using it.	Transmitted by droppings of infected birds.
Reported in the U.S., Canada, Scotland, England, and Germany. Hexamitiasis is not of major importance today.	The disease does not respond well to treatment. Furazolidone, aureomycin, and terramycin have been used with some success.	Segregation of age groups, and sanitation.	
	Several deworming drugs are on the market. Select and use according to manufacturer's directions.	Sanitation and rotation of range and yards. Careful use of old litter. Cage rearing and housing.	See Figure 10.29 for life cycle.
The disease occurs most frequently in southern and southeastern U.S.	Drug treatment of leucocytozoonosis has had limited success. Clopidol may be used as a treatment or preventive.	Exterminate black fly population, and do not raise turkeys near running streams. Segregate breeding and brooding operations. Brooding in houses with cheesecloth over openings during black fly season. Sulfadimethoxine and sulfaquinoxaline prevent infection of certain types of the parasite.	
Worldwide. Heavy infestations affect bird performance.	Select an approved insecticide and use according to manufacturer's directions.	Buy louse-free birds and never add lousy birds to clean stock.	
Worldwide. Can cause reduced performance.	Select an approved insecticide and use according to manufacturer's directions.	Sanitation. Examine birds frequently for signs of mites. Preventive insecticide treatment of quarters. Control sparrows.	
The northern fowl mite, *Ornithonyssus sylviarum*, is rated as the most important permanent parasite in all major producing areas of the U.S.	Select an approved insecticide and use according to manufacturer's directions.	Examine birds frequently for evidence of mite infestation. Preventive insecticide treatment of quarters. Control sparrows.	
	Butynorate is the most widely used product for treatment. Use according to manufacturer's directions.	Eliminate the intermediate insect hosts. Control snails, earthworms, beetles, and flies.	See Figure 10.30 for life cycle.
Worldwide.	Sanitation; remove infected birds; and use one of the recommended drugs according to manufacturer's directions.	Sanitation, clean feed and water. Eliminate recovered or carrier birds.	

Coccidiosis can affect young chickens and turkeys with a major loss of growth rate. The *Eimeria* pathogens invade cells of the intestine with, for instance, *Eimeria tenella* focused in the ceca of the chicken. The *Eimeria* are very species-specific, affecting only the host and possibly very closely related species. Costs of prevention are more than $90 million in the United States and more than $300 million worldwide.

Anticoccidial agents, such as the coccidiostats, are routinely added to both starter and grower diets.

- In 1999, salinomycin was reported to be used in both starter and grower diets in over 50% of broiler chicken units in the United States, with nicarbazin also used extensively.
- Ionophores such as monesin, narasin, and lasalocid are also employed.

Vaccines against *Eimeria tenella* are available (e.g., Coccivac® from Schering-Plough Animal Health) and are effective. They are used particularly with broiler breeder flocks.

Other Internal Parasites

Poultry can also be infected with other internal parasites. If found in the intestine, these essentially "rob" the bird of nutrients. They can also invade other organs (e.g., blood and eye), destroying tissues. These parasites include:

- Roundworms (nematodes) (organs affected: digestive tract, eye, other tissues)
- Tapeworms (cestodes) (organs affected: digestive tract)
- Flukes (trematodes) (organs affected: digestive tract, liver, and other tissues)

Wild birds are frequently the host to these parasites, and thus they can be spread. There are FDA-approved medications against many of these internal parasites.

The multicellular internal parasites are classified as follows:

Roundworms

Phylum Nematoda

Tapeworms

Phylum Platyhelminthes
Class Cestoda

Fluke

Phylum Platyhelminthes
Class Trematoda

Examples of internal parasites of poultry are:

- *Nematodes,* or *roundworms*
  - *Ascaridia galli.* This is a large worm >2 in. (>50 mm) long that has been found in the intestine of all poultry. The life cycle of *A. galli* is shown in Figure 11.13.
  - Hair worm (various species of *Capillaria* can be found in the intestine of all poultry along with some in the eye).
  - Cecal worm (*Heterakis gallinis* or *H.gallinarum*). This worm is about 1/2 in. long (male 7–13 mm; female 10–15 mm). It can be found in the ceca of all poultry.
- *Tapeworms* are segmented and flat with a long ribbon-like appearance. They frequently infect chickens and turkeys in backyard flocks or on range conditions. The head or scolex is embedded into the tissue of the small intestine. The segments extend down the intestine. The lowest segments are called gravid and are filled with the infective eggs (or onchospheres). They are shed with the excreta. There may be an intermediate host such as beetles. The life cycle of tapeworms is shown in Figure 11.14. Among the eight tapeworm species that infect chickens in North America and Europe is *Davainea proglottina.*
- *Trematodes,* or *flukes,* can infect poultry. These flukes can infest organs such as the liver, pancreas, eye, and the oviduct (e.g., the oviduct fluke—*Prosthogonimus* sp.). Flukes are not particularly species-specific and are found in multiple wild birds and poultry. Their intermediate hosts are snails and other molluscs. Since water is viewed as a media for infection, free-range poultry and particularly ducks are vulnerable to fluke infestation.

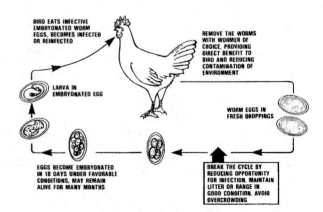

Figure 11.13 Life cycle of the common, large roundworm, *Ascaridia galli,* the most prevalent of the worm parasites of chickens and the cause of heavy economic losses. *(Original diagram from E. Ensminger)*

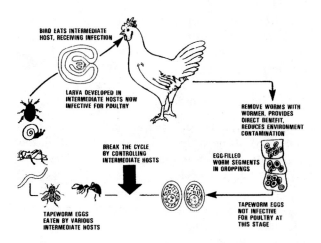

Figure 11.14 Life cycle of tapeworms, several species of which infect chickens and turkeys. Poultry become infected by eating the intermediate host. (*Original diagram from E. Ensminger*)

External Parasites and Bloodsucking Insects

Poultry are affected by external parasites (insects such as lice and the arachnids, ticks and mites) together with bloodsucking insects (bugs, mosquitoes, and beetles). These cause reduced weight gain and egg production; mar the skin, resulting in downgrading of carcass quality; and act as vectors for the spread of infectious diseases. Heavy infestations can cause high mortality among young poults.

Insects

There are more species of insects on the planet than any other class of species. Among the numerous species are bees and butterflies, together with ectoparasites and bloodsucking insects. They are classified as:

> Phylum Arthropoda
>
> Class Insecta
>
> Order Mallophaga (lice) (see Figure 11.15)
>
> Order Siphonoptera (fleas)
>
> Order Coleoptera (beetles)
>
> Order Hemiptera (bugs)
>
> Order Diptera (flies and mosquitoes)

Insects impacting poultry include the following: lice, flies, beetles (e.g., darkling or lesser mealworm), fleas, mosquitoes, and bugs (e.g., the bloodsucking bedbug). Mosquitoes can act as vectors for the spread of diseases. This may be from wild birds to poultry, or zoonotic diseases (e.g., West Nile virus), to people from animals and birds.

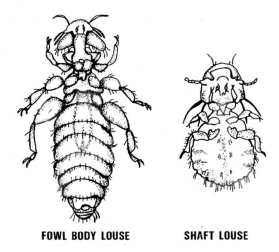

FOWL BODY LOUSE   **SHAFT LOUSE**

Figure 11.15 Lice. (*Original diagram from E. Ensminger*)

Arachnids

Arachnids include free-living animals such as spiders and the ectoparasites: mites (see Figure 11.16) and ticks. They are classified as:

> Phylum Arthropoda
>
> Class Arachnida

Examples of arachnids affecting poultry are the chicken or red mite (*Dermanyssus gallinae*), northern fowl mite (*Ornithonyssus sylviarum*), tropical fowl mite (*O. bursa*), feather mites, scaly leg mites (*Knemidocoptes mutans*), and depluming mite (*Knemidocoptes gallinae*), together with fowl ticks.

Controlling External Parasites and Bloodsucking Insects

Control procedures include Integrated Pest Management (IPM) and spraying, dusting, and misting with insecticides. According to *Diseases of Poultry*

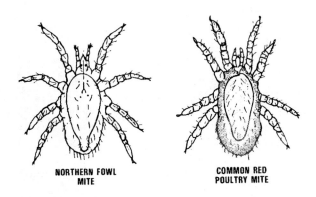

NORTHERN FOWL MITE   **COMMON RED POULTRY MITE**

Figure 11.16 Northern fowl mite, *Arnithonyssus sylviarum* (left), and common red mite, *Dermanyssus gallinae* (right). (*Original diagram from E. Ensminger*)

(1997), the following insecticides can be used in poultry houses:

- Cyromazine (trade name: Larvadex)
- Carbaryl (Sevin)
- Chlorpyrifos (Dursban)
- Coumaphos (Co-Ral)
- Dichlorvos (Vapona)
- Dimethoate (Cygon)
- Fenthion (Baytex)
- Fenvalerate (a pyrethroid) (Ectrin)
- Malathion (Malathion)
- Methomyl (Malrin)
- Naled (Dibrom)
- Permethrin (a pyrethroid) (Ectiban, Atroban)
- Propoxyr (Baygon)
- Stirofos (Rapon)

Spread of Infection

Infectious diseases are spread in poultry by the following:

- Contagion from diseased or carrier birds. (This is most important within an individual house or flock.)
- Contagion from other animals (insects, rodents, and free-flying birds) (within and between individual houses or flocks). This is unlikely with good management and pest control measures.
- Airborne. Some microorganisms can be spread in the wind and airflow (within and between individual houses or flocks). Transmission between individual houses or flocks can be very important in localities with heavy poultry populations.
- Contact with contaminated materials (e.g., litter, feed) or objects (crates, feeders, feed bags, and waterers). This is unlikely with good management and disinfection.
- Egg transmission. A number of diseases are transmitted from the hen to the chick through the egg.
- Carcasses of dead birds that have not been disposed of properly.
- Impure water, such as surface drainage water (unlikely with good management and closed water systems).
- Shoes and clothing of people who move from flock to flock (unlikely if biosecurity measures are taken).

Strategies to prevent pathogenic diseases include:

- Biosecurity (e.g., see Figure 11.17)
- Superior animal management
- Vaccination
- Breeding for disease resistance
- Surveillance and monitoring by producers and veterinarians
- Judicious use of antibiotics

Other Sources of Pathogens

Live-bird markets can represent a source of pathogenic organisms in view of lack of uniform disinfection and the continuous presence of birds as disease reservoirs. Similarly, backyard flocks may represent reservoirs of pathogens. Although illegal in many states, roosters are reared for cockfighting. The unregulated nature of this lends itself to the possibility of being a reservoir for pathogens and a means to transmit poultry diseases via the poultry unit personnel.

Impact of Other Animals (Beetles, Rodents, and Flies)

Beetles

Beetles are found in the litter of poultry houses. They have either beneficial or detrimental effects on poultry production. The rove and hister beetles have a beneficial effect, consuming fly eggs and

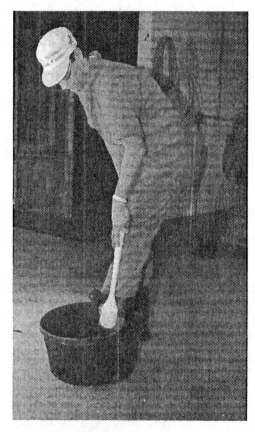

Figure 11.17 Worker scrubbing, rinsing, and disinfecting boots before entering a poultry facility to prevent the transmission of diseases. *(ISU photo by Bob Elbert)*

larva. The darkling beetle may cause structural damage and/or disease. While on the other hand, the larder beetle aids in the decomposition of feathers and such, but may damage the facilities.

Pesticides can be used to control the population of beetles. Darkling beetles (*Terebrionidae*) in the litter can serve as a reservoir for pathogens, including *Salmonella* (with consequences for food safety) and viruses (e.g., Marek's disease).

Rodents

Rodents create severe negative consequences for the poultry industry by spreading diseases of poultry, including zoonotic diseases that affect people. Rats and mice are sources and reservoirs of disease-causing microorganisms. The pathogens may be transmitted via the feces or external parasites.

The major rodents that infest poultry facilities are brown rats, roof rats, and house mice. Rodent populations must be controlled, and the success of these efforts monitored. Approaches include physical exclusion from facilities and feed, rodenticides, and trapping. (See Chapter 8 for details.)

Flies

The adult fly lays her eggs in wet manure and other damp, rotting matter. From the eggs, the larva or maggot emerges. Larvae grow and develop and ultimately enter the pupa stage. Following metamorphosis, the adult fly leaves the pupal case. Flies can be *vectors* for poultry diseases, including the common tapeworm and Newcastle disease.

The problem of flies can be greatly reduced by good management:

- Striving to eliminate sites where maggots can grow and adult flies can feed
- Frequently cleaning facilities and drying manure
- Using pesticides (carbamates, organophosphates, and pyrethroids) (see above)
- Applying biological control with beneficial insects such as the parasitic wasp or some beetles (rove and hister beetles, particularly in layer houses)

Reducing fly numbers is not only a good-neighbor policy, but it also may reduce diseases in the flock.

Wild Birds

Wild birds are a source of disease-causing microorganisms, including emerging diseases such as West Nile virus (infectious to poultry, wild birds, horses, and people, and transmitted by mosquitoes).

IMMUNITY AND OTHER DEFENSES AGAINST DISEASE

The body has a well-developed defense mechanism that must be understood and utilized in controlling infectious diseases. Immunity is the ability to resist infection. An animal has two types of protective mechanisms:

1. Those that hinder or prevent invasion of organisms.
2. Those that combat agents that invade the body.

The former includes the intact skin and mucous membranes, creating a direct barrier and mucous secretions that "wash out" invading organisms and inflammation responses. Mast cells release histamine, serotonin, and prostaglandins, which then cause inflammation.

Mechanisms combating infections include the white blood cells that engulf microorganisms (cellular immunity) (see Figure 11.18) and circulating antibodies (see Figure 11.19) that bind microorganisms or their toxins (humoral immunity). T-lymphocytes also produce proteins called cytokines that aid the body's defense against pathogenic microorganisms. These include interferons that help the body resist viral infections.

The avian immune system has been extensively employed in research on immunology. There are two primary lymphoid tissues:

1. **Thymus (in the neck)**—where T immune cells originate (T cells being responsible for cell-mediated response to foreign organisms)
2. **Bursa of Fabricius (by the cloaca)**—where B immune cells originate (B cells being responsible for the humoral or antibody-mediated response to foreign organisms)

There is no bursa of Fabricius in mammals, instead, B cells originate in the bone marrow. The secondary lymphoid tissues contain both B and T cells and include:

1. The spleen (adjacent to the liver)
2. Lymph nodes
3. Harderian gland (close to the eyeballs)
4. Peyer's patches (close to the junction of the ileum and ceca)
5. Meckel's diverticulum (remnants of the yolk stalk)

The purpose of vaccines is to stimulate an active production of antibodies (see Figure 11.19). Active immunity depends upon the production of antibodies, which are glycoproteins (immunoglobulins)

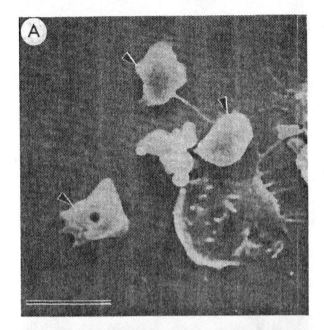

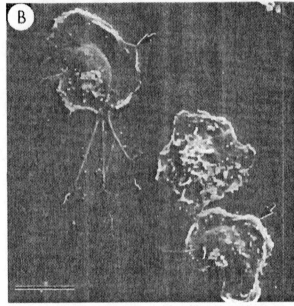

Figure 11.18 Scanning electron micrograph of heterophils, the avian equivalent of neutrophils in mammals, following stimulation by lymphokines showing extended pseudopodia. *(Reprinted from Kogut, M. H., Lowry, V. K., Moyes, R. B., Bowden, L. L., Bowden, R., Genovese, K., and DeLoach, I. R. (1998). Lymphokine-augmented activation of avian heterophils.* Poultry Science *77, 964–971 with permission from* Poultry Science*)*

in the blood serum and other body fluids. Antibodies are specific for the organism that stimulated their production. Thus, immunity to one disease does not provide immunity to others. All antibodies have the following characteristics:

- They have at least two binding sites for the antigen.
- They have two (or multiples of two) heavy chains and two (or multiples of two) light chains.

- They have regions in the protein that are constant and regions that are variable, being different for each antibody for a single clone of antibody-producing cells.

There are five classes or isotypes of immunoglobulins (Ig) (antibodies). These are:

1. **IgM.** (This is produced during the early response to an infection or vaccine. It is less specific. It has five multiples of the two heavy and two light chains.)
2. **IgG, or gamma globulin.** (This is produced during the later response to an infection or vaccine. It is very specific. It has one multiple of the two heavy and two light chains.) It is the most important antibody in the blood.
3. **IgD.** (An antibody found on the surface of B-cells.)
4. **IgA.** (This is found in the intestine.)
5. **IgE.** (This is involved in allergies.)

Passive immunity is the transfer of antibodies from the individual in which they are produced to another individual. This may be done by the injection of serum collected from an immunized animal. Antibodies are transferred from the hen to the offspring through the egg yolk. Passive immunity is of short duration with a marked decline in the antibody levels within 21 days. Passive protection against infection usually lasts no longer than 5 weeks.

INFECTIOUS DISEASE PREVENTION AND CONTROL

An effective disease prevention and control program should embrace:

1. A flock health program
2. Biosecurity
3. Hatchery management and sanitation (see Figure 11.20)
4. Use of disinfectants
5. Use of antibiotics and coccidiostats
6. Vaccination (see Figure 11.21)
7. Critical role of veterinarians and diagnostic laboratories
8. Planned responses to disease outbreaks

Specific Pathogen Free (SPF) birds may be considered.

Flock Health Program

There are certain basic principles that should always be observed:

Color Plate 1   Brooding chicks. *(ISU photo by Bob Elbert)*

Color Plate 4   Young chicks. *(Courtesy, Hy-Line International)*

Color Plate 2   Hy-Line white egg layer (W-36). *(Courtesy, Hy-Line International)*

Color Plate 5   Wild turkeys. *(Courtesy, USDA)*

Color Plate 3   Hy-Line brown egg layer. *(Courtesy, Hy-Line International)*

Color Plate 6   Turkey spermatozoan. The light micrograph (labeled with antibody to turkey sperm mitochondria counter stained with ethidium bromide) shows the acrosome (A), nucleus (N), mid-piece containing mitachondria (M) and tail or flagella (F). *(Reprinted from Korn, N., Thurston, R. J., Pooser, B. P. and Scott, T. R. (2000). Ultrastructure of spermatozoa from Japanese quail. Poultry Science 79, 407–414 with permission from Poultry Science)*

Color Plate 7 Examples of traditional breeds of chickens. A. Barred Plymouth Rock rooster, B. Buff Orpington hen, C. Buff Orpington rooster, D. Silkie Bantum, hen, E. Golden Polish hen, F. Golden Polish rooster, G. Rhode Island rooster, I. Red Rhode Island hen rooster. *(ISU photo by Bob Elbert)*

Color Plate 8 Muscovy ducks. *(Courtesy, USDA)*

Color Plate 9 Young turkey poults. *(Courtesy, Gretta Irwin, Iowa Turkey Federation)*

Color Plate 10 (A) Leghorn strains used for production of table eggs lay white-shelled eggs. Strains of chickens used for the production of brown-shelled table eggs and broiler strains produce eggshells in a variety of shades of brown. These vary from light cream through dark brown. (B) Each breed has a characteristic eggshell color. Some breeds of birds lay white-shelled eggs. Others lay eggshells in shades of brown from light cream through dark brown. A few breeds lay yellow shelled eggs and a few lay blue or green shelled eggs. (1) White Leghorn, (2) Partridge Cochin, (3) White Laced Red Cornish, (4) Silver Laced Wyandotte, (5) Barred Plymouth Rock, (6) Buff Orpington, (7) Aracauna.

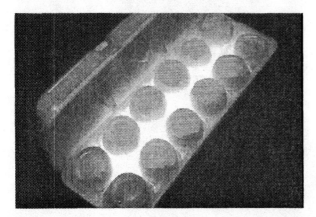

Color Plate 11 Carton of eggs with various color yolks. *(Courtesy, Kemin Industries)*

Color Plate 12 Eggs with additional nutrients/vitamins. Sparboe Egg Carton. *(Courtesy, Kemin Industries)*

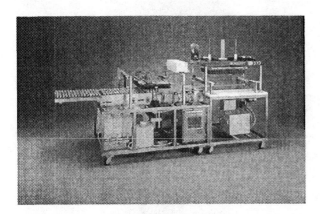

Color Plate 13 The Inovojet® system allows automated vaccination of embryos and is being adapted for automated candling, sexing of embryos and *in ovo* feeding. *(Courtesy, Embrex, Inc.)*

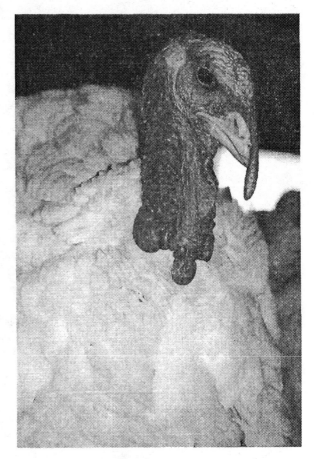

Color Plate 14 Adult male or tom turkey. *(Courtesy, Gretta Irwin, Iowa Turkey Federation)*

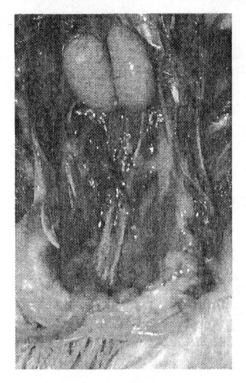

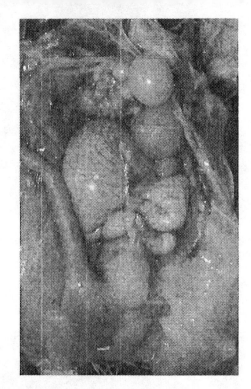

Color Plate 15 Reproductive anatomy of male chicken.

Color Plate 16 Reproductive anatomy of female chicken.

Color Plate 17 Types of combs. (A) rose comb, (B) pea comb (*Photo by George Brant*), (C) V comb, (D) single comb *(ISU photo by Bob Elbert)*

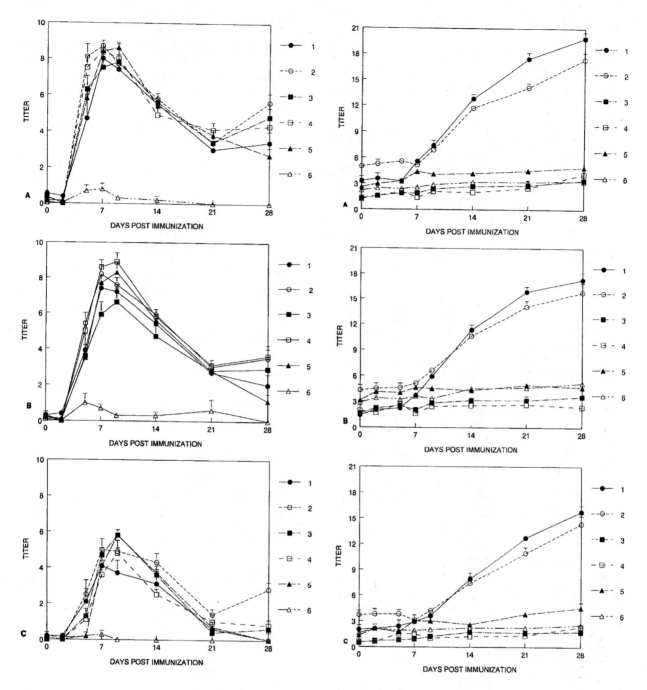

Figure 11.19 Example of production of antibodies in 5-week-old chickens in response to a foreign protein or immunogen (administered in different ways). *(Reprinted from Permentier, H. K., Walraven, M., and Nieuwland, M. G. B. (1998). Antibody responses and body weights of chicken lines separated for high and low humoral responsiveness to sheep red blood cells. 2. Effects of separate application of Freund's complete and incomplete adjuvants and antigen.* Poultry Science *77, 256–265 with permission from* Poultry Science)

- Employ a biosecurity program. Personnel other than essential SHOULD NOT visit poultry units.
- Purchase only day-old chicks or hatching eggs from a source with excellent genetics and poultry health.
- Regulate temperature, humidity, and ventilation during brooding and growing to ensure comfort of the birds. Prevent overheating, chilling, or buildup of ammonia in the air and moisture in the litter.

- Keep birds separate according to source and age groups. Follow an "all-in, all-out" program.
- Don't crowd birds. Crowding reduces growth and feed efficiency while increasing cannibalism, feather picking, and other stress-related problems.
- Maintain hatchery supply flocks on separate premises from other birds.
- Provide a good commercial feed (or a carefully formulated home-mixed feed).

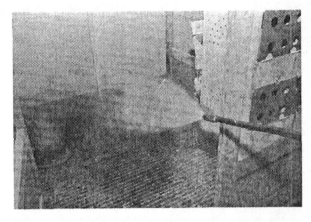

Figure 11.20 Sanitation. Cleaning represents 95% of sanitation involving washing, rinsing, and drying. Disinfecting is important but represents 5% of the sanitation program. *(ISU photo by Bob Elbert)*

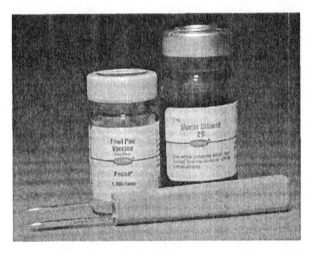

Figure 11.21 Examples of poultry vaccines. *(ISU photo by Bob Elbert)*

- Provide an adequate supply of clean, quality water, preferably in a closed system. Monitor water quality for bacteria, particularly coliform, and other contamination (minerals that adversely affect poultry such as nitrates, zinc, sodium). For smaller-scale producers: avoid watering from surface tanks, streams, or ponds.
- Develop the vaccination program in consultation with a poultry veterinarian.
- Control internal and external parasites.
- Control vermin and screen out free-flying birds.
- If a disease problem develops, obtain an early veterinary diagnosis and apply the best treatment and eradication measures.
- Dispose of all dead birds as follows: incineration, composting, or pit or deep burial.
- Maintain good records relative to flock health including vaccination history, disease problems, and medication employed.
- Avoid keeping chickens and turkeys on the same premises.

Biosecurity

Biosecurity programs are developed on an individual farm or company basis in an attempt to prevent transmission of diseases by people. The programs always involve:

- Carefully thought-out standard operating procedures (SOPs)
- Educational programs for workers, visitors, and service people
- Monitoring and policing

The operating procedures involve use of coveralls, boots, head coverings, disinfectants, and so on. There may be shower-in, shower-out and/or disposable shoe and head coverings. Vehicles may also need to be subjected to intense washing. Personal hygiene is a must. Service personnel (deliveries of feed, supplies, transportation of birds or eggs to the processor, dead bird pickup, etc.) should be restricted to specific areas on the poultry unit.

Hatchery Management and Sanitation

Hatching eggs should be collected at frequent intervals using the following practices. Use clean and disinfected containers to collect the eggs and prevent contamination from organisms on hands or clothing. Maintain the identity of all eggs relative to the breeder flocks of origin. Do not use dirty eggs for hatching. Collect dirty eggs in a separate container from hatching eggs and clean. Store eggs in a cool place in properly cleaned and disinfected racks following fumigation. Hold the storage period to as short a time as possible before setting. Use new or fumigated cases to transport eggs to the hatchery. Burn soiled egg case fillers.

An effective program for the control of infections includes the following. Arrange the hatchery building with separate rooms each with separate ventilation for egg receiving, incubation and hatching, chick holding, and disposal of shells/mebranes/dead embryos, plus cleaning of trays. Use biosecurity with admission permitted only to authorized personnel who have taken proper precautions (e.g., shower-in, shower-out). Clean thoroughly and disinfect frequently. Incinerate all hatchery waste. Clean thoroughly, then fumigate the hatching compartment of incubators, including hatching trays, after each hatch. Transport day-old chicks and poults in clean, new boxes or in disinfected plastic cartons. After each use, clean and disinfect all crates and vehicles used for transporting started or adult birds. Maintain the identity of all chicks and poults relative to the breeder flock of origin. Avoid mixing the progeny of different breeder flocks.

Cleaning and Disinfecting

In poultry houses, cleaning and disinfecting should include these steps: Remove all litter and droppings. Scrub the walls, floors, and equipment with hot water containing detergent. Rinse. Spray with a suitable disinfectant following the manufacturer's directions.

In hatcheries, cleaning and disinfecting should include the following:

- Remove trays, fans, and so forth for separate cleaning.
- Thoroughly clean the ceiling, walls, and floors. Rinse.
- Replace cleaned fans, trays, and so forth and bring the incubator to the normal operating temperature.
- Fumigate the hatcher before inserting eggs.

If eggs are hatching and incubating in the same machine, clean the entire machine after each hatch.

Disinfectants

Disinfectants are chemicals that kill bacteria (bactericidal activity) and protozoa along with destroying viruses, molds, and the transmittable forms of internal parasites. Cleaning and disinfection become extremely important in breaking the cycle of infection-reinfection. Also, in the case of a disease outbreak and depopulation, the premises must be disinfected. Thorough cleaning should precede disinfection. Organic matter can protect microorganisms. Heat (steam, hot water, etc.) is an effective disinfectant. In addition, a chemical disinfectant is used. A good chemical disinfectant should have the following characteristics:

- Kill pathogenic microorganisms
- Remain stable in and penetrate organic matter rapidly
- Dissolve in water and remain in solution
- Be nontoxic to birds and humans
- Be economical to use

The number of available disinfectants is large. Table 11.3 gives a summary of the limitations, usefulness, and strength of some common disinfectants (also see Figure 11.22). When using a disinfectant, always follow the manufacturer's directions.

Vaccination or Immunization

Vaccination is the injection of some agent (vaccine) into an animal to develop a competent immune response and hence prevent disease (see Figure 11.21). Vaccines contain large numbers of the disease-causing organism (either live but nonpathogenic/nonvirulent or killed) or potential surface proteins. Most poultry vaccines are of the live-virus type, produced by growing laboratory strains of virus in embryonated chicken eggs or in cell culture systems. Bacterial vaccines, called bacterins, are killed or inactivated preparations of bacteria, produced by growing selected strains of bacterial organisms in artificial media.

So-called vaccination outbreaks do occur. The factors that influence vaccine-response in poultry are many, mainly depending on the host and environment. Ordinarily some protective immunity is conferred when birds are vaccinated, although the vaccine within itself cannot guarantee it. A sound vaccination program is part of a good management program, but not a substitute for a sanitation program.

Diseases for which Vaccines Are Available

Viral diseases that can be controlled by vaccination include:

- Avian encephalomyelitis (epidemic tremors)
- Fowl pox
- Gumboro (infectious bursal disease)
- Infectious bronchitis
- Laryngotracheitis
- Marek's disease
- Newcastle disease

    Bacterins available commercially include:

- Erysipelas bacterin
- Fowl cholera bacterin
- Mixed bacterins

Vaccination Programs for Chickens and Turkeys

In a concentrated poultry area, various diseases are a constant threat to the producer's profits. Vaccination is cheap insurance against heavy losses from some of these diseases. Vaccinate chickens, turkeys, and other poultry only after getting expert advice from a poultry veterinarian.

Figure 11.22 Examples of disinfectants. *(ISU photo by Bob Elbert)*

TABLE 11.3 Disinfectant Guide[1]

Kind of Disinfectant	Usefulness	Strength	Limitations and Comments
Alcohol (ethyl-ethanol, isopropyl, methanol)	Primarily as skin disinfectants and for emergency purposes on instruments.	70% alcohol—the content usually found in rubbing alcohol.	They are too costly for general disinfection and are ineffective against bacterial spores.
Boric acid[2]	As a wash for eyes, and other sensitive parts of the body.	1 oz in 1 pt water (about 6% solution).	It is a weak antiseptic. It may cause harm to the nervous system if absorbed into the body in large amounts. For this and other reasons, antibiotic solutions and saline solutions are fast replacing it.
Chlorines (sodium hypochlorite, chloramine-T)	Used (1) for egg dipping and washing, (2) in processing plants, (3) for sanitizing poultry drinking water, and (4) as a deodorant. Chlorines will kill all kinds of bacteria, fungi, and viruses, providing the concentration is sufficiently high.	Generally used at about 200 ppm for disinfection and 50 ppm for sanitizing.	They are corrosive to metals and neutralized by organic materials. Not effective against TB organisms and spores.
Cresols (many commercial products available)	Recommended for disinfecting houses, equipment, and footbaths. Cresols in fuel oil are the best disinfectants for dirt floors. Effective against tuberculosis and the red mite.	Cresol is usually used as a 2 to 4% solution (1 cup to 2 gal of water makes a 4% solution). Cresols can be incorporated in water, kerosene, or fuel oil.	Effective on organic material. Cannot be used where odor may be absorbed.
Formaldehyde (may be used as a gas or as a liquid)	Effective against viruses, bacteria, and fungi. It is often used to disinfect buildings following a disease outbreak. It is commonly used in fumigating hatchery eggs, prior to and during incubation.	As a liquid disinfectant, it is usually used as a 1 to 2% solution. As a gaseous disinfectant (fumigant), use 1½ lb potassium permanganate plus 3 pt of formaldehyde. Also, gas may be released by heating paraformaldehyde.	It has a disagreeable odor, destroys living tissue, and can be extremely poisonous. The bactericidal effectiveness of the gas is dependent upon having the proper relative humidity (about 75%) and temperature (about 86°F).
Heat (by steam, hot water, burning, or boiling)	In the burning of rubbish or articles of little value, and in disposing of infected body discharges. The steam "Jenny" is effective for disinfection (example: poultry equipment) if *properly employed*, particularly if used in conjunction with a phenolic germicide.	10 minutes' exposure to boiling water is usually sufficient.	Exposure to boiling water will destroy all ordinary disease germs but sometimes fails to kill the spores of such diseases as anthrax and tetanus. Moist heat is preferred to dry heat, and steam under pressure is the most effective. Heat may be impractical or too expensive.
Iodine[2] (tincture)	Extensively used as skin disinfectant, for minor cuts and bruises.	Generally used as tincture of iodine, either 2% or 7%.	Never cover with a bandage. Clean skin before applying iodine. It is corrosive to metals.
Iodophor ("tamed" iodine)	Effective against all bacteria (both Gram-negative and Gram-positive), fungi, and most viruses. Used for (1) egg dipping, (2) hatchery or poultry house disinfection, and (3) sanitizing processing plants, footbaths, and poultry drinking water.	Usually used as a disinfectant at concentrations of 50 to 75 ppm titratable iodine, and as a sanitizer at levels of 12.5 to 25 ppm. At 12.5 ppm titratable iodine, it can be used as an antiseptic in drinking water.	They are inhibited in their activity by organic matter. They are quite expensive. They should not be used near heat. When the characteristic iodine color fades, effectiveness is gone.
Lime (quicklime, burnt lime, calcium oxide)	As a deodorant when sprinkled on manure and animal discharges, or as a disinfectant when sprinkled on the floor or used as a newly made "milk of lime" or as a whitewash.	Use as a dust; as "milk of lime"; or as a whitewash, but *use fresh*.	Not effective against spores. Wear goggles when adding water to quicklime.

184

Lye (sodium hydroxide, caustic soda)	On concrete floors.	Lye is usually used as either a 2% or 5% solution. To prepare a 2% solution, add 1 can of lye to 5 gal of water. To prepare a 5% solution, add 1 can lye to 2 gal water.	Damages fabrics, aluminum, and painted surfaces. Be careful, for it will burn the hands and face. Not effective against organism of TB. Lye solutions are most effective when used hot. Diluted vinegar can be used to neutralize lye.
Lysol (the brand name of a product of cresol plus soap)	For disinfecting surgical instruments. Useful as a skin disinfectant and for use on the hands before surgery.	0.5 to 2.0%.	Has a disagreeable odor. Does not mix well with hard water. Less costly than phenol.
Phenol (carbolic acid): 1. Phenolics—coal tar derivatives 2. Synthetic phenols	Commonly used for (1) egg dipping, (2) disinfection of hatcheries, poultry houses, and equipment, and (3) footbath solutions. Effective against bacteria and fungi.	Both phenolics (coal tar) and synthetic phenols vary widely in effectiveness from one compound to another. Note and follow manufacturer's directions. Generally used at 100 ppm for disinfecting and 50 ppm for sanitizing.	Organic materials have a diluting effect, but do not inactivate them. Ineffective against viruses.
Quaternary ammonium compounds (QACs) "quats"	Effective against bacteria and fungi. Used for (1) egg washing and dipping, (2) disinfecting hatcheries, poultry houses, and equipment, and (3) sanitizing poultry drinking water.	For sanitizing, use 200 ppm. For disinfecting, use 400 to 800 ppm.	They can corrode metal. Adversely affected by organic matter. Not very potent in combating viruses. Not effective against TB organisms and spores. Inactivated by soaps, calcium, magnesium, iron, and aluminum salts.
Sal soda	It may be used in place of lye against certain diseases.	10 1/2% solution (13 1/2 oz to 1 gal of water).	
Sal soda and soda ash (or sodium carbonate)	They may be used in place of lye.	4% solution (1 lb to 3 gal of water). Most effective in hot solution.	Commonly used as cleaning agents, but have disinfectant properties, especially when used as a hot solution.
Soap	Its power to kill germs is very limited. Greatest usefulness is in cleansing and dissolving coatings from various surfaces, including the skin, prior to application of a good disinfectant.	As commercially prepared.	Although indispensable for sanitizing surfaces, soaps should not be used as disinfectants. They are not regularly effective; staphylococci and the organisms which cause diarrheal diseases are resistant.

[1]For metric conversions, see the Appendix.
[2]Sometimes loosely classified as a disinfectant, but actually an antiseptic that is useful only on living tissue.

For replacement chicks (commercial egg or breeders), purchase or use day-old chicks vaccinated at the hatchery against Marek's disease and later vaccinate for Newcastle disease, infectious bronchitis, infectious bursal disease, avian encephalomyelitis, and fowl pox. Laryngotracheitis may be vaccinated for if there are local outbreaks.

Broilers may be vaccinated against Marek's disease, Newcastle disease, infectious bronchitis, and infectious bursal disease. Laryngotracheitis may be vaccinated for if there are local outbreaks. Turkeys may be vaccinated against Newcastle disease, hemorrhagic enteritis, and fowl cholera, and against coryza if there are local outbreaks. Turkey breeders may be additionally vaccinated against fowl pox and erysipelas.

Role of Veterinarian and Diagnostic Laboratory

Large integrators have poultry veterinarians together with pathology laboratories. Small and medium producers may hire a consulting veterinarian with poultry expertise. Diagnostic laboratories (commercial or state-operated at a fee) are available to producers. Accurate identification of poultry diseases is essential for treatment. A wrong diagnosis can result in depopulation when not required or improper medication. These can be costly. When a disease outbreak is suspected, live birds showing typical symptoms should be immediately submitted to a poultry diagnostic laboratory for examination. Such laboratories are equipped to identify the disease problem and make recommendations for control. Practicing veterinarians, industry service people, and trained extension personnel working with producers and a diagnostic laboratory can bring about a reduction in losses due to disease.

Treatment of Disease Outbreaks

Treatment of disease should be initiated only after a reliable diagnosis from a poultry veterinarian assisted by a poultry diagnostic laboratory. Diseases can be treated by:

- Allowing them to run their course
- Culling sick birds
- Depopulation of the house or facility
- Administering drugs

Misdiagnosis can result in considerable losses. Drugs should normally be used according to the instructions on the label. Drugs may produce toxic effects if used improperly.

Prescription Antibiotics and Judicious use of Antibiotics

In the United States, the Food and Drug Administration requires that a valid veterinarian-client-patient re-lationship "be established before the use of any prescription antibiotic or any antibiotic used not in accordance to labeled direction." However, veterinarians in integrated poultry production systems are generally averse to using antibiotics in an extralabel manner.

It is advisable that prescription antibiotics are used with the minimal number of diseased birds and those in contact with them (i.e., the same house). This latter is to prevent secondary infection while reducing the cost of treatment and also minimizing environmental contamination and the risks of antibiotic resistance. The concept of the judicious use of antibiotics comes from an American Veterinary Medical Association (AVMA) initiative.

Specific Pathogen Free (SPF) Program

The Specific Pathogen Free (SPF) program is a combination of breeding, testing, sanitation, and management practices designed to establish and maintain breeder flocks free of specific known infectious diseases. This program is directed toward eggs produced for vaccine production. An SPF program uses regularly cleaned and disinfected caged birds in locked rooms constructed to exclude vermin and wild birds. Feed (medicated with coccidiostat) should be in bulk-feed facilities with outside filler pipes. Facilities must be at least 100 ft (30 m) from public highways and at least 1,000 ft (300 m) from other poultry houses on adjacent premises. Shoe covers should be available. All units must have a footbath with an approved disinfectant (changed daily) and a stiff brush next to the door in the entry room. All personnel must wear covers and clean their footwear upon entering or leaving. Only pullorum-typhoid-free day-old chicks should be received, arriving directly from the hatchery in new or disinfected shipping equipment. Workers should shower-in, shower-out and should not visit the premises where other poultry are kept or poultry products are processed, or have pet birds. Maintain good records. If birds become positive for diseases, depopulate the facility immediately, disposing of carcasses by incineration.

Genetic Resistance to Disease

Selected lines of chickens have been found to vary in their resistance and susceptibility to various diseases. Thus, genetics shows promise in reducing the impact of diseases. Genomics will allow identification of the genes responsible for disease resistance and these can then be introgressed by primary breeders.

METABOLIC DISEASES

Ascites

Ascites is the accumulation of fluid in the body cavity leading to heart failure and ultimately death of the bird. It is also known as pulmonary hypertension (high blood pressure in the lung) syndrome. This disease results in over $1 billion per year in losses worldwide. Ascites reduces livability by contributing to mortalities in apparently healthy flocks and significantly to live-bird condemnations (culls).

Environment can induce ascites (e.g., low atmospheric pressure/oxygen availability at high altitudes and cold stress). Other predisposing factors are aspergillosis and reduced oxygen-carrying capacity of the blood due to carbon monoxide. Predisposition for ascites is genetic. Rapidly growing broiler chickens may not be able to supply the oxygen needs of tissues, and there may be a defect in the mitochondria.

Bone Diseases

Leg weakness and Tibial Dyschondroplasia

Leg weakness profoundly affects broiler chickens and turkeys, leading to splaying of the legs, poor performance, and on-farm condemnations. Leg weakness or tibial dyschondroplasia is a disease of the long bone (the tibia). The growth plate fails to calcify in the cartilage (transitional zone). Tibial dyschondroplasia is particularly found in fast-growing broiler chickens (see Figure 11.23). This disease leads to loss of mobility, lack of access to feed and water, and is associated with very poor performance. Results of research with addition of the metabolite of vitamin D (25-hydroxyvitamin D_3) to the feed are promising. Experimentally,

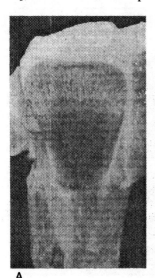

A   B

Figure 11.23 Comparison between growth plate (cartilage) of control (A) and chickens with dyschondroplasia (B). Note: There is a lesion under the growth plate in the birds with dyschondroplasia. This is opaque and not supplied with blood vessels. *(Reprinted from Farquharson, C., and Jefferies, D. (2000). Chondrocytes and longitudinal bone growth: the development of tibial dyschondroplasia. Poultry Science 79, 994–1004 with permission from Poultry Science)*

tibial dyschondroplasia can be induced by a low calcium diet, and this can be overcome with 25-hydroxyvitamin D_3.

Caged Layer Fatigue, Osteoporosis, and Bone Fracturing

With the calcium demands of egg production, it is not surprising that laying hens can show abnormalities of bones. Associated with caged layer fatigue are the symptoms of osteoporosis, namely, bones become more porous but brittle, bend under pressure (e.g., the ribs), and also have the tendency to break. Fracture of fragile limbs, bones, and ribs can lead ultimately to death. Bone fracture can also occur prior to death during transportation to a processor or during processing. Moreover, the fracturing of the bones leads to problems of processing old or spent hens.

Muscle Diseases and Myopathy

Muscle myopathy is observed in growing turkeys. This deep pectoral myopathy is also known as green muscle disease, or Oregon muscle disease. There is a marked reduction in meat quality and performance.

In fast-growing broiler chickens, particularly in stressed birds (e.g., heat stress and/or transportation stress), muscle can show unusual characteristics. Under histological examination, some of the muscle fibers die and the dead tissue is replaced by invading white blood cells. This is referred to as Growth Associated Idiopathic Myopathy. Birds with this have lower meat quality, exhibiting pale, soft, and exudative meat (PSE).

Reproductive Diseases

The laying hen can become infertile or die due to diseases of the reproductive tract. This includes the following:

- Salpingitis—the oviduct becomes filled (engorged) with yolk and possibly egg white
- Egg bound—the calcified egg is retained in the oviduct
- Peritonitis following internal ovulation into the abdominal cavity followed by inflammation

There can also be secondary bacterial infection.

ENVIRONMENTAL STRESS

Heat Stress

High environmental temperatures have been associated with mortalities in broiler chickens. Although this has been greatly improved with tunnel ventilation, growth rates and feed efficiency are reduced in the hottest months of the year. In laying hens, heat stress is associated with the following:

- Reduced egg production
- Reduced egg weight
- Reduced shell weight

Cold Stress

Cold stress impairs performance of fast-growing broiler chicks. Moreover, it may lead to ascites (see p. 187).

Low Atmospheric Pressure

At high altitudes, fast-growing broiler chickens are more susceptible to ascites (see p. 187) because ascites is fairly common in broiler chickens raised at high altitudes. Experimentally, ascites can be induced by low oxygen pressure, for example, in hypobaric chambers (i.e., with low atmospheric pressure).

BEHAVIORAL PROBLEMS

There are a number of behaviors that impact poultry production.

Cannibalism/Persecution

Cannibalism/persecution can take multiple forms including:

- Vent picking (or the area below the vent) is a very severe form of cannibalism that is found in laying hens.

- Toe picking is most commonly seen in chicks or young game birds.
- Head picking usually follows injuries to the comb or wattles caused by freezing or by fighting between males.
- Blueback is caused by feather picking in turkeys, followed by exposure to sunlight. Blueback may result from overcrowding in the brooder, keeping the poults on the sunporch too long, and lack of sufficient fiber in the ration.
- Egg eating is costly but can be reduced by beak trimming.

Hysteria/Fright

Excessive fright can be observed in growing pullets or layers. With floor-housed birds, there is flight and piling up in the corners of the house, resulting in suffocation. Caged layers may attempt to fly, resulting in injuries to wings and legs. Care should be taken not to frighten birds.

 ## USEFUL WEBSITES

General Information
http://netvet.wustl.edu/birds.htm

Breeding Company Sites
Arbor Acres: http://www.aaf.com/home.htm

Hybrid Turkeys (Canada): http://www.hybridturkeys.com

Nicholas Turkey Breeding Farms: http://www.nicholas-turkey.com

---

BOX 11.1　New Issues in Poultry Health

IN OVO VACCINATION
Commercial equipment is available through Embrex (North Carolina, USA) for vaccination of fertile/embryonated eggs (see Figure 11.24). Automated vaccination is performed into the amnion at day 18 of incubation. This gives satisfactory protection, along with being a labor saving process for vaccination.

BIOTERRORISM
In the United States, Canada, and Western Europe, the industry and its employees may face animal vulnerability to bioterrorism from international terrorists, extreme animal rights activists, and criminal enterprises. Diseases might be introduced to and severely impact an industry, particularly where animals are located together in very high concentrations. The importance of biosecurity and trusted employees cannot be overestimated. This could potentially greatly damage the reputation of the product for domestic consumers and exports.

ANTIBIOTICS
Antibiotics have been used as growth "promotants" in poultry for about 50 years because they can improve growth rate and feed efficiency. Among the major growth-promoting antibiotics used by the industry are:

- Bacitracin (in methylene, disalicide, and zinc forms)
- Virginiamycin
- Barbermycin

(continued)

BOX 11.1 Continued

Figure 11.24 *In ovo* vaccination of eggs by the Inovojet® system. *(Courtesy of Embrex, Inc.)*

In the absence of antibiotics, there are reductions in growth, livability, and feed efficiency together with less uniformity in the birds. Uniformity is very important to processing and marketing.

Antibiotics act by changing microbial populations in the gastrointestinal tract and reducing subclinical infections. These antimicrobial agents require regulatory approval by government agencies/departments to establish their efficacy (do they work?) and safety (are they safe to animals and people consuming the livestock product?). In the United States, antibiotics (individual and combinations) must meet Federal Drug Administration (FDA) standards and approval process. Concerns are raised that the use of antibiotics is leading to antibiotic-resistant bacteria. This is hotly debated. Europe is moving to ban the use of antibiotics as growth promotants. This issue is covered in more detail in Chapter 18.

PROBIOTICS AND PREBIOTICS

Probiotics are live beneficial microorganisms (e.g., lactic acid bacteria) that aid digestion and/or reduce pathogenic bacteria (disease-causing in poultry or food-borne pathogens for people) in the gastrointestinal tract. Prebiotics are feed ingredients that modify the population of microorganisms in the gastrointestinal tract. Prebiotics include either nutrients used preferentially by benign bacteria or dead bacteria (e.g., *Aspergillus*), which enhance the gastrointestinal tract's immune responses. Grit that is high in calcium, oystershells, and other grit are still provided for broiler breeders (parent stock).

BOTANICS/BOTANICALS

It may be possible to use botanicals partially to replace growth-promoting antibiotics. These are extracts of specific plants (e.g., garlic, rosemary, anise, horseradish, juniper, yarrow, cinnamon, etc.). An example in poultry feed is Apex R3010 (from Braes Feed Ingredients). The company has reported improved performance in broiler chickens with their product and that it meets FDA GRAS (generally recognized as safe) criteria. Some botanicals can be used in the European Union and with producers of natural and organic poultry in North America.

BOX 11.2 Glossary of Poultry Health Terms

Active immunity This is immunity (antibody production) to disease acquired by the animal in response to the disease or to vaccination.

Acute A disease with a short and relatively severe course.

Allergy A severe reaction, or sensitivity, that occurs in some individuals following the introduction of certain antigens into their bodies.

Anemia A condition in which the blood is deficient in red blood cells and/or hemoglobin.

Anthelmintic A chemical that kills or removes worm parasites.

Antibiotic A chemical (usually produced by molds or bacteria) that kills or inhibits growth of bacteria.

Antibody A protein produced by immune tissues that binds antigens (e.g., pathogenic organisms).

Antigen A substance that stimulates formation of antibodies by the body.

Antiserum Serum containing specific antibodies. This may be used to treat a specific disease.

Antitoxin An antibody that will neutralize a toxin.

Atrophy Wasting away or diminution in size.

Autopsy Inspection (dissection, histology) of a dead animal to determine the cause of death.

Bacterin Killed suspension of bacterial organisms used for immunizing.

Bacteriostat A product that retards bacterial growth.

Broad-spectrum antibiotic An antibiotic killing both Gram-positive and Gram-negative bacteria.

(continued)

BOX 11.2 Continued

Carrier An apparently healthy animal that harbors disease organisms and is capable of transmitting them to susceptible animals.

Chronic A disease of long duration.

Coccidiostat A drug incorporated to prevent coccidiosis.

Contagious An infectious disease that may be transmitted readily from one animal to another.

Cyanosis Bluish discoloration of the skin, particularly the comb and wattles in birds.

Debilitating Weakening.

Edema Presence of abnormal amounts of fluid in tissues.

Emaciated A severe loss of weight.

Exudate A fluid oozing from tissue.

Hemorrhage Escape of blood from vessels; bleeding.

Intradermal An injection into the skin, between the layers.

Intramuscular An injection into a muscle.

Intraperitoneal An injection into the body or peritoneal cavity.

Intravenous An injection into a vein.

Lesion Gross or visible changes in size, shape, color, or structure of an organ.

Metabolism The chemical changes to macronutrients after absorption including (1) the processes by which nutrients are used in the growth/development or repair of body tissues, and (2) the breaking down processes in which nutrients are oxidized to provide energy for the animal.

Morbidity Sick rate.

Mortality Death rate.

Necrosis Death or dying of local tissue.

Oral Given by mouth.

Pathogenic Disease-producing.

Postmortem, or necropsy Examination after death.

Prophylaxis Preventive treatment against disease.

Rigor mortis Stiffening of the body after death.

Sporadic A disease outbreak occurring here and there.

Spore Bacterial or fungal stage of life capable of resisting unfavorable environmental conditions.

Subcutaneous An injection under the skin.

Symptoms Detectable signs of disease.

Therapy Treating disease.

Toxin Poison (e.g., produced by microorganisms).

Vaccine A suspension of dead microorganisms (bacteria or virus) or microorganisms that have had their pathogenic properties removed or attenuated but retained their antigenic properties.

Virulence The degree of pathogenicity of microorganisms.

FURTHER READING

Diseases of Poultry. Edited by B. W. Calnek, H. John Barnes, Charles W. Beard, Larry R. McDougald, and Y. M. Saif. Ames, IA: Iowa State Press, 1997.

Pattison, M. *The Health of Poultry.* Blackwell Science, 1993.

Sainsbury, D. *Poultry Health and Management: Chickens, Ducks, Turkeys, Geese, and Quail.* Blackwell Science, 2000.

12

Food Safety

> **Objectives**
>
> After studying this chapter, you should be able to:
>
> 1. Understand the importance of food safety.
> 2. List the major pathogens related to food safety issues in poultry.
> 3. Understand how irradiation improves food safety.
> 4. List the major government agencies regulating the safety of food.
> 5. List the different kinds of floors used in egg production.
> 6. Give recommendations on consumer handling of eggs and poultry meat.
> 7. Define *HACCP*.

Undoubtedly, food is several degrees safer today than it was 100 years ago, particularly in developed countries. Indeed, many countries boast of having "the safest food in the world." Policy makers, farmers, food processors, restaurants, and consumers have long recognized the importance of food safety—"from farm to fork." Regulations are in place and stringently enforced. Where problems have arisen, food items have been recalled.

There have been, however, significant outbreaks of diseases from food in North America (e.g., *Escherichia coli* O157:H7 in undercooked beef and unpasteurized apple cider and *Salmonella* serotype Typhimurium in poultry meat) and Western Europe (bovine spongiform encephalitis [BSE, or mad cow disease] in beef and *Salmonella* serotype *Enteriditis* in eggs).

There is a growing realization that the issue of food safety is still with us. Food-borne pathogens give rise to diseases that range in severity from simple diarrhea to life-threatening situations requiring hospitalization and even leading to death. Food-borne disease has a major impact in both developed and developing countries. For instance, the high estimates from the U.S. Centers for Disease Control (CDC) are that food-borne pathogens cause about 76 million illnesses, 325,000 hospitalizations, and 5,000 human deaths in the United States per year. Many food-borne illnesses are the result of animal products. For instance, *Escherichia coli* O157:H7 is as-

sociated with cattle. Pathogens associated with poultry and eggs include:

- *Listeria monocytogenes*
- *Campylobacter jejuni*
- *Salmonella* species such as:
  - *Salmonella* serotype Enteriditis (most frequently associated with eggs—see later in the chapter)
  - *Salmonella* serotype Typhimurium (the most commonly isolated serotype, which can colonize the crop of broiler chickens and contaminate chicken meat)

The impact of these food-borne pathogens in the United States is summarized in Table 12.1. Most at risk from food-borne disease are people whose immune systems are not functioning adequately (immuno compromised) such as the elderly, young children, people receiving chemotherapy, and individuals infected with HIV.

GOVERNMENT AND FOOD SAFETY

Most countries have consumer education programs about food safety and rigorously enforce regulations together with regulatory agencies that "police" the enforcement of regulations. Some countries have moved to a single department with responsibility for food safety (e.g., the United Kingdom). In others,

TABLE 12.1 Pathogenic Organisms in Food and Their Significance in the United States

Pathogen	Poultry Meat	Egg	Other	Food	Incidence of Food-Borne Diseases per Year	Hospitalizations per Year	Deaths per year	Economic Cost
Campylobacter jejuni or coli	Major	Minor	Minor	Hamburger, water, milk, mushrooms, cheese, pork, shellfish	1.96 million	11 thousand	99	$2.0 billion
Clostridium perfringens	Major	—	Major Minor	Meats, beans, seafood	0.25 million	41	7	—
Escherichia coli	Minor	Minor	Major Minor	Beef (particularly ground), apple cider, dairy, vegetables, melon	62 thousand[1]	2 thousand	52	$659 million
Listeria monocytogenes	Minor	—	Major Minor	Cheese, ground meat, dairy, seafood, vegetables	2.5 thousand	2.3 thousand	499	$2.3 billion
Salmonella (nontyphoid)	Major	Major	Major Minor	Meat, milk, fruits, vegetables, shellfish	1.34 million	16 thousand	553	
Staphylococcus aureus	Major	—	Major Minor	Sliced meats and dairy	0.18 million	1.7 thousand	2	

[1] Other E. coli; 106 thousand cases per year for a cost of $330 million.

Source: USDA Economic Research Service, 2000.

for instance the United States, there are independent agencies of the federal government with responsibility for food safety. These include the following:

1. USDA's Food Safety and Inspection Service (FSIS)
2. USDA's Animal and Plant Health Inspection Service (APHIS)
3. The Food and Drug Administration (FDA)
4. The Centers for Disease Control (CDC)

APHIS in its entirety or APHIS's Agricultural Quarantine and Inspection Program (which inspects food imports) may become part of the Department of Homeland Security.

PATHOGENS AND MICROORGANISMS AND FOOD SAFETY

Food contains microorganisms or has them on its surface, particularly when sealed packages are left open. Most microorganisms in food are *benign* (although they may cause spoilage of food). Some are *pathogenic*, that is, they can cause human or animal diseases. These pathogens include bacteria and viruses together with the less common protozoa.

Pathogens and other microorganisms are introduced into the food, as purchased from grocery stores, in the following manner:

1. During production on the farm
2. During transportation
3. During processing

Moreover, pathogens can be introduced into the food via other routes:

1. The store
2. The consumer's home
3. Restaurant kitchens
4. Community events (e.g., family reunion picnics and business, office, or church potluck meals)

If food is prepared correctly, which usually involves adequate cooking, the vast majority of the microorganisms will be killed or removed. It should be emphasized that microorganisms in food multiply rapidly at warm temperatures and in an environment with plentiful nutrients for bacterial growth. The danger zone is between 40° and 140°F (4.5°–60°C), with the rate of bacterial growth becoming higher as temperatures rise.

Food Safety Educational Programs

There are numerous food safety educational programs with very similar content (instructions) to en-

TABLE 12.2 Tips for Assuring the Safety of the Food We Eat

- Thoroughly *wash* hands and surfaces used for food preparation (using soap and antibacterial cleansers).
- Keep completely *separate* foods that are to be cooked (vegetables, meats and eggs, etc.) from foods to be eaten cold (e.g., salads).
- *Cook* at the correct temperature with the internal temperature of the food over 160° to 170°F (71°–77°C) depending on the food. Use of a cooking thermometer is recommended.
- *Store food* at temperatures that slow bacterial growth (i.e., in a refrigerator or freezer). Thawing of frozen food or defrosting should also be done at low temperatures, for instance, in cold water that is changed every 30 minutes.

sure food safety. One example is the Fight BAC™ (fight bacteria) program that was developed by the USDA's Food Safety and Inspection Service.

Food Handling for Safety

The requirements are so simple but so many of us forget. What is needed is covered in Table 12.2 but can be summarized as:

1. Washing hands and cleaning utensils and work surfaces including cutting boards to prevent cross-contamination.
2. Keeping foods separate during storage and during their preparation to prevent cross-contamination.
3. Cooking at sufficiently high temperature to kill pathogens.
4. Chilling food (storing refrigerated or frozen) to slow bacterial growth.

Eggs and poultry meat may harbor pathogens that give rise to food-borne diseases in the human population (see Table 12.1). In virtually all cases of pathogens from poultry or eggs, the organisms will be killed with adequate cooking (to over 160°F or 71°C). However, pasteurized eggs may be consumed raw, particularly in salad dressings (e.g., Caesar salad, mayonnaise).

FOOD SAFETY ASPECTS OF EGGS AND POULTRY MEAT

Pasteurization

Eggs are used in cooking either as table eggs or as egg products. The safety of egg products is greatly enhanced by their pasteurization (rapid heating and cooling in a process analogous to that with milk). This has been a legal requirement in the United States for egg products since 1970. The USDA's Food Safety and Inspection Service advises consumers that even when using egg products to heat them.

Salmonella and Poultry and Eggs

The U.S. Centers for Disease Control considers *Salmonella* to be a threat to public health in the United States with over 1.3 million cases per year and 16 thousand people hospitalized per year in the United States alone. Poultry meat and eggs are significant sources of infection (see Table 12.1). Figures 12.1 and 12.2 show *Salmonella* colonizing pig intestinal tissue. This situation is similar in poultry.

Salmonella and Poultry Meat

Salmonella contamination is found on both the outside surface and in the eviscerated body cavity of chickens, turkeys, and other poultry. This is the result of the following:

- Spilling gut contents, particularly from the colon and ceca (equivalent of the large intestine and rectum in mammals)
- Regurgitation of the contents of the crop

The number of salmonella in the crop contents is increased when feed is not present in the crop. Paradoxically, feed and water are withheld for a period before slaughter in an attempt to reduce release of feces.

Salmonella and Eggs

There is increasing evidence that eggs can be a significant source of *Salmonella*.

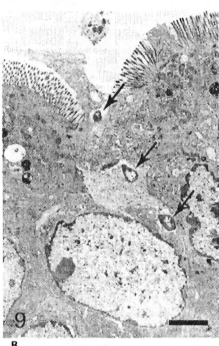

Figure 12.2 *Salmonella* typhimurium colonizing intestinal tissue. Transmission electron micrograph of ileum 25 minutes (A) and 60 minutes (B) after administration of *Salmonella* typhimurium. Organisms shown by arrows invading intestinal cells. *(Courtesy, Tom Stabel, USDA ARS National Animal Disease Center)*

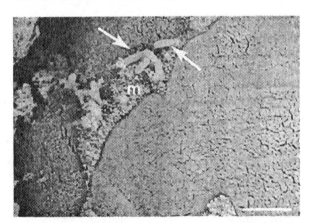

Figure 12.1 *Salmonella* typhimurium on intestinal tissue. Scanning electron micrograph of pig ileum 5 minutes after administration of *Salmonella* typhimurium. Organism shown by arrow adhering to intestinal cells. *(Courtesy, Tom Stabel, USDA ARS National Animal Disease Center)*

SALMONELLA CONTAMINATION OF THE EGG SURFACE. *Salmonella* has been isolated from the ceca of most old hens (between 24–65% in different flocks). It is not surprising that the surface of eggs may be contaminated with *Salmonella* and other bacteria. Thus,

Grade A eggs must be sold washed and disinfected. Unlike most food, eggs do not need to be washed by the consumer.

INCIDENCE OF *SALMONELLA* SEROTYPE ENTERITIS INFECTION.

Salmonella serotype Enteritis, one subtype of *Salmonella*, has emerged as an important cause of disease in people. The rate of *Salmonella* serotype Enteritis infection in the United States is rising in the human population. The CDC estimates the following rate of infection:

- 0.6 cases per 100,000 people in 1976
- 3.9 cases per 100,000 people in 1996

Someone infected with the *Salmonella* serotype Enteritidis bacterium usually has fever, abdominal cramps, and diarrhea beginning 12 to 72 hours after consuming a contaminated food or beverage. The illness usually lasts 4 to 7 days.

SALMONELLA CONTAMINATION OF THE EGG CONTENT.

Uncooked eggs are a real hazard. The CDC concluded that eggs, and specifically Grade A shell eggs, are the predominant source of *Salmonella* serotype Enteritidis infections, with uncooked or undercooked eggs being a particular threat to public health. The causative link between *Salmonella* serotype Enteritis and eggs was first recognized in the United Kingdom. Illnesses related to contaminated eggs have been increasingly documented in Europe and North America.

NOTE: *Salmonella* serotype Enteritidis can be found inside perfectly disinfected normal-appearing eggs. If the eggs are eaten raw or undercooked, the bacterium can cause illness.

Salmonella serotype Enteritidis can infect or colonize the ovaries and also the oviducts of hens and hence contaminate eggs (the yolk, respectively, which is much more likely, and the egg white) before the shells are formed. Only about 0.01% of eggs in the northeastern United States are internally contaminated with *Salmonella enteritidis.* In other parts of the United States, this is even less common. Few hens are infected at any one time, and they only occasionally lay contaminated eggs. *Salmonella* Enteritidis has been isolated from cecal cultures from about 0.3% of hens.

The Food and Drug Administration has implemented several actions to reduce the risk of illness and death caused by *Salmonella* enteritidis. The FDA requires that the retailer include the following statement on shell eggs that have not been specifically processed to destroy all live salmonellae:

SAFE HANDLING INSTRUCTIONS:

To prevent illness from bacteria:

- Keep eggs refrigerated.
- Cook eggs until yolks are firm.
- Cook foods containing eggs thoroughly.

Moreover, the FDA requires that the retailer refrigerate shell eggs promptly when they are received and to store the eggs at 45°F (7.2°C) or cooler.

Listeria

It is increasingly recognized that contamination with *Listeria* is a problem with chicken and turkey meat and products. The organism responsible is *Listeria monocytogenes* (see Figure 12.3). People eating food contaminated with this pathogen can develop listeriosis. This is an unusual but potentially very serious (or fatal) disease with symptoms such as fever and severe headache that may cause miscarriage (spontaneous abortion) or death. The latter is particularly a problem with the very young, the elderly, and others with immune impairment such as AIDS or people on chemotherapy.

The USDA's Food Safety and Inspection Service has recalled chicken and turkey products on multiple occasions. This leads to major financial losses to processors and reduces public confidence in poultry. Examples of products recalled in the United States include ready-to-eat chicken and turkey, chicken salad, chicken egg rolls, smoked turkey, turkey ham, chicken chow, and roast ducklings.

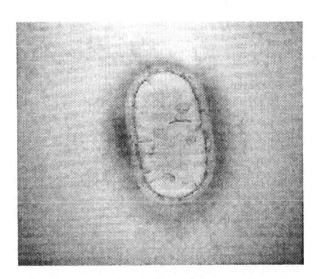

Figure 12.3 Electron micrograph of *Listeria monocytogenes.* (Courtesy, Irene Wesley, E. M. by Al Ritchie, USDA National Animal Disease Center)

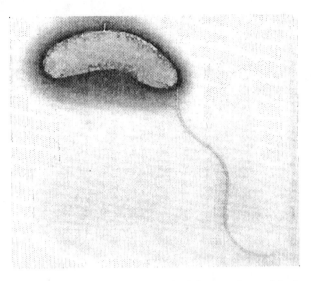

Figure 12.5 USDA inspection of poultry. *(Courtesy, USDA)*

Figure 12.4 Electron micrograph of *Campylobacter jejuni*. *(Courtesy, Irene Wesley, E. M. by Al Ritchie, USDA National Animal Disease Center)*

Campylobacter

Another emerging food-borne disease is *Campylobacter jejuni* (see Figure 12.4).

ROLE OF USDA'S FOOD SAFETY AND INSPECTION SERVICE

The Food Safety and Inspection Service (FSIS) protects public health by inspection of meat and poultry (see Figure 12.5). For instance, the FSIS recommends not using raw eggs in such recipes as:

- Hollandaise sauce
- Caesar salad
- Ice cream
- Eggnog
- Milkshakes
- "Health" drinks produced in a blender, etc.

    Also see Table 12.2.

    A satisfactory alternative is pasteurized eggs from processors who break and package liquid eggs. According to the FSIS, eggs with intact shells should be stored refrigerated because bacteria multiply rapidly at room temperature. The shelf life for table and shell eggs under these circumstances is 3 to 5 weeks.

    Again, according to the FSIS, it is not a good idea to purchase cracked eggs. If eggs are found to be cracked, the white and yolk should be removed from the shell, stored in a refrigerator, and consumed within 2 days. The importance of refrigeration is illustrated in Figure 12.6.

Figure 12.6 Importance of refrigeration in post-harvest food safety. *(Courtesy, Iowa Turkey Federation)*

HACCP-Based Inspection

The Pathogen Reduction and Hazard Analysis and Critical Control Point (HACCP) System was adopted as the new system for meat and poultry inspection in the United States in the late 1990s. This science-based system replaced traditional meat inspection (see Figure 12.5). The system includes:

- Establishment of standards for food safety and nonfood safety defects with the goal of reducing food-borne pathogens
- Ongoing verification of implementation
- Educational/training programs for HACCP teams

    The system is to safeguard the consumer from food-borne pathogens in poultry and other meat. It identifies what it calls Critical Control Points (CCPs) and uses various system approaches to reduce hazards. Examples of CCPs include (1) scalding freshly killed chickens in a common bath to loosen feathers; (2) defeathering many birds with the same rubber fingers; (3) eviscerating many birds with an improperly adjusted machine; and (4) cooling defeathered chickens in a common chill bath.

IRRADIATION

Irradiation of foods greatly reduces the contamination in microorganisms that are either pathogens (cause disease) or food spoilage organisms.

Irradiated foods are identified by the RADURA symbol (see Figure 12.7). The ionizing radiation can be generated from a radioactive source (e.g., cobalt 60 or cesium 137) or a linear accelerator (see Figure 12.8). There is abundant evidence that both processes are safe with no contamination of the food. Radiation pasteurization of meat extends the shelf life and represents a major step to reducing food-borne disease.

Radiation treatment of food has been used for spices and for food for astronauts for over 30 years. In the late 1990s, irradiation of meat was approved in the United States by the federal government: for poultry (1990), beef (1999), and pork (1999). It is beginning to be used. In the United States, irradiation of food is regulated jointly by the FDA and USDA's FSIS under a memorandum agreement.

The "jury is still out" as to whether the consumer will accept what should be a boon for public health. The name *radiation* conjures up "visions of

TREATED BY

IRRADIATION

DO NOT IRRADIATE AGAIN

LOT #

LINEAR ACCELERATOR FACILITY
IOWA STATE UNIVERSITY

Figure 12.7 RADURA—the *green* symbol for food that had been irradiated.

nuclear weapons and nuclear power stations." Some have argued to use the names *cold pasteurization* or *cold sterilization* instead.

Irradiated Meat

Labeling is required for irradiated meat. Some environmental and consumer activists have opposed irradiation treatment based on the stated concerns

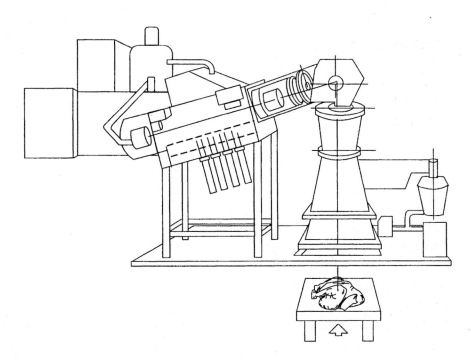

Figure 12.8 Schematic diagram showing irradiation of meat using electrons from a linear accelerator.

that it is a way of masking unsanitary conditions in processing plants. The meat industry is slow to adopt radiation treatment based on high capital costs for the equipment, energy costs, and concerns of a consumer "backlash." Some fast-food companies are beginning to sell irradiated meat in test markets.

ANTIBIOTICS

There is a growing movement to ban antibiotics from production—to improve the safety of the product. Antibiotics are used therapeutically to fight bacterial diseases in people and domestic animals. They have been used as growth promotants in poultry and also pigs since the 1950s. There is concern that this nontherapeutic use of antibiotics in healthy animals to promote growth is leading to antibiotic-resistant bacteria. The issue is debated with scientists on both sides. Many scientists and physicians working in the human health field are concerned that the use of antibiotics in livestock will lead to bacteria becoming resistant to the antibiotics. Hence, these antibiotics will no longer be useful to fight diseases in people. Other possible causes of antibiotic resistance include:

- Overprescription by physicians
- People not taking all the antibiotics prescribed

In the United States, the National Research Council and Institute of Medicine have concluded that "*the benefit to human health in the proper use of antibiotics in food animals is related to the ability for these drugs to combat infectious bacteria that can be transferred to humans by either direct contact with the sick animal, consumption of food contaminated with pathogens from animals, or proliferation into the environment.*"

Irrespective of the cause of antibiotic resistance, it is undoubtedly real. Concern about the agricultural use of antibiotics is leading to changes in usage. Sweden was one of the first countries to eliminate the nontherapeutic use of antibiotics. The European Union has announced that it will be banning the nontherapeutic use of antibiotics. In the United States, the Food and Drug Administration continues to allow the nontherapeutic use of antibiotics. Nongovernment organizations (NGOs), coalitions, and activists are discussing with large producers about the reduction or elimination of antibiotics as growth promotants. Some companies reportedly have been reducing their use of antibiotics while others will not be feeding medically important antibiotics to poultry. Judicious use of antibiotics for poultry diseases is covered in Chapter 11.

 USEFUL WEBSITES

For Poultry Inspection and HACCP— see USDA FSIS

http://www.fsis.usda.gov

13

Poultry Houses and Equipment

Objectives

After studying this chapter, you should be able to:

1. List and discuss major factors in determining building placement.
2. Define *vapor barrier* and give its function.
3. List the major reasons for insulating buildings.
4. Discuss the purpose of lighting in buildings.
5. Give recommendations for lighting of different classes of poultry.
6. Discuss heat production and loss by a bird, giving reasons for its consideration in building construction.
7. Define *insulation*, list desirable properties, and give common examples of insulating materials.
8. Define *R-value.*
9. List and describe the main types of building floors.

Poultry housing and equipment are needed for the comfort, protection, and efficient production of eggs and meat. The jungle fowl, ancestor of the modern chicken, was protected from the sun and the wind by vegetation. The natural laying and mating season of many birds is in the spring of the year, stimulated by the increasing day length. It follows that for maximum egg production and growth during other seasons of the year, spring-like conditions should be emulated. This is achieved by housing, which provides protection from weather, plus providing artificial light. Poultry houses also make it possible to care for the flock in accordance with recommended practices, using a minimum of effort.

No standard set of poultry buildings and equipment can be expected to be well adapted to such diverse conditions and systems of poultry production as exist in the United States or elsewhere. This is evidenced by a map showing three U.S. winter temperature zones (Figure 13.1) as related to the amount of insulation needed in poultry houses. Extremes in temperature, and the duration of these extremes, along with other climatic factors that are closely associated with temperature, such as amount of sunlight, rainfall, latitude, and elevation, determine the type of housing and construction necessary. Thus, detailed plans and specifications for poultry houses in a given area should be obtained from a local architect or from the college of agriculture of the specific state.

LOCATION

It is important that poultry buildings be located and planned correctly for efficient, profitable, and pleasant operation. To this end, consideration should be given to the following (also see discussion for nuisance to neighbors in Chapter 9 and biosecurity issues in Chapter 11):

1. **Water supply.** High-quality water must be available and plentiful.
2. **Roads.** It is preferable that poultry buildings be located near an all-weather road or highway that is well maintained (but not used for transportation of other poultry). Normally, a location along an all-weather road has better access to electric and telephone lines.
3. **Telephone and electricity.** The units should be near well-maintained telephone and electric lines.
4. **Topography.** The topography should be high and level with no abrupt slopes. A relatively

CLIMATE	TOTAL RESISTANCE (R)	
ZONE	WALLS	CEILINGS
MILD 9		12
MODERATE 9–14		16
COLD 14		23

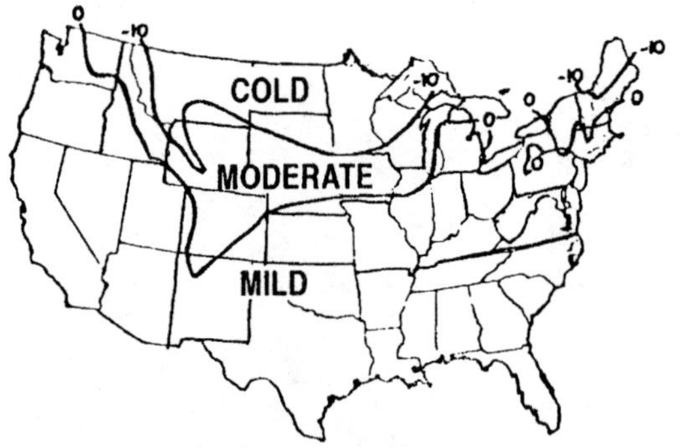

Figure 13.1 Recommended insulation, showing three U.S. winter temperature zones as related to the amount of insulation needed in poultry houses. (*Adapted from* Structures and Environment Handbook, *7th ed., Midwest Plan Service, Iowa State University*)

level area requires less site preparation, thereby lowering building costs.

5. **Drainage.** The soil should be porous and the slope gentle, which contributes to dryness. Construction should not be close to waterways or wetlands.

6. **Layout of operations.** Prior to starting construction at a new site, poultry producers may avoid much subsequent difficulty and expense by first planning (with a specialist as consultant). First, sketch out the buildings and equipment required to meet the needs in the most efficient and economical manner, with room for expansion. Consideration should be given to the following:

   a. **Plans for the flow of all materials.** Producers should develop detailed plans for the flow of all materials, with primary consideration given to maximum automation and disease prevention. These plans should include provision for (1) delivering the feed to the birds at the desired time and place, (2) providing a water supply, (3) delivering and distributing litter, (4) removing manure, and (5) marketing. All these considerations, and more, enter into the handling of materials to, within, and from buildings.

   The layout of operations should first be put on paper in sketch form. From this, the architect and/or engineer can design, or recommend for purchase, buildings and equipment that most effectively and eco-

nomically meet the production requirements of the enterprise.

   b. **Landscaping (trees, and so on).** Landscaping is a valuable attribute. Trees and bushes provide privacy from neighbors, disrupt wind pattern and disperse odors, and reduce local temperatures. A grassy area aids rodent detection and makes the facility look well cared for.

   c. **Orientation.** Fortunately, the poultry farm need not be oriented with the compass. Although in general the farmstead plan will be developed to present the front to the road, most buildings can be turned, quarter-turned, or reversed, as may be necessary to take advantage of the prevailing winds, sunlight, and view. In general, poultry houses are placed with the long axis north and south.

   d. **Fire protection.** Poultry buildings should be far enough apart so that fire will not spread easily from one building to another. In general, this means at least 100 ft (30 m) apart in the case of large buildings. In acquiring added fire protection through spacing buildings farther apart, avoid extreme distances that reduce efficiency; fire insurance is probably cheaper than labor.

   e. **Appearance.** Careful attention to the arrangement of buildings adds to the attractiveness of the entire unit. Unsightly objects (e.g., lagoons, etc.) should not be visible from the highway; fences and buildings should be repaired and painted regularly; and yards and driveways should be kept free of rubbish, scattered equipment, and so forth.

   f. **Expansion.** Provision should be made for easy expansion. Many times buildings can be expanded in size by extending their length, provided no other buildings or utilities interfere.

SPACE REQUIREMENTS OF BUILDINGS AND EQUIPMENT

The intensification of poultry production has created the need for scientific, yet practical, information relative to space requirements. Moreover, the recommendations have changed from time to time, as a result of new experiments and experiences. One of the first and frequently one of the most difficult problems confronting the producer who wishes to construct a building or item of equipment is that of

TABLE 13.1 Space Requirements for Poultry

Type of Bird	Type of Facility	Age of Birds (wk)	Space/Bird (sq ft)	Space/Bird (sq cm)	Feeder Space/Bird Linear (in.)	Feeder Space/Bird Linear (cm)	Water Space/Bird Linear (in.)	Water Space/Bird Linear (cm)
Commercial layers	Floor	0–4	0.30	279	1.0	2.5	0.2	0.5
	Floor	4–8	0.60	557	1.0	2.5	0.4	1.0
	Floor	9–16	1.25	1,161	1.5	3.8	0.6	1.5
	Floor	16	1.50	1,394	1.5	3.8	1.0	2.5
	Cage, individual	over	0.50	465	3.0	7.6	1.5	3.8
	Cage, colony	Adult Adult	0.50	465	3.0	7.6	1.5	3.8
Broiler—breeder pullets	Floor	0–8	0.80	743	1.0	2.5	0.5	1.3
	Floor	9–16	1.30	1,208	3.0	7.6	0.6	1.5
	Floor	16 over	2.00	1,858	4.0	10.2	1.0	2.5
Broiler—breeder hens	Floor	Adult	2.50	2,322	4.0	10.2	2.0	5.0
	Slat floor	Adult	2.00	1,858	4.0	10.2	2.0	5.0
Broilers	Floor	0–4	0.30	279	1.0	2.5	0.2	0.5
	Floor	4–8	0.75	697	1.0	2.5	0.2	0.5
Roasters	Floor	0–8	0.80	743	1.0	2.5	0.5	1.3
	Floor	9–16	1.50	1,394	2.0	5.0	1.0	2.5
Turkey—breeders	Confinement on floor	Adult	4.50	4,181	3.0	7.6	1.0	2.5
	Confinement and range	Adult	2.50 in house plus range	2,322	3.0	7.6	1.0	2.5
Turkeys—market	Confinement on floor	0–4	1.25	1,161	1.0	2.5	0.5	1.3
	Confinement on floor	4–16	2.50	2,322	2.0	5.0	1.0	2.5
	Confinement on floor	16–29	4.00	3,716	2.5	6.4	1.0	2.5
Duck—breeders	Confinement and yard	Adult	2.50 in house plus yard	2,322	2.0	5.0	1.5	3.8
Ducklings—market	Floor	0–3	1.00	929	1.0	2.5	0.5	1.3
	Floor	3–5	1.50	1,394	1.5	3.8	1.0	2.5
	Floor	After 5	2.00	1,858	2.0	5.0	1.5	3.8
Geese	Floor	0–2	1.25	1,161	1.5	3.8	1.0	2.0
	Floor	After 2	2.50	2,322	2.5	6.4	2.0	5.0

Source: "Report of the Committee on Avian Facilities," Poultry Science, 1974, 53: 2257, with turkeys, ducks, and geese added.

arriving at the proper size or dimensions. Suggested space requirements are given in Table 13.1.

Poultry Houses

Birds of good genetics, no matter how well fed, will not perform optimally unless well housed. Good housing is essential to the comfort and health of the birds. It also allows for convenience and ease of management for the poultry producer at a reasonable cost. In order to design any building properly it is necessary to know the purpose for which it is to be constructed. In addition, the designer must know the nature and number of units to be housed, their size and the space required for each, and the conditions required for best operation or production. The designer must also be aware of the normal weather conditions of the area and have a knowledge of available materials that are best suited for use under the given conditions.

Requirements of Poultry Houses

There are certain general requirements of all poultry buildings that should always be considered; among them, reasonable construction and maintenance costs, minimum labor costs, and minimum utility costs. For poultry, however, increased emphasis needs to be placed on the following features of buildings:

learned about environmental control, but the gap between awareness and application is becoming smaller.

The critical temperature is that temperature at which the heat created by digestion and body metabolism just equals that which the bird dissipates by convection, evaporation, radiation, and conduction. The comfort zone is the range in temperature within which the bird may perform with little or no discomfort. At temperatures below the comfort zone, additional nutrients need to be converted to heat to keep the body warm; and at temperatures above the comfort zone, nutrients are needed to help keep the bird cool. The optimum temperature is the temperature at which the bird responds most favorably, as determined or measured by maximum rate of gains or production, feed efficiency, and/or production (see Figure 13.5).

The critical temperature and comfort zone vary with different species, ages, breeds, and the physiological and productive status. The species' differences result primarily from the kinds of thermoregulatory mechanism provided by nature, such as type of coat (feathers, hair, wool), sweat glands, and so on.

The temperature varies according to age, too. For example, the comfort zone of baby chicks is about 95°F (35°C), whereas the comfort zone of adult layers is 60° to 65°F (15° – 18°C) and for broilers, 70° to 75°F (21°–24°C).

Stresses at both high and low temperature are increased with high humidity. The respired air has less of a cooling effect. As humidity of the air increases, discomfort at any temperature increases, and nutrient utilization decreases.

Air movement (wind) results in body heat being removed at a more rapid rate than when there is no wind. In warm weather, air movement may make the birds more comfortable; but in cold weather, it adds to the stress of temperature. At low temperatures, the nutrients required to maintain body temperature are increased as the wind velocity increases. In addition to the wind, a drafty condition where the wind passes through small openings directly onto some portion or all of the body will usually be more detrimental to comfort and nutrient utilization than the wind itself.

Environmentally controlled buildings are costly to construct, but they make for the ultimate in bird comfort, health, and efficiency of feed utilization. Also, they lend themselves to automation, which results in a saving in labor; and, because of minimizing space requirements, they effect a saving in land cost. Today, environmental control is rather common in poultry, and it will increase.

Before an environmental system can be designed for poultry, it is important to know the (1) heat production, (2) vapor production, and (3) space requirements (see Table 13.1). This information is as pertinent to designing poultry buildings as nutrient requirements are to balancing diets.

Heat Production of Poultry

The heat production of a laying hen is given in Table 13.2. This may be used as a guide, but in doing so, consideration should be given to the fact that heat production varies with age, body weight, diet, breed, activity, house temperature, and humidity at high temperatures. Table 13.2 gives both total heat production and sensible heat production. Total heat production includes both sensible heat and latent heat combined. Latent heat refers to the energy involved in a change of state and cannot be measured with a thermometer; evaporation of water or respired moisture from the lungs are examples. Sensible heat is the portion of the total heat (measurable by a thermometer) that can be used for warming air and compensating for building losses.

Since ventilation also involves a transfer of heat, it is important to conserve heat in the building to maintain desired temperatures and reduce the need for supplemental heat. In a well-insulated building, mature birds may produce sufficient heat to provide a desirable balance between heat and moisture, but newly hatched poultry will usually require supplemental heat. The major requirement of summer ventilation is temperature control, which requires moving more air than in the winter.

TABLE 13.2 Heat Production of a Laying Hen

Heat Source	Unit		Heat Production, BTU/Hr			Heat Production, Kcal/Hr		
			Temperature	Total	Sensible	Temperature	Total	Sensible
	(lb)	**(kg)**	**(°F)**			**(°C)**		
Layer hen	4.5	2.04	55	40	28	12.8	10.1	7.1

Source: Adapted from Agricultural Engineers Yearbook, *St. Joseph, MI, ASAE Data Sheet D-249.2, p. 424.*

Vapor Production of Poultry

Birds give off moisture during normal respiration; and the higher the temperature the greater the moisture. This moisture should be removed from buildings through the ventilation system. Most building designers govern the amount of winter ventilation by the need for moisture removal. Also, cognizance is taken of the fact that moisture removal in the winter is lower than in the summer; hence, less air is needed. However, lack of heat makes moisture removal more difficult in the wintertime. Table 13.3 gives the information necessary for determining the approximate amount of moisture to be removed for layers.

Recommended Environmental Controls

The comfort of poultry is a function of temperature, humidity, and air movement. Likewise, the heat loss from birds is a function of these three items.

Temperature, humidity, and ventilation recommendations for layers, broilers, and turkeys are given in Table 13.4. This table will be helpful in obtaining a satisfactory environment in confinement poultry buildings, which require careful planning and design.

The prime function of the winter ventilation system is to control moisture, whereas the summer ventilation system is primarily for temperature control. If air in poultry houses is supplied at a rate sufficient to control moisture, that is, to keep the inside relative humidity in winter below 75%, then this will usually provide the needed fresh air, help suppress odors, and prevent an ammonia buildup.

Lighting

The intensity of light required by poultry (about 1 footcandle) is much less than is required for plant growth or for an everyday activity of people such as reading.

1. Layers. Artificial light in the laying house should give a 16-hour day. If natural daylight is longer, artificial light should maintain the longest daylight period, although a light regimen in excess of 17 hours is of doubtful value. Use intensity of 1 foot-candle at bird height. (Approximately 40-watt bulb with reflector every 12 ft, 7 ft above the floor in floor systems.) Never decrease day length or light intensity during the laying period. A photoelectric cell may be used in connection with a time clock to turn lights off at dawn and on at twilight.

2. Replacement pullets. During the growing period, pullets should not be exposed to increasing amounts of light prior to 21 weeks of age if best egg production is to be attained. Rather, short or decreasing light during the growing period is desirable. The following lighting regimen is recommended for pullets: 0 to 3 weeks, 1.0 foot-candle, 20 hours of light and 4 hours of darkness; 3 to 12 weeks of age, 0.1 to 0.5 foot-candle, 16 hours of light and 8 hours of darkness; 12 to 21 weeks of age, 0.5 footcandle, continue 16 hours of light and 8 hours of

TABLE 13.3 Vapor Production of Laying Hens

Vapor Source	Unit		Temperature		Vapor Production		Vapor Production	
	(lb)	**(kg)**	**(°F)**	**(°C)**	**(lb/hr)**	**(BTU/hr)**	**(kg/hr)**	**(kcal/hr)**
Layer hen	4.5	2.04	50	10	0.012	12	0.005	3.0

Source: Adapted from Agricultural Engineers Yearbook, *St. Joseph, MI, ASAE Data Sheet D-249.2, p. 424.*

TABLE 13.4 Recommended Environmental Conditions for Poultry[1]

| Kind of Poultry | Temperature | | | | | Commonly Used Ventilation Rates[1] | | | Drinking Water | | | | |
	Comfort Zone		Optimum		Acceptable Humidity	Basis	Winter	Summer	Winter		Summer	
	(°F)	**(°C)**	**(°F)**	**(°C)**	**(%)**		**(cfm)**	**(cfm)**	**(°F)**	**(°C)**	**(°F)**	**(°C)**
Layers	50–75	10–24	55–70	13–21	50–75	per bird	2	5	50	10	60–75	15–24
Broilers	95[2]	35	70[3]	21	50–75	per lb body weight	0.5	1	50	10	60–75	15–24
Turkeys	95–100[2]	35–38	70[3]	21	50–75	per lb body weight	0.5	1	50	10	60–75	15–24

[1]Two different ventilating systems may be provided; one for winter and one for summer.
[2]First week after hatching.
[3]After sixth week.

darkness to 16 weeks of age, followed by decreasing to 8 hours of light and 16 hours of darkness at 21 weeks of age. Long or increasing light should begin at 21 weeks of age, with the light increased at the rate of 15 minutes per week until 17 hours, then maintained. One 25-watt bulb with reflector every 12 ft, 7 ft above the floor will give approximately 0.5 foot-candle intensity.

3. Broilers. Lighting in lighttight houses might be as follows: 0 to 5 days, 3.5 foot-candle continuous light; 6 days to market, 0.35 foot-candle, 23 hours of continuous light and 1 hour of darkness. Or from 6 days to market, intermittent light may be used; 1 hour of dim light (feeding time) to 3 hours of darkness (resting time). The following lighting is recommended for broilers in open-sided houses: 0 to 2 days, 3.5 foot-candle of continuous light; 2 days to market, 0.5 foot-candle, 23 hours of continuous light and 1 hour of darkness.

4. Turkeys. Market turkeys: 0 to 2 weeks, 10 to 15 foot-candle, continuous light; 2 weeks to market, 0.5 to 1.5 foot-candle, 16 hours of continuous light and 8 hours of darkness. Turkey breeders: same lighting as market turkeys to 29 weeks of age. After 29 weeks, 5 to 7 foot-candle, 14 hours of light and 10 hours of darkness in spring and summer (or 8 hours of light and 16 hours of darkness in the fall).

Ventilation

A laying hen breathing 40 times per minute uses air at the rate of 0.019 cu ft per minute (0.03 m^3/hour). Compared with each other, the air breathed in is 5% richer in oxygen and the air breathed out 5% richer in carbon dioxide. But respiration needs are greatly overshadowed by the rate of air change necessary to remove moisture from a tightly built house. In most instances, if ventilation is adequate to keep the house, particularly the litter, dry, it will more than meet the air requirements of the birds.

The factors affecting and the procedure for calculating ventilation requirements of poultry are discussed later in this chapter in the several subsections under Environmental Factors. Only the ventilation equipment is presented at this point. Forced-air ventilation is used in most commercial poultry houses. Most systems use exhaust fans and air intakes. Poultry specialists should be consulted in planning and installing a fan system for ventilation.

Design and Construction

The types, uses, and plans of poultry houses are very diverse. Poultry housing may be one of the following types: (1) open shed, (2) semienclosed, or (3) enclosed with light and temperature control.

Open sheds are suited for free-range birds together occasionally with breeder turkeys, plus ducks, geese, and game birds.

Semienclosed houses give more protection to the birds and provide more environmental control.

Enclosed houses are environmentally controlled, with the light, temperature, and ventilation regulated.

Poultry enterprises have buildings adapted to their specialized purposes. Thus, chicken houses are designed for the brooding of chicks, replacement pullets, layers, or broilers; and turkey houses are designed for breeder turkeys, brooding poults, or market turkeys. Although types, purposes, and plans of poultry houses differ widely, certain principles are similar; that is, certain principles are observed relative to roof types, floor types, cage types, and building materials.

Floor Types

Floors may be either solid or slotted. **Solid floors** should be dry, easily cleaned, rat-proof, and durable. They may be and are constructed of numerous materials including the following: clay, clay with a concrete border, plank, concrete, concrete with board surfacing, cork brick, creosoted wooden blocks, cinders, or various combinations of these materials. Regardless of the type of flooring material, for a good dry bed there should be a combination of surface and subsurface drainage, together with a cover provided by a suitable absorbent litter. After considering both the advantages and the disadvantages of the many types of flooring material, most producers are agreed that where a solid floor is desired, concrete is usually the most satisfactory. Concrete floors are sanitary, durable, and rat-proof. If properly constructed, they are dry; and if bedded with litter, they are not cold.

Dirt floors and deep litter are sometimes used in broiler houses. They are much less expensive than other types of floors, and many growers are well satisfied with them. In some cases, the entire accumulation of litter and manure is removed after each lot of broilers is sold; in other instances, the litter is used for three or four lots before changing.

Slotted floors are floors with slots through which the feces pass to a storage area below or nearby (see Figure 13.6). Such floors are not new; they have been used in Europe for over 200 years. The advantages of slotted floors are (1) less space per bird is needed, (2) bedding is eliminated, (3) manure handling is reduced, (4) sanitation is increased, and (5) labor savings are increased.

Figure 13.6 Example of slotted floor. *(ISU photo by Bob Elbert)*

The disadvantages of slotted floors are (1) higher initial cost than conventional solid floors, (2) less flexibility in the use of the building, (3) any spilled feed is lost through the slots, and (4) environmental conditions become more critical.

Slotted floors may be used for the entire floor area, or they may cover only about half the floor, usually a strip down the center of a long house. In both cases, a pit is provided below the slotted floor for the collection of manure. Such pits are often equipped with mechanical scrapers for periodic removal of manure. In some installations, the pits are deep enough to permit removal of manure by the use of a tractor with a scraper blade.

Cage Types

The vast majority of layers are maintained in cages. Also, replacement pullets can be raised in cages.

Building Materials

Several types of building materials are used in poultry house construction; among them, wood, concrete blocks, hollow tile, and metal (steel or aluminum).

Wood houses are most common because of ease of construction and availability of material. The outside wall is generally drop siding. There is considerable heat loss through a single layer of wood; hence, the use of sheathing or insulation board on the inside, to provide double-wall construction, is recommended. Wood houses require frequent painting.

Concrete block houses are not widely used for poultry because they are costly and there is considerable heat loss through the walls. Some houses are built of blocks made of cinders and concrete, which are lighter and less costly than concrete.

Hollow tile houses are popular in areas where tile is produced and is cheaper than other con-

struction material. The tile is durable, does not require paint, inside insulation is not necessary, the walls are rat-proof, and the house is fireproof.

Metal houses (steel or aluminum) are durable and easy to keep clean, but expensive. Double-wall construction or insulation is recommended, because single-wall houses are difficult to keep warm during winter.

Environmental Factors

Effective environmentally controlled poultry houses have a variety of features designed to provide an optimum environment of temperature, moisture, air movement, and gas content—features such as tight construction, insulation, vapor barriers, and ventilation. To understand the basis of all the environmental factors, however, it is first necessary to know how heat is produced and eliminated by the bird.

Heat Production and Elimination by the Bird

During the process of body metabolism, heat is liberated as a by-product. But the process is not constant; it is affected by the following:

- Oxygen intake
- Feed intake
- Activity of the bird
- Production parameters (growth rate, muscle protein synthesis, egg production, etc.)

Increases in any of these factors will lead to greater heat production in the deep-body section of the chicken. This heat must be liberated; otherwise, the body temperature will rise. Because of the variation in the heat of metabolism, the body temperature of the chicken fluctuates between 105° and 107°F (40.5°–41.5°C).

Deep-body temperature varies; it is higher in small birds than in large birds, higher in males than in females, higher in active birds than in inactive ones, and so on. Also, it rises with the presence of the food in the digestive tract, with increased activity, and with higher ambient temperature. Also, deep-body temperature is highest in the morning and lowest in the evening (about 3 hours after darkness).

Heat is moved from the body of the bird to the air by the following four pathways:

1. **From the skin**—(known as sensible or direct heat loss) by (a) radiation, (b) convection, and (c) conduction.
2. **By vaporization**—of moisture in the respiratory tract (insensible or indirect heat loss).

3. By excretion—(not great).
4. By production—of eggs (minor).

The amount of heat lost from the skin surface to the surrounding air by radiation depends on the spread between the two temperatures. When air temperature is low, heat loss by radiation is great; when air temperature is high, heat loss by radiation is low. At ambient temperatures that are optimum for the bird's well-being, about 75% of the total heat lost from the skin is by radiation and is the major means of dissipation.

When cool air comes in contact with the surface of the bird, the air is warmed, expands (becomes lighter), and rises. As heat is carried away with the warm air, cooler air moves in; and the process is repeated. Heat lost in this manner is by convection.

Loss of heat from the surface of the bird by conduction is that type that occurs when the body surface of the bird comes in contact with any cooler surrounding object, such as the floor or soil. Generally, this type of heat loss is minimal.

As room temperature rises, less sensible heat is lost from the body and the bird resorts to vaporization—the changing of water from liquid to vapor in the respiratory tract—as a means of removing heat. Heat is necessary to change water from liquid to vapor. At ambient temperatures of about 86°F or above, the bird pants in an endeavor to cause more air to (1) move through the respiratory system, (2) become vaporized, and (3) carry off more heat; and the higher the ambient temperature, the faster the panting. If the inhaled air is completely saturated (has 100% relative humidity), it cannot absorb moisture in the respiratory tract; hence, there can be no vaporization, and the bird has lost its last chance to survive.

Unlike heat lost through the skin, heat lost by vaporization does not raise the room temperature. It does cause the birds to drink more water in order to prevent dehydration, often resulting in wet droppings. Table 13.5 gives estimates of a bird's heat loss.

TABLE 13.5   Poultry Heat Production

Type of Heat	5-lb (2.3kg) Bird at Building Temp. of 55°F (13°C)	
	BTU/hr/bird	**(Kcal/hr/bird)**
Sensible heat	31.2	(7.9)
Insensible heat	13.0	(3.3)
Total heat	44.2	(11.2)

Source: Structures and Environment Handbook, *7th ed., Midwest Plan Service, Iowa State University, p. 165, Table 214–7.*

Insulation

Insulation is any material that reduces the rate at which heat is transferred from one area to another. Although all building materials have some insulation value, the term *insulation* is generally reserved for those products that provide this one service.

Insulation has several functions, the most important of which are:

1. **To reduce the rate of heat loss from buildings during cold weather.** Insulation helps conserve heat during periods of cold weather. Conserving bird heat helps maintain desirable housing conditions without the addition of great amounts of supplemental heat.
2. **To reduce the rate at which heat passes into buildings during hot weather.** Insulation helps reduce the rate of heat gain in hot weather, an important consideration, for instance, in the South (United States).
3. **To control moisture condensation by making the wall and ceiling surfaces warmer.** In a poorly insulated building, the inside ceiling and wall surfaces become cold in the winter, bringing discomfort to birds. When the surface temperature is low enough, the air next to the surface becomes saturated and moisture condenses. If the surface temperature is below freezing, frost forms. Well-insulated buildings reduce condensation by keeping the walls and ceilings relatively warm.
4. **To reduce frost heaving.** Correct foundation insulation (insulating the outside of the foundation wall if possible) reduces frost heaving, keeps floors warmer, and reduces heat losses.

Most insulation materials are bulky, porous, and lightweight, with countless air spaces. Generally speaking, the lighter the material, the better its insulating properties; and the more air pockets in the material, the better it insulates. Some building materials, like wood, are good insulators, while others, like concrete and metal, are poor insulators. Figure 13.7 shows how thick building materials must be to have the same insulating value as glass wool batt.

Insulation, as well as other building materials, is rated according to its ability to resist the flow of heat. This is commonly referred to as the R-value. The R, or insulation value, may be given per inch (2.5 cm) of thickness, or for the total thickness of a material. The relative insulation values of some common building materials are given in Table 13.6. Table 13.7 shows insulation values for some common types of construction, including roofs and ceil-

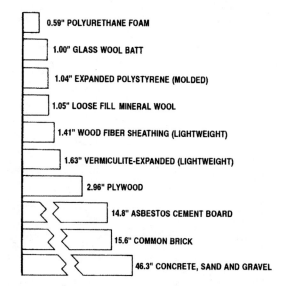

Figure 13.7 Thickness of materials required to equal the insulation value of 1 in. glass wool batt. *(Structures and Environment Handbook, 7th ed., Midwest Plan Service, Iowa State University, p. 162, Fig. 214-6. To convert inches to centimeters, multiply by 2.54)*

ings. Tables 13.8, 13.9, and 13.10 show heat output of a bird, sensible and insensible heat loss, and recommended airflows, respectively.

Vapor Barrier

In poultry buildings, and in other buildings where the relative humidity is high, insulation should be protected from moisture. Moisture, in the form of water vapor, tends to move from warmer moist areas to the cooler outside. So, moisture enters the wall, moves outward, then condenses when it reaches a cold enough area. Condensed water in the wall greatly reduces the value of the insulation and may damage the wall. To eliminate this flow of moisture, a vapor barrier should be placed near the warm side of the wall, preferably immediately beneath the interior lining. Common vapor barriers are aluminum foil, 4 mil plastic film, and some of the asphalt-impregnated building papers.

Insulation Needs

The amount of insulation needed in poultry houses varies according to winter temperature (see Figure 13.1).

Airflow

When designing a large poultry house, the air exchange needs to be calculated. The ventilation requirements in a poultry house are given in Table 13.10.

Prefabricated Houses

Pre-engineered and prefabricated poultry houses—complete with insulation, ventilating system, and all equipment—are important. Fabricators of such buildings have the distinct advantages of (1) price savings due to purchase of materials in quantity lots, (2) economical and controlled fabricating, and (3) well-trained personnel for developing the best in plans and specifications.

From the standpoint of the poultry producer, a prefabricated unit is often appealing because (1) the full cost of such a building is known in advance, rather than a contractor's estimate; (2) financing may be easier to arrange as a result of the entire transaction being carried out with one supplier; and (3) construction will likely take less time than when a local contractor is involved.

Some poultry producers, however, still prefer to design and build their own houses and to select and install the necessary equipment. By so doing, they may be able to effect a savings.

Requirements of Poultry Equipment

The size and design of poultry equipment may differ; that is, not all self-feeders are the same. Yet there are certain fundamentals of poultry equipment that are similar regardless of the kind of equipment, the design, or the size. These requirements are:

- Equipment should be useful, practical, and efficient.
- Poultry equipment receives heavy use. Thus, it should be strongly and durably built.
- Poultry equipment should be dependable so that it will function without getting out of order. Overly complicated equipment sometimes requires more of the operator's time than practical, noncomplex equipment.
- Low annual cost and upkeep.
- Some poultry equipment should be movable so that it may be shifted from one location to another (subject to biosecurity).
- Poultry equipment should be accessible.
- Save feed. Much feed may be saved when fed in properly constructed equipment. When such equipment is used, birds eat their feed without throwing it out of the feeder.
- Modern equipment reduces the labor required in caring for the birds.
- Poultry equipment should prevent pollution.

Automation

Automation is a coined word meaning the mechanical handling of materials. Producers automate to reduce labor costs. Modern equipment has eliminated much

TABLE 13.6   Insulation R-Values of Some Common Materials

Material	Insulation Value[1,2]	
	Thickness	For Thickness Listed
	(in.)	**(in.)**
Batt or blanket insulation:		
Glass wool, mineral wool, or fiberglass	3.70	
Fill-type insulation:		
Glass or mineral wool	3.00–3.50	
Vermiculite (expanded)	2.13–2.27	
Shavings or sawdust	2.22	
Paper or wood pulp	3.70	
Rigid insulation:		
Wood fiber sheathing	2.27–2.63	
Expanded polystyrene, extruded	4.00–5.26	
Expanded polystyrene, molded	3.57	
Expanded polyurethane (aged)	6.25	
Glass fiber	4.00	
Ordinary building materials:		
Concrete, poured	0.08	
Plywood, ⅜ in.	1.25	0.47
Plywood, ½ in.	1.25	0.63
Hardboard, ¼ in.	1.00–1.37	
Cement asbestos board, ⅛ in.		0.03
Lumber (fir, pine), ¾ in.	1.25	0.94
Wood-beveled siding, ½″ × 8″		0.81
Asphalt shingles		0.44
Wood shingles		0.94
Window glass, includes surface conditions:		
Single-glazed		0.89
Single-glazed plus storm windows		1.79
Double-pane insulating glass		1.45–1.73
Air space (¾ in. or larger)		0.90
Surface conditions:		
Inside surface		0.68
Outside surface (15 mph wind)		0.17

[1]Mean temperature of 76°F (24°C).
[2]To convert inches to centimeters, multiply by 2.54.
Source: Structures and Environment Handbook, *7th ed., Midwest Plan Service, Iowa State University, p. 162, Table 214–1.*

labor needs such as feeding, watering, adding litter, cleaning, and handling eggs, which have all been mechanized. Producers are using self-unloading trucks, self-feeders, feed augers and belts, laborsaving processing equipment, automatic waterers, and manure disposal units. Automation of the poultry industry will increase.

Types of Equipment

Good equipment is essential for satisfactory poultry production. Examples of feeders and waterers are illustrated in Figures 13.8, 13.9 and 13.10. It should be simple in construction; trouble-free; adequate for the intended function; but not excessive; movable and easily cleaned; and laborsaving.

Feeding Equipment

Since feed cost is the major item in poultry production, it is necessary that there be adequate feeder space (for feeder space requirements of different species and ages of poultry, see Table 13.11). They should be easy to fill, easy to clean, built to avoid waste, arranged so that the birds cannot roost on them, high enough so that the birds cannot scratch litter into them, and so constructed that as long as they contain any feed, the birds can reach it.

Automated feeders are standard equipment on commercial layer and broiler farms. They save labor and keep feed fresh. It is important, however, that standby generating equipment be available for their operation in case of power failure.

TABLE 13.7   Insulation R-Values for Some Common Construction (Surface Conditions Included)

Type of Construction	"R"-Value
Roofs and Ceilings:	
Asphalt shingles, wood sheathing, vented attic space, ½ in. insulating board ceiling	4.95
Wood shingles, wood sheathing, vented attic space, ½ in. insulating board ceiling	5.45
Metal roofing on nailing girts, wood sheathing, vented attic space, ½ in. insulating board ceiling	3.53
Metal roofing on nailing girts, vented attic, 3 in. blanket insulation (mineral wool), ½ in. plywood ceiling	13.94
Metal roofing on nailing girts, vented attic, 4 in. fill insulation (glass or mineral wool), ½ in. plywood ceiling	16.88
Metal roofing on nailing girts, vented attic, 6 in. blanket insulation (mineral wool), ½ in. plywood ceiling	25.04
Metal roofing on nailing girts, hay mow, 12 in. of hay or straw	20 (approx)
Doors:	
Wood siding, beveled, ¾" × 10"	1.90
Plywood, ¾ in. blanket insulation between two sheets ½ in. plywood	4.88
Plywood, 1½ in. blanket insulation between two sheets ½ in. plywood	7.65
Floor perimeter (per foot of length of exterior wall):	
Concrete, without perimeter insulation	1.23
Concrete, with 2" × 24" perimeter insulation	2.22

Source: Adapted from Structures and Environment Handbook, *7th ed., Midwest Plan Service, Iowa State University, p. 163, Table 214–5.*

TABLE 13.8   Total Heat Output by a Chicken at 70°F (21°C)

Average Body Weight		Total Heat Output per Hour			
		Per Lb Body Weight	Per Kg Body Weight	Per Bird	
(lb)	**(kg)**	**BTU**	**(Kcal)**	**BTU**	**(Kcal)**
1	0.5	20.0	(11.1)	20.0	(5.1)
2	0.9	14.5	(8.3)	29.0	(7.3)
3	1.4	11.5	(6.4)	34.5	(8.7)
4	1.8	10.0	(5.6)	40.0	(10.1)
5	2.3	9.0	(5.0)	45.0	(11.4)
6	2.7	8.2	(4.6)	49.2	(12.5)

Source: M. O. North, "Basics of Poultry House Ventilation," Poultry Tribune, *March 1978, p. 12.*

TABLE 13.9   Sensible and Insensible Heat Production by a Single, Resting Leghorn Layer as Influenced by Ambient Temperature

Ambient Temperature		Sensible Heat	Insensible Heat	Output of Sensible Heat per Hour of Body Weight (lb) in BTU (or kg in Kcal)	
(°F)	**(°C)**	**(%)**	**(%)**	**(lb)**	**(kg)**
40	4	90	10	9.0	(5.2)
60	16	80	20	7.9	(4.4)
80	27	60	40	6.1	(3.4)
100	38	40	60	4.3	(2.4)

Source: M. O. North, "Basics of Poultry House Ventilation," Poultry Tribune, *March 1978, p. 12.*

TABLE 13.10   Recommended Airflow through a Chicken Unit

Air Temperature in House		Cu Ft of Air per Minute per Lb of Body Weight	M³ of Air per Minute per Kg of Body Weight
(°F)	**(°C)**		
40	4	0.48	0.03
60	16	0.72	0.04
80	27	0.96	0.06
100	38	1.20	0.07
110	43	1.32	0.08

Source: M. O. North, "Basics of Poultry House Ventilation," Poultry Tribune, *March 1978, p. 12.*

Figure 13.8 Example of an automatic waterer—nipple waterer. *(ISU photo by Bob Elbert)*

Figure 13.10 Automated feeder and waterer.

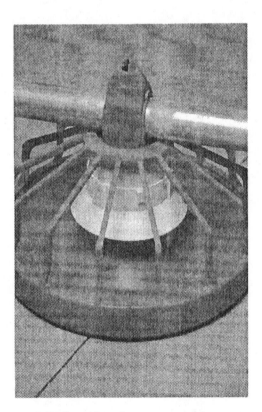

Figure 13.9 Example of an automated waterer—pan waterer. *(ISU photo by Bob Elbert)*

TABLE 13.11   Cage Sizes, Birds per Cage, Feed and Water Space[1]

Cage Size	No. Birds	Total Sq. In.	Sq. In. per Bird	Inches Feed and Water Space per Bird
8" × 16"	1	128	128	8
8" × 16"	2	128	64	4
12" × 16"	3	192	64	4
12" × 18"	3	216	72	4
12" × 18"	4	216	53	3
24" × 18"	7	432	62	3.5
24" × 18"	8	432	54	3
12" × 20"	4	240	60	3
30" × 24"	10	720	72	3
48" × 36"	20	1,728	86	2.4
48" × 36"	25	1,728	69	1.9

[1] 1 in. = 2.5 cm, 1 sq. in. = 16.4 cm^2.

To keep the water clean, the container(s) should be high enough that litter will not be scratched into it, and located so that the birds cannot contaminate it with droppings. Automatic watering devices are commonly used on commercial layer, broiler, and turkey operations. For large flocks, watering facilities running the entire length of the house are satisfactory. For cages, drip-type or small-cup waterers minimize cleaning and spillage (see Table 13.11).

OTHER POULTRY FACILITIES AND EQUIPMENT

In modern large-scale poultry operations, two other types of facilities and equipment, which for the most part are used in all types of poultry production operations, merit mentioning. Namely, (1) facilities and equipment for handling excreta, and (2) an emergency warning system.

Waterers

The importance of high-quality water is evidenced by the fact that laying hens drink 2 to 3 units (lb or kg) of water for each unit (lb or kg) of feed that they consume.

Watering devices should keep the water clean, be easily cleaned, prevent spillage of water around the vessels or containers, and keep the water cool in warm weather and safe from freezing in cold weather.

Facilities and Equipment for Handling Excreta

The facilities and equipment for handling poultry waste will vary according to the disposition made of it. Among the common manure disposal systems and equipment are the following:

1. **Dry spreading.** The spreading of dry excreta or composted excreta with litter on cropland is still a very common and economical way to dispose of it. The major problems or drawbacks to this method are the land acreage required and the odors produced. A manure spreader, a manure pit where storage is planned, and a field constitute the necessary facilities and equipment for this system.

2. **Wet spreading.** This refers to excreta to which water is added to create a slurry and facilitate handling. Liquid manure is stored in large watertight storage pits or tanks. For conveying the manure away from the storage tank or pit, either a watertight manure spreader or an irrigation pipe must be available.

3. **Lagoon.** Outdoor anaerobic lagoons should be >3 ft (1 m) deep, and up to an acre in size. They should have outlet mechanisms to control water depth (such that there is not an overflow into waterways with major disastrous environmental consequences) and perhaps provision for supplying oxygen at the surface. Lagoon effluent is then land-applied.

4. **Incinerators.** Incinerators are rather expensive, require the use of considerable fuel to consume manure, and may create an odor problem. The use of incinerators for poultry waste is being used in Europe.

Emergency Warning Systems

The confinement of large flocks of birds in a mechanically controlled environment entails considerable risk because of the possibility of (1) power or equipment failure or (2) fire or abnormal temperatures. To guard against such troubles, an emergency warning system should be installed. Such a warning system may save its cost during just one power interruption, fire, or undesirable temperature in an incubator, brooder house, layer house, broiler house, egg storage room, or furnace room. The continuation of the poultry enterprise as the result of a timely warning can be far more valuable than any monetary insurance settlement after the operation has failed.

Cleaning and Spraying Equipment

Thorough cleaning should precede disinfecting. For big operations, a power sprayer should be available for use in applying both disinfectants and insecticides for fly control.

Mortalities

Mortalities can be disposed of by composting (recommended), incineration (also recommended), rendering, and burial in a pit. Either composting or incineration is a satisfactory form of flesh disposal. If done properly, there is no odor, disease, rodent, fly, or water pollution problem. Composters or incinerators may be home-constructed. Incinerators (e.g., a steel-jacketed commercial unit) may also be purchased. Propane, gas, or oil may be used as the fuel for the incinerator.

SPECIALIZED BUILDINGS AND EQUIPMENT FOR SPECIALIZED PURPOSES

The large and highly specialized poultry enterprises have developed buildings and equipment adapted to their highly specialized purposes. Certain principles are observed in the facilities and equipment for layers, for brooding of chicks, for replacement pullets, for broilers, for breeder turkeys, for turkey poults, for market turkeys, and for ducks and geese. Building and equipment requirements for each of these specialty areas follow.

HOUSING AND EQUIPMENT FOR LAYERS

The design and construction of houses and equipment for layers should be such as to provide for top performance of the layers, optimum environmental control, functional arrangement of equipment, maximum labor efficiency, satisfactory waste disposal, and minimum housing and care costs per dozen eggs produced.

Houses

Layer houses may be colony houses, multiple-unit houses, or multiple-story houses; some are permanent and others are movable. The starting point in designing a layer house is the selection of the type of laying system.

1. **Cage system.** Early in human development wild hens were caught and kept in cages in order to facilitate egg gathering. Then, beginning in the 1930s, a limited number of commercial layers were kept in cages. Today, more than 90% of the layers in the United States are kept in cages.

Concrete floors are widely used in caged layer houses. In the cage system, small wire cages placed side by side, in long rows, hold the birds. These cages stand in rows on each side of an aisle (usually about 30 to 36 in. wide), placed at a convenient working height for the operator. The cage arrangement may range from single-deck to five decks. Where multiple decks are used, the second row (and subsequent rows) of cages may be located directly above the lower row, or placed so that the bottom row projects forward about one-half the cage's depth, giving a stair-step effect.

Originally, one or two hens were kept in individual wire cages. To reduce the cost per bird, however, colony cages, which hold up to 20 to 25 hens, have been developed. Three or four hens per cage appears to be the most popular arrangement at the present time. Cage sizes are not standardized, but Table 13.11 will serve as a useful guide.

While the cage system (a) accommodates more birds in a given floor area than the litter floor system and (b) eliminates many internal parasite troubles, it does give rise to problems. High initial investment and high labor requirements have been experienced in the hen-per-cage system. With the colony-type cage arrangement, cannibalism and similar social problems have often been encountered. Controlling flies and removing manure have been particularly difficult with the cage system.

The most important things are to design the building to fit the local climate; to provide adequate protection, ventilation, and cooling for the birds; and to offer good working conditions for the operator. It is further recommended that anyone planning to construct a layer house should inspect all types or systems and confer with those who have used them; then, reach a decision.

In addition to the layer house, a building for handling, cooling, and holding eggs under refrigeration on the egg farm is essential. Most poultry feed is purchased in bulk and delivered in bulk tanks from which it is withdrawn as used, by gravity or mechanical conveyors. Where whole grain is used, it may be practical to have sufficient storage facilities to permit purchase of grain at harvest time and/or when prices are usually more favorable. A service building for storing supplies and small equipment, and for making repairs, is also needed. It may be a separate building, or it may be a part of the garage or egg room.

Equipment

Good laying-house equipment is essential for satisfactory production. It should be simple in construction, movable, and easily cleaned.

1. Feeders. Since feed cost is the major item in egg production, it is necessary that there be adequate feeder space and that the feeders be good. They should be easy to fill, easy to clean, built to avoid waste, arranged so that the birds cannot roost on them, high enough so that the birds cannot scratch litter into them, and so constructed that as long as they contain any feed the birds can reach it. Automatic (mechanical) feeders are standard equipment in large commercial egg operations.

2. Waterers. Laying hens drink 2 to 3 units (lb or kg/liters) of water for each unit (lb or kg) of feed they eat. Watering devices should keep the water clean, be easily cleaned, and prevent spillage of water around the vessels or containers. Also, it is important that the waterers be distributed throughout the laying house so that a hen never has to travel more than 15 ft (3 m) for a drink. Many cage installations for layers are equipped with either drip-type or small-cup waterers that minimize cleaning problems as well as spillage.

Care of the House

Good husbandry and housekeeping are essential for optimum production and high egg quality. Also, it is necessary to minimize the spread of diseases. Accordingly, the following practices should be a part of the regular chores of the employee: inspect the birds daily, clean waterers daily, keep lightbulbs clean, clean windows regularly, control flies and rodents, and inspect all equipment routinely. Additionally, the following management and environmental factors should receive special attention.

1. Cooling. In warm areas, summer heat may cause retarded egg production and even result in death losses. Well-ventilated houses help, but when temperatures become extreme, artificial cooling is necessary. During extremely hot weather, the house can be kept more comfortable by increasing the movement of air, cleaning the fan blades and screens, painting the roof white, and sprinkling the roof. Foggers, controlled by thermostats, may also be used to produce a fine spray. For small producers, in an emergency, sprinkling water over the hens with a garden hose, using a fine spray, will help cut down death due to heat prostration if enough breeze is available to evaporate the water and cool the birds.

2. Excreta removal. A manure removal or holding system should be planned before the operation is started. The importance of this becomes apparent when it is realized that 100,000 layers will produce over 12 tons of manure a day, or well over 4,000 tons a year.

BROODER HOUSES AND EQUIPMENT FOR CHICKS

Wherever chicks are raised, whether for broilers or replacement pullets, and whether for continued rearing in confinement or on the range, artificial brooding of some kind is necessary. No phase of the poultry business is so important as brooding, as it is the part that makes for a proper start in life.

Brooder Houses

Until recently, brooder houses for chicks were, for the most part, either portable or stationary buildings that were used for other purposes the balance of the year—part of the laying house, or the garage, or perhaps one end of the machine shed was used for brooding purposes. Housing arrangements of this type are still common among small farm flock owners. However, in large commercial installations, where it is not uncommon to find 30,000 to 50,000 or more chicks being brooded together as a unit, special brooder houses or arrangements are common.

Brooding Equipment

Heating, feeding, and watering equipment are the three main items of equipment needed for the brooding of chicks.

1. **Heating equipment.** The heat requirements for artificially brooding chicks may be supplied by a wide variety of devices and methods, among them the following:

   a. **Portable brooders.** These units are, as indicated by the name, portable or movable. Although they come in a wide variety of styles and sizes, they generally consist of a central heating unit surrounded by a hover. They may be heated by gas, oil, or electricity. Portable brooders cost less to install than central heating systems, but they cost more per chick to operate. Also, they require more attention and labor, and there is more fire hazard from using them than central systems.

   b. **Infrared lamps.** In this method, infrared lamps are suspended 18 to 27 in (~60 cm). above the floor litter. These lamps do not heat the surrounding air, but they warm the chicks in the same manner as direct rays from the sun. One 250-watt infrared bulb will suffice for 60 to 100 chicks.

   c. **Battery brooders.** Commercial battery brooders may be either (1) unheated brooders made for use in warm rooms, or (2) those equipped with heating units and warm compartments for use in rooms held at 60° to 70°F (16° to 21°C). Most batteries of today have heating units warmed by electricity and equipped with thermostat regulators.

   d. **Central heating.** Large commercial operations, which handle 5,000 to 25,000 or more chicks per house, need a central heating system. Several different heating systems have been developed for such use. Most of them are highly automated, thermostatically controlled, and fueled by oil, gas, or electricity. Central heating may provide warmth through either hot water (pipes or radiant heating) or hot air (direct or indirect).

2. **Feeders and waterers.** Feeding and watering equipment that is suited to 6-week-old birds is not satisfactory for day-old chicks. Hence, special feeding and watering equipment must be provided, and the chicks must learn to eat and drink when they are first placed in the brooder.

HOUSING AND EQUIPMENT FOR REPLACEMENT (STARTED) PULLETS

Regular replacement of the laying flock is one of the most important and most expensive essentials of egg production. Usually, it is wise to provide pullets with levels of environment, housing, feed, and general management, considerably above minimum standards, but consistent with realistic cost consideration. This calls for housing that gives reasonable protection from heat and cold and provides good ventilation; and it means adequate feed and water space and facilities.

Housing

Effective isolation of growing stock from older birds is important for disease control, particularly during the early stages. Hence, separate housing should be provided for replacement pullets, completely separated from layers. Housing that is suitable for brooding chicks, or for housing a laying flock, will be satisfactory for rearing pullets. Usually, they are started out in the brooder house, then switched to the layer house.

Feeders and Waterers

Water space is less important than a good distribution of waterers through the house with a constant supply of fresh water. The same principle applies to some degree to feeders. Both feeders and waterers should be located to permit birds in any part of the house to feed or drink conveniently, without having to find their way around barriers or to travel more than 15 feet (5 m).

Other Equipment

Hoppers for grit should be provided for replacement pullets throughout the growing period, and, as they near maturity, additional hoppers should be provided for oystershell or other calcium supplement.

Roosts are not always used where pullets are reared. When used over enclosed pits, roosts aid in sanitation. Also, they add to the comfort of the birds in hot weather and help to establish desirable roosting habits for the laying period that follows. Some producers also feel that flocks with good roosting habits are less likely to present severe problems of floor eggs.

Nesting equipment should be available by the time the pullets begin to lay. Of course, only a few nests are necessary if pullets are moved from rearing quarters to their laying quarters before production becomes heavy.

Except for automatic feeders and waterers, little mechanization exists in the vast majority of pullet-rearing facilities, especially when compared to layer facilities.

HOUSES AND EQUIPMENT FOR BROILERS

In modern commercial broiler production, the bird spends its entire life in one house; that is, it is not brooded in a special brooder house, then moved to a house for growing because broiler raising is basically a brooding operation. Instead, brooder houses are thoroughly cleaned and disinfected between flocks, preferably with the quarters left idle 1 or 2 weeks before starting a new group. Today, the vast majority of the nation's broilers are produced in large production units. Hence, in the discussion that follows, only the buildings and equipment common to these will be described.

Housing

A broiler house should provide clean, dry, comfortable surroundings for birds throughout the year. The house should be kept warm enough, but not too warm; the litter should be kept reasonably dry; provisions should be made to modify the air circulation as broilers grow; and fresh air should be circulated, but the house should be free from drafts. In short, the broiler house should not be too cold, too hot, too wet, or too dry.

Most of the new broiler houses being constructed today are 24 to 40 ft (8 to 13 m) wide, with gable-type roofs. Truss-type (of either wood or steel) construction is replacing pole-type due to lower labor costs in construction and greater ease of cleaning with a tractor. The length varies from 200 to 600 ft (65 to 200 m), with most of them averaging 300 to 400 ft (~100 m). Capacity varies, but in the newer houses they generally range from 7,200 to 20,000 broilers. All of the birds may be in one large pen, but the newer trend is to pen units of 1,200 to 2,500 birds. There is increasing interest in controlled environment housing for broilers. When the broiler house is insulated and environmentally controlled, a centrally heated brooding system may be difficult to justify. However, where broiler houses are not environmentally controlled, a central heating system, of either hot or cold air or hot water, is preferred. Where a chick hover is used, large (1,000 capacity) units are most popular.

Although maintaining adequate temperature is of great importance, constant attention is needed to ventilate the broiler house properly. This calls for a building that is properly insulated and in which there is a forced ventilation system.

Feeders

Baby chicks should be started eating from new, cut-down chick boxes or box lids placed at floor level, allowing one feeder lid per 100 chicks. These feeders allow the chicks to find the feed easily.

Following the box lid stage, broilers may be fed from trough feeders, hanging tube-type feeders, or mechanical feeders. Bulk feed bins and mechanical feeders are the most costly of the various types of feeding equipment, but they also make for more saving in labor. As a rule of thumb, installation of a mechanical feeder or feeders is worthwhile if the investment is no more than five times the labor saved per year.

Waterers

Clean water at about 55°F (~13°C) should be available at all times. Gallon fountains should be provided for baby chicks, but these should be replaced by automatic hanging waterers as soon as chicks have learned to drink. All waterers should be cleaned and washed daily.

HOUSES AND EQUIPMENT FOR TURKEYS

Although there are some species differences between turkeys and chickens, primarily because of a difference in size, the general principles relative to buildings and equipment for turkeys and chicks are very similar.

Facilities for Breeder Turkeys

Breeding turkeys may be kept (1) on restricted range, (2) in confinement housing, or (3) in semi-confinement.

Open range without shelters should not be used unless the flock can be protected from cold winds by trees and/or sloping hillsides. Turkey range should be enclosed by a well-constructed, permanent fence that is at least 5 ft (1.5 m) high, preferably with an electric wire around the outside to further discourage predators. A 4-acre (2 hectare) area will accommodate about 600 breeders, with provision for some rotation of pens and two night pens. Night pens should be equipped with roosts and lights arranged to ensure the entire area is lighted. Day pens should be equipped with waterers, feeders, and nests.

Buildings for strict confinement are usually 40 to 50 ft (12 to 15 m) wide and covered with wire on all sides. Most operators enclose three sides of the shelter with plastic to give protection during the winter months. Strict confinement lends itself to automatic feeding and watering.

Semiconfinement gives the protection advantage of complete confinement during the bad weather, with the added yard space for improved sanitation and mating. Buildings for semiconfinement are usually of lower cost construction than those used in strict confinement; generally they consist of an open-front, pole-type building, with a fenced yard.

Facilities for Brooding Turkey Poults

Turkey poults should be brooded and reared separately from all ages of chickens and adult turkeys. In large commercial turkey operations, they are usually brooded in permanent-type houses. Some growers prefer to brood in batteries for the first 2 or 3 weeks before placing the poults on the floor. Regardless of the method of brooding employed, at least 1 sq ft (0.09 sq m) of floor space is required to 8 weeks of age. More floor space must be provided as the poults grow older.

Any of the types of eating equipment commonly used for baby chicks can be used for brooding poults. To begin with, poults should be maintained at a temperature of 95° to 100°F (~38°C), with the temperature lessening following the first week. A guard should be placed around the brooder for the first 2 weeks to prevent crowding and smothering in the corners of the house.

The watering and feeding facilities used by turkey poults are very similar to those used by baby chicks, but generally they are somewhat larger and, of course, more space needs to be provided per bird.

Facilities for Market Turkeys

For the first 8 to 10 weeks of life, the brooding of turkey poults is the same regardless of whether they are subsequently to be raised in confinement or moved to the range. Although a limited number of market turkeys are reared on the range, there is an increasing tendency to grow them in confinement in pole-type shelters, or even in environmentally controlled buildings.

Where turkeys are to be range-raised under a rotation system, portable range shelters, roosts, feeders, and waterers are necessary. These may be placed on runners or wheels and moved where and when needed.

When confinement reared, turkeys are generally provided with a pole-type shelter, although more costly environmentally controlled units are used. When raised in strict confinement, 3 to 5 sq ft (0.27 to 0.45 sq m) of floor space should be provided to carry each bird to market age and weight, the amount depending on the size of the strain of turkeys. Automatic waterers and feeders, or bulk feeders, help reduce labor costs. Care should be taken to provide adequate ventilation and occasional additions to the litter to reduce dampness and dirty litter conditions.

FURTHER READING

ASAE. *Design Values for Emergency Ventilation and Care of Livestock and Poultry.* In ASAE Standards, EP282.2. St. Joseph, MI: American Society of Agricultural Engineers, 1993.

Bell, D. D., and W. D. Weaver. *Commercial Chicken Meat and Egg Production,* 5th ed. Norwell, MA: Kluwer Academic Publishers, 2002.

Leeson, S., and J. D. Summers. *Commercial Poultry Nutrition.* Guelph, Ontario: University Books, 1997.

Lesson, S., and J. D. Summers. *Broiler Breeder Production.* Guelph, Ontario: University Books, 2000.

14

Eggs and Layers

Objectives

After studying this chapter, you should be able to:

1. List the major components of eggs.
2. List the major components of egg yolk, egg white, and egg shell.
3. List the top egg-producing states.
4. List the major breeders of commercial white-egg and brown-egg layers.
5. List the major companies producing eggs.
6. List the major kinds of contracts made with egg producers.
7. List the different kinds of floors used in egg production.
8. Give recommendations on the rearing and production of egg layers and replacement pullets.
9. Define *phase feeding* and give reasons why it is important.
10. List the major costs of egg production in order of importance.
11. List procedures for collection and handling of table eggs to maximize quality.

The commercial production of chicken eggs is of great importance worldwide and is growing (see Figures 14.1 to 14.7 for examples for layers and eggs). Little demand for duck eggs exists in the United States. There is limited quantitative data on duck egg production or consumption in the world. However, duck eggs appear to have a large market in China and other countries in Asia together with significant demand in Western Europe. *Coturnix* (or quail) eggs have a significant niche market, particularly in Europe where high prices are paid for these as shell eggs or pickled hard-boiled eggs.

This chapter addresses the following topics related to the chicken egg:

- The composition of the egg.
- The biology of the production of the egg by the hen.
- The commercial production of eggs.
- Handling and marketing eggs.

THE COMPOSITION OF THE EGG

Eggs are a rich source of nutrients including:

- Essential amino acids
- Vitamins including vitamin A, choline, B complex, and others
- Minerals (e.g., potassium and zinc)

In view of the nutritional value of eggs, consumption of eggs by children and the elderly is often recommended.

Composition of a hen's egg (%):

Yolk

Water, 49

Protein, 16

Fat, 33

Carbohydrate, 1

Mineral, 1

Figure 14.1 What it's all about! A layer and an egg. *(Courtesy, Hy-Line International)*

Figure 14.3 Hy-Line white egg layers (top, W-36 and bottom, W-98). *(Courtesy, Hy-Line International)*

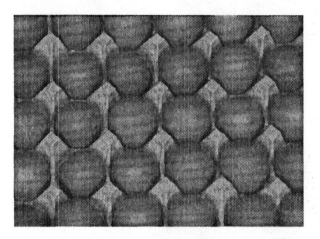

Figure 14.2 Commercial brown eggs. Some consumers prefer brown eggs. *(Courtesy, Hy-Line International)*

Figure 14.4 Hy-Line brown egg layer. *(Courtesy, Hy-Line International)*

Figure 14.5 Hubbard Isa laying hens. A, Babcock B-3000—a layer breeder; B, Isawhite—a layer producing large eggs; C, Shaver white—a layer with good stress resistance). *(Courtesy, Hubbard Isa)*

Albumen (or egg white)

Water, 87

Protein, 11

Fat, 0

Carbohydrate, 1

Mineral, 1

Shell

Water, <1

Protein, 3

Fat, 2

Carbohydrate, 0

Mineral, 95

The Composition of the Yolk

The yolk contains the following:

1. Triglycerides (70%) and other lipids including cholesterol (5%) and phospholipids (25%).
2. The phosphoproteins and lipoproteins and phosphovitin and lipovitellin in granules.

Figure 14.6 Hendricks Poultry Breeders laying hens that produce white eggs with a feed efficiency of about 1:2.0 (A, DeKalb White, and B, Bovans White). *(Courtesy, Centurian Poultry)*

Figure 14.7 Hendricks Poultry Breeders laying hens that produce brown eggs with a feed efficiency of about 1:2.1 to 2.4 (A, DeKalb Brown; B, Hisex Brown; C, Bovans Goldline; D, Bovans Brown; E, Bovans Nera). *(Courtesy, Centurian Poultry)*

3. Soluble proteins such as α-livetin (plasma albumen), β-livetin (α_2-plasma glycoprotein), and γ-livetin (antibodies or immunoglobulins).
4. Binding proteins for nutrients (e.g., riboflavin, biotin, thiamin, vitamin A, vitamin D, iron).

NOTE: The transfer of antibodies to yolk makes it attractive to produce antibodies in chickens (for vaccines, testing, etc.) because these can be harvested from eggs.

C

D   E

Figure 14.7   Continued

Fat Composition

Much of the fat in eggs is in the form of triglyceride (~70%) together with some phospholipid (~25%) and cholesterol (~5%). The fatty acid composition of egg triglycerides is:

- Monounsaturated about 44% (predominantly 18:1 oleic acid)
- Polyunsaturated about 10% (predominantly 18:2 linoleic acid)
- Saturated about 40% (predominantly 16:0 palmitic acid)

The composition of egg phospholipids is:

- Lecithin (~75%)
- Cephalin (~20%)
- Sphingolipids (2.5%)

CHOLESTEROL CONTENT.   The cholesterol content of a large egg is 213 mg (this compares to the recommended maximum consumption of 300 mg per day).

Yolk Pigmentation

The color of egg yolk is dependent on pigments in the diet of the bird. In the United States, the light yellow yolk is due to the consumption of pigments such as lutein and β-carotene coming from marigold petals and yellow corn, respectively. These natural pigments are also nutrients for people, with lutein related to health of the eye and β-carotene converted to the active forms of vitamin A by the body.

The Composition of the Egg White

The major proteins in the egg white (or albumen) are:

1. Ovalbumin 54%
2. Ovotransferrin 13% (binds iron, zinc, and copper)
3. Ovomucoid 11% (inhibits proteases, e.g., of bacteria)
4. Ovoglobulins 8% (antibodies)
5. Lysozyme 3.5% (an enzyme that lyses or breaks down bacteria)
6. Ovomucin about 2% (antimicrobial)

There are also minor components, many of which bind vitamins such as avidin (binds biotin), flavoprotein (binds riboflavin), and thiamine-binding proteins.

The Composition of the Egg Shell

The shell (composed of calcium carbonate in a protein matrix) provides protection from microorganisms and physical damage. In addition, the developing embryo obtains significant amounts of calcium from the shell. This is critical for bone development.

BIOLOGY OF THE PRODUCTION OF THE EGG

Yolk

The liver synthesizes the precursors of yolk, namely:

1. The phospholipoprotein—*vitellogenin*
2. The very low-density lipoproteins (VLDLs) rich in triglycerides

Estrogens, and possibly other hormones, are required for the synthesis of the yolk precursors. They pass via the bloodstream to the developing follicles in the ovary (see Chapter 2) where they are deposited in the ovum (as yolk). The vitellogenin molecules are split to yield *lipovitellins* and *phosvitin*.

Egg White (Albumen)

The oviduct produces the egg white proteins. Most of this occurs in the magnum region. The process is predominantly under the control of estrogens (e.g., estradiol). An exception to this is avidin, where progesterone is the major controlling hormone.

Shell

The oviduct produces the shell (made up of calcium carbonate). This occurs in the shell gland. The process involves transfer of calcium from the feed in the intestines to the shell and from the bones. Hormones are important to the process.

COMMERCIAL PRODUCTION OF EGGS—LAYERS

The commercial production of eggs is important worldwide. The industry has the following characteristics in the United States:

- The per capita consumption of eggs in the United States is now rising steadily following declines for more than 20 years and a low point in 1991.
- The top egg-producing states are Iowa, Ohio, California, Pennsylvania, Indiana, Texas, Nebraska, Minnesota, Georgia, and Florida.
- Sixty-five U.S. firms each own more than a million layers.
- Eight firms own more than 5 million birds.
- The largest U.S. egg company is Cal-Maine Foods, Inc. (see Table 14.1).
- In North America, over 90% of the layers are in cages or on wire; and more and more of all replacement pullets are brooded and reared in

TABLE 14.1   The Top 10 Egg Companies in the United States

1. Cal-Maine Foods, Inc.	Jackson, MS	19.9 million birds
2. Rose Acre Farms	Seymour, IN	16.2 million
3. Micheals Foods	Minneapolis, MN	15.0 million
4. De Coster Egg Farms	Turner, ME	12.6 million
5. Buckeye Egg Farm	Croton, OH	10.5 million
6. Fort Recovery Equity	Fort Recovery, OH	8.0 million
7. Moark Productions, Inc.	Neosho, MO	6.5 million
8. Midwest Poultry Sciences, L. P.	Mentone, IN	5.7 million
9. Sparboe Companies	Litchfield, MN	5.0 million
10. ISE America, Inc.	Newberry, SC	4.8 million

Source: From the Egg Industry, Jan. 2000.

cages from day-old. The reasons for this are reduced labor requirements and greater efficiency.
- There is an increase in the production of free-range and organic eggs in North America and particularly Western Europe.
- There is greater processing of eggs in the United States, with 58% of eggs sold retail in 2001, 30% further processed (up from 19% in 1988), 11% going to food service/restaurants, and about 1% exported.

There is a tendency of egg producers to do the entire job of production, processing, and marketing; with their eggs trucked directly from their egg-holding room facilities to the store-door delivery. All characteristics of egg production operations should be considered when planning a new egg production enterprise.

Business Aspects of Egg Production

Today's commercial egg production units are large and intricate, and they are destined to get even bigger and more complicated. It is unusual for anyone to start a commercial egg enterprise with fewer than several hundred thousand birds. With the increase in size and complexity of units, the business aspects have become more sophisticated. It is important, therefore, that commercial egg producers be knowledgeable in financing, labor requirements, contracts, record keeping, and costs and returns. This will be briefly addressed here but also more extensively in Chapter 17.

Location

Determining where to locate laying units requires consideration of the competitive advantages and disadvantages of the location. Factors include the costs of the following:

- Land
- Buildings
- Feed (proximity to corn and soybean production will likely result in lower costs and the ability to contract for long-term supplies while having access to land for application of animal waste)
- Labor
- Meeting requirements of local, state, and federal regulations

In addition, access and proximity to markets is a consideration.

Financing

One of the first things to consider when thinking of becoming a commercial egg producer is how much money it will take, and where the money will come from.

Let's assume starting a layer enterprise with 100,000 birds. For the house and equipment, this will call for an investment of $8.00 per bird, or an outlay of $800,000. The 20-week-old pullets may cost as much as $2.75 each, or $275,000. Land is needed for the physical plant, with the value determined by location and prevailing price. Operating capital is needed, including 75 to 85 lb (~36 kg) of feed per hen per year plus supplies. Thus, in total, 100,000 layers will require an investment of more than $1 million.

Labor Requirements

Caring for layers is a full-time job, 7 days a week, and 52 weeks a year. It calls for reliable employees. Assuming automation, one worker can care for at least 100,000 hens.

Egg Contracts

Narrowing profit margins, fluctuating egg prices, expanded flock size, large investment, and a desire for a stable income caused producers to contract production and/or marketing. Over 90% of the nation's eggs were produced under contracts or in integrated operations (see Chapter 17). Contracts may cover the following:

- Management of layers (feed, space, animal health program, biosecurity, etc.)
- Strain of birds and replacement pullets
- Months of lay (production and quality)
- Quantity and quality of eggs to be delivered

The types of egg contracts include producer contracts, credit contracts, and marketing contracts.

In egg contracts, producers usually own the buildings and equipment and furnish labor together with management and utilities in return for a fixed rate per layer per week or per dozen top-grade eggs produced. Contractors usually furnish hens, feed, animal health program, supervision, and a market for the eggs. In some areas, the contractors haul the eggs; in others the producers do the hauling. Generally, the producer only washes the dirty and badly stained eggs.

The terms of contracts vary among areas and are affected by the amount of competition and the level of commercial egg prices. The following types of egg production contracts are rather typical:

- **A fixed fee per dozen eggs.** The producer furnishes housing, equipment, labor, utilities, and sometimes litter and receives a payment per dozen eggs produced. The contractor furnishes ready-to-lay hens, feed, medication, and owns all eggs as well as the salvage hens at the end of the production period. There may be bonuses for good feed conversion, egg quality, or low mortality.

- **A fixed fee per hen per month.** The producer furnishes the same items as in the above contract, but is paid per hen rather than per unit eggs.
- **Percentage of returns.** The producer furnishes the items listed in the first type of contract and receives a percentage of total egg returns (usually averages 15 to 18%). The feed supplier usually receives 50 to 55% of the return, while the supplier of ready-to-lay pullets gets 26 to 28% (usually retains ownership of birds). The percentage received by any party must be proportional to the contribution in the form of material and/or services.

Records and Accounting

Keeping and using performance records are vital to success in egg production. The following records are important:

1. **Feed consumption.** This is critical to determine profitability. Also, a drop in feed consumption may precede a drop in egg production.
2. **Number of eggs produced** per day, week, month, and so on.
3. **Egg quality and egg size.** The goal should be 90% or more Grade As, fewer than 5% small eggs, and fewer than 20% medium eggs. Large or extra-large Grade A or AAs (see Chapter 16 for egg grading) have the most potential for profit.
4. **Mortality.** The producer should know whether mortality is excessive (signs of a disease problem).

Costs and Returns

As in any business, profits in egg production can be increased by either (1) lowering costs or (2) increasing returns, or both. Costs of egg production include both cash and noncash items. Feed purchased is by far the largest cash cost. Among the noncash items are labor, interest, and depreciation. Eggs are the chief source of returns on a commercial egg farm. Spent hens sold for meat are of very minor importance.

COST OF PRODUCING EGGS. The cost of producing eggs varies with location primarily based on the cost of feed ingredients. For instance, the cost of feed is lower in the Corn Belt of the United States. Costs will vary from between 40 to 50¢ per dozen depending on the commodity prices of corn and soybeans.

Estimated costs related to 42¢/dozen eggs (see Bell and Weaver, 2002):

1. Feed, 48%
2. Replacement pullets, 22%
3. Housing and equipment, 11%
4. Labor, 8%
5. Miscellaneous, 11%

LOWERING COSTS. Some production costs can be lowered. The place to start is the highest-cost items—feed, pullets, and labor. These accounted for about 80% of the cost of production.

To lower feed costs:

a. The producer should be located in an area where suitable major feed ingredients are produced (eliminating transportation costs) and near a feed supplier who has low operating costs, and a willingness to pass some of this efficiency on to the producer. For every $5.00 per ton saved on feed, the cost of production is lowered 1¢ a dozen.
b. Least-cost formulation of diets and forward contracting/hedging for feed ingredients are important.
c. Optimal formulation and feed mixing for maximal production and reduction in other costs (e.g., costs of animal waste disposal).

To reduce pullet costs, pullets may be raised for $2.00 a bird. Every 20¢ extra paid for pullets is equal to 1¢ a dozen increase in the cost of production. This may, however, reduce the quality of pullets or divert effort from the primary business. Labor costs can also be lowered by good personnel management and automation.

INCREASING RETURNS. Returns can be increased! Egg prices fluctuate widely and wildly! Therefore, profits have ranged from nonexistent to very high. Returns from commercial egg production can be improved by more eggs per bird, improved egg quality, access to markets and niche markets, reducing costs (see above), and efficient business and poultry management (pullet rearing, feed mixing, and egg processing).

- **Increase eggs per bird.** The one thing that will do more to increase returns than all else is to increase the number of eggs produced per bird. Birds eat little more feed and use no more lights, ventilation, or any other cost item when they produce more than 20 dozen eggs per bird in 12 months in the laying house or less than 20 dozen eggs.

  Increased production does not just happen; it requires planning and work. It may mean adjusting the lighting, feeding an appropriate diet, animal health programs, and so on. Increased production of 24 eggs per bird in a 300,000-bird unit will mean $360,000 (at 60¢ a dozen) more income per flock. Yet costs will be increased very little.
- **Set high goals.** These spur high achievement and increase returns. Examples for a commer-

cial egg operation might be: (1) eggs per hen, 12-month cycle, >275, feed per dozen eggs produced, < 3.5 lb (1.6 kg), and (2) mortality over 12 months, < 10%.

Breeding and Genetics

Selection of superior genetics is extremely important. Continuous improvement is being made in breeding. A single company, Hy-Line International, dominates the U.S. market with more than 80% market share. On the other hand, three companies dominate the world market. (for examples of hens see Figures 14.3 to 14.7.

The world market shares for primary breeders of layers (2000):

Chickens—White-Egg Layers
- Lohmann-Hy-Line International, 68% of market share
- Hendrix Poultry Breeding, 22% of market share
- Hubbard-ISA, 10% of market share

Chickens—Brown-Egg Layers
- Hubbard-ISA, 57% of market share
- Hendrix Poultry Breeding, 23% of market share
- Lohmann-Hy-Line, 17% of market share

Breeders produce and sell "name lines" such as the following (see Chapter 4):

- Babcock (Hubbard-ISA)
- Bovans (Hendrix Poultry Breeding, produced by Centurian Poultry in Georgia for the North American market)
- Hy-Line (Lohmann-Hy-Line)
- Hisex (Hendrix Poultry Breeding)
- ISA (Hubbard-ISA)
- Shaver (Hubbard-ISA)

In-bred lines are crossed to produce the commercial chick or breeders. Each line or strain of a breeder carries a brand name and number for identification similar to makes and models of automobiles.

Issues to consider in purchasing pullets (point-of-lay hens) include:

- Egg production and quality
- Feed efficiency (smaller hens tending to high feed efficiencies)
- Livability (or low mortality)
- Service from the primary producer/supplier

Replacement Pullets

Poultry production involves specialization within a specialization. Some commercial egg producers only produce eggs, leaving the rearing of replacement pullets to others. The commercial egg producer has two alternatives in procuring replacement stock:

1. Day-old chicks.
2. Ready-to-lay pullets (or point-of-lay) (~20 weeks of age). Predominantly, large commercial egg producers buy pullets rather than raise them.

Animal health programs should start with the purchase of replacement stock. Economic losses can be prevented much more easily by not introducing a disease than by acting after the disease has gained a foothold.

Day-Old Chicks

Price should be one of the least-important considerations when buying day-old chicks for raising replacement pullets. Some searching questions to be asked are:

- Do the breeder and hatchery firms have reputations for reliability?
- Will the hatching eggs come from healthy breeders and be vaccinated against Marek's disease?
- Will the entire order be filled by chicks of only one strain coming from just one hatchery and from eggs produced in a single location?
- What will be the age spread, if any, between the youngest and oldest chicks delivered?
- **Is the strain of stock relatively free of mortality problems?**
- Does the strain's body size permit housing at the cage density planned?
- **Do the egg number, size, and quality characteristics of the strain meet the marketing requirements?**
- **What is the feed efficiency?**
- Will the chicks be delivered in clean, uncontaminated boxes and equipment, without a prolonged period of transit? If new boxes are not used, how are the boxes cleaned and disinfected?

Separating Male From Female Chicks

In white egg-laying strains, male chicks are distinguished from female by using the sex-linked rate of feathering locus discussed in Chapter 4. Male chicks Kk exhibit slow feathering while female chicks kk show fast feathering. This is determined by inspection of the relative lengths of the primary wing feathers and primary wing feather coverts on newly hatched chicks. Alternatively, chicks can be sexed by inspection of the vent. Male chicks of laying strains are euthanized.

Ready-to-lay Pullets

Quality birds are essential regardless of the age at the time of purchase. However, when purchasing started pullets, more information is needed than when purchasing day-old chicks. With ready-to-lay pullets, it is important to make sure that they are well developed, healthy, and capable of expressing their full genetic potential as layers.

The choice of a supplier of started pullets is critical. The supplier should be a specialist with a good reputation for producing high-quality stock. When contracting for pullets, the buyer and the pullet grower should agree on some or all of the following management practices that the grower will be using:

- Strain and source of stock
- Number of birds to be housed in each unit (depending on size of unit)
- Diets
- Animal health and biosecurity measures
- Lighting programs

Written instructions should specify what the buyer will receive. These should include information on the following: body weight, livability, feed consumption, production in test trials, the brand and serial number of each vaccine used, with vaccination dates and, perhaps, copies of reports from the monitoring veterinarian. There may also be the opportunity to visit the pullet grow out operation (obviously in clean protective clothing) during the growing period. The buyer should also reserve the right to reject individual birds or the entire lot if they do not meet the quality standards specified.

The manner in which the pullets are loaded, transported, and placed in the cages on the buyer's premises will significantly affect their subsequent performance. Thus, the contract should specify who is to deliver the birds, and that all equipment be thoroughly sanitized. Delivery instructions should be sufficiently detailed to include method and time of loading, type of hauling equipment to be used, number of birds per crate or cage, the route of travel, and arrival time. The owner or manager should be on hand to supervise the distribution of the pullets into the laying cages.

Raising Pullets—Nutrition of Pullets

Before beginning to lay eggs, pullets consume about 16.5 lb (7.5 kg) of feed per 100 pullets per day.

Alternative Replacement Programs

When the average rate of lay falls below 65–70%, commercial egg producers can choose either of the following: culling and using replacement pullets or forced molting.

The replacement program that egg producers choose is one of the most important management decisions. It should be based on the performance and marketing arrangement of the flock. In addition, there is the issue of welfare (discussed in more detail later). The choice may be made from among many alternative programs, but not all of them give equal financial returns. Forced molting is likely to be an integral part of the program chosen.

Excluding capital costs, replacement pullets represent the second largest cost of producing eggs (feed is first). Replacement cost is reduced primarily by increasing the number of dozens of eggs a given flock produces during its lifetime, which is achieved by keeping the flock for a longer period of lay. When the savings exceed increases in other costs or reductions in income, the extended period of lay is more profitable. A sound replacement program should be well thought out and should be based on the flock's normal performance, realistic estimates of costs and prices, and marketing arrangements.

All-Pullet Flocks

An all-pullet flock is made up of birds that are less than 6 months old when the production year starts. In comparison with hens, pullets have the following advantages:

- Pullets lay 30% more eggs than hens will lay during their second year of production.
- Pullets require less feed to produce a dozen eggs.
- Pullet eggs have a higher interior quality (a high percentage of thick white compared to thin white).
- Pullets lay eggs with stronger shells.

The all-pullet system is still used in North America. There is a growing interest in the system in Europe where welfare regulations are moving to limit forced molting. All-pullet flocks are easy to manage. Once a cutoff age is selected, new pullets can be ordered at routine intervals. The only real problem is to determine the selling age at which total profits will be maximized.

Egg producers choose a wide range of cutoff ages of all-pullet programs—from as short as 50 weeks of age at sale to as long as 108 weeks, with an average of about 82 weeks. At the 82-week (20-month) age, most flocks are at 50 to 55% production; the number of "undergrade" eggs approaches 20%, and the flock size is approximately 80% of the number originally housed.

Molting

Molting is a normal process of birds, occurring in both sexes. In the wild state, birds shed and renew plumage, for instance, following migratory flights and/or following breeding. Chickens in commercial egg production have a different molting pattern.

They have been bred for high egg production and their environment (light and temperature) is controlled to remove major seasonal influences. A natural molt does not normally occur until about 8 to 12 months of egg production when egg production is decreasing. High producers tend to molt late; but once production ceases, molting is rapid. Molting is under hormonal control from the pituitary gland, gonads, and thyroid glands. It is associated with a drop in estrogen levels. Several factors affect the onset and length of molting; among them (1) weight, physical condition, and nutrition of the bird and (2) length of day (long-day length) exposure. Thus, by drastically reducing the length of light per day and/or starving the birds, molting can be induced or speeded up while terminating egg production.

Decreasing day length is a normal trigger for molting. Therefore, lighting programs for layers should provide either constant long or increasing day length. Stresses such as temporary feed or water shortage, disease, cold temperature, or sudden changes in the lighting program can also initiate a partial or premature molt, evidenced by the number of feathers on the floor of the poultry house. In these cases, chickens lose some head and neck feathers. If the molt continues beyond this point, a more severe drop in performance can be expected.

Egg producers molt hens by forced molting, also called forced resting, or induced molting or recycling. Chickens are molted when egg production falls below 65% compared to peak rates of lay which are >90%. Egg production declines to this arbitrary figure most typically at around 70 weeks of age or about 38 weeks of egg production.

Forced Rest or Forced (Induced) Molting

Under forced resting, a layer flock is induced to shed and replace its feathers at a time selected by the flock manager near the end of a normal laying cycle. A flock may be force rested earlier for other reasons. An induced molt causes all of the hens in a flock to go out of production for a period of time. During this period, regression and rejuvenation of the ovary and reproductive tract occurs. This is accompanied by first the loss followed by replacement of feathers. After a molt, the egg production rate increases markedly and egg quality is improved. Egg production usually peaks slightly below the previous peak rate.

Hens can be recycled once or even twice. The forced resting effectively rejuvenates the hen's reproductive system, with the ovary and oviduct decreasing to immature size during molting and then recovering. Well over 60% of laying flocks in the United States are recycled. The most important reason for induced molting is that it improves egg production and quality and, hence, returns (and ultimately profits).

Moreover, it reduces costs of replacement pullets. The question of whether it pays to recycle depends primarily upon the relative performance of all-pullet versus recycled flocks. There are also welfare issues that can affect consumer confidence.

In the industry, molting most frequently involves withholding feed (and in some cases, also water) for a period. Feed can be withheld for 3 to 14 days with the hours and intensity of lighting reduced. This results in a 25 to 30% decrease in body weight, loss of feathering, and the reproductive organs becoming immature or juvenile. Following restriction, the hens are placed on a maintenance diet for regrowth of the ovaries and oviduct.

Alternative methods of molting include diets low in energy (or high in fiber), low in sodium, low in calcium, and/or high in zinc (20,000 ppm). Poultry scientists are attempting to develop flock-friendly methods of molting that do not involve feed removal, do not cause a major loss in body weight, and do not increase economic losses.

TYPES OF RECYCLING. Egg-producing hens may be molted one or more times in the following types of recycling programs:

Two-cycle molting program—one molt and two cycles of egg production. The two-cycle molting program extends productive life by 6 to 8 months beyond the standard age at which all-pullet flocks are sold; and each hen produces an extra 100 to 150 eggs, thus reducing hen replacement costs per dozen eggs.

Three-cycle molting program—two molts and three cycles of egg production. The hens are first molted after about 9 months of production, then held through a shorter production period, molted again, followed by an even shorter period of lay, and then replaced.

ADVANTAGES OF FORCED RESTING. Forced resting has the following advantages:

- Greatly reduced replacement costs of pullets
- Greater production (peaking at ~85%)
- Improved egg quality and shell quality
- Lack of the small or medium eggs produced by new hens early in the first cycle

DISADVANTAGES OF FORCED RESTING. Disadvantages of forced resting include:

- Increased mortality (~1.5% of the flock during the resting period)
- Considerable interest from people concerned about animal welfare and consequently potential loss of consumer confidence in the product

DOES FORCED RESTING PAY? There is no simple answer to this question, since it depends on a variety of economic factors. The following general guidelines should be considered:

1. Recycling becomes less profitable as egg prices increase and as price differential between egg sizes decreases.
2. As the cost of replacement pullets increases, it becomes more advantageous to recycle.
3. The profitability of recycling increases as the price paid for spent hens decreases. The market for spent hens continues to diminish at a rapid rate. Spent hens are increasingly disposed of by composting (preferred) or by rendering.
4. Using replacement pullets instead of recycling ties up capital.
5. There are obvious welfare issues related to the methods of forced resting.

Housing and Equipment

Layer houses and equipment come in many styles and sizes. Floor houses are generally 36 to 70 ft (6–12 m) wide and 200 to 600 ft (65–200 m) long, with a refrigerated area for storage of eggs. The most common type of manual cage house is 9 to 12 ft (3–4 m) wide and 250 to 300 ft (~100 m) long. There is a concrete aisle down the center and a refrigerated area at one end in which the layers are fed by manually driven electric feed carts, and eggs are collected by hand. Mechanical cage houses vary from 8 to 32 ft (2.5–11 m) wide and 270 to 450 ft (90–150 m) long, and they are equipped with mechanical feeders and egg belts (also see Chapter 11).

Housing Requirements

Poultry housing should be reasonable in construction and maintenance costs, reduce labor, have utility value, and confine the birds. Additionally, poultry houses are frequently environmentally controlled for temperature (requiring insulation and a vapor barrier), ventilation, and lighting (see Chapter 11).

Housing Systems

Housing systems for egg production include (1) cages, (2) floor, (3) slats, and (4) partial floor and partial slats. Each system can be varied as to size, construction material, and extent of mechanization. There is no one best system or arrangement. All of them must take into account:

- Bird comfort
- Operator efficiency
- Operational costs
- Egg handling
- Durability
- Initial cost
- Service availability

BIRD COMFORT. The hen must be reasonably comfortable if she is to produce a large number of high-quality eggs with good feed efficiency while staying healthy. The house must be arranged efficiently for the operator. The operating costs of all fans, feeders, and animal waste removal systems must be held down or their use may not be justified. Provision must be made for handling eggs rapidly with minimum breakage. The total cost must be competitive with other producers.

CAGE SYSTEM. The cage system is the major method of producing table eggs (see Figure 14.8 for various designs). The cage system involves many small wire compartments each equipped with its own feeder and waterer and sloping floors so the eggs roll out of the cage for easy gathering by hand or egg belt. Figure 14.9 shows a cage setup with collection belts. Droppings fall through the cage floor into pits or onto dropping boards, where they are scraped into pits. Dropping pits range from 6 in. to more than 8 ft deep. With shallow pits, the droppings are scraped or washed into a holding chamber or directly into spreaders

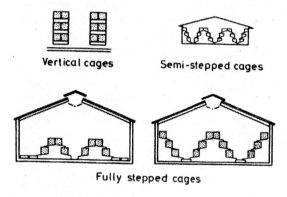

Vertical cages   Semi-stepped cages

Fully stepped cages

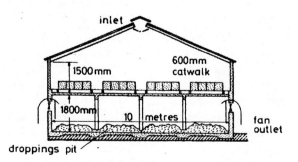

Flat deck cages over deep pit for droppings

Figure 14.8 Cage Systems. (Note: 10 meters is equivalent to 33 feet; 600 mm is 2 feet, 1500 mm is 5 feet, and 1800 mm is 6 feet. *(Reprinted from Sainsbury, D. (2000).* Poultry Health and Management: Chickens, Ducks, Turkeys, Geese, Quail. *Published by Blackwell Science, Oxford with permission)*

Figure 14.9 Automated egg collection by a moving belt with eggs conveyed by a revolving collecting device. *(Reprinted from Sainsbury, D. (2000).* Poultry Health and Management: Chickens, Ducks, Turkeys, Geese, Quail. *Published by Blackwell Science, Oxford with permission)*

every few days. In the deeper pits they can accumulate for prolonged periods of time (a year or more).

Feed may be supplied and eggs picked up by hand with ventilation adjusted by the operator. These jobs are usually mechanized so that the feed, air, water, eggs, and excreta are all moved mechanically.

Cages vary in size and bird capacity. Each will hold from 2 to more than 15 birds. Up to the present time, the most successful cage has 4 birds or less. Each bird should have from 54 to 64 sq in. (~400 cm²) of cage floor space. The dimensions of the smallest cage (2 birds) are 8 in. (20 cm) wide, 16 in. (40 cm) deep, and 14 in. (36 cm) high. Cages for more than 2 birds are simply wider and/or deeper.

Many cages have been installed in a stair-stepped arrangement with three, four, and sometimes five rows of cages. There are service walks down each side of the house and between each row of cages. Walks are 26 in. to 30 in. (~0.7 m) wide, while cage rows may be up to 90 in. wide. Some stair-stepped cages have been modified so that the bottom rows of cages are separated by approximately 8 in. (20 cm). This necessitates the use of a dropping board on the top of the bottom cages to prevent excreta from the birds in the top cages from falling through the cages below. The dropping boards have to be scraped frequently. Cages are suspended from the ceiling of a house. The bottom of the lower cage is usually 24 to 28 in. (~70 cm) above the aisle walkway. This puts the top cage at shoulder height. The excreta drops into the pit below the cages.

Triple-decked, narrow-aisle cage systems offer capacity. The interior aisles are 12 to 14 in. (~35 cm) wide, but the outside aisles are approximately 24 in. (60 cm) to provide work space. The bottom cage is level with the aisle. The top of the bottom cage is about knee high, the middle cage a little over waist high, and the top cage head high. This type of house is com-

pletely mechanized. Four-deck systems also exist. It appears that five, six, or more decks can be used, but several problems need to be solved to make them feasible. Inspection of birds and equipment in a "tall house" is more difficult since some birds are higher and some lower than eye level.

Another cage arrangement is known as the flat-deck system. Two arrangements are possible, one a wall-to-wall plan (without aisles) and the other with aisles. The difference between the two is that an overhead personnel carrier is needed to service the birds in the wall-to-wall system. The carrier enables the caretaker to move over the cages to inspect and care for the birds and equipment.

Cage-system equipment costs more per bird than floor-system equipment. This is offset by housing more birds in a given size building, making the overall cost of house and equipment quite similar. Cages offer the following advantages compared to floor systems:

1. Eliminates the need for pressure on the hens by breaking the flock up into many small societies.
2. Eliminates the floor-egg problem.
3. Produces an egg that is more acceptable to egg receivers and processors.
4. May reduce some diseases.

Some of the disadvantages of cage systems compared to floor units include:

1. Hens tend to have a very rough and ragged appearance.
2. The entire house must have an acceptable total environment since the hen cannot move to a more comfortable location.
3. There may be more odor because the droppings are not mixed with litter and there is more dust from the hens because they are housed closer together.

FLOOR SYSTEM. This is the oldest of the systems. It consists of litter, varying in depth from light to built-up, covering the entire floor. Feeders and waterers are located on the litter, and nests line one or two sides of the house. This arrangement takes a low investment, since it needs a minimum of equipment. But it requires a well-insulated house with a properly designed and well-managed ventilation system. It may be appropriate for small producers who are producing eggs for the organic and noncaged niche markets.

SLAT OR WIRE FLOOR. This involves slats or wire over the entire floor. Wire floors are subject to rust and mechanical damage, and sometimes they produce foot problems in the chickens. Slat floors are made by arranging strips of wood, metal, or concrete parallel to each other about 1/2 in. (1 cm) apart so that the droppings are pushed between the slats by the birds' feet. Droppings then accumulate in a pit below. The depth of the pit may vary from 2 or 3 ft (0.6–1 m) up to 8 or 10 ft (~3 m). The floor must be removed at least once a year to clean shallow pits. With deep pits, droppings may accumulate for several years if no water is permitted to enter the dropping area. Breeder flocks are frequently housed on slat floors.

Mechanical cleaning with a cable-type cleaner beneath slats or wire floors frequently presents problems because of the large area and volume of manure. The operator of slat or wire floor houses should maintain a constant program of rat and mouse control. Feather picking and cannibalism may be a problem on slat or wire floors. There may be flightiness if pen size, bird concentration, or noise levels become intolerable.

PARTIAL FLOOR, PARTIAL SLAT. This consists of litter covering about 60% of the floor area with a raised slat or wire floor over the remaining 40%—usually down the middle of the house. This raised floor provides a place for feed, water, and roosts. About 70% of the droppings collect below the raised floor and are removed by a mechanical cleaner about twice a week. Removal of droppings also gets rid of considerable moisture, thus reducing the ventilation problem. The floor litter is easier to manage since it contains fewer droppings and less moisture.

Pullet Housing and Equipment

Regular replacement of the laying flock is one of the most important and most expensive essentials of egg production. It can rank next to feed in the cost of egg production and it exceeds labor, housing, equipment, interest, taxes, and other costs. Producers may choose between recycling existing hens or pullets, growing their own pullets, or purchasing started pullets. In any event, the building and equipment requirements are the same, whether pullets are raised for sale or raised by the one who will retain them as layers.

It is wise to provide pullets with an environment (housing, feed, and general management) above minimum standards but consistent with economic costs. This calls for housing that gives protection from heat and cold and provides good ventilation; and it means optimal feed, water, and space.

HOUSING. Effective isolation of growing stock from older birds is important for disease control, particularly during the early stages. Hence, separate housing should be provided for replacement pullets, completely separated from layers. Housing that is suitable for brooding chicks, or for housing a laying flock, will be satisfactory for rearing pullets.

Usually, they are started out in the brooder house, then switched to the layer house.

FEEDERS AND WATERERS. It is critically important to have a good distribution of waterers through the house with a constant supply of fresh water. The same principle applies to some degree to feeders. Both feeders and waterers should be located to permit birds in any part of the house or cage to feed or drink conveniently, without having to travel more than 15 ft (5 m).

Layer Housing and Equipment

The design and construction of houses and equipment for layers should be such as to provide for top performance of the layers, optimum environmental control, functional arrangement of equipment, maximum labor efficiency, satisfactory waste disposal, and minimum housing and care costs per dozen eggs produced.

HOUSING AND LAYER EQUIPMENT. The starting point in designing a layer house is the selection of the type of laying system. Presently, layer houses are being arranged in the following ways, or according to the following systems: cage system (by far the most common), floor or litter-type house, slat or wire floor, or combination floor and slat (see Chapter 11).

Ventilation. Of the various ventilation systems, a mechanical system is best. There are two basic types of mechanical ventilation: (1) exhaust fans and (2) pressure or intake fans.

In an exhaust system, fans force air out of the building. This creates a slight vacuum in the building. Air then comes in through intake openings to equalize pressure. Where fresh air inlets appear uniformly around the building, fresh air distribution should occur uniformly.

In a pressure fan system, fans draw fresh air from the outside and build up enough pressure to push stale, moisture-laden air out through exhaust ports and any other openings. Fan capacity is an important factor. The correct capacity depends primarily on the total weight of poultry in the house. Approximately 1 cfm (cubic foot per minute) handles 1 lb of bird (0.06 m^3/min/kg bird). Many commercial systems have a 7 cfm per bird (0.2 m^3/min/bird) rating for added safety. For 1,000 4-lb birds, 4,000 cfm is needed (for 1,000 2-kg birds, 100 m^3/min is needed). A fan or fans should then be selected to deliver this quantity of air at 1/8 in. (0.3 cm) static pressure.

Small houses, 24 ft (8 m) wide or less, with birds housed on the floor may be satisfactorily ventilated with the gravity system. This is based on the principle that warm air rises. A gravity system needs draft-free inlets and controlled outlets.

Adequate summer ventilation is extremely important. Egg production drops when temperatures exceed 84°F (29°C), due to decreased feed consumption. During the winter, fans are operated at much lower capacity (one-fourth or more cfm per bird), depending on the outside temperature. Individual capacity thermostats on each fan are set to run at different temperatures to provide fresh air regularly while minimizing heat loss.

Feeders. Automated feeding distributes feed to the feed trough's serving cages. Mechanical feeders are used almost universally in large enterprises. Feed is delivered by a chain, cable, or traveling feed hopper to the feed trough in each cage. Feed distribution is activated by a time clock. In some small operations, a motorized feed cart is used. Feed is delivered to the farm in bulk feed trucks, then put into bulk feed bins outside the house. From the bins, it is augered into a feed cart. Then, a motorized feed cart is driven by the employee to distribute feed to the multidecked cages.

Waterers. There are three ways of delivering water to birds in cages.

1. The nipple waterer is hung from the top of each cage. Birds drink by pushing the nipple up, which allows the water to flow into their mouths.
2. A cup drinker is placed in each cage or between every two cages. The hen depresses a small trigger with her beak that causes water to flow into the cup. Birds must be taught to use these drinkers.
   Cup and nipple drinkers usually do not have to be cleaned frequently.
3. The continuous flow V-shaped water trough—water enters one end and flows down a slight slope or grade to the other end. Clean, fresh water is always available, but a lot of water is used and frequent cleaning is required (preferably daily).

Egg-Handling Equipment. Egg collection belts are used in larger units. In some installations, eggs are cross-conveyed to a packing station within the house. In very large production units, eggs may be conveyed through a tunnel from all of the laying houses to an egg-processing room. In the processing room, they can be directed through an egg washer, dryer, candler (to check interior quality), grader (to determine size), and packer (to carton) without being handled except to remove defective eggs.

Eggs are collected by hand in smaller houses. They are usually put into filler flats on a cart that is pushed or driven up and down the aisles. Next, they are cased and then held in a cooler while awaiting delivery to a processing plant.

Refrigeration. Eggs are a perishable product and should be handled with care. When they are being candled, washed, and sized, the room temperature should be < 72°F (< 22°C). Eggs should be cooled to < 60°F (< 16°C) as soon after production as possible, then held at that temperature until used. An egg cooler room should have sufficient capacity to hold at least one week's production.

Nutrition of Laying Hens

Feed is by far the greatest expense item in producing eggs. Therefore, the efficient producer uses a feeding program that maximizes profit.

A layer ration must furnish an adequate supply of nutrients. The essential nutrients are:

1. Adequate energy, furnished by starches and fats
2. Protein (with the optimal balance of amino acids, particularly limiting essential amino acids)
3. Necessary minerals
4. Sufficient vitamins

The diet must be formulated to meet or exceed the breeders' stated or established requirements and be broadly consistent with National Research Council (NRC) recommendations.

Producers can do one of the following:

- Buy their feed commercially prepared.
- Mix locally grown grain either with commercial supplements or individual ingredients.
- Producers, especially the large ones, do on-farm mixing.

It should be remembered that feed mixing is complex and exacting and should be attempted only by those persons who are knowledgeable. Also, regardless of how poultry feed is obtained, it should be formulated and mixed correctly (also see Chapters 6 and 7).

Beginning of Egg Production

There are both quantitative and qualitative changes in nutrition as laying hens come into lay. These include the following:

- Food intake increases immediately prior to egg laying and while laying.
- Calcium concentrations should be increased to about 3% 10 days before the expected time of first egg.

Feeding Programs

Egg producers use an all-mash diet. Commercial egg producers use this feeding program exclusively. Under this program the operator uses a complete feed—

a must when a mechanical feeder is used. It may be fed as:

- Meal
- Granules
- Crumbles of various sizes

An all-mash program offers the greatest assurance of good production. Also, it comes closest to ensuring uniform yolk color. A 14 to 17% protein mash is recommended for layers.

To minimize waste, feeders and waterers should be adjusted to shoulder height of the birds, and hoppers should be filled only one-third full. In non-mechanized houses, hoppers should be placed on a 4 in. (10 cm) block or suspended at an equivalent height so that birds reach out rather than down to feed. Feeders on 18 or 20 in. (48 cm) legs should be avoided because their use predisposes the flock to cannibalism as a result of the birds standing at eye level to each other. Allow feed hoppers to empty occasionally to avoid mold development. Palatable, fresh mash increases consumption.

A 4 ft (1.2 m) water trough will accommodate 100 birds. A laying flock drinks about 2 lb (1 kg) of water per pound (0.45 kg) of feed consumed; more in warm weather. This amounts to about 50 gal (46 l) per day per 1,000 hens. The water supply must be kept clean. Flock health demands clean, fresh high-quality water. It is critical for high rates of egg production.

Feed Conversion (Efficiency)

Feed conversion refers to the units (pounds or kilograms) of feed required to produce a dozen eggs, without regard to the size of eggs produced, (i.e., medium, large, or jumbo).

Rate of production is a very important factor in determining efficiency. Divide the units (pounds or kilograms) of feed by the dozens of eggs to obtain the feed conversion, or units (pounds or kg) of feed required to produce a dozen eggs. This should be < 4.0 (for lb) for profitable production. The strain of birds, season of year, management, disease level, and feed influence feed consumption.

Phase Feeding

Phase feeding involves changing the laying hen's diet adjusted for the following: (1) age and state of production, (2) season and temperature, and (3) the strains of birds. For example, a hen laying at the rate of 60% has different nutritional requirements than one laying at the rate of 80%. Therefore, the main objective of phase feeding is to reduce the waste of nutrients caused by feeding more than what a bird

actually needs under different sets of conditions. In this way, feed efficiency can be improved and the cost of producing eggs reduced.

Flock Health

Good livability and health are requirements for layer profits. Deaths, medications, and condemnations make for tremendous losses. But even greater economic losses result from decreased egg production.

More than 1% mortality per month is excessive. Any abnormal drop in egg production or in daily feed consumption indicates the need to check flock health and corrective treatment. All chickens are susceptible to diseases resulting from two causes: (1) stress and/or (2) infections. Stress factors include poor nutrition and environment. Infections may occur in a single individual or be widespread, depending upon the pathogen involved, the resistance of the flock, and the environment.

Layer Health Program

There is no such thing as a standard health program for all layer farms. The following points are basic:

1. Purchase healthy, disease-free replacement stock (-day-old chicks or ready-to-lay replacement pullets).
2. Minimize stress, including ensuring sanitation and proper nutrition.
3. Develop and follow a precise vaccination program.
4. Ensure biosecurity. Keep visitors away from poultry buildings. Do not mix birds of different ages. Never permit contaminated equipment from other poultry farms on facility. Screen or cover all poultry house openings so wild birds and rodents cannot enter.
5. Remove sick, injured, and dead birds as soon as noticed (burn, compost, or bury dead birds immediately).
6. Ensure clean water of a high quality. Test water quality. Clean trough-type waterers daily. Check operation of individual cups and direct-action valves or nipples.
7. Check operation of feeders, fans, and lights.
8. Obtain reliable diagnosis from poultry veterinarian before administering drugs or biologics. Check birds regularly for internal and external parasites.
9. Maintain good records on flock health.

The following five components are keys in any disease prevention program:

1. Sanitation
2. Nutrition

3. Environmental quality
4. Vaccination
5. Parasite control

SANITATION.   There is no substitute for sanitation. This includes everything the birds or eggs come in contact with. Clean, disinfect, and air out a poultry house before putting birds of any age in it. Disinfecting is not a substitute for cleaning. Disinfectants are only effective on clean surfaces. When preparing a previously occupied poultry house, follow these steps:

- Remove dust, dirt, crusted manure, litter, feed, and so forth.
- Clean the building and all equipment thoroughly, including air intakes, overhead ledges, and fans.
- Disinfect by selecting the proper disinfectant and following the manufacturer's directions. All disinfected surfaces must be dry and the building aired out before placing birds.
- Clean automatic waterers regularly.

NUTRITION.   Nutrition plays a major role in the health of the flock and the birds' ability to resist disease.

ENVIRONMENTAL QUALITY.   Environmental conditions contribute directly to flock health. A poor environment makes birds more susceptible to disease and contributes to the spread of many diseases. Adequately lighted, well-ventilated, relatively dry quarters are desirable. A good environment provides:

- Shelter, protection, and comfort.
- Convenient and adequate supplies of proper feed, clean water, and fresh air.
- Equipment and facilities that are conveniently arranged for both birds and personnel.
- Waterers should be within easy reach of all birds and should not overflow.

If litter is used, it should be absorbent, relatively dust-free, and resistant to matting. Litter under waterers and feed troughs should be replaced when it becomes damp or matted. Moisture in litter fosters parasite development and growth of molds and bacteria. With cage systems, the operator must control the environment uniformly to get satisfactory performance.

Poultry Health for Layers

Many of the serious diseases to which layers are susceptible can be prevented by animal health programs including vaccination and by proper management (see Chapter 10).

MANAGEMENT OF EGG PRODUCTION

Key Management Decisions

Management involves decision making. Many of the decisions egg producers make are only made once; hence, they had better be right. Among the layer management decisions that must be made are:

1. Whether to go into egg production, and where.
2. Method of financing.
3. Contract versus independent production.
4. The breed and breeding; and whether to buy or raise replacements.
5. Type and size of house and equipment.
6. The diet to use; and whether to buy commercial feed or home-mix.
7. The flock health program.
8. How and where to market eggs.
9. What is ahead.

Additionally, the capable manager sets high production goals and is quick to sense when all is not well in the hen house and how to rectify the situation.

Production Goals

Producers of commercial eggs strive for the maximum of egg production at a minimum cost, yet they recognize that net returns are more important than costs as such. The largest item of cost in the production of eggs is feed. It normally constitutes 50 to 70% of the total cost. Labor costs and cost of replacement pullets are the other two major cost items in production.

Generally speaking, higher egg production means lower costs per dozen eggs. This is because the feed required for maintenance is constant for hens of any given weight and bears no relation to the number of eggs laid. Hence, it is important that commercial producers strive for high egg production.

The following production goals are suggested for the commercial egg producer:

1. Production of 270 eggs per hen per year.
2. Feed conversion of less than 3.6 lb of feed per dozen eggs.
3. A laying house mortality of less than 10%.
4. More than 75% extra-large and large Grade A eggs.
5. More than 95% marketable eggs.
6. On-farm egg breakage under 2%.
7. Mortality of less than 5% from 1 day old to 5 months.

Size does not assure success and profitability in commercial egg production. Rather, it makes it im-

BOX 14.1 Cost of Producing Replacement Pullets

The cost of producing replacement pullets varies with the costs of the following:

1. Feed
2. Vaccination
3. Labor

Estimated cost related to $2.37/20-week-old pullet:

- Feed, 42%
- Chicks, 24%
- Labor and management, 15%
- Buildings and equipment depreciation/interest, 10%
- Miscellaneous including utilities, fuel, and vaccination, 9%

(Based on D. D. Bell and W. D. Weaver, Commercial Chicken Meat and Egg Production, Kluwer Academic Publishers, 2002.)

perative that there be superior management and adequate ventilation.

Overview of Egg Production

The production of table eggs is divided into two phases. The first phase is the production of started pullets. This involves brooding and rearing of chicks until they are old enough to start actual egg production. Started pullets are also called replacement pullets since once you've been through one egg production cycle you will be using these birds to replace the old birds. The second phase involves the actual production of eggs. These two phases of production are usually carried out in different locations that are some distance apart in order to reduce the incidence of infectious diseases.

Started/Replacement Pullet Management

Pullets should have outstanding genetics and be healthy and vigorous. Proper feeding, care, and management during the pullet starting period develop the potential productive egg capacity of the birds, whereas poor practices and management may reduce the genetic potential. The following man-

agement pointers in raising starter pullets are recommended. The costs of producing replacement pullets are outlined in Box 14.1.

Brooding of Replacement Pullets

Conditions for brooding replacement pullets include the following:

- **Floor.** Replacement pullets may be grown on solid floors or on wire floors. In addition to being convenient, wire floors alleviate later problems of adjustment to wire floors of laying cages. Successful pullet rearing requires adequate floor space. Overcrowding may result in a breakdown of the social order, feather picking, cannibalism, high mortality, and excessive culls. Optimum insulation and forced-air circulation can reduce floor space requirements by as much as 50% over uninsulated conventional housing. Suggested floor space requirements are shown in Table 14.2.
- **Brooder.** First week brooding temperatures should be at least 90° to 95°F (32°–35°C), with a 5°F (2.75°C) weekly reduction until 70°F (21°C) is reached, or until heat is no longer required. Chilling or overheating young birds should be

TABLE 14.2 Minimum Floor Space Requirements for Pullets

	Space/Bird							
	Controlled Environment				Uncontrolled Environment			
Age of Pullets	In Cages		On Floor		In Cages		On Floor	
	(sq ft)	(sq m)	(sq ft)	(sq m)	(sq ft)	(sq m)	(sq ft)	(sq m)
1 day to 7 weeks	0.20	0.018	0.50	0.045	0.30	0.027	0.75	0.067
7 to 11 weeks	0.25	0.022	0.70	0.063	0.40	0.036	1.00	0.090
11 to 22 weeks	0.40	0.036	1.20	0.108	0.55	0.049	1.80	0.162

Source: Adapted from Production of Commercial Layer Replacement Pullets, *University of Georgia.*

TABLE 14.3   Recommended Feeder Space for Pullets

	Feeder Space/1,000 Birds[1]			
	Cages		Floor	
Age	Trough Feeders	Trough Feeders	Pan Feeders
1 day to 10 weeks	92 ft (1.1 in./bird)	10 feeder lids	10 feeder lids
10 days to 10 weeks	92 ft (1.1 in./bird)	78 ft (.94 in./bird)	22 (46 birds/pan)
10 to 22 weeks	167 ft (2.0 in./bird)	117 ft (1.4 in./bird)	31 (32 birds/pan)

[1]To convert feet to meters, multiply by 0.305.
Source: Adapted from Production of Commercial Layer Replacement Pullets, *University of Georgia.*

TABLE 14.4   Recommended Waterer Space for Pullets

	Waterer Space/1,000 Birds			
	Cages			Floor
Age	Trough Waterers		Cup	Trough Feeders
	(ft)	(m)		
1 day to 10 weeks	—	—	45 cups	15 1-gal fountains
10 days to 10 weeks	40	12.2	55 cups	20-ft (6.1-m) trough
10 to 22 weeks	80	24.4	84 cups	40-ft (12.2-m) trough

Source: Adapted from Production of Commercial Layer Replacement Pullets, *University of Georgia.*

avoided. Cage rearing requires that the entire house be uniformly comfortable. Any one of several types of commercial brooders will work satisfactorily if properly used. Gas, oil, or electric (hover-type) brooders, or infrared, are most common. Each of the brooding systems offers certain advantages that make it desirable. The availability and cost of fuel must be considered also in choosing the type of brooder to use. Chicks should not be crowded during the brooding period. Where hovers are used, 8 to 10 sq in. (~64 sq cm) of hover space per chick should be provided.

- **Litter.** Wood shavings, straw, rice hulls, or other commercial litter may be used. The litter selection should be on the basis of what is most convenient and economical.
- **Feeders.** It is important that adequate feeder space be provided. The amount of space will vary according to age, cage versus floor feeding, and trough versus pan feeders. Table 14.3 gives recommended feeder space per 1,000 birds.
- **Waterers/drinkers.** The value of adequate water space is evidenced by the fact that water is consumed at about twice the rate of feed. Even sick—off feed—birds will continue to drink, provided that water is easily accessible to them. Inadequate water space can cause or accelerate health problems, particularly for growing birds. Recommended water space is given in Table 14.4.

- **Feeding.** Since growth rate is not critical in White Leghorn pullets, the dietary energy level of the ration is usually lower than that of broiler diets. The young bird eats mainly to satisfy its energy requirements; hence, the intake of other nutrients is dependent upon the energy level of the feed.
  - The feeding objective in commercial pullet production is to manage nutrient intake so that the birds develop their full genetic potential. This is primarily accomplished by avoiding excessive body fat and early maturity. Obesity may reduce the total number of eggs, while early maturity tends to cause more small eggs and thus fewer large market eggs.
  - Feeding during the growing and development stage (8 to 22 weeks) usually follows either one or two basic programs:
    1. Full feed with a controlled lighting program.
    2. Controlled feed intake, using a high-energy feed and appropriate lighting control.
- **Vaccination.** A critical component to hen health is vaccination (See Table 14.5, also see Chapter 11).

Sanitation

A strict sanitation program is a must in raising starter pullets. Recommended rules are:

TABLE 14.5 Outline of Suggested Vaccination Program

Vaccination	Age	Method of Administration
Marek's disease	1 day	Subcutaneous
Newcastle disease[1]	14 days	Spray, eyedrop, drinking water
	4 weeks	Spray, eyedrop, drinking water
	14 weeks	Drinking water, internasal spray
Bronchitis		(Given along with Newcastle)
Fowl pox[2]	1 day	Wing web
	8–10 weeks	Wing web
Epidemic tremor	12–18 weeks	Drinking water, inoculate 10% of birds
Laryngotracheitis	12–14 weeks	Modified eyedrop
Fowl cholera	12–18 weeks (optional)	Subcutaneous

[1]Potential velogenic viscerotrophic Newcastle challenge would require modification of the above program.

[2]In endemic areas, pigeon or modified fowl pox may be used.

Source: Adapted from Production of Commercial Layer Replacement Pullets, *University of Georgia.*

a. Do not mix ages or strains of birds in the same house.
b. Do not allow visitors in the house.
c. Screen out birds and control rats, mice, and other rodents.
d. Do not permit contaminated equipment to come on the farm.
e. Use composting, incinerator, or disposal pit for dead birds.
f. Keep birds free from external and internal parasites.
g. Necropsy all dead birds.

Vaccination

The vaccination program should be in keeping with the needs of the area. Also, it should be completed far enough ahead of moving the pullets so that they will not be going through a vaccination reaction during or soon after moving.

Moving the Pullets

The pullets should be moved in such a manner that there will be a minimum of stress. This calls for clean (preferably steamed) coops and trucks, not moving in inclement weather, providing fresh air but avoiding drafts during the move, and gentle handling.

Layer Hen Management Practices

Even though routine, certain management practices are prerequisites to success in egg production; among them, beak trimming, lighting, animal waste/excreta handling, culling, nesting, broodiness, and dubbing.

Beak Trimming

Beak trimming is the removing of a portion of the upper (and a lesser portion of the lower) beak of the fowl. It is an effective way to control cannibalism. This can be done at anytime from day old on. It is done with an electrically heated blade that both cuts and cauterizes to prevent bleeding. Chicks that have been beak trimmed do not need further attention for at least 1 year.

Lighting

All birds, including laying hens, are sensitive to light. They usually reproduce in the spring as day length increases. They cease egg production, molt, and regrow feathers as day length decreases (fall). Poultry producers have taken advantage of this phenomenon and, with lighting, given the laying hen a condition of perpetual spring. The result is year-round egg production.

Windowless houses make exact control of day length a simple matter. Operators with windowed houses can get the same effect if they consider the present day length and what it will be at the various times in the bird's life, then make adjustments accordingly.

Many pullets are grown under short or decreasing daylight to 19 or 20 weeks of age. The day length is then increased to >14 hours of light per year (also see Chapter 8).

Excreta (Animal Waste)

One of the major problems on a large egg farm is disposal of the excreta. The best solution is to spread it on fields as fertilizer, although the possibility of using manure as a feed for ruminants should be considered. Some producers handle excreta as a liquid. The animal waste is naturally degraded in anaerobic lagoons with a loss of nitrogen (nitrogen gas plus some ammonia) and carbon (carbon dioxide and methane) to the air. The liquid can then be land applied. Alternatives exist for producers who encounter odor or regulatory problems. These include:

- In-house drying, since excreta dried to 30 to 50% moisture has little odor
- Composting
- Incineration

A combination of regular removal, hauling, and spreading is a good policy. Spreading in the winter may not be possible when land is frozen. Similarly, summer application may be precluded by the presence of growing crops.

Culling

With improvements in genetics, there is a reduced need to cull. Laying hens may be culled for the following reasons:

1. To discard the few injured or unhealthy pullets at the time of placing in layer caging/houses.
2. To remove injured or overtly unhealthy hens from caging (preventing cannibalism and practicing humane management).

It is difficult to identify poor-producing birds except by those molting.

Nesting

When housed pullets come into production, encourage nesting by making frequent trips through the house. On each trip, pick up the floor eggs and place them in a nest. Level off the floor nests birds have made in the litter. Persistence will pay off—pretty soon the pullets will get the "idea."

Place nest pads in wire floor nests for the first 3 months or until the flock has reached peak production. Arrange nests in the pen convenient to the birds, and with consideration given to established social orders.

Nests are used primarily for the production of hatching eggs and by small scale producers.

Broodiness

Broodiness has been largely bred out of laying strains by selective breeding. It is rarely seen in hens in cages. Some broodiness may be observed in floor-managed flocks. Broodies should be removed as soon as observed and placed in a slat- or wire-floored broody coop. Feed them a complete ration. They should be ready to return to the laying pen in about 7 days.

Dubbing

Dubbing is the removing of the comb and, in some instances, the wattles from chickens. It may be done to either males or females. Usually, dubbing is done when chicks are a day old, using curved manicure scissors. At this age very little bleeding or discomfort is noted. The reasons for dubbing are:

- It reduces the possibility of injury to the large, tender comb surface.
- It may allow birds with large combs to eat and drink more easily.

The comb is a secondary sex characteristic. It may also function to reduce temperature (by heat exchange).

There are both real and perceived welfare issues. Dubbing may be stressful to the bird and has a significant labor cost. Hence, it should be done only if it serves a definite need, makes economic sense, and can be defended.

Dubbing is rarely used in commercial settings. It is sometimes used in research.

Behavioral Problems

With caging of hens, abnormal behaviors are evident including cannibalism and egg eating.

CANNIBALISM. Many types of cannibalism occur; among them, the following:

- **Vent picking.** Picking of the vent, or the area below the vent.
- **Toe picking.** Toe picking is most commonly seen in chicks. It may be brought on by hunger.
- **Head picking.** Head picking may follow injuries to the comb.

EGG EATING. Egg eating is costly to producers. If one bird acquires the habit, it usually spreads quickly throughout the flock. Egg eating can lead to vent picking. Egg eating is predisposed by factors favoring egg breakage. Once the egg-eating habit has started, it is very difficult to stop. If the birds have not been beak trimmed, this should be done immediately. Also, nests should be darkened and eggs should be collected frequently.

Producing Quality Eggs

Eggs are perishable and they should be handled as such. It is of utmost importance, therefore, that producers, packers, and retailers maintain superior quality. Consumers want eggs with fresh-laid appearance, good flavor, and high nutritive value. The shells should be strong, regular, and clean (see Figure 14.10 for how shell thickness can be determined); the white (or albumen) should be thick, clear, and firm; and the yolk should be light-colored.

Contrary to popular belief, not all freshly laid eggs are high quality. However, a high percentage of

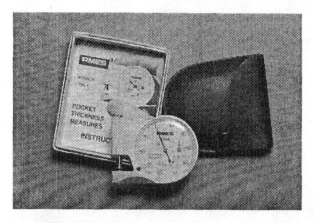

Figure 14.10 Shell thickness is an important factor considered by breeders. Thickness is correlated with shell strength. The photo shows a tool for measuring shell thickness. *(ISU photo by Bob Elbert)*

top-quality eggs can be produced by adopting the following practices:

- Selecting birds with the genetic background to lay high-quality eggs.
- Feeding well-balanced rations with adequate calcium, phosphorus, manganese, and vitamin D_3.
- Keeping the flock disease-free.
- Replacing or recycling birds in the laying flock when they are about 70 weeks old—the highest-quality eggs are laid by recycled hens. Older hens (late in the cycle of producing eggs) lay eggs lacking in acceptable shell quality and albumen firmness.

Handling Table Eggs

Observing the following rules will aid in maintaining high-quality eggs all the way to the consumer:

- Gather eggs frequently or continuously.
- Produce clean eggs—eggs are usually cleanest when they are laid.
- Wash all eggs.
- Clean soiled eggs.
- Cool eggs properly. Cooling eggs immediately after gathering removes the animal heat and retards any reaction that might be conducive to deterioration of quality. The egg cooler should be large enough to accommodate the daily production plus eggs held until they are marketed. An egg cooler operating at 55°F (18°C) or lower with a relative humidity of 75 to 80% is considered adequate. Eggs cooled prior to packing will sweat if removed from the cooler for candling or sizing. Therefore, eggs should be processed in coolers or in an adjoining air-conditioned room.

- Candle eggs. Candling is the most practical way to determine the interior quality of shell eggs. The object of candling is to discover and cull out eggs with blood spots and checks. The shell should be sound and free from checks or cracks, and it should be of good texture.
- Separate eggs into weight classes. See Chapter 16.
- Pack eggs properly. Eggs should be packed with small ends down. It is possible for the air cell to break loose and move to the small end when the large end of the egg is packed down. It is also important that cartons and cases be kept clean. This prevents the formation of mold that may pass off-flavors to the eggs.
- Eggs should be moved to market as quickly as possible, keeping all eggs cool en route to market (see Figure 14.11).

After eggs are delivered, maintenance of quality is the retailer's responsibility. Generally speaking, retailers have adequate facilities to protect the high-quality product that the producer has delivered to them. Refrigeration is a must for both short-term holding and displaying of eggs.

Handling Hatching or Embryonated Eggs

The fertile egg is at a fairly advanced stage of development from an early embryological standpoint at the time of laying. Accordingly, ideal handling would consist of setting the egg at once so that development could proceed without being checked. Obviously, this is not practical under most conditions, for hatching eggs must be held for varying lengths of time. The practical problem, therefore, is to hold these eggs in a suspended state of development without destroying the developing embryo. To this end, specific handling practices are essential (see Chapter 3).

Figure 14.11 Eggs sold in cartons.

Figure 14.12 Alternative egg products meet the needs of consumers on low-fat diets.

MARKETING EGGS

Eggs are marketed (see Chapter 17) either through processors who pick them up at the farm—they grade the eggs for size and quality and pay producers according to the quality and size—or through large producers who grade and package eggs on their establishments due to the lower costs involved.

There is greater processing of eggs. In 2001, the following situation existed:

- 58% eggs were sold retail,
- 30% further processed (up from 19% in 1988) (see Figures 14.12 and 14.13 for examples of further processed egg product),
- 11% going to food service/restaurants, etc.
- ~ 1% exported.

Shell Eggs

Pasteurization of shell eggs has been used experimentally. It has not been adopted by the industry.

Liquid Eggs

The attraction of liquid eggs (with the shell removed) includes:

- Convenience
- Uniformity of product
- Food safety, because liquid eggs are pasteurized

Liquid eggs have replaced shell eggs for fast-food restaurants (for breakfast egg sandwiches, such as Egg McMuffin) and many institutional users of eggs. Liquid eggs and other egg products now represent more

Figure 14.13 Rapidly cooked egg product.

than 30% of eggs produced in the United States (compared with ~6% in 1950). Liquid eggs are manufactured by breaking virtually all peewee and small eggs together with many medium eggs. In some cases, an egg breaker will use all the eggs produced on a farm. Liquid eggs may be sold to a food manufacturer as liquid or frozen or dried eggs. These may be further processed for retail products (frozen French toast and waffles, ice cream, egg noodles, mayonnaise, baked goods, etc.).

White and Brown Eggs

There is no difference in the nutritional value of white- or brown-shelled eggs. They come from different breeds of chickens. There are marked consumer preferences for the color of the eggshells. In the United States, the vast majority of eggs (>95%) consumed have white shells. In New England and elsewhere in North America, there is still demand for brown-shelled eggs. In Western Europe, consumers tend to prefer brown eggs. In the past, an exception to this was Germany but there is now a trend toward brown eggs there also.

Embryonated Eggs

Fertilized or embryonated eggs are used in biomedical research and for teaching (K–12 and college). This represents a small niche market. The health status of the parent stock is critical.

 USEFUL WEBSITES

Hendrix Poultry Breeding: http://www.hendrix-poultry.nl

Hubbard-ISA: http://www.hubbard-isa.com/

Lohmann-Hy-Line International: http://www.hyline.com

BOX 14.2 What's Ahead for Eggs and Layers

- Science and technology have made great changes in the egg industry. Further advances lie ahead.
- The egg industry will be concentrated in areas with high corn and soybean production and hence lower feed costs.
- Energy costs will be higher in the years to come, resulting in those areas having the lowest energy cost and/or requiring the least energy being favored from the standpoint of energy costs for environmentally controlled buildings and so forth.
- Pollution control (primarily due to excreta and dead bird disposal, odor, and flies) will continue to be a critical factor in the site selection for an operation of large layer units in the years ahead, favoring locations with relative ease of compliance with state and federal regulations.
- Commercial production units will be larger. Flocks of layers greater than 1,000,000 will be more common.
- More vertical coordination will come with stronger links between retailers and producers in the supply chain.
- There will be increases in niche markets for eggs including organic, free-range, and eggs with specific nutritional properties.
- There will be increased interest by consumers in brown eggs.
- There will be improved housing and automation.
- Egg production and its efficiency will continue to increase.
- Egg consumption will increase as medical opinions allay consumer fears of cholesterol and heart disease from eating eggs, and as new products appear on the market.
- Spent layers will increasingly be composted, rendered, and even incinerated.

FURTHER READING

Barbut, S. *Poultry Products Processing: An Industry Guide.* Boca Raton, FL: CRC Press, 2002.

Bell, D. D., and W. D. Weaver. *Commercial Chicken Meat and Egg Production,* 5th ed. Norwell, MA: Kluwer Academic Publishers, 2002.

Diseases of Poultry. Edited by B. W. Calnek, H. John Barnes, Charles W. Beard, Larry R. McDougald, and Y. M. Saif. Ames, IA: Iowa State Press, 1997.

Pattison, M. *The Health of Poultry.* Blackwell Science, 1993. Sainsbury, D. *Poultry Health and Management: Chickens, Ducks, Turkeys, Geese, and Quail.* Blackwell Science, 2000.

Sturkie's Avian Physiology, 5th ed. Edited by G. C. Whittow. San Diego, CA: Academic Press, 2000.

15

Chicken Meat and Its Production (Broilers)

Objectives

After studying this chapter, you should be able to:

1. List the major weights and ages at which broilers are marketed.
2. List the basic composition of broiler meat.
3. List the major breeders of commercial broilers.
4. Give recommendations on brooding, rearing, and production of broilers.
5. List the major diseases encountered in broiler production and give recommendations for prevention or control.
6. Give reasons why several programs of lighting can be used to increase growth of broilers.
7. List the top broiler-processing companies in the United States.
8. List the typical procedures performed in the conversion of broilers to meat.

Production of chicken meat is of great importance worldwide and is growing (see Chapter 1). Chicken meat can be divided into broilers (the vast majority of production), roasters, and Rock Cornish game hen, with capon an additional but very minor contributor.

This chapter addresses the following topics related to chicken meat:

- Marketing of chicken meat
- Composition of chicken meat
- The biology of meat production
- The commercial production of broiler chickens
- Broiler breeders and reproduction
- Other chicken meat (roasters and capons)

MARKETING OF CHICKEN MEAT

Chicken meat is popular because of consumers' perceptions of the following:

- Low price
- Consistent quality product

- Healthful, low-fat meat
- The availability of multiple products at retail (see Figures 15.1, to 15.11 for examples of chicken products) and fast-food units.

Chicken meat is marketed predominantly as broiler chickens. These are marketed at 6 to 7 weeks old with a dressed weight of 2.6 to 3.7 lb (1.2–1.7 kg) and a liveweight of about 5 lb (2.3 kg). Broiler chickens can be subdivided into small (< 4.4 lb or 2 kg), medium (4.4–5.5 lb or 2–2.5 kg), and heavy (> 5.5 lb or > 2.5 kg). In the United States, the average liveweight size of broiler chickens going to the processor increased to 5.2 lb (2.4 kg) in 2001.

In the United States, according to the USDA, young chickens are classified by weight with different markets.

1. < 4.2 lb (<1.9 kg) food service/fast food (predominantly 3.6–4.0 lb or 1.6–1.8 kg)
2. 4.21–5.25 lb (1.9–2.4 kg) retailer (dressed)
3. > 5.26 lb (2.4 kg) for deboning

Figure 15.1   Ready to cook dressed chicken.

Figure 15.2   Ready to cook chicken breast fillets.

Figure 15.3   Pre-cooked chicken for fajitas.

Figure 15.4   BBQ style chicken wings.

Figure 15.5   Prepared chicken enchilada meal.

Figure 15.6   Prepared grilled chicken quesadillas as snacks or a complete meal.

Figure 15.7 Easy to cook chicken primavera skillet dinner.

Figure 15.10 Easy to prepare lemon peppered chicken.

Figure 15.8 Ready to cook chicken breast tenderloins.

Figure 15.11 Prepared chicken enchilada suiza.

Figure 15.9 Canned chicken breast.

In 2002, these represented the following percentages of the number of birds slaughtered, respectively:

1. < 4.2 lb, 29%
2. 4.21–5.25 lb, 29%
3. > 5.26, 42%

Larger broiler chickens for deboning represent somewhat over 50% of chicken meat.

In addition, chickens are marketed as:

1. Roaster chickens at about 8 weeks with a dressed weight of 6.6 lb (3.0 kg)
2. Rock Cornish game hens at 3 to 4 weeks with a dressed weight of 1.3 lb (0.6 kg)

The dressing percentage for broiler chickens is about 70%.

Of chickens produced in the United States, the following is a breakdown of retail markets:

1. 48% are sold to retail outlets (supermarkets, etc.).
2. 35% are sold to food service such as fast-food restaurants (McDonald's has > 27,000 restaurants worldwide), and institutions (schools, hospitals, prisons).
3. 17% are exported.

Increasingly, chickens are processed prior to sale as illustrated by the following:

- 20% are sold as whole birds.
- 50% are cut up.
- 30% are further processed (e.g., as nuggets—McNuggets—and frozen meals).

Further processing includes mechanically deboned chicken meat. Ground chicken meat products in an emulsion include frankfurters. Drying is another step in further processing. Poultry meat can be dried (using hot air) as a method of preservation, to reduce weight, and/or to accommodate the food to which it is added. Perhaps the best example of dried chicken is in dried chicken soup mixes.

COMPOSITION OF CHICKEN MEAT

The carcass of poultry consists of the following tissues: muscle, adipose tissue, or fat, and bone.

A major advantage of chicken meat, particularly breast, is the low-fat content. Uncooked chicken breast has the following composition:

- Water, 69%
- Protein, 20%
- Fat, 11%

With removal of the skin, the fat content is decreased to < 2%.

Uncooked chicken dark meat has the following composition:

- Water, 65%
- Protein, 17%
- Fat, 18%

Removal of the skin reduces the fat to 4%. Not only does skinless chicken have a much lower fat content than red meat but there is also proportionately less saturated fatty acids. Chicken fat has a profile of fatty acids with less stearic (a saturated fatty acid) and more unsaturated fatty acids (e.g., linoleic) than beef or pork. Because saturated fatty acids are associated with elevated arterial cholesterol and cardiovascular disease, skinless poultry meat has this double advantage as long as it is not cooked with animal fat (e.g., butter).

Muscle

Protein makes up about 19% of chicken meat (see Figure 15.12 for a schematic representation of the details of the stuctured muscle). There are multiple proteins including the following:

- Myofibril proteins. These proteins cause muscle contraction (they represent 61% of muscle protein and include actin, myosin, and tropomyosin).
- Sarcoplasmic soluble proteins such as myoglobin and the glycolytic enzymes (representing 29% of muscle protein).
- The stromal proteins including collagen (found in connective tissue) and mitochondria (representing 10% of muscle protein).

Skeletal muscle myofibril proteins are arranged parallel to each other in myofibrils.

Proteins represent about 19% of the chemical composition of muscle, which is shown as follows:

Myofibril proteins (11.5%) including

Myosin	5.5
Actin	2.5
Tropomyosin	0.6
Troponin	0.6

Sarcoplasmic proteins (5.5%) including

Myosin	0.2
Hemoglobin	0.6
Glycolytic enzymes	2.2

Stromal proteins (2.0%) including

Collagen	1.0
Mitochondria	0.95

Myoglobin concentrations are much higher in dark meat than red meat.

There are changes in the meat after slaughter. These include pH and rigor mortis (stiffness of death due to muscle contraction as glycogen is depleted). The pH in meat declines after death. Components of the criteria for the quality of meat include the pH and the rate at which pH declines.

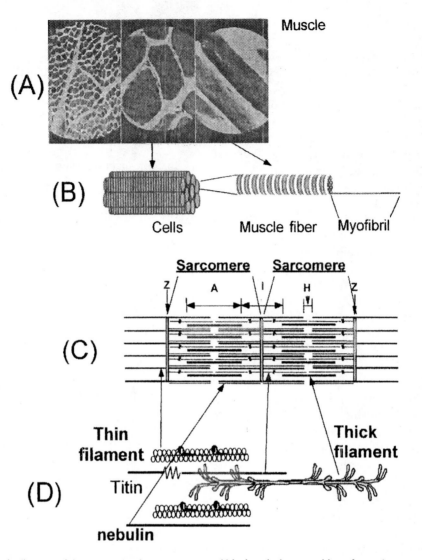

Figure 15.12 Schematic diagram of the structure, ultra-structure, and biochemical composition of muscle at progressively greater levels of magnification. Note: the muscle cells are made up of muscle fibers which are in turn made up of myofibrils. The thick filaments are myosin and the thin filaments are predominantly actin. (A)–(D) represent increasing magnification. *(Reprinted from Wick, M. (1999). Filament assembly properties of the sarcomeric myosin heavy chain.* Poultry Science *78, 735–742 with permission from* Poultry Science)

Adipose Tissue

There are two major sites for adipose tissue in poultry: subcutaneous, or under the skin, and abdominal, or leaf fat (this is the body cavity and mostly removed during evisceration). Most adipose tissue is of one cell type—the adipocyte, or fat cell. This is a large cell with a droplet of fat making up most of the cell. In turn, the vast majority of the fat stored in the fat cell is triglyceride—molecules containing three fatty acids combined with glycerol.

THE BIOLOGY OF MEAT PRODUCTION

The rapid growth rate of broiler chickens reflects superior genetics (intense selection for growth rate) and improved nutrition. The effects are manifest by nutrient availability to growing tissues and

endogenous hormones and factors. In chickens, as in other livestock, growth and development is controlled by numerous hormones and growth factors.

Hormones and Growth

Growth of muscle and bone requires normal levels of hormones including the following:

- Thyroxine (T_4) and its active metabolite, triiodothyronine (T_3)
- Insulin-like growth factor-I (IGF-I)
- Growth hormone (GH)

Muscle Development

Skeletal muscle is the major component of meat. Thus, muscle development and growth is the basis of meat production. Muscles develop in the chick embryo. The cells that are to become muscle are called **myoblasts.** In the early embryonic stage, the myoblasts divide repeatedly, forming many of these cells. These myoblasts then come together to form **myotubes.** The cell membranes break down, creating a myotube with multiple nuclei and a common cytoplasm. The muscle proteins are synthesized in the myotube by ribosomes. The major muscle proteins are actin and myosin.

Muscle Growth

Muscle growth includes increases in the size of muscle fibers due to accumulation of specific proteins and the incorporation of nuclei from **satellite cells.** Surprisingly, muscle protein is being both synthesized and broken down (or degraded) during growth. The rate of accumulation or accretion of muscle protein is the difference between the rate of synthesis and the rate of degradation.

> Rate of muscle protein accretion = rate of muscle protein synthesis − rate of muscle protein degradation

Adipose Tissue

The adipose tissue stores fat as triglyceride. Unlike the situation in many mammals, in poultry the liver produces the fatty acids. These travel to the fat cells via the bloodstream combined with lipoproteins. In the adipocytes, the fatty acids are linked (esterified) to glycerol to form triglyceride. At times when energy is needed, the triglyceride can be broken down in the process of lipolysis. Hormones change the rates of fatty acid synthesis and lipolysis. For instance, glucagon (released at times when food is not available) acts to inhibit fatty acid synthesis and increase triglyceride breakdown.

Lipogenesis, or Fatty Acid Synthesis

In chickens and other poultry, fatty acids are synthesized in the liver. They are then transported to adipose tissue via the bloodstream. Fatty acids are stored as triglyceride. Synthesis of fatty acids occurs when energy supply is greater than demand.

$$\text{Glucose} \Rightarrow \text{Liver} \Rightarrow \text{Fatty Acids} \Rightarrow \text{Adipose Cell} \Rightarrow \text{Triglyceride}$$

Triglyceride Synthesis by Adipose Tissue

Triglycerides are synthesized in the adipocytes from fatty acids and **glycerol-3-phosphate,** in a ratio of 3:1. The glycerol-3-phosphate is derived from glucose by the process of glycolysis. Triglyceride syn-

thesis occurs when energy supply is greater than demand.

$$3 \text{ Fatty Acids} + \text{Glycerol-3-phosphate} \Rightarrow \text{Triglyceride}$$
$$\Uparrow$$
$$\text{Glucose}$$

Triglyceride Breakdown by Adipose Tissue

Triglyceride breakdown, or **lipolysis,** is the enzymic cleavage of triglyceride to yield fatty acids and glycerol. Lipolysis occurs when energy supply is less than demand.

$$\text{Triglyceride} \Rightarrow \text{Hormone sensitive lipase} \Rightarrow$$
$$3 \text{ Fatty Acids} + \text{Glycerol}$$

Adipose Tissue Growth

The amount of fat or adipose tissue increases during growth by increases in the number of fat cells and an increase in the size of adipocytes due to triglyceride accumulation. As growth progresses, the relative importance of increasing the fat cell size becomes progressively greater. Some breeds of poultry have more fat cells. Females have more fat cells and larger fat cells than males. The size of the fat cells is also affected by the nutrition the bird is receiving.

Impact of Adipose Tissue on the Poultry Industry

Adipose tissue and particularly excess fat has negative consequences for the poultry industry for the following reasons:

1. Fat is high in calories and energy dense, and therefore takes a considerable amount of feed (lowering feed:gain efficiency).
2. Abdominal fat is removed from the carcass as a waste by-product.
3. Excess fat is not a desirable trait for the consumer.

While subcutaneous fat adds to the eating quality of chicken, it also has disadvantages for people concerned with their weight (due to the calories) or cardiovascular disease (saturated/animal fat content).

COMMERCIAL PRODUCTION OF BROILERS

Chapter 1 provides an overview of broiler production. The increase in broiler production can be attributed to genetics/breeding, nutrition, disease control, management, processing, marketing, and

TABLE 15.1   **Top Broiler Companies in the United States**

			Weekly Million lb Average	
Rank	Company	Million Head	Liveweight	Average Weight
1	Tyson	40.6	192	4.7
2	Gold Kish	14.8	77.4	5.2
3	Con Agra Poultry	15.2	68.5	4.5
4	Perdue Farms	11.8	62.2	5.3
5	Pilgrim's Pride*	11.4	54.8	4.8
6	Wayne Farms	5.1	33.0	6.5
7	Sanderson Farms	5.0	27.0	5.4
8	Cagle's	4.2	23.1	5.5
9	Mountaire Farms	3.8	23.0	6.0
10	Foster Farms	4.2	21.2	5.0

*Pilgrim's Pride acquisition early in 2001 of WLR (including Wampler Foods) brings it to #3 ranking.
Source: Rankings from WATT Poultry USA.

vertical coordination in the supply chain. Broiler production has changed from small, independent farm flocks and small processors scattered across the country to a highly integrated, efficient industry that is concentrated in fewer areas. Table 15.1 lists the largest broiler companies in the United States.

Business Aspects

A typical broiler unit might consist of a hatchery, a feed mill, a processing plant, a field service and management staff, and 150 to 300 growers. With the increase in size and integration of broiler enterprises, the business aspects have become more important. More capital is required, knowledge of contracts has become important, competent management is in demand, and data/information management is essential (see Chapter 18).

Financing

Banks and the Farm Credit Association finance the broiler industry through credit extended to feed manufacturers, dressing-plant processors, hatcheries, and integrated operators. Money is loaned at the local level to growers with mortgages as collateral.

Labor Requirements

Efficiency of labor is an important factor in determining profits in the broiler business. The more broilers raised per person, the lower the production costs, and the bigger the operation, the more feasible it is to automate. Automatic feeding and watering systems may reduce labor costs by 60%. A farm family, devoting full time to the operation, can care for more than 120,000 broilers at one time in five to six houses, and grow out at least six flocks per year.

Cost of Producing Chicken Meat

The cost of producing chicken meat has declined over the last 50 years. For instance, in the United States, this is evident both in nominal dollars and inflation-adjusted dollars.

From:

- 26.5¢/lb liveweight (nominal dollar in the 1940s) equivalent to $2.28 (in 2002 inflation-adjusted dollars).

To:

- 25¢/lb liveweight (nominal and inflation-adjusted dollars in 2002).

Analyzing Outcome of the Broiler Contract

Good management requires data on the performance of each flock. Box 15.1 summarizes the cost of chicken production. Good records are exceedingly important. Three types of broiler records should be kept: (1) those related to the flock, (2) those pertaining to the contract settlement, and (3) those pertaining to profit or loss. Costs and returns in the broiler business are critical to decision making, when deciding whether or not to enter the business, and in determining how well an established enterprise is doing. Costs and returns of broiler production differ from year to year and from area to area.

Annual net returns to contract poultry producers are relatively modest ($3,000–$5,000 for each house), at least until the unit is paid for. Labor requirements are also modest, but the capital requirements are great. Returns from contract broiler production are consistent and dependable, while labor requirements are low, making this type of en-

BOX 15.1   Economics of Broiler Production

Fixed payment per pound (e.g., 4.25¢/lb or 9.3¢/kg)

Variable-bonus-payment based on grower relative to other growers

Costs for houses	$720,000
Birds placed	160,000
Livability	95 %
Number of chickens produced	152,000
Weight of chickens produced	760,000 lb (345 metric tons)
Cycles per year	6
Total weight of chickens produced	4,560,000 lb (2,070 metric tons)
Revenue per year ($0.0425/lb)	$193,800

Operating expenses (e.g., litter, utilities, etc.) $60,000

Grower net return excluding loan repayment, $133,800

For a 7-year loan of $100, annual repayments are the following at different interest rates:

　　6%—$17.61
　　7%—$18.56
　　8%—$19.21
　　9%—$19.87
　　10%—$20.54

Loan repayment on buildings over, for example, 7 years (corresponding to the length of contract). For a 10-year loan of $100, annual repayments of $14.90 are for an interest rate of 8% assuming an interest rate of 8% – $138,312 per year. ($19.21 × 7200)

Net loss to the grower, $4,512! ($135,800 – 138,312)

This emphasizes the importance of negotiating for as low an interest rate as possible and for loan terms that give the grower a positive cash flow.

During the period when loans are being repaid, the following are critical to profitability:

1. Bonus payments
2. Reducing mortality
3. Lower-cost construction
4. Reducing expenses (e.g., litter)

Arguably, there will be moves to increase the number of flock cycles to seven per year. Once the loan is repaid, net returns will improve tremendously. Costs for building maintenance will, however, be rising progressively. There may be a case for expansion halfway through the loan or at its end such that the loan repayment of these new buildings is amortized over more birds.

(Recalculated from Ag Decision Maker, 1995, Iowa State University Extension; and Aho, 2002 in Commercial Chicken Meat and Egg Production (Bell and Weaver), Kluwer Academic Publishers.)

terprise attractive to many farm families. Moreover, some contract broiler growers have only one to two houses, which they care for on a part-time basis, and which they consider as supplementary income to a full-time job off the farm.

As is true in any business, profits in broiler production can be increased by lowering costs and increasing returns. Net returns depend on a satisfactory contract, reasonable broiler prices, efficiency of production, and volume.

LOWERING COSTS. Because most commercial growers raise broilers in large volume, small differences in production costs have a great impact on profit.

- **Feed.** Because feed is the highest production cost item, any change in feed cost and feed efficiency will cut the cost of production.
- **Mortality.** Mortality increases feed consumption per unit of broiler produced.

- **Building and equipment costs.** Facility costs are high. Comparison should be made between (1) package deals consisting of houses and equipment on an installed basis, (2) buying the houses and equipment from different suppliers, and (3) homebuilt houses. Also, growers should visit successful growers, expositions, and so forth to learn what is new and working well.

INCREASING RETURNS. Contract payments vary with the contractor and with such factors as mortality, feed conversion, average weights, and condemnations. Growers have increased returns by producing market birds of higher quality and at a younger age.

Housing and Equipment

In commercial broiler production, the bird spends its entire life in one house. Thus, brooder houses are cleaned and disinfected between flocks. There are, of course, many different styles and designs of houses, and even more variations in equipment. The important thing is that broiler houses and equipment provide for bird comfort and performance (also see Chapter 13).

Housing

The broiler house should provide clean, dry, comfortable surroundings throughout the year. It should be kept warm enough, but not too warm, and the litter should be kept dry. There should be ventilation with fresh air circulated, but the house should be free from drafts. In short, the broiler house should not be too cold, too hot, too wet, or too dry. The importance of housing is apparent from the seasonal effects on broilers: growth is best in the fall, with hot weather adversely affecting growth and feed efficiency.

LOCATION. The broiler house should be located to comply with state regulations concerning the required separation distance to neighbors' residences (complying with the law is a good-neighbor policy) together with the distance to other poultry units (~1 mile, to safeguard broiler health). Planting rapidly growing trees around the unit reduces neighbor complaints ("out of sight, out of mind"). The trees can also influence local wind patterns, air currents, and air or odor from houses being diluted further and moved to higher altitudes.

There should be adequate water drainage, but broiler houses should be set back a reasonable distance from streams, creeks, rivers, and lakes to prevent polluting surface water. There needs to be a good access road from the houses to the highway because of feed and bird hauling. The poultry unit should be set back from roads used to transport birds for biosecurity. The cost of bringing in utilities such as water, gas, and electricity should be considered.

CONSTRUCTION. Broiler houses in different areas are of many types and dimensions, and are constructed of various materials. Housing should be economical, yet substantial enough to last many years with a minimum of maintenance. The house should be sufficiently insulated to heat easily in winter and have sufficient ventilation for cooling birds in the summer. From 1/2 to 1 sq ft (0.04 to 0.09 sq m) of floor space should be provided per bird. Many broiler houses that are being constructed have gable-type roofs, truss-type (either wood or steel) construction, and single floors; and they are 32 to 40 ft (10 to 12 m) in width and of sufficient length to give the desired bird capacity. Most of them have a minimum capacity of 20,000 to 50,000 birds or more.

Most broiler houses have concrete floors. This is advantageous for rodent control. However, whenever the soil is dry, porous, and sandy, some houses have dirt floors. Basic house designs give satisfactory results such as environmentally controlled houses, which are lighttight and ventilation-controlled according to the requirements of the birds, and open houses, usually with curtains or windows.

The trend is toward environmentally controlled housing such as tunnel-ventilated buildings. These are insulated, windowless, and fan-ventilated. This provides a more uniform temperature, along with the fresh air. The final result is an improved market-quality bird and higher income.

VENTILATION. The main functions of ventilation are to maintain oxygen, keep carbon dioxide levels low, remove dust or moisture and ammonia from the building, and maintain suitable temperatures. Proper ventilation requires considerable management, because of large variations in exterior temperatures from time to time and increasing requirements of broiler flocks as they grow.

Air movement requirements are best determined by observing bird comfort, litter condition, and odor buildup. If the house is fan-ventilated, air movement of 3 to 4 cu ft (0.084 to 0.112 cu m) per bird per minute is sufficient under most conditions. However, total fan capacity of 7 cu ft (0.196 cu m) per bird per minute should be installed to handle air on excessively hot, humid days. During the first week, excessive ventilation should be avoided.

Ventilation is controlled by the use of high-speed or multispeed fans operated both continuously and thermostatically. The ventilation system must let air into the building in a way that avoids drafts on the chicks, and yet promotes air mixing, air change, and dust removal. A smaller air intake may be required in winter as cold air entering the building has a much larger expansion ratio than warmer air entering at other times of the year. Most ventilation systems may be classified as either of two types: negative pressure or positive pressure. In the negative pressure system, exhaust fans expel air that has been drawn into the pen through intakes, usually located in the opposite wall. The positive pressure system uses fans to force air into the pen and air escapes out through ventilation openings; this makes it easy to filter the incoming air—a decided advantage from the standpoint of disease control. As previously mentioned, it is important to have a system that provides uniform air change without drafts.

Some provision should be made for emergency ventilation in the event of an electric power failure. This can be done by either providing sufficient auxiliary electric power to operate fans, or installing doors in the sides of the building that can be opened to allow natural airflow and thus prevent smothering if the fan ventilation system fails.

RELATIVE HUMIDITY. It has been demonstrated that the performance of a broiler flock is affected by the humidity level. A relative humidity of 60 to 70% appears optimal. Birds may show discomfort by huddling when the relative humidity is 45% or less and the room temperature is 60° to 70°F (16° to 21°C).

When dry, dusty conditions prevail, it may be advantageous to spray the walls and ceiling with a fine mist, using fogging nozzles to bring relative humidity up to acceptable levels. Without optimizing humidity, litter moisture levels can rise and reduce broiler performance.

LIGHT. Lights are used to encourage feed consumption and optimum growth, and to prevent chickens from piling or stampeding when scared. Broilers may be lighted most of the night or intermittently through the night. To avoid piling in case of power failures, it is desirable to allow at least 1 hour of darkness in every 24-hour period.

The following lighting systems are commonly used for broilers:

- **Continuous light in open-sided houses.** Start with 48 hours of continuous light, then supply dim artificial light during all dark hours except for 1 hour at night (so that the birds will not panic in case of power failure). At floor level, the light intensity should be 0.5 foot-candle (5.4 lux), which can be supplied by one 150-watt bulb for each 1,000 sq ft (90 sq m) of floor space.

- **Light in lighttight houses.** Provide 3.5 foot-candles (37.7 lux) of continuous light at floor level for the first 5 days, until the chicks develop their eating habits and know the location of feeders and waterers. On the sixth day, reduce the light intensity to 0.35 foot-candle (3.8 lux), which can be supplied by one 125-watt bulb for each 1,000 sq ft of floor space. Then beginning on the eleventh day, use one of the following lighting programs:
- **Continuous dim light.** Provide 23 hours of light with an intensity of 0.35 foot-candle (3.8 lux) and 1 hour of darkness.
- **Intermittent dim light.** Provide 1 hour of 0.35 foot-candle (3.8 lux) light intensity (feeding time), followed by 3 hours of darkness (resting time), then repeat. This provides 6 hours of eating time each 24 hours. It is noteworthy that intermittent light produces better growth than continuous light. Although the reasons for this are not known, it is thought to be due to better feed utilization.

CAUTION: Do not change the lighting program once either continuous or intermittent lighting is started, and when an intermittent light system is used, up to 50% more feeder and water space may be required since most of the birds must be able to eat at one time.

Incandescent bulbs are considered superior to other light sources. Bright white light may be a contributing cause of feather picking, which can lead to cannibalism. Blue lights may be used when catching birds, and red lights may be used in case of cannibalism. Some interest has been shown in using continuous light the first week, followed by an intermittent lighting program (such as 3 hours of light followed by 1 hour of darkness) for the remainder of the growing period. In order to use an intermittent lighting program effectively, it is essential that the building be blacked out to prevent entry of light through doors or ventilation openings.

AUXILIARY POWER. A generator is required for auxiliary electric power to maintain heating, lighting, and ventilating operations in case of power failure. Additionally, a battery-operated alarm system should be installed in the broiler house and wired to the grower's residence. This should be activated by a power failure or when extreme temperatures occur.

Equipment

The two chief items of equipment are feeders and waterers (see Figure 15.13 for examples of automated feeders and waterers).

Figure 15.13 The automatic waterers and the moving chain of feed seem to stretch into "infinity" in this big broiler house. *(Courtesy, USDA)*

Figure 15.14 Broilers on a computer-formulated ration. *(Courtesy, USDA)*

FEEDERS. Inadequate feeder space causes uneven and slow growth. No point in the house should be more than 10 to 15 ft (~4.5 m) from a feeder. Broiler chicks are started with additional temporary feeders—trays with feed for the first 5 to 7 days. This allows the chicks to find feed easier. Feed should be placed in the feed troughs at the same time. Temporary feeders may be removed as soon as the chicks are eating from the troughs.

Mechanical feeders are generally used widely, with 2 to 2 1/2 in. (~7 cm) of feeder space provided per bird. These reduce labor and allow the grower to care for more birds. However, mechanical feeding does not reduce the necessity for frequent and regular observation of the birds. Some growers use a trough or hanging circular-type feeder. For the trough-type feeder, provide 2 in. (5 cm) of trough space per bird through 6 weeks of age, and 3 in. thereafter to market time at 7 to 8 weeks of age. When circular-type feeders are used, 20% less feeder space per bird will be sufficient.

WATERERS. It is important that broilers have an adequate supply of clean, high-quality water. Four types of waterers are used in broiler production:

- Nipple
- Cup
- Bell
- Trough

Nipple and cup drinkers are attached directly to water pipes and are triggered by birds to release water. These provide clean water in an automated system. Troughs are available in several lengths and sizes. Troughs should run lengthwise to the house and be equally spaced on both sides of the house. Bell-type waterers are similar in principle to troughs, and should also be equally spaced in the

house. Bell and trough waterers should be cleaned and washed daily. In most broiler houses, the water is metered. Daily water consumption is a good indicator of bird health.

Nutrition and Feeding

Since feed constitutes about 70% of the cost of producing broilers, it is important to give special attention to it (see Figure 15.14).

Most broiler feeds are produced by a feed manufacturing company, the feed-mixing unit of an integrated operation, or a large independent producer. The processed feeds are usually mixed in a centrally located mill and delivered in bulk by truck to the broiler farm. There is no "best" formula for the efficient feeding of a broiler flock. Rather, there are many very good formulas; hence, the grower should select the formula that will give the highest net returns. It should be noted that feed ingredients vary considerably in nutrient value, thus introducing an undesirable variable to feed formulation. As a safeguard against nutritional deficiencies, feed manufacturers usually blend ingredients from a wide range of sources.

Broiler feeds are high in both protein and energy. The importance of protein quality and of the protein-energy ratio in broiler rations is well recognized. Dietary demands of rapidly growing broiler chicks require that all nutrients be accurately balanced for optimum performance.

Although feeding systems vary, an effective broiler feeding program consists of the following:

1. Starter crumbles and/or pellets
2. Grower feed (1 and 2)
3. Finisher feed

Feed conversion is superior for birds on crumbles and pellets compared with mash. The feeding of crumbles and pellets tends to reduce the amount of feed lost in the litter compared with feeding mash.

Crumbled or pelleted feed is usually purchased in bulk and stored in upright metal tanks. The feed is moved by auger from the storage tank to the automatic feeder or feeders. Care should be taken to ensure that bulk feed tanks and augers are watertight to prevent the accumulation of moldy feeds. It is also beneficial to locate feed tanks on the leeward side of the building (also see Chapters 5–7).

Growth Rate and Feed Consumption

With improved breeding, nutrition, and management, feed conversion and growth rate of broilers are constantly improving. Tremendous gains in production efficiency have been realized in recent years. With narrow margins between production cost and selling price, it becomes increasingly important for the grower to pay strict attention to every step in the management program that will help improve production efficiencies.

In Ovo Feeding

Research has found that injection of amino acids and starch into the amnion results in more rapid growth posthatch. The chick embryos swallow nutrients, which are then digested and absorbed from the intestine. With adaptation, it may be possible to use the technology for *in ovo* vaccination to do this in an automated manner.

Broiler Health

The general issues of poultry health are covered in detail in Chapter 11. See Figure 15.15 for an example of healthy broiler chickens in spacious grow out houses. Specific concerns related to broiler produc-

Figure 15.15 Broiler chickens are free to roam in large, spacious grow out houses. State of the art ventilation systems keep the birds comfortable for year round production. *(Courtesy, USDA)*

tion are discussed here. The emphasis should be on disease prevention. Presently, average cumulative mortality runs about 4%.

The primary objective of a preventive program is to avoid the mechanical spread of pathogens. An effective preventive program is the all-in, all-out system, in which only one age of birds is on the farm at the same time. Between flocks there is a period of time when no birds are on the premises, during which any cycle of infectious diseases is broken. The next group of birds has a clean start, without contracting any disease from older birds on the farm. Advances in isolation and disease control have made it more practical to keep chicks of several ages on the same farm.

The following are key points in any broiler health program (also see Box 15.2).

1. Biosecurity
   a. Sanitation
   b. Disinfection
   c. Removal of dead birds
   d. Rodent control
   e. Access control (people)
   f. Access control (wild birds)
2. Good nutrition
3. Good environment
4. Vaccination
5. Medication
6. Parasite control

Good Nutrition

Nutrition plays a major role in the health of the broiler flock. The nutritional needs of healthy broilers are normally satisfied by the feed balanced for the required nutrients.

Good Water Quality

It is critically important for broiler health that the water supply is regularly checked for quality (excess minerals and the unlikely possibility of pathogens).

Good Environment

Environmental conditions contribute directly to broiler flock health. A poor environment makes birds more susceptible to disease and contributes to the spread of diseases. The following are very important:

- Adequate ventilation (good air quality with low ammonia and moisture)
- Proper lighting
- Dry litter

BOX 15.2   Broiler Health Program

A health program is fundamental to successful broiler production. It should include the following:

1. Isolate the broiler farm. Enclose the premises with a tight fence, and lock all entrance gates. Be aware of feed and supply trucks entering the enclosure.
2. Start with disease-free chicks.
3. Vaccinate chicks or embryos against Marek's disease at the hatchery, and against other common diseases.
4. Use effective drugs in the feed to prevent coccidiosis. Nonmedicated feed may be required the last 3 to 5 days before slaughter; check the feed tag.
5. Keep feed and water clean.
6. Screen house to keep out birds and rodents.
7. Do not allow visitors or service people inside the broiler house unless they wear disinfected boots and clean clothing.
8. Avoid contact with other flocks.
9. When there are several age groups on the farm, always care for the youngest birds first.
10. Obtain a laboratory diagnosis when disease problems arise.
11. Clean the house completely between each flock—ceiling, rafters, walls, floor, and surrounding premises. Also, repair, scrub, and disinfect all equipment—feeders, waterers, and brooders.
12. Rework litter. All caked and wet litter should be removed and replaced with fresh, clean litter before chicks arrive.
13. Add clean litter as necessary at least 3 in. deep after each clean-out. Wood shavings, processed pine bark, cane litter, and rice hulls are suitable litter materials. Avoid moldy or musty litter to prevent aspergillosis (mold growth in the bird's respiratory tract).

Poultry house ventilation requires constant attention. Most broilers are housed in environmentally controlled buildings that are mechanically ventilated. However, some broiler growers in warmer regions of the United States still use nonmechanical ventilation—curtains, panels, and windows.

Wet litter is never desirable. It can be the cause of disease outbreaks or breast blisters. However, excessive dust may trigger respiratory trouble. Both conditions can be avoided with proper ventilation made possible by adequate heat. Dust has sometimes been a major problem with fan ventilation, perhaps because of lower humidity and overdrying of the litter.

Disinfection

The producer should know what constitutes proper cleaning, why it is necessary, and how to do it. Most pathogens are killed by disinfection following thorough cleaning. Viral diseases are best controlled by strict cleanliness. Effective cleaning begins with the removal of litter and manure. The cleaning process includes scrubbing with brushes until surfaces are visibly clean, using a good detergent, flushing with clean water, then applying an approved disinfectant. One advantage of a concrete floor is that lye (caustic soda) can be used effectively. The presence of organic matter (dirt and excreta) in the disinfecting solution or on the surfaces to be disinfected may prevent the solution from destroying disease organisms so that no disinfection results (also see Chapter 11).

Vaccination

Protecting birds by vaccinating them against diseases common to the area is good insurance against costly disease losses. Most integrators have a recommended vaccination program, specifying the time to vaccinate and the type of vaccine to use. Broiler chickens are usually vaccinated against Marek's disease and Newcastle-bronchitis. The age of vaccination for Newcastle disease virus (NDV) or infectious bronchitis virus (IBV) will vary. Other vaccines may be used such as infectious bursal disease (IBD) and laryngotracheitis (LT). Vaccination against Marek's disease can be performed at the hatchery or injected by an automated process to 18-day-old embryonated eggs. Embryos can also be vaccinated against infectious bursal disease, Newcastle disease, and infectious bronchitis with an automated machine. Broiler chicks also receive protection via antibodies from the hen in the yolk. For instance, broiler breeder hens receive inactivated vaccines such as Newcastle disease virus, infectious bronchitis virus, reovirus, and infectious bursal disease virus subcutaneously. Booster vaccines for Newcastle bronchitis may also be given at about 10 days old and about 4 weeks via the drinking water, or sprays should be used.

Vaccines should be kept cool and used immediately after they are opened. Improper handling of vaccines can render them useless. Costly disease losses may result from improperly handled vaccines.

Antibiotics and Coccidiostats

Coccidiosis, a protozoan disease, is an ever-present hazard when raising young chicks. In order to prevent outbreaks, coccidiostats are included in feeds. Care must be taken to ensure that any feed containing a coccidiostat is removed well ahead of slaughtering time.

Antibiotics can be administered in either the feed or water to enhance growth or prevent subclinical or clinical infections. The specific antibiotic to be used is recommended by poultry veterinarians or poultry specialists. These may also have different withdrawal periods prior to slaughter of the flock. Detailed information on withdrawal times can be obtained from poultry advisors or by reading the directions on the label.

Parasite Control

A wide variety of external and internal parasites affect broilers. The prevention and control of these is one of the quickest, cheapest, and most dependable methods of increasing production. Flocks should be checked for the presence of tapeworms and intestinal roundworms (see Chapter 11).

Lice, mites, and flies are the most common external parasites of broilers. Heavy infestations reduce performance. The control of these parasites is simple if the right insecticide is used properly.

BROILER MANAGEMENT

Management is the key to success in the broiler business. The geneticist can increase the potential of production, and the nutritionist can formulate a ration that can reach this potential. Without proper management, the broiler will never get there. Management can make or break a broiler enterprise. Fortunately, a broiler operation responds favorably to good management.

Broiler growers should set production goals such as the following:

- Birds weighing 5.0 lb (2.3 kg) at 6 weeks of age
- A feed conversion of 1:1.75
- Mortality under 3.0%

Management Practices

Good management practices are summarized as follows:

- Start with quality, healthy, beak-trimmed chicks from reliable sources.
- Keep houses and equipment clean.
- Keep litter clean, dry, and free from mold.
- Brood birds carefully; have good sanitary management.
- Supply adequate heat and ventilation.
- Provide enough floor space.
- Give adequate space for feeders and waterers.
- Adapt the poultry health program to local needs.
- Dispose of dead birds promptly.
- Keep visitors out of houses; lock doors.

Management practices involve multiple details that add up to make broiler management very important.

Beak Trimming

Cannibalism can become a serious problem in a broiler flock. It begins by chicks picking tail feathers, toes, vents, and can progress to other parts of the body.

Cannibalism is caused by a combination of conditions such as overcrowding, excessive light intensity, insufficient ventilation, overheating, inadequate feeder and waterer space, and nutritional deficiencies. It is simple to alleviate these. Unfortunately, once cannibalism has started in a flock it is difficult to bring under control. Beak trimming greatly reduces feather picking and cannibalism, and has little detrimental effect on bird performance. It may be done at 1 to 8 days old by trimming off less than one-third of the beak with a commercial hot-blade-type trimmer. In addition, it can improve the percentage of Grade A birds marketed (also see Chapter 8).

Raising Males and Females Separately

Broilers can be sexed at the hatchery. Sexing may be by vent inspection or rate of feathering; the latter only if parental stocks were specially selected and mated. Separation of sexes allows the males to be marketed first. Males have the following characteristics:

- At least 1% heavier than females at hatching time.
- Grow faster than females and weigh more at a given age.
- Less fat—prior to market weight, male birds are less fat.
- Males convert feed to meat more efficiently than females.

As strains are bred for increased growth, the weight variation between the males and females increases. Intermingling the sexes does not detrimentally affect growth, provided that at least 1 sq ft of floor space is provided per bird. If the space is smaller than this, the growth rate of females may be depressed.

Temperature

The recommended temperature for broilers is as follows:

Age of Chick	Temperature
First week	92–95°F (33–35°C)
Second week	88–90°F (31–32°C)
Third week	83–85°F (28–29°C)
Fourth week	78–80°F (26–27°C)
Fifth week	73–75°F (23–24°C)
Sixth to eighth week	70°F (21°C)

The thermometer reading should be taken at the level of the chicks under the heater. With infrared lamps, readings are difficult to take. It is suggested that the lamp be hung a minimum of 18 in. (50 cm) above the floor so that the chicks form a circle. Comfortable chicks will bed down evenly or form a doughnut-shaped ring under the light. If they scatter, it is too hot. If they huddle, it is too cold.

Lighting

Light management for broilers is very important. The type of lighting system is determined by the housing—either an open-sided or lighttight house. However, the objective is the same:

- To provide sufficient light to enable the birds to see, eat, and drink.
- To avoid light intensity such that there will be cannibalism, excess activity, and piling.

These objectives are met when the intensity of illumination at bird level is about 0.35 to 0.50 foot-candle.

Disposal of Dead Birds

Dispose of all dead birds promptly by composting, incineration, deep pit burial, or rendering.

Litter

Provide a dry, absorbent litter material 2 to 4 in. (~8 cm) deep. Dry sawdust, shavings, peat moss, vermiculite, straw, rice hulls, and other products are available. Use the one that is most convenient and economical.

Litter is now not usually removed following the harvesting of a broiler flock. Rather, there is natural composting. Excessive height of litter dictates the removal of some (also see Chapter 8).

Excreta (Animal Waste) Handling

One of the major problems of broiler production is the disposal of animal waste. The best solution is to spread the animal waste/litter mixture on fields as fertilizer.

Brooding Broiler Chickens

In a typical grower contract, the contractor (the company) supervises the growing of the broilers and furnishes chicks, feed, and medication. The grower provides housing, equipment, and labor. The contractor decides when to market the birds and to whom they shall be sold. The contractor also outlines the brooding program to be followed. (This is covered in detail in Chapter 8.) *Brooding is critical to the success of the grower.*

- **Housing.** The first requirement for growing broilers is adequate housing. Since broiler production is essentially a chick-brooding operation, the house must control such factors as temperature, moisture, air movement, and light. Whatever the location, houses should be capable of maintaining a temperature of >70°F (>21°C) throughout the year.
- **Brooders.** The type of brooder to use varies (see Chapter 8). Brooder guards to keep chicks confined to the brooding area are recommended by some manufacturers. Corrugated cardboard 14 to 18 in. (~40 cm) high may be used particularly in winter, not only to confine the chicks but to aid in preventing floor drafts. Poultry netting may be used to replace cardboard for summer brooding. The guards should extend out about 3 ft (1 m) from the edge of the brooder to allow the chicks to move out from under the brooder if they become too hot. The guard may be removed after about the first week.
- **Feeding equipment.** Broilers are started with supplementary feed to allow them to find the feed more easily. Temporary feeders may be removed as soon as chicks are eating from the troughs. Mechanical feeders are generally used.
- **Waterers/drinkers.** Adequate watering space is essential (see Chapter 8).
- **Vaccination.** Protecting broilers by vaccinating against diseases common to the area is good insurance against costly disease losses. Marek's vaccine is usually administered at the hatchery (*in ovo* or at 1 day old). (Also see Chapter 11.) Immunization against both Newcastle and infectious bronchitis should be between the first and fourteenth day of age. Intraocular or intranasal dust, spray, and water-type vaccines are available; hence, the method that has given the most suitable results should be used. Booster vaccinations for Newcastle may be given when the birds are 4 to 5 weeks old.

- **Litter.** Broilers be may provided with fresh or reused dry, absorbent material 2 to 4 in. (~8 cm)deep. Wood shavings, sawdust, straw, rice hulls, and other products may be used. The choice of the litter should be made on the basis of what is most convenient and economical.
- **Light.** All-night lights (0.35 to 0.50 foot-candle at bird level) will give a big assist in keeping the birds from crowding in corners or piling up in case of disturbance during the night, and will tend to increase feed consumption, especially during hot weather.
- **Temperature.** The brooder temperature should be 95°F (35°C) at the edge of the hover, 2 in. (5 cm) above the litter. In controlled-environment houses, however, satisfactory results can be obtained with starting temperatures as low as 88°F (31°C).

In case of trouble. If the birds go off feed or if there is a sudden increase in mortality, immediate steps should be taken to determine the trouble. It is best to have a positive disease diagnosis before any treatment is started. To this end, birds should be sent to a diagnostic laboratory.

BROILER BREEDERS AND REPRODUCTION

Normally, adult males and females are housed together so that insemination can occur.

Breeding and Genetics

The key to the success of broiler production worldwide is the tremendous improvements in the growth rate and, consequently, feed efficiency achieved by poultry breeders. The selection of commercial broiler stock (for growth, feed conversion, livability, and rate of condemnation) can be the determining factor between profit and loss.

Broiler breeder flocks consist of a female line and a male line. The male lines have dominant white feathers and are selected for the following:

- Rapid growth
- Meat characteristics, such as breast width and carcass yield
- Livability
- Rapid feathering

Female lines used in producing broilers must have outstanding growth rate, high hatchability, and good, but not outstanding, production of eggs of desirable size and texture (see Figure 15.16 and 15.17).

If male and female lines are produced by different breeders, as they often are, each line must cross

Figure 15.16 Hatchery flocks of hens and roosters provide fertile eggs that are hatched into broiler chickens. Eggs are collected on a regular basis to help maintain optimum productivity. *(Courtesy, USDA)*

Figure 15.17 Broiler breeders. *(Courtesy, Dr. M. J. Wineland, North Carolina State University)*

well with other lines if it is to remain competitive. The breeder must determine the ability of male and female lines to cross well, whether the lines are developed by separate breeders or the same breeder.

Primary Breeders

Modern primary breeders are large international companies. The major primary breeders of broiler chickens are listed in alphabetical order as follows:

1. Arbor Acres (part of the Aviagen Group)
2. Cobb-Vantress (a wholly owned subsidiary of Tyson Food)
3. Hubbard (under Hubbard-ISA, which is a subsidiary of Merial Animal Health) (see Figure 15.18 for examples of Hubbard-ISA birds)
4. ISA (under Hubbard-ISA, which is a subsidiary of Merial Animal Health)
5. Ross Breeders (part of the Aviagen Group)
6. Shaver (under Hubbard-ISA, which is a subsidiary of Merial Animal Health)

Figure 15.18 Hubbard-ISA broiler breeder genetic lines. (A) The Hubbard Classic package of male and female broiler breeders. These have a balance of reproductive performance with broiler efficiency and are feather sexable. The broiler chicks produced are targeted for markets with lighter body weights 3.3 to 4.5 lbs (1.5 to 2.0 kg); (B) The Hubbard HI-Y package of male and female broiler breeders. The broiler chicks produced are targeted for markets with heavier body weights 4.4 to 6.6 lbs (2.0 to 3.0 kg); (C) The Hubbard Ultra Yield broiler breeder. The chicks produced are targeted for the market with heavy body weights > 5 lbs (2.5 kg). *(Courtesy, Hubbard ISA)*

Broiler Breeder Females

The broiler breeder hen produces eggs for a shorter period than laying hens. Moreover, in the older broiler breeder (>1 year old), the daily egg-laying cycle becomes irregular and the number of soft-shelled eggs increases. Egg production in the broiler breeder shows the following characteristics:

- Peaks during weeks 6–10 of production at about 85%
- Declines thereafter (e.g., at 40 weeks of production) at 60%

Photostimulation of Broiler Breeders (Lighting)

Broiler breeders, like layers, are brought into egg production by increasing the length of the day (length of lighting). This is achieved by extending the day length to 14 or greater hours of light per day (frequently 16 hours of light per day). Broiler breeder pullets are on short days prior to photostimulation. Broiler breeders are not photostimulated until at about 21 to 22 weeks of age to delay sexual maturation. This delay gives greater uniformity of egg production, with all the flock coming into lay and producing larger eggs.

Induced Molting

Induced molting can also be used with broiler breeders. There is the potential to increase the following:

- Egg production
- Egg quality (improved albumen based on Haugh units)
- Embryonic and, consequently, posthatching growth

Egg Storage

The longer eggs are stored, the greater the percentage of embryonic mortalities, the lower the overall hatchability, and the reduction in albumen quality (see Figures 15.19 and 15.20).

Egg Incubation

The temperature for incubation of chicken eggs including broilers is 100°F (37.8°C).

Egg Size

Since broiler chickens go to market at progressively younger ages, growth of the embryo (or *in ovo* growth) becomes increasingly important. Growth in the egg is more than 30% of growth (3 weeks or 21 days as a percentage of 3 weeks' embryonic growth plus 6.5 weeks posthatch growth). Broiler chicks from large eggs are larger at hatching than

Figure 15.20 Baskets of day-old chicks delivered to broiler house and placed in brooder pen in Mississippi. *(Courtesy, USDA)*

those from small eggs. As broilers grow, the effect of this relationship continues.

Sexing (Separation of Male from Female Chicks)

Chicks can be sexed (gender determined) by the following methods: vent inspection and sex-linked pattern of feathering on wings. It is possible that an experimental method of automated sexing will be adopted. This involves taking a small sample of allantois fluid at about day 15 of incubation. Fluid with high levels of estrogen conjugate comes from female chick embryos. The gender of the embryo is thus determined and the egg sorted by gender before hatching.

Broiler Breeder Males

Full semen quality as determined by the sperm quality index is achieved at about 30 weeks of age in broiler breeders. There are age-related diseases in male fertility/semen quality, leading to the frequent practice of replacing males by approximately 1 year of age. Some primary chicken breeders use artificial insemination (also see Chapter 4).

Artificial Insemination

Artificial insemination has been used with some broiler breeders for over 20 years. The advantages include:

- Increased fertility
- Reduced damage to female during mating
- Discontinued use of large males that may have difficulty mating
- Allows separation of males and females
- Semen can be evaluated for quality (e.g., spermatozoa motility and number of abnormal spermatozoa)
- Semen can be diluted or extended

Figure 15.19 Removing newly hatched chicks from an incubator. *(Courtesy, USDA)*

Insemination can occur every 10 to 14 days as spermatozoa are stored in the oviduct. The optimal dose of sperm per insemination has been estimated as about 150 million every 5 to 7 days.

Semen Quality

The sperm quality index (SQI) is a measure of semen quality. It is determined by deflection of a beam of light due to movement of the spermatozoa.

Broiler Breeder Nutrition

Nutritional requirements and practices for feeding the growing and adult broiler breeder show differences for both the broiler and laying chicken.

Nutrition During Growth of Broiler Breeder Pullets

Feed intake is restricted during growth of broiler breeder pullets. This is done to produce smaller, leaner hens with lower maintenance requirements. Feed is restricted by either of the following approaches:

- Reducing the feed provided each day
- Feeding on alternate days ("skip-day" feeding)

In either system, broiler diets can be employed as:

- Starter (19% protein ME 1,380 kcal/lb or 3,035 kcal/kg) 1–6 weeks
- Grower (15% protein ME 1,335 kcal/lb or 2,935 kcal/kg) 7–20 weeks
- Breeder (16% protein ME 1,320 kcal/lb or 290 kcal/kg) 21 weeks plus

The restriction on feeding begins at 4 weeks of age.

ADVANTAGES OF FEED RESTRICTION. Advantages of feed restriction during growth include:

- Reduced feed costs during growth
- Reduced feed costs during egg production
- Reduced mortality during rearing and egg production
- Little effect on number of eggs produced but larger eggs

DISADVANTAGES OF FEED RESTRICTION. Disadvantages of feed restriction during growth include:

- Delay in initiation of egg production
- Welfare issues and public perception

If a broiler breeder pullet received a high-energy/high-protein diet ad libitum during growth, body weight would be 9.5 lb (4.3 kg) at 24 weeks old in comparison with the ideal with feed restriction of about 5 lb (2.3 kg)!

Nutrition of the Broiler Breeder Hen

The broiler breeder hens have a calcium-rich (3.0%) feed freely available. In addition, calcium-containing crushed oystershell may be available. The calcium is for eggshell production and building calcium stores in the medullary bone (stimulated by the hormone estradiol and other estrogens together with androgens) prior to egg production.

Nutrition During Growth of Broiler Breeder Males

The diet for growing male broiler breeders is restricted to prevent obesity, using skip-day and daily restriction approaches.

Broiler Breeder Males (Roosters)

If the males are kept separate from the hens (if artificial insemination is practiced), the diets for broiler breeder roosters should be lower in protein plus calcium and somewhat lower in energy than hens.

- Protein, 16%
- Calcium, 0.90%
- ME 1,275 kcal/lb (2,805 kcal/kg)

HARVESTING AND PROCESSING

Most broilers are marketed when they are between 6 and 7 weeks of age. Marketing involves moving the birds from the house(s) in which they are produced to the processing plant. Improper handling of broilers immediately prior to and during shipment will result in deaths, excess bruises, and lowered quality.

The feeders should be removed about 2 hours before the catching crew arrives. Removing or elevating feeding and watering equipment will prevent bruises during catching. (Water should be removed just before the birds are caught; too early removal will result in excess dehydration.)

Catching Birds/Harvesting

Chickens are caught by hand and placed into crates by a loading crew of seven to ten people. Each catcher lifts about 1,000 to 1,500 broilers per hour. Typically, catchers pick up the birds by their legs. It is possible to catch them without injuring or damaging the birds. Catching may result in up to 20 to 25% of birds being bruised. Welfare and economic considerations dictate that the crew be well supervised with incentives to do the job right.

Catching requires the following:

a. Using an experienced well-supervised crew.
b. Working under a dim blue light at night.
c. Corralling birds in small groups (of about 200) to prevent smothering and undue injury.
d. Grasping birds by the shanks, with no more than four or five being carried at a time.
e. Placing birds in the crates gently.
f. Handling the crates carefully, preferably on pallets that can be moved easily with a hoist.

Catching may become automated with mechanical harvesting. This involves the use of a "catcher or harvester" of rubber paddles or fingers together with a conveyer belt to the transporter. Advantages of mechanical harvesting include:

- Lower stress to birds
- Fewer bruises
- Reduced labor costs
- Improved working conditions for catching crew

Transportation/Hauling

Poultry are hand-loaded into loose crates (plastic), fixed crates, or modular containers and then transported from the production unit to the processor in trucks. Each crate or container can hold about 14 birds. Transportation may represent a 3- to 5-hour journey and up to 12 hours in crates before, during, and after transportation. Birds are stressed during transportation due to vibration, lack of food and water, environmental temperatures, and lack of space. It is important to protect the in-transit birds from extremes in weather such as chilling in cold weather or overheating in hot weather.

Shrinkage, or weight loss from the time feed and water are removed until the birds are weighed at the processing plant, varies according to temperature and length of time involved; it ranges from 2% for a 3-hour period to 6% for a 15-hour period. Some bird deaths occur during transportation.

Slaughter and Processing

The slaughter and processing of broilers is an assembly line operation conducted under sanitary conditions. Birds are inspected twice: (a) live (antemortem) and (b) after slaughter (postmortem), when the carcasses and viscera are examined. The following are steps in slaughter and processing:

- Unloading birds from transport vehicle
- Shackling to a conveying chain
- Stunning

- Killing by cutting throat and by severing the jugular vein
- Bleed out
- Scalding
- Feather removal
- Removal of oil gland and feet
- Evisceration
- Removal of crop, head, and lungs
- Washing
- Cooling

The processing of a 5-lb (2.3-kg) (liveweight) broiler will result in about a 12% loss in blood and feathers, carcass or eviscerated weight of 70% of liveweight, and chilling gains of 9%; with variations according to weight and sex. The heavier the bird, the lower the percentage of blood and feathers; males have a higher percentage loss of blood and feathers than females; and the chilling gains on females are greater than on males (Figures 15.21 to 15.23 illustrate processing of broiler chickens).

Figure 15.21 Harvesting chickens to be transported to the processing plant. *(Courtesy, USDA)*

Figure 15.22 Broiler chickens being processed and checked out by USDA inspectors. *(Courtesy, USDA)*

Figure 15.23 Chicken processing plant. *(Courtesy, USDA)*

Processing Plant Prior to Slaughter

Birds are moved to the shackling line from the crates or containers and placed head down with the shackles under their knee joints. Unloading birds is again usually done by hand. Automated unloading has been developed by use of conveyer belts. Care needs to be taken during unloading to prevent bruising.

Slaughter

The most common system involves shackled birds receiving the following:

- Stunning (e.g., electric or gas comet) to immobilize and render unconscious and to minimize pain associated with slaughter
- Cutting the neck to kill and bleed out over 2 to 5 minutes

Postslaughter

After bleed out, the birds are moved usually via the shackle line through the following:

- Scalding at 122–127°F (50–53°C). The birds are immersed in warm water for 1 minute to loosen the feather follicles.
- Feather removal is achieved by mechanical pickers and pluckers. These consist of rotating drums with rubber fingers.
- Next, the oil gland and feet are removed under rotating blades and bars/rails position the bird for cutting.
- The carcasses are rehung on another shackle line automatically.
- Evisceration is done automatically. The viscera are inspected and parts (e.g., gizzard, liver, heart) may be salvaged and cleaned for the giblets.

- The lungs, head, and crop are removed.
- Both the inside and outside of the carcass are washed.
- The carcass passes through a cooling tank containing ice water before being graded, sorted by weight, and prepared for delivery to consumer markets and/or further processing.

Condemnation During Processing

Broiler condemnations average about 2.5% with variation between processing plants. Condemnations may be due to respiratory diseases, overt signs of leukosis complex, and bruises. Also, bruises can result in downgrading or in only partial condemnation; for example, a severe breast or wing bruise may be cut out or off as unacceptable for human consumption. Bruises are responsible for over half of the downgrading. They are due to mishandling of the birds during catching, loading, and transporting to and at the processing plant. Improper bleeding, eviscerating, and overscalding in processing also result in condemnations. Broiler growers and processors strive to minimize such condemnations because they represent a monetary loss.

Kosher and Halal Slaughtering

Poultry may be slaughtered following Islamic or Jewish religious laws, respectively, Halal and Kosher. Neither practice allows stunning of the birds before killing. In Kosher slaughter, the blood vessels (carotid artery and jugular vein) in the neck are cut with a knife by someone trained and approved by a rabbi. This leads to death and exsanguination/ bleed out. An analogous system is used for Halal slaughtering. It is unlikely that concerns about animal welfare will affect the exception in civil law (made for religious freedom) that allows Kosher and Halal slaughtering.

Standards, Grades, and Classes

The U.S. standards of quality and grades for poultry are used for trading purposes (see Chapter 17). Some retailers require that all ready-to-cook poultry be identified with an official grade mark.

Other Products and By-Products

Pet Food

Pet food represents a considerable and expanding market for poultry meat and poultry by-products— the rendered material from offal in pet food. Dog and cat food come in three forms:

1. Canned soft food
2. Dry food
3. Semisoft dry food

Dollar sales of cat and dog food are increasing by about 4% in the United States, with similar growth elsewhere. Pet food uses poultry meat and poultry by-products. An estimate of use of poultry by-products worldwide is approaching 1 million tons per year (~1 million metric tons/year).

Feathers

Feathers are a useful by-product of the production of poultry meat, with feathers being 7% of the liveweight of poultry. Feathers can be classified as:

1. Down (fluff only with no shaft)
2. Partial down with some shaft (e.g., plumbs, fluff, three-quarter, half-fluff)
3. Hard and saddle feathers

Major uses are for insulation and comfort or as feather meal for animal feed.

INSULATION AND COMFORT IN BEDDING AND WINTER CLOTHES. Feathers are very useful for bedding and winter clothes.

1. They are a natural, durable product.
2. They are biodegradable.
3. Because of the amount of air trapped, they are excellent lightweight insulators.
4. They are comfortable.

Down feathers are separated from the rest of the feathers by flowing with their low density in air. They can travel farther while the other feathers settle out. The best feathers are used in sporting equipment (e.g., arrows and darts). Feathers must be washed (5–8 times) in soap to eliminate animal waste, but some oils should be left attached. They may also be treated with a salt solution and dilute acid.

FEATHERS AS ANIMAL FEED. While feathers are principally made of the protein keratin, they are not digestible. However, with high-pressure cooking at >212°F (>100°C), hydrolysis occurs, increasing the availability of amino acids. Feather meal is produced following drying and grinding of the hydrolyzed feather. It can be added to poultry or swine or ruminant feeds, being a good source of arginine, cysteine, and threonine. Feather meal has 75–90% crude protein.

Offal and Bone

Bone and offal consist of the following:

- Edible offal (liver, hearts, and gizzard—collectively called giblets)

- Inedible offal such as the rest of the gastrointestinal tract (head, feed, bone from deboned poultry meats), lungs, and kidneys
- Blood may be considered as either edible or inedible offal

Inedible offal is rendered by cooking at high temperatures (212°F; 100°C) under pressure to destroy pathogenic organisms. The resulting material can be used for livestock feed or pet food.

ROASTERS

Roasters are young chickens that are grown similar to broilers, but they are older and heavier, ranging in weight from 6 to 8 lb (2.7–3.6 kg). The management program for roasters is quite similar to that of broilers up until about 6 weeks of age, following which some changes in feeding and management should be made. Pointers pertinent to roasters follow:

- Broiler chickens of some genetic backgrounds are well suited to increasing to roaster weight.
- Roasters should be given the kind and amount of feed recommended for broilers during the first 6 weeks.
- Roasters may be of either sex. However, males grow more rapidly than females, with the result that they reach market weight more quickly. It takes only about 9 weeks for males to reach the required weight, whereas it takes about 11 weeks for females. For this reason, it is strongly recommended that chicks intended as roasters be sex-separated at hatching.

A 7-lb (3.2-kg) roaster requires about 1.5 sq ft (0.13 m²) of floor space in comparison with the 0.8 sq ft (0.07 m²) required by a 4-lb (1.8-kg) broiler. Likewise, roasters need more room to eat and drink than broilers. They should have 2 in. (5 cm) of feeder space and 1 in. (2.5 cm) of water space per bird after 8 weeks of age. Also, roasters, being heavier than broilers, require more ventilation.

Breast Blisters

A higher incidence of breast blisters is encountered in roasters than in broilers. Heavy birds seem to sit more than light birds; and males sit more than females. Thus, the incidence of breast blisters is related to body weight, sex, and the material with which the breast comes in contact.

Breast blisters in roasters contribute to economic losses. The blisters must be cut out at processing, and this downgrades the carcass. Since roasters are sold as whole-body birds, breast blisters

BOX 15.3   What's Ahead for Chicken (Broiler) Meat

- The broiler industry will continue to be concentrated in some states with feed availability, appropriate infrastructure (processing, transportation etc) and a mild climate (lower housing and energy costs).
- Animal waste-excreta or dead birds, odor, and flies) will be a critical factor in site selection for broiler operations, favoring remoteness from neighbors and compliance with state and federal regulations.
- Labor costs and environmental regulations may lead to greater development in Mexico, Brazil and Argentina.
- Larger units and more integrated broiler companies.
- Breeding stock will be selected for greater livability.
- Per capita consumption of broilers will increase with more broilers further processed.
- Packing of chicken and processed products under brands will continue to increase, with some brands consistently selling at premium prices.
- Greater marketing from the processor to the retailer or consumer.
- Restaurants and fast-food outlets use more chicken.
- The cost of production will be further reduced by improved genetics and management etc.
- Feed and energy costs will show fluctuations
- Costs of packaging and labor will increase but offset by increased efficiency and productivity.

are more serious than in broilers. The incidence of breast blisters in roasters may be reduced by (1) using females instead of males, (2) keeping the litter dry, (3) keeping deep litter, (4) feeding more often (when birds are eating, they are not sitting), and (5) stirring the birds occasionally.

CAPONS

In the past when broiler growth was much slower than today, capons were popular chicken meat. Capons are surgically castrated male chickens. The objective of caponizing is to improve the quality of the meat produced. Capons do not develop the normal male characteristics such as the head, comb, and wattles. They do not crow or fight like roosters,

and the dark meat of capons is lighter-colored than meat of males of the same age. When marketed, they range up to 20 weeks of age and weigh >12 lb (5.5 kg). The breeds and crosses used for broiler production make satisfactory capons. Diets for broilers are also satisfactory for capons.

 USEFUL WEBSITES

Cobb-Vantress: http://www.cobb-vantress.com/

Hubbard-ISA: http://www.hubbard-isa.com/

Ross Breeders: http://www.rossbreeders.com/

FURTHER READING

Arbor Acres Breeder Manual. 2002.

Barbut, S. *Poultry Products Processing: An Industry Guide.* Boca Raton, FL: CRC Press, 2002.

Bell, D. D., and W. D. Weaver. *Commercial Chicken Meat and Egg Production,* 5th ed. Norwell MA: Kluwer Academic Publishers, 2002.

Diseases of Poultry. Edited by B. W. Calnek, H. John Barnes, Charles W. Beard, Larry R. McDougald, and Y. M. Saif. Ames, IA: Iowa State Press, 1997.

Hunton, P. *Poultry Production.* Amsterdam, Netherlands: Elsevier, 1995.

Lesson, S., and J. D. Summers. Broiler Breeder Production. Guelph, Canada: University Books, 2000.

Pattison, M. *The Health of Poultry.* Harlow, England: Longman, 1993.

Sainsbury, D. *Poultry Health and Management: Chickens, Ducks, Turkeys, Geese, and Quail.* Blackwell Science: Oxford, England, 2000.

Sturkie's Avian Physiology, 5th ed. Edited by G. C. Whittow. San Diego, CA: Academic Press, 2000.

16

Turkeys and Turkey Meat

Objectives

After studying this chapter, you should be able to:

1. List the major weights and ages at which turkeys are marketed.
2. List the basic composition of turkey meat.
3. List the major breeders of commercial turkeys.
4. List the major varieties of exhibition turkeys popular in North America.
5. Give recommendations on the rearing and production of turkeys.
6. List the major diseases encountered in turkey production.
7. Define *photostimulation* and give recommendations on utilizing this concept to improve egg production in turkey breeders.
8. Give reasons and recommendations for artificial insemination in turkeys.
9. Give suggestions for maximizing fertility and hatchability of turkey eggs.
10. List the top turkey processing companies in the United States.

Turkeys are native to the New World (see Chapter 2 for details on their domestication). At one time in American history, the turkey vied with the bald eagle as the national bird, but the latter won the honor by a small margin of the United States Congress. Production of turkey meat is of great importance in North America and is growing in Europe (see Chapter 1). See Figure 16.1 for a photograph of an adult male turkey. This chapter addresses the following topics related to the production of turkey meat:

- Marketing of turkey meat
- Composition of turkey meat
- The biology of turkey meat production
- The commercial production of turkeys
- Turkey breeders and turkey poult production (reproduction)

MARKETING OF TURKEY MEAT

Turkey meat is marketed as:

- Broiler hens at 10 weeks old with a dressed weight of about 9.2 lb (4.2 kg)

- Young hens at 16 weeks with a dressed weight of about 15.4 lb (7.0 kg)
- Young toms at 17 to 18 weeks with a dressed weight of about 27.5 lb (12.5 kg)

The dressing percentage for broiler turkeys is about 77%.

Further Processing of Turkey

Further processing of turkey includes the following:

- Turkey ham and smoked turkey
- Deli sliced turkey, turkey ham, and smoked turkey
- Turkey bacon with alternating white and dark meats
- Mechanically deboned turkey meat
- Ground turkey meat products such as turkey hamburger, sausage, and chili
- Ground turkey meat products in an emulsion (e.g., frankfurters)

Examples of prepared turkey are illustrated in Figures 16.2 and 16.3.

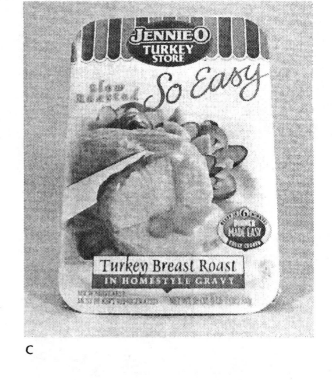

C

Figure 16.1 Adult tom (male) turkey. *(Courtesy, USDA)*

A

D

B

E

Figure 16.2 Foods from turkey.

A

B

Figure 16.3 Foods from turkey.

COMPOSITION OF TURKEY MEAT

Uncooked turkey breast has the following composition:

- Water, 70%
- Protein, 22%
- Fat, 7%

Uncooked turkey dark meat is 71% water, 19% protein, and 9% fat. Removal of the skin reduces the fat considerably. Not only does skinless turkey have much less fat than pork and beef but there is also proportionately less saturated fatty acids.

Muscle and Meat

Protein makes up about 19% of turkey meat. There are multiple proteins that are the same as in

the chicken (see Chapter 15). They include the following:

- Myofibril proteins including actin, myosin, and tropomyosin
- Sarcoplasmic soluble proteins such as myoglobin and the glycolytic enzymes
- The stromal proteins including collagen (connective tissue) and mitochondria

THE BIOLOGY OF MEAT PRODUCTION

The rapid growth rate of turkeys reflects superior genetics (intense selection for growth rate) and improved nutrition. The biology of growth of meat is similar to that in the chicken (see Chapter 15).

COMMERCIAL PRODUCTION OF TURKEYS

Today, turkey production and marketing is a highly efficient process, and turkey meat consumption is substantial throughout the year. For an overview of United States and world turkey production, see Chapter 1. The leading turkey producing and processing companies are shown in Table 16.1.

Business Aspects

With the increase in size and integration of turkey enterprises, the business aspects have become more important. More capital is required, knowledge of contracts is important, competent management is in demand, and good records and computers are essential. (See Chapter 18 for details.) This section will focus on specific turkey issues.

Financing

Banks and the Farm Credit Association finance the turkey industry through credit extended to feed

TABLE 16.1 Top Turkey Companies in the United States in 2001

Rank	Company	Annual Production/ Processing Billion lb Liveweight
1	Jennie-O Turkey Store	1.25
2	Cargill	1.20
3	Butterball Turkey	0.80
4	Carolina Turkeys	0.55
5	Pilgrim's Pride	0.54
6	Louis Rich	0.29
7	Bil Mar Foods	0.26
8	House of Raeford	0.25
9	Foster Farms	0.23
10	Perdue Farms	0.21

Source: Rankings, based on WATT Poultry USA.

manufacturers, processors, hatcheries, and integrated operators. Much of the money is loaned at the local level to growers with mortgages as collateral.

Labor Requirements

Efficiency of labor is important to profitability in turkey production. The more turkeys raised per worker (productivity), the lower the production costs. Also the bigger the operation, the more feasible it is to automate. Greater productivity is a primary force increasing the number of turkeys per farm. Labor saving equipment includes automatic feeding and watering systems.

Turkey Contracts

When turkeys are grown under contract, contractors generally provide poults, feed, and services, while producers provide labor and facilities.
Types of contracts include:

> **Profit-sharing contract.** The contractor provides inputs (poults, feed, services, medication, etc.) on an "at-cost basis." The cost of these inputs, plus hauling charges, is first deducted from the gross receipts from the sale of turkeys. Fixed percentages of the remaining receipts are distributed to the contractor and to the grower.
>
> **Base pay and bonuses contract.** The contractor provides poults, feed, medication, and services, and pays the grower based on the number of weeks to market plus bonuses for livability, feed conversion, and grade yield.

Records/Costs and Returns

Good records are important for flock and business management. These should include feed consumption, mortality, and vaccination dates.

Costs and returns in the turkey business are pertinent to decision making, when deciding whether or not to enter the business, and in determining how well an established enterprise is doing. Costs and returns of turkey production differ from year to year and from area to area. The differences in production costs between growers can account for the success or failure of the enterprise. So, persons planning to become turkey growers should set goals to produce turkeys for less than the average of their competitors.

Costs can be lowered! The place to start is the highest-cost item. Because feed represents about two-thirds the cost of production, any change in feed cost or efficiency will dramatically influence the cost of production. Feed costs can be lowered through (1) locating the turkey operation in an area where major feed ingredients are produced, (2) using improved feed formulations, (3) selecting efficient turkey strains, and (4) lowering mortality (high mortality increases feed consumption per unit weight of turkey produced). Turkey growers buy 10% more poults than they expect to raise, simply because this figure represents the average mortality; thus, a savings in the cost of poults is achieved by lowering mortality.

Labor costs can be lowered by clear expectations such as standard operating procedures (SOPs), good supervision, and automation. The turkey producer should also be aware that facilities and equipment change rapidly. It is important that the grower visit successful operations and trade shows, and observe scientific presentations to learn what is new and working well.

BREEDING AND GENETICS

Although raising turkeys for market is the largest phase of the turkey industry, the breeding of turkeys is the foundation. Primary breeders provide turkey poults and parent breeding stock while retaining grandparent, great-grandparent, and pedigree lines (also see Chapter 4). The hatchery is a critical component of a primary breeder, hatching eggs from grandparent, great-grandparent, and pedigree lines and possibly, at a separate location, eggs from parent lines.

Turkey selection has involved the following:

- Mass selection
- Specific breeding index characteristics including
  - Growth rate
  - Feed efficiency
  - Carcass yield and breast yield
  - Livability
- BLUP (best linear unbiased prediction)
- DNA markers for marker-assisted selection (beginning to be used)

Primary Breeders

The major primary breeders of turkeys in the world are listed in alphabetical order as follows:

- British United Turkeys (BUT)/British United Turkeys of America (BUTA) (a subsidiary of Merial Animal Health)
- Hybrid Turkeys (a subsidiary of the parent company, Nutreco)
- Nicholas Turkey Breeding Farms (part of the Aviagen Group)

Figure 16.4 BUTA 37 roaster turkey. *(Courtesy, Hubbard ISA)*

The major primary breeders of turkeys in the United States are:

- British United Turkeys of America (BUTA) (a subsidiary of Merial Animal Health) (see Figure 16.4 for an example of BUTA turkey.)
- Nicholas Turkey Breeding Farms (part of the Aviagen Group)

Hybrid Turkeys (a subsidiary of the parent company, Nutreco) is also a primary breeder of turkeys. It is located in Canada but provides breeding birds for U.S. producers.

Breeds and Varieties

For an understanding of breeds of turkeys, it is important to know the parts of a turkey (see Figure 16.5). Seven standard varieties, popularly called breeds, of domesticated turkeys are recognized by the American Poultry Association and described in detail in *The Standard of Perfection*, namely:

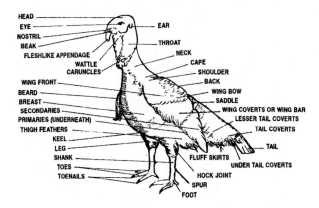

Figure 16.5 External anatomy of a turkey. *(Original diagram from E. Ensminger)*

Figure 16.6 Bourbon Red turkeys. *(ISU photo by Bob Elbert)*

- Bronze
- White Holland
- Bourbon Red (see Figure 16.6)
- Narragansett
- Black
- Slate
- Beltsville Small White

Including wild turkeys, there are 10 other varieties, chief of which are the Broad-Breasted Bronze and the Broad-Breasted White. Presently, only the Broad-Breasted White (also called Large White) variety of turkeys is important commercially. (Also see Chapter 4, especially for color genes.)

Broad-Breasted Bronze

The Broad-Breasted Bronze turkey originated in England. It was imported into North America about 1930. Within a few years after its introduction, it became the most widely grown of all varieties. It was used to produce the Broad-Breasted White and the Beltsville Small White.

The Broad-Breasted Bronze resembles the almost-extinct standardbred Bronze in color. Its basic plumage color is black, which results in dark-colored pinfeathers, a disadvantage that contributed to its replacement by the Broad-Breasted White.

The Broad-Breasted Bronze and the Broad-Breasted White are inferior to the Beltsville Small White, producing fewer eggs, with lower fertility and hatchability. Artificial insemination is standard practice in breeding Broad-Breasted Bronze and other heavy broad-breasted turkeys since it is required for acceptable fertility.

Broad-Breasted White

The Broad-Breasted White or Large White turkey was developed in the early 1950s through pedigree breeding and selection at Cornell University, from crosses of Broad-Breasted Bronze and White Holland. Private breeders developed other large white strains through crossbreeding. A few lines were developed using naturally occurring white mutations in Broad-Breasted Bronze flocks. Selection has resulted in rapid growth, large size, and broad-breasted conformation, which are the bases of commercial turkey production.

Beltsville Small White

The Beltsville Small White was developed by pedigree breeding and selection at the USDA Agriculture Research Center, Beltsville, Maryland (between 1941 and 1962). It resembles the Broad-Breasted White in color and body type but is smaller. Egg production, fertility, and hatchability tend to be higher, and broodiness lower in the Beltsville Small White than in heavy varieties. It is, however, not widely used. Because one person can care for the same number of big turkeys as small turkeys, the labor costs per unit weight (pound) are higher on the small variety.

Breeding Operators and Systems

Two types of operators are involved in reproducing turkeys: foundation breeders and multipliers. The foundation breeder follows an intensive genetic program designed to achieve increased fertility, hatchability, livability of poults, uniformity of growth, better feed conversion, shortening the number of weeks to market, and more uniformity in type, size, and weight. The multiplier works very closely with the breeder (same company) and multiplies turkeys from a definite line or strain from the breeder.

Housing and Equipment

Either of three types of housing systems may be used for breeder turkeys: (1) confinement breeding facilities, (2) range breeding facilities, and (3) combination range and confinement. These latter two are now uncommon in North America, but there is a move back to free-range systems in Europe, particularly in the UK. Formerly, range breeding was commonplace; the breeders were allowed to roam in open pens with range-type or open-shed shelters. The space requirements for turkey quarters, feeders, and waterers are important; and they are affected by age, size (large-type versus small-type), and kind of rearing (brooder house, confinement, semiconfinement, or range). Table 16.2 shows the floor space and land recommendations per bird for brooding, rearing, and breeding turkeys. Table 16.3 gives the feeder and waterer space recommendations (also see Chapter 13 and Figures 16.7 to 16.10).

Litter

Deep litter, usually consisting of soft wood shavings or wheat straw, is usually used in pole houses (see Figure 16.11). Four to six inches (~12.5 cm) of litter are usually placed on the floor before the birds arrive. With the addition of excreta, there is natural composting. It should be kept dry and clean by adding to it when necessary and removing prominent accumulations of droppings and wet litter that contribute to the development of leg problems and fungus diseases. Adequate ventilation is the key to good litter management, and preventing buildup of pathogens.

Feeders

Either large-capacity metal hopper-type or automated feeders should be used. Allow the linear inches of feeder space per bird shown in Table 16.3. Feeders should be well distributed throughout the area, available to the birds, and easily filled by the operator.

TABLE 16.2   Floor Space and Land Recommendations per Bird for Brooding, Rearing, and Breeding Turkeys[1]

	Rearing							
	Brooder House		Confinement		Semiconfinement		Range	
Age	Floor		Floor		Floor		Land[2]	
	(sq ft)	(sq m)	(sq ft)	(sq m)	(sq ft)	(sq m)	(sq ft)	(sq m)
0 to 8 weeks	1.25	0.116	—	—	—	—	—	—
8 to 12 weeks	—	—	1.5	0.1	1.5	0.1	6	0.6
12 to 16 weeks	—	—	2.5	0.2	1.5	0.1	16	1.5
16 weeks to market	—	—	4.0	0.4	1.5	0.1	21	2.0
Breeders	—	—	4.5	0.4	1.5	0.1	26	2.4

[1]Growers who plan to raise turkeys to market weight must have building space and provide land area at the maximum recommendations for older birds even though there is some wasted space earlier in the growing period. Intermediate growers who plan to grow birds only 8 or 12 weeks, then sell to a finish grower, may find the intermediate breakdown helpful.

[2]Space for house, pen, and a minimum of 100 ft between pens.

TABLE 16.3   Feeder and Waterer Space Recommendations

Age	Space per Turkey			
	Feeder[1]		Waterer[2]	
	(lin in.)	(lin cm)	(lin in.)	(lin cm)
0 to 2 weeks	1.0	2.5	0.5	1.25
2 to 4 weeks	1.0	2.5	0.5	1.25
4 to 6 weeks	2.0	5	1.0	2.5
6 to 8 weeks	2.0	5	1.0	2.5
8 to 12 weeks	2.0	5	1.0	2.5
12 to 16 weeks	2.0	5	1.0	2.5
16 to 20 weeks	2.5	6.5	1.0	2.5
20 weeks to market	2.5	6.5	1.0	2.5
Breeders	3.0	7.5	1.0	2.55

[1]Location of feeders is just as important as amount of feeder space. Both pan or feeding space and feed storage capacity are important, especially with older turkeys.

[2]Locate waterers so that each poult is always within 10 to 15 ft (3–5 m) of water. More water space than is recommended above may be needed during hot weather. A sudden drop in water consumption often forewarns of a possible disease problem.

Figure 16.7   Young turkey poults. *(Courtesy, Gretta Irwin, Iowa Turkey Federation)*

Figure 16.9   Young turkeys in modern grow out house. *(Courtesy, P. R. Ferket, North Carolina State University)*

Figure 16.8   Turkey confinement building. *(Courtesy, Gretta Irwin, Iowa Turkey Federation)*

Figure 16.10   Breeding turkeys. *(Courtesy, P. R. Ferket, North Carolina State University)*

Figure 16.11 Moving fresh litter for raising turkeys. *(Courtesy, Gretta Irwin, Iowa Turkey Federation)*

Waterers

Water should always be in good supply, high quality, and well located. Allow the linear inches of water space per bird shown in Table 16.3. Large-type turkeys will consume 16 to 20 (~70 liters) gal of water per day per 100 birds, depending on the rate of production and environmental temperature. In cold climates, it is necessary to keep the water from freezing. Waterers should be clean. When cleaning, water should not be poured out on the litter.

NUTRITION AND FEEDING

Nutrition is critically important. Feed is about two-thirds of the cost of producing turkeys. Feeding turkeys for maximum profit requires the simultaneous use of knowledge of nutrition and a judicious assessment of prevailing feed prices. A successful program must provide the turkeys with the required nutrients in the amounts and proportions to satisfy their needs, and, at the same time, remain well within the economic limits necessary for a profit potential to be high.

The basic nutrient categories are proteins, carbohydrates, fats, minerals, vitamins, and water. Distributed within these categories are some 40 separate nutrients that turkeys require (also see Chapters 6 and 7). With improved genetics, together with nutrition and management, feed conversion and growth rate of turkeys are constantly improving.

Weather conditions always play a role in the performance of turkeys from year to year and season to season. Generally speaking, extremely hot weather depresses growth rate and extremely cold weather lowers feed efficiency.

Turkey Health

Turkeys seem to be more susceptible to pathogenic disease and require a higher level of management than chickens. One severe outbreak of a disease is usually all that is needed to turn a flock from a profit to a loss. Many diseases are at least partly under the control of the turkey producer, and they are not inevitable—they may be controlled or prevented by the following (for details see Chapter 11):

- Proper management
- Biosecurity including good plans and well-trained employees
- Vaccination
- Professional services from poultry veterinarians and pathology laboratories

Pathogens can be spread from bird to bird or from vectors (see Chapter 11 for details). Of the common pathogens, only the *Salmonellas*, the paracolons, and the *Mycoplasmas* are transmissible vertically through the egg.

Professional Services from Veterinarians and Laboratories

From one sample submitted to a laboratory, blood may be tested for the presence of pullorum disease, fowl typhoid, paratyphoid, infectious sinusitis (*Mycoplasma gallisepticum* infection), and possibly Arizona and *Mycoplasma meleagridis* infections.

INFECTIOUS SINUSITIS. Infectious sinusitis is caused by *Mycoplasma gallisepticum* (MG) infection. There may be secondary infections from *E. coli* or *Pasteurella multocida* (fowl cholera). Condemnations can reach 70%. While eradicated from turkeys, the pathogen can come from infected chickens or other birds (emphasizing the importance of biosecurity, keeping turkeys and chickens separate, and enforcing a policy that employees do not keep birds, poultry, or otherwise). While vaccines are not available, infected birds can be treated with the antibiotic, tylosin. Following a 30-day withdrawal period, turkeys can be marketed. It is then critically important to ensure that the premises are completely depopulated and thoroughly cleaned since the pathogen does survive well outside the bird's body. Also, the facility should be isolated from other turkey units because the pathogen can be transported in feces.

PNEUMONIA/RESPIRATORY DISEASES. Infection with *Ornothobacterium rhinotracheale* (ORT) causes respiratory disease in turkeys, with depression in growth and increased mortality in commercial turkeys, and reduced egg production in turkey breeders. Infection with this Gram-negative bacteria can be overcome with antibiotic treatment.

Vaccinations

Vaccines help build resistance to specific diseases. No vaccine is completely effective. There may be adverse reactions or lack of effectiveness. Vaccine producers are responsible for producing safe, potent, and effective vaccines. Users also have responsibilities such as ensuring that vaccines are refrigerated or stored properly, used within the expiration date, mixed according to directions, applied to birds that are in good health when vaccinated, and administered according to the directions. The vaccination program should be based upon the requirements of the flock and the advice of turkey veterinarians. Vaccination for breeder flocks includes the following diseases: fowl pox, Newcastle disease, avian encephalomyelitis, erysipelas, fowl cholera, fowl typhoid, paratyphoid, and Arizona paracolon. A poult vaccination program may include protection against the following diseases: avian encephalomyelitis, erysipelas, fowl cholera, fowl pox, and Newcastle disease (also see Chapter 11).

ERYSIPELAS. In some areas, erysipelas is a problem. In addition to causing death, the disease can affect the fertilizing capacity of male turkeys. Also, marketing losses may result from condemnations or downgrading losses due to postmortem evidence of septicemia or to lack of finish. Turkeys should be immunized against erysipelas in areas where the disease is known to be a problem.

FOWL CHOLERA. Vaccination should be considered in areas where fowl cholera is present. Commercially produced bacterins and live vaccines are available. This is not a substitute for good sanitation.

FOWL POX. Vaccination against fowl pox has been used for many years, as a means of protecting the flock and preventing a drop in egg production and fertility. Turkeys do not develop as long an immunity as chickens. Initially, turkeys are vaccinated when they are 2 to 3 months old, but those to be used as breeders should be revaccinated before production and at 3- to 4-month intervals.

HEMORRHAGIC ENTERITIS. This is a significant problem in some turkey-producing areas. To control hemorrhagic enteritis, avirulent isolates can be successfully used as live, water-administrated vaccines. If flocks are less than 100% protected by vaccination, they are subsequently protected by lateral transmission of vaccine virus within 2 to 3 weeks.

NEWCASTLE DISEASE. Breeder turkeys should be vaccinated since the disease could cause a serious disruption of egg production. A common practice is to vaccinate in the drinking water with a B1 type of Newcastle disease vaccine. Other possibilities include injection of LaSota strain and a tissue culture strain.

Antibiotics and Coccidiostats

Antibiotics and coccidiostats are widely used in the turkey industry for the prevention of disease and as growth stimulants. Turkey producers may:

1. Dip the hatching eggs in antibiotics to control *Arizonosis* and *Mycoplasma* infection.
2. Start the poult on a coccidiostat and an antibiotic.
3. Give increased drug treatment at times of stress.
4. Use drugs to treat specific infectious diseases such as fowl cholera.

Combating Disease

Before a disease strikes, the turkey producer should have a plan for combating the situation. This involves detection of diseases, defensive measures, diagnosis, and postmortem examination.

Detection (Signs of Disease)

One of the first signs of disease outbreak is reduced feed consumption. Birds may be nervous or they may appear droopy and listless. Coughing and sneezing indicate respiratory infections; abnormal droppings suggest intestinal disorders. Young poults may crowd together and seem to be cold even though heat is available. These warning signals indicate the need to check the birds' health. When sick birds are detected, they should be isolated.

Diagnosis

Once sickness is indicated, an accurate diagnosis should be obtained as soon as possible. It is important to know what is wrong to treat the flock properly and to prevent further losses. Work with your veterinarian to obtain an accurate diagnosis. Sick or dead birds should be subject to necropsy with samples going to a pathology laboratory.

Parasites

The turkey producer should be alert for parasites, both external and internal. The most common external parasite is the fowl mite. This small, gray mite is distinguishable from a fleck of dust only by the fact that it moves quite rapidly. Fowl mites live mostly in the tail feathers and in the fluff feathers at the rear of the keel, and infestations of these tiny parasites can build up very quickly in a flock, especially in toms, often before being noticed. The insemination crew should report mites.

Roundworms are a common internal parasite of turkeys. Once a flock has roundworms, or has run on litter or ground that has been used with turkeys with roundworms, the flock are continually infested. A regular, once-a-month worming with a wormer will reduce worms to a harmless level.

Abnormalities and Injuries

A number of abnormalities and injuries are found in turkeys. A few of the more important will be discussed.

Breast Blisters and Calluses

Breast blisters and calluses are much more common in toms than in hens, and are less common in dimple-breasted birds than in those without the dimple. They may be caused by irritation of the skin that covers the breastbone. They can cause serious losses due to downgrading of the carcass.

Cannibalism

Feather picking is a mild form of cannibalism found in turkeys, especially during the growing period. It results in unsightly appearance and potentially further injury (flesh picking) and can retard growth rate. Management practices that reduce feather picking include (1) avoiding overcrowding, and (2) feeding an adequate diet of pellets rather than loose mash.

Pendulous or Drop Crop

Pendulous crop is sometimes called baggy, sour, or dropped crop. There is a weakening of the crop wall such that feed and water accumulate. These become foul smelling. There is likely to be secondary bacterial and/or fungal infections. Feeding continues but because the injesta passes out of the crop too slowly, digestion is impaired and without sufficient absorbed nutrients, emaciation occurs. Seriously affected birds seldom recover.

Some turkeys, particularly types that are not broad-breasted, possess a genetic weakness that predisposes them to this. When these susceptible birds are exposed to hot weather with consequent heavy consumption of liquids, the pendulous crop may develop. Preventive measures include (1) selecting strains not carrying the genetic factor, (2) avoiding exposure of turkeys to excessive heat or dehydration, (3) giving continuous and easy access to cool drinking water, and (4) providing shade for turkeys on the range.

Rickets (Including Tibial Abnormalities)

Abnormal development of the bones can occur with severe consequences to growth and greatly increased mortality. Remedies include:

- Feeding a diet with optimal calcium (1.5%), phosphorus (0.75%), and vitamin D (2,000 IU/lb feed or 4,500 IU/kg feed).
- Ensuring that feed is not contaminated by fungal toxins (mycotoxins).
- Reducing animal fat in the feed.
- Using fresh feed without rancid fats that can cause enteritis.

Stampeding

Turkeys are subject to fright, especially at night. Severe losses from injury and deaths by predatory animals may result from stampedes. Stampeding may be reduced by avoiding disturbances and by the provision of low-intensity night lighting.

TURKEY MANAGEMENT

Management is the key to success in the turkey business. The geneticist can lift the potential of production; the nutritionist can formulate a ration that can reach this potential; but without proper management, the turkey producer will never get there. Management gives point and purpose to everything else—to breeding, feeding, housing, health, and marketing. It can make or break a turkey enterprise. Fortunately, a turkey operation responds favorably to good management. High goals spur high achievement. Turkey producers have specific goals, for example the following:

- **Fertility of eggs.** Lowering incubated egg infertility to < 5%.
- **Embryonic fertility.** Lowering embryonic deaths to < 5%.
- **Mortality of poults.** Lowering poult mortality to < 4%.
- **Feed efficiency.** < 2.2 and < 2.4 units of feed per unit weight for turkey hen and tom growth.

Brooding

The successful brooding of poults is critical to turkey production. It depends upon satisfying all the requirements of the young birds at this early

Figure 16.12 Setting-up for brooding poults. *(Courtesy, Gretta Irwin, Iowa Turkey Federation)*

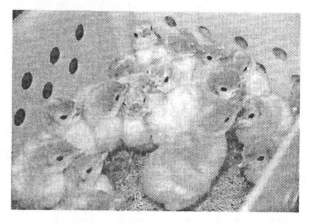

Figure 16.14 Newly hatched turkey poults. *(Courtesy, Gretta Irwin, Iowa Turkey Federation)*

Figure 16.13 Brooding poults. *(Courtesy, Gretta Irwin, Iowa Turkey Federation)*

Figure 16.15 Sexing turkey poults. *(Courtesy, Gretta Irwin, Iowa Turkey Federation)*

stage in their development (see Figures 16.12 to 16.14).

Selecting Poults

Start with high-genetic quality sexed poults from pullorum-typhoid clean (note that between 1993 and 1999 no turkeys tested positive for either pullorum or typhoid) and PPLO tested breeder flocks from a hatchery. The sexed poults should be raised separately.

Separating the Sexes

The sex of day-old poults may be determined by examining the vent (see Figure 16.15). Commercially sexed poults are available at many hatcheries.

Where accurately sexed day-old poults are available at reasonable cost, raising toms and hens separately is practicable. The advantages of separating the sexes are:

- Injuries to the hens are reduced.
- Hens can be marketed at an earlier age than the toms.
- Sex-separated flocks can be fed different diets and thus be more efficient.

Some producers buy straight-run poults, then separate the sexes at 12 to 14 weeks of age.

Brooding Turkey Poults

Large-scale turkey production calls for large-scale brooding; and, to be successful, it must be properly done. The successful brooding of turkey poults necessitates that attention be given to the same factors and principles as are involved in the successful brooding of chicks, with consideration given to difference in size and certain species peculiarities—for example, the difficulty in getting poults to eat and drink.

- **Temperature.** Young poults must be kept warm and dry. For the first 2 weeks, the temperature 3 in.

(7.5 cm) above the floor at the edge of the hover should be 95° to 100°F (35°–37.8oC); and temperature may be reduced about 5°F (2.75°C) each week until heat is no longer needed. Temperature near the floor outside the brooding area should be about 70°F (21°C). A guard should be placed around the brooder for the first 2 weeks to prevent crowding and smothering in the corners of the house.

- **Ventilation.** Mechanical ventilation is highly desirable in all brooder houses, especially in those 40 ft (12 m) or more in width. Trouble from respiratory diseases is invited when the air in the brooder house is low in oxygen and contaminated with exhaust gases from the brooder stoves.

- **Floor space.** For brooding poults to 8 weeks of age in naturally ventilated brooder houses, allow 1.25 sq ft (0.11 m^2) of floor space for large-type poults and 1 sq ft (0.09 m^2) for small-type poults. In insulated force-ventilated houses, 1 sq ft (0.09 m^2) will be adequate for large-type poults to 8 weeks of age.

- **Floor.** Poults may be started on covered or uncovered litter, on asphalt roofing (not tar paper), on wire or slat floors, or in batteries. To start the poults on covered litter, proceed as follows:

  - Distribute about 2 in. (5 cm) of suitable litter material on the brooder house floor.
  - On top of this litter, under the hover and extending 3 ft (1m) beyond, place a layer of strong, rough-surface paper.
  - Enclose the paper with a brooder ring or poult guard.
  - After 5 or 6 days, remove the paper and increase the litter to a depth of 4 in. (10 cm), then allow the poults on the litter.

- **Litter.** For starting poults on uncovered litter for the first 5 or 6 days, either use softwood shavings or bright straw that is free from chaff. Following this, litter may be chosen largely according to availability and price, with consideration given to shavings, wheat straw, beardless barley straw, peat moss, shredded cane, rice hulls, and processed flax straw. If free of harmful molds, peanut hulls, crushed corncobs, and shredded corn stover make good litter. The following materials are unsatisfactory for use as poult litter: splinty shavings, sawdust, oat hulls, cottonseed hulls, dried beet pulp, and the straw of rye, oats, and bearded barley. Good litter should be put down to begin with. Then wet or caked litter should be removed at intervals and replaced with clean litter. To promote sanitation, feeders and waterers should either be moved frequently or placed on wire- or slat-covered platforms after the poults are about 3 weeks old.

- **Feeders and waterers.** The watering and feeding facilities for turkey poults are very similar to those used for chickens, but generally they are somewhat larger, and, of course, more space is needed per bird.

Start with small trough-type metal or wooden feeders placed like spokes in a wheel, part under the hover and part outside, but within the brooder ring. Provide 2 linear in. (5 cm) of feeding space per bird. In order to encourage eating, heap the trough with a little mash in a few paper plates or egg flats for the first day or two, and place some glass marbles on the feed to attract the poults. After 7 to 10 days, use larger feeders, but retain about 2 linear in. (5 cm) of feeder space per bird during the remainder of the brooding period.

For watering, start with one circular 1- to 2-gal glass (4–8 l) or metal baby-poult-size waterer, with a narrow drinking space not over 1 in. (2.5 cm) wide and 1 in. (2.5 cm) deep, for each 50 poults; or as an alternative, start with one baby-poult-size automatic water trough 4 ft (1.2 m) long for each 80 poults. Place the waterers around the edge of the hover and put a few colored glass marbles in each one. As an extra precaution, some growers add one quart-size (one liter) waterer per 50 poults for the first 2 days.

After about 2 weeks, replace the baby-poult-size waterers with larger waterers that provide a water depth of 1 in. (2.5 cm) and about 1/2 (1.25 cm) linear in. of watering space per bird. Gradually move the small waterers near the larger ones to assist the poults in making the change. Wash and rinse waterers and drinkers daily.

- **Lighting.** For the first 2 weeks of brooding in all types of houses, the room should be well lighted day and night at 10 to 15 foot-candles at poult level. After 2 weeks, only dim light at 0.5 foot-candle is needed at night, none in the daytime. In windowless houses, the light intensity can be reduced gradually to about 1 foot-candle during the 16-hour day and 0.5 foot-candle during the 8-hour night. The dim light at night may help to prevent piling and stampeding.

Cleaning the Brooder House

The brooder house should be completely cleaned prior to a new brood of poults. All equipment should be scrubbed, repaired, and disinfected.

Starting the Poults

The poults should be placed in the brooding quarters and given feed and water within 24 hours after hatching. Poults are usually started on litter or wire (plastic-coated) or slat floors. It is possible to use battery cages.

Brooder Ring or Poult Guard

Poults are placed within a brooder ring. This should be 16 to 18 in. high, and it may be made of cardboard or lightweight hardware cloth or lightweight chicken wire. This protects the poults, keeping them close to heat, water, and feed (see Figures 16.12 and 16.13).

Litter

For the youngest poults, the litter inside the brooder rings should be covered with rough-surfaced and absorbent paper. This is for firm footing (reducing the development of leg problems), and also to prevent young poults from eating enough litter to clog their digestive tract. Do not use newspapers or magazines since these can lead to leg problems. Remove the paper litter cover as soon as poults are on feed, at least by the end of the first week. The cover also prevents moisture accumulation. This combined with the heat from the brooder would provide a favorable environment for mold growth.

Space Requirements

The space requirements for brooding are shown in Tables 16.2 and 16.3.

Temperature

Young poults must be kept warm and dry. For at least the first 2 weeks, the temperature 3 in. above the floor at the edge of the hover should be 95°F for dark poults and about 100°F for white poults. The temperature outside the brooding area should be maintained at about 70°F.

Ventilation

Ventilation is always important with a well-designed system of fans and intakes preferable.

Brooder Houses

Before constructing a brooder house, the turkey producer should check plans based on successful experiences by other producers, consultants, and from County Extension Services. Brooder houses often are used both for brooding and for rearing to market age. Width usually varies from 24 to 40 ft (7.5 to 12 m), and length from 100 ft up to 600 ft (30 to 180 m). Clear-span, rigid-frame, gable-roof houses 40 ft (12 m) wide and 300 ft (100 m) long are popular. For cold-weather brooding, insulation of sidewalls and ceiling is needed, especially if the brooders are the type that do not provide an abundance of heat. Mechanical ventilation is highly desirable in all brooder houses and necessary in those over 40 ft wide. Ventilating systems must follow sound engineering principles.

Partitions are not necessary in small houses used for brooding and rearing, but some operators install them at intervals in long houses so that no more than 1,000 to 1,500 birds are allowed in a pen. This arrangement is useful in separating the turkeys by age and sex and can help to control stampedes and disease outbreaks. Pole-type breeding and rearing houses also may be used for moderate-weather brooding by enclosing the open sides on the pens with clear plastic, adding electricity, and installing brooders and water supply.

Brooder houses as well as breeding and rearing houses should be built to exclude small wild birds and rats as well as larger birds and animals. Some of the requirements are:

- Concrete foundations
- tight-fitting doors and windows
- all openings covered by 3/4-in. (1.8 cm) (or smaller) mesh wire.

Slat Floors and Wire Floors

Removable, 4-ft square sections composed of plaster lath or narrow lathlike slats mounted on metal or wooden supports make good brooder house floors. Wire floors can also be used but should be plastic-covered to reduce leg problems. These, like slats, do not use litter and, therefore, there is a cost reduction. They may reduce pathogen-borne diseases but increase leg problems. Roosts during the brooding period may be desirable.

Types of Brooders

Poults need a dependable source of artificial heat within the first few weeks of life. During the first week, they should be provided a uniform temperature of about 95° F. Thereafter, the temperature may be lowered 5° F per week. Portable brooders come in a number of different styles and sizes. In commercial operations, heating is by gas, oil, or electricity.

Handling Practices

BEAK TRIMMING. Poults should be beak trimmed (only the tip of the beak removed) to control feather picking. This is done either by the hatchery at 1 day old or at 3 to 5 weeks old.

DESNOODING. Removal of the snood may help to prevent head injuries from picking or fighting and may reduce the spread of erysipelas, but it is not a common practice today. At day-old age, the snood can be removed by thumbnail and finger pressure.

TABLE 16.4   Comparison of Large-Type Tom Turkeys Raised Under Either Range or Confinement

Rearing System	Age	Body Weight		Feed/Gain	Livability
		(lb)	**(kg)**		**(%)**
Range	19 weeks–3 days	24.3	11.0	2.90	91.3
Confinement	19 weeks–1 day	22.4	10.2	2.76	93.0

Source: Data from L.S. Jensen, University of Georgia.

After the age of about 3 weeks, it can be cut off close to the head with sharp, pointed scissors.

TOE CLIPPING.   Clipping the toes of each foot of day-old poults is used by some turkey growers to prevent scratched and torn backs—an important cause of downgrading carcasses. Clipping is best done at the hatchery with surgical shears.

CATCHING TURKEYS.   For catching turkeys, a darkened room is best. In the dark or semidarkness, turkeys can be picked up with both legs without confusion or injury. Portable catching chutes, preferably with a conveyor, can be used for catching and loading range turkeys.

Range or Free Range Turkeys

Some turkeys are raised in range conditions. Table 16.4 compares performance on range and confinement.

Moving Turkeys

When poults are about 8 weeks of age, they are ready to be moved to the range. Movement to range is most critical and should be carefully planned. Two things should be checked before moving poults to range: (1) the 5-day or 1-week weather forecast, and (2) the readiness of equipment. Avoid moving poults when the weather is threatening. Cool, wet, and rainy weather place a severe stress on poults that have just been moved to range. It is best to move in the early morning, because this allows poults plenty of time to locate water, feed, and shelter before nightfall. Never move in the heat of the day, for this is hard on both birds and handlers. Some growers move about one-third of their poults the first morning, then skip a day or two and move the remainder of the flock. Large wire-enclosed four-wheel trailers are best for moving turkeys to ranges. Moving trailers may be constructed so that the birds can be loaded and unloaded by driving or herding.

Handling Turkeys

Range turkeys can be driven from place to place. One person can control a good-size flock. Dogs can be used for driving turkeys if well trained and gentle.

Predator Control

A tight fence around the range area will reduce predator entry. A perimeter electric fence 10 to 18 in. (25 to 45 cm) above ground level and 2 to 3 ft (0.6 to 1.0 m) outside the poultry fencing will further reduce predatory entry. When an electric fence is used, it should be properly constructed; otherwise, it may be hazardous. Some growers ward off predators with devices that produce loud noises or by keeping a dog leashed near the range area. Federal, and possibly state assistance, should be obtained when predator losses become severe.

PROCESSING TURKEYS

Most producers market their birds through integrated firms, processors, or cooperative organizations. The turkeys are collected and trucked to processing plants (see Figures 16.16 to 16.19 for illustrations of processing). Some turkey growers sell their turkeys directly to consumers or to local retail dealers. These growers may have their own processing and even freezing facilities.

Until the 1960s, fresh and frozen turkey was readily available only around the Thanksgiving and Christmas holiday season. Today, turkey is available

Figure 16.16 Processing turkeys at West Liberty Foods. *(Courtesy, Gretta Irwin, Iowa Turkey Federation)*

Figure 16.17 Refrigerated storage of turkey hams at West Liberty Foods. *(Courtesy, Gretta Irwin, Iowa Turkey Federation)*

Figure 16.18 Further processing turkeys into hams at West Liberty Foods. *(Courtesy, Gretta Irwin, Iowa Turkey Federation)*

Figure 16.19 Processing turkey meat into hams. *(Courtesy, Gretta Irwin, Iowa Turkey Federation)*

year-round. About 40% of the per capita consumption of turkey meat still occurs in the fourth quarter (October, November, and December), reflecting Thanksgiving and Christmas sales. The remaining 60% is distributed fairly equally throughout the rest of the year. Marketing of turkeys is covered in more detail in Chapter 17.

Slaughtering and Processing

Turkeys are slaughtered and processed in a manner very similar to that in chickens (see Chapter 15). They may also be slaughtered following Islamic or Jewish religious laws, respectively, Halal and Kosher (for details see Chapter 15). Most turkeys are now processed in plants that process turkeys only. The usual steps in processing are:

1. **Suspended by legs.** The live turkey is suspended by the legs, the feet held in a steel shackle.
2. **Stunning.** An electric stunner is usually used to prevent struggling and relax the muscles that hold the feathers.
3. **Bleeding.** The head is held in one hand, and a cut is made across the throat so that both branches of the jugular vein are severed at or close to their junction.
4. **Scalding.** The feathers are loosened, usually by the subscald method in which the bird is immersed in agitated water at 140°F for about 30 seconds.
5. **Feather removal (picking).** The feathers are removed by a rubber-fingered machine.
6. **Evisceration.** Usually, the turkey is suspended by the head and both hocks, a crosswise cut is made between the rear end of the keel and the vent, and the intestines are removed.
7. **Chilling.** The eviscerated carcass is chilled in ice water or ice slush to an internal temperature of 35° to 39°F.
8. **Packaging.** After chilling, the carcass is drained, a plastic wrap is applied, the air in it is exhausted by vacuum, and the wrapper is sealed. The bird is then readied for freezing or for marketing fresh chilled, unfrozen. Alternatively, further processing occurs.

Inspection

Inspection is the examination of birds for indication of disease or other conditions that might make them unfit for human consumption. In the United States, all ready-to-cook poultry and poultry products are federally inspected. Antemortem inspection of poultry is required before it enters the processing plant. This is a spot check of each load of birds, not of each individual bird. Postmortem inspection of each bird can occur when the abdominal cavity is opened and the visceral organs are removed (also see Chapter 12).

Grading

Grading consists of identifying individual birds according to class, quality, and weight. Final grading of ready-to-cook poultry is after the birds are eviscerated and cooled and before they are packaged, frozen, or further processed, but not while they are still warm (see Chapter 17).

Other Products and By-Products

Offal, bone, and feathers from turkeys are marketed in a manner very similar to that in chickens (see Chapter 15 for details).

TURKEY BREEDERS (TURKEY POULT PRODUCTION)

Turkey breeders are the equivalent of broiler breeders, with broiler chick production instead of the production of turkey poults.

Turkey Breeders and Reproduction

Reproduction in turkeys is highly seasonal, requiring long-day length for reproduction. Under natural conditions, turkey hens start egg production in spring, when the natural light gets to about 14 hours per day after being as low as 8 hours per day in the winter months (see Figures 16.20 and 16.21 for adult turkeys).

To bring turkey hens into lay, the hours of light are increased. This is known as photo stimulation or

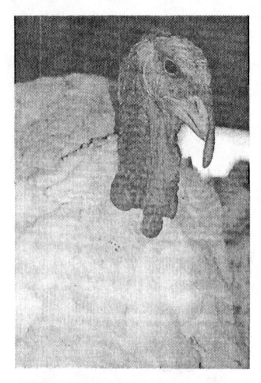

Figure 16.21 Adult tom turkey. *(Courtesy, Gretta Irwin, Iowa Turkey Federation)*

lighting of the turkeys. The increase in day length, or photoperiod, also initiates a process that leads to the birds being unresponsive to these long-day lengths. This is photorefractoriness. It may involve the effects of thyroid hormones.

After an initial peak in eggs produced, egg production declines. After about 20 weeks of laying eggs, the rate of egg production falls to a low level. Moreover, turkey hens start to show incubation/nesting behavior. If they are permitted to sit on the eggs, laying stops altogether. The breeder hens can be recycled or brought back into lay by restricting day length to 6 to 8 hours/light per day for about 6 weeks, then increasing the day length again.

Reproduction can be divided into the following phases:

* Place immature hens on short-day length.
* Increase day length to 14 to 15 hours/light per day.
* Development of the ovary and reproductive tract (oviduct).
* Egg production peaks.
* Egg lay declines due to photorefractoriness and/or shift to incubation physiology with nesting, and so on.
* Transfer to short days to recycle (or "break" photorefractoriness) and prevent nesting and incubation.
* Increase day length for reproductive resurgence.

Figure 16.20 Laying turkey hen. *(Courtesy, Gretta Irwin, Iowa Turkey Federation)*

TABLE 16.5   Lighting Schedule for Commercial Turkeys

| | Age | Hours of Light | Light Intensity | |
			Lux	Foot-candles
	1 day	24	100	10
	2 days	23	100	10
	3–5 days	20	100	10
	6–9 days	18	80–100	8–10
Toms/males	9 days–15 weeks	16	60–100	6–10
Hens/females	9 days–17 weeks	16–18	60–100	6–10
Toms/males	15–18 weeks	16–18	60–100	6–10
Hens/females	13–16 weeks	16–18	60–100	6–10

Source: Adapted from Hybrid Turkeys information. http://www.hybridturkeys.com/Media/PDF_files/Management/Mng_lhgt_com.pdf

TABLE 16.6   Lighting Schedule for Turkey Breeder Hens

| Age | Hours of Light | Light Intensity | |
		Lux	Foot-candles
1 day	24	75	7
2–4 days	23	75	7
1–17 weeks	14	54	5
17–21 weeks	8	54	5
21–27 weeks	7	54	5
27–30 weeks	6	54	5
30–36 weeks	14	107	10
36 weeks–end of cycle	15	107	10

Source: Adapted from Hybrid Turkeys information. http://www.hybridturkeys.com/Media/PDF_files/Management/Mng_lighting_brd.pdf

TABLE 16.7   Lighting Schedule for Turkey Breeder Toms

| Age | Hours of Light | Light Intensity | |
		Lux	Foot-candles
1 day	24	75	7
2–4 days	23	75	7
1–20 weeks	12	43	4
20–24 weeks	13	43	4
24 weeks to end	14	32	3

Source: Adapted from Hybrid Turkeys information.

For possible lighting schedules for breeding hens and toms, see Tables 16.5 to 16.7. Following insemination, spermatozoa are stored in sperm ducts (glands or nests) for up to several weeks in the hen's vagina. After each egg is laid, some of the sperm leave the sperm nest and travel to the top of the oviduct to fertilize the next egg being laid. This is the only time an egg can be fertilized. As soon as the

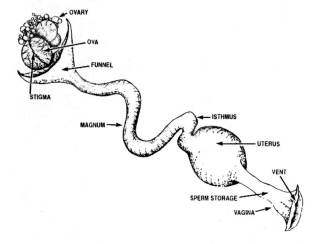

Figure 16.22   Reproductive system of turkey hen. *(Original diagram from E. Ensminger)*

yolk is coated with egg white, the sperm cannot get to the ovum (see Figure 16.22).

Egg Production

Egg production in turkeys shows the following characteristics:

- Peaks at week 3 at about 72%
- Declines gradually to about 50% at week 26

(Nicholas Technical Information, 2002)

Hatchability

Hatchability also shows changes during egg production:

- Week 1, 74%
- Week 5/6, 83%
- Week 26, 73%

(Nicholas Technical Information, 2002)

Lighting for Breeder Turkeys, or Photostimulation

Commercial turkey hens are placed on short-day length lighting (< 8 hours of light per day) at about 18 weeks of age to prevent reproductive development when they are too young or small for optimal production. Feed may be restricted to ensure that a target body weight is not exceeded. At about 30 weeks of age, the day length is increased to a long-day length (14 or 15 hours of light per day) to bring about rapid sexual maturation. This is equivalent to the increases in day length of spring that bring about breeding in wild birds. Breeder hens should not be subjected to lighting earlier than 29 weeks of age. The recommended light intensity is 5 to 7 foot-candles, 12 in. off the ground. A uniform light pattern is also important (see Table 16.5).

How is this controlled?

Light indirectly via eye or directly on hypothalamus
⇓
Increased release of gonadotropin-releasing hormone (GnRH)
⇓
Anterior pituitary gland
⇓
Increased release of luteinizing hormone (LH) and follicle stimulating hormone (FSH)
⇓
Ovary

- Develops and matures
- Releases estradiol ⇒ oviduct grows, matures, and supplies egg white and shell
- Follicle grows and ovulates

Artificial Insemination

Today's broad-breasted turkeys would not be practical without artificial insemination. Turkeys with a conformation that pleases the consumer and a meat yield that satisfies the processor do not have the agility to mate often and successfully. Males (toms) are much larger than females (hens). Mating is not only problematic but may physically injure the hen.

The turkey industry in the United States and Canada has used artificial insemination for breeding for about 50 years. Because turkey semen cannot be frozen, the storage time for turkey semen is limited. The semen may be diluted with an extender (containing isotonic salts and proteins, including yolk) to increase the number of hens that can be inseminated and to extend the life of the semen (see Figure 16.23). The reason for almost complete utilization of artificial insemination for breeding turkeys is animal safety. Insemination can occur every 7 to 14 days because spermatozoa are stored in the oviduct.

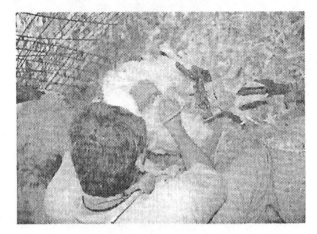

Figure 16.23 Inseminating hen. *(Courtesy, Gretta Irwin, Iowa Turkey Federation)*

Advantages of Artificial Insemination

The advantages of artificial insemination include:

- Increased fertility
- Reduced damage to female during mating
- Reduced use of large males that have difficulty mating
- Allows separation of males and females
- Semen can be evaluated for quality (e.g., spermatozoa motility and number of abnormal spermatozoa)
- Semen can be diluted or extended

Semen Characteristics

Each drop of good semen contains millions of very active sperm. Under the microscope, they can be seen swimming about very rapidly, but they die after a few minutes. Toms produce semen with different characteristics:

- Semen volume (average, 0.2 ml).
- Spermatozoa concentration as determined by centrifugation or spectrophotometer.
- Number of live (motile) spermatozoa. (A good sample should contain 8 billion spermatozoa per ml with > 85% normal cells and > 90% motile.)
- Fertilizing capability.

Note: High-quality semen should be thick and pearly white.

Sperm produced early in the season appear to be much longer lived than sperm produced late in the season. Despite the ability of sperm to survive in the vagina, they are very delicate. Sperm have thin cell walls and they are easily killed by hyper- and hypotonic salt solutions. Sperm survive best at 33° to

37°F (1°–3°C). At temperatures above this, spermatozoa show reduced survival.

Semen Collection

Semen is collected by stimulation. Only two cloacal strokes should be necessary to extract the semen. An extender is added to the semen (to a >1:1 ratio) and then it is brought to storage temperature.

Semen Storage

Semen is stored at 33° to 37°F (1°–3°C) because sperm survive best at this temperature range. Semen is then stored with the extender and with agitation (farm agitator).

Semen Abnormalities

In the yellow semen syndrome, there are increases in the number of abnormal spermatozoa, along with reduced hatchability and reduced fertility. The yellow coloration may be due to high levels of riboflavin.

NO SEXING OF SEMEN. In cattle, X (female offspring) and Y (male offspring) spermatozoa are beginning to be separated. Semen can then be sold with a "guarantee" of a bull or heifer calf. Sexing semen in turkeys is not possible because the male is homozygous (ZZ) and the female gamete determines gender of the poult.

Insemination

Based on the concentration of spermatozoa, inseminate at a dose of 18 hens per ml.

TIME AND METHOD OF INSEMINATION. The usual time for artificial insemination is about 2 weeks after lighting the hens. The progress of the hens should determine when the flock should be inseminated. Artificially inseminate the flock every 5 to 10 days. The straw method of insemination is recommended. Use one straw to inseminate each hen, reducing the danger of spreading disease from hen to hen.

Natural Mating

In the natural process of fertilization, mating takes place when the hen is receptive. Hens have a stronger propensity to mate a few days just before egg production starts and a few days thereafter. By the end of the season, they may mate less frequently. In mating, the hen everts the vagina beyond the lips of the cloaca and the tom deposits semen directly on the everted tip of the vagina. The hen immediately withdraws, carrying the semen far up the vagina. Sperm swim into tiny structures in the walls of the vagina, called sperm nests or glands. A few swim out and are carried by the action in the hen's oviduct to where fertilization takes place.

Male:Female Ratios

In single-male natural mating, the following ratios are recommended:

- Small-type turkeys, 20 hens to 1 young tom
- Medium-type turkeys, 18 to 1
- Large-type turkeys, 16 to 1

In natural flock matings, the following ratios are recommended:

- Small-type turkeys, 14 to 1
- Medium-type turkeys, 12 to 1
- Large-type turkeys, 10 to 1

Additionally, a few extra toms, preferably late hatched or out-of-season, should be maintained separately or as replacements for those that become injured or die.

Management of the Breeding Turkey

Tom Management

Tom housing can be very similar to hen housing, except that feed, water, and roosting space should be larger. It is recommended that there be 6 sq ft (0.75 sq m) per tom.

Preferably, toms should be in a separate building because of the need for different management. If the toms are housed in the same building with the hens, the fence separating them should be solid. However, if they are separated by 20 ft (~6 m), wire fencing can be used. Up to 25 toms may be kept together in a pen.

Toms should be lighted 3 weeks prior to lighting hens to assure semen being ready at the time of mating (or semen collection for artificial insemination). Toms that are to be used for the summer flock may not require this since they mature on a long-day length.

Toms should be "premilked" (ejaculated) at least once before collecting semen for artificial insemination. If they are not premilked, they may not give enough semen for the first artificial insemination.

Hen Management

NESTS. Nests with tip-up fronts, or partial trapnests, one for each 4 to 5 hens, are recommended. They may be made of wood or of more durable welded iron or heavy plastic. The tip-ups, or gates, must be carefully constructed and balanced or equipped with springs so that the tip-up closes behind the hen when she enters and remains closed until she leaves. As she leaves, the tip-up is automatically reset so another hen may enter. Nests should be arranged in rows, preferably back-to-back to facilitate efficient gathering. Nests should be installed and made operational well in advance of lighting the hens. Otherwise, hens may lay on the ground and it will be hard to break them of this habit. Individual nests should be at least 2 ft high, 1 1/2 ft wide, and 2 ft long. Nests should have enough depth so as to allow at least 4 in. of nest bedding to be used. Nests should be placed on the floor or about 6 in. off the floor. A variety of materials are satisfactory for nesting litter, including sawdust, wood shavings, and others.

Broody Hen Control

Broodiness in turkey hens is their natural desire to incubate and hatch eggs. In nature, broodiness is essential for survival of the species. Since the incubator came into use, however, broodiness is a problem, with much reduced egg production, together with broken and dirty eggs.

Broodiness should be halted as soon as it is detected in a flock. One of the first signs of the onset of broodiness is that the hen spends more and more time on the nest, after laying, until she no longer leaves the nest. Broody hens should be removed from the nest in the evening. Different techniques have been practiced to eliminate or "break up" broodiness including breeding, environment, and lighting.

Out-of-Season Breeding

In most of the Northern Hemisphere, turkeys hatched between August 1 and April 1 can be considered out-of-season birds. Out-of-season hens require a light control program in order to lay well and long.

To obtain high egg production, the out-of-season hens, but not the toms, should be preconditioned by placing them under light restriction between 18 and 20 weeks of age. When put under light restriction at various ages, the following light schedule should be maintained: up to 30 weeks—<8 hours of light per day. Hens should be observed carefully during the period of light restriction. If they show signs of sexual activity, such as squatting, lower the period of light from 8 hours to 6 or even 4 hours.

The darkout period should be as dark as possible. During the light period, 1 foot-candle is ample. When properly done, white birds can barely be seen

and it is barely possible to read newspaper headlines (brownout). The density of birds in a darkout facility should not exceed 3 sq ft per hen. Out-of-season breeding stock should be fed and housed in the same manner as normal-season birds and the intensities of light used should be the same. Failure to obtain efficient out-of-season production usually stems from (1) starting the light-restriction period too late—after production starts, or (2) allowing too much light during the brownout hours of the restriction period.

Forced Molting or Forced Rest

A satisfactory second round of egg production during the off-season usually can be obtained from normal-season breeding hens by forced molting (forced resting). Reduction in day length causes egg production to cease and a molt to start. Forced molting is accomplished by placing the birds in a completely dark house. This means that air intakes must be darkened so that no light comes in. The exhaust fans must have hoods that are baffled to prevent light from entering, or light control shutters on the inside of the fans to prevent any light from entering. In order to have good ventilation, it is recommended that 2 cfm of air movement per pound of bird be provided, especially during hot weather.

During the first 72 hours of forced molting, the birds should not receive any feed or water. After 72 hours of fasting, they should be given feed and water for only 3 hours. During this 3-hour period, a dim light should be used so that the birds can see the feed and water. After 2 or 3 days of 3-hour daily feeding, the birds may be fed and watered for a 6-hour period.

If the program is functioning satisfactorily, the birds will shed a lot of feathers and there will be complete stoppage of egg production within a few days. This program of light restriction should be continued until the end of the 12-week period. This is recommended because it takes that long for the birds to molt and grow the new primary feathers as well as the body feathers. At the end of the 12-week period, the birds are put under stimulatory light and brought back into production. Preferably, the males used for second-season breeding should be young. If necessary, the original males may be used.

Nutrition of the Breeding Turkey

The nutritional requirements and practices for breeding turkeys are different from those for turkeys raised for meat. This is analogous to the situation with broiler breeders (see Chapter 15). There is also a seasonal effect on feed consumption.

For every 10°F (6°C) decrease in temperature expect a 6% increase in feed consumption.

Feeding Breeder Turkey Hens

The nutritional requirements vary gradually with age, except for the abrupt change when they come into egg production. At that time, there are markedly increased requirements for protein and calcium to meet the needs of egg production. Examples of stepwise changes in turkey breeder hen diets during growth and as they come into lay are as follows:

Starter	0–3 weeks
Grower 1	3–7 weeks
Grower 2	7–11 weeks
Developer	11–16 weeks
Holder	16–29 weeks
Breeder	29 to end

(*Nicholas Technical Information*)

Starter

Protein	26.5%
Energy	2,670 kcal/kg
Calcium	1.4%

Grower 1

Protein	22.5%
Energy	2,740 kcal/kg
Calcium	1.35%

Grower 2

Protein	20.5%
Energy	2,780 kcal/kg
Calcium	1.3%

Developer

Protein	19%
Energy	2,820 kcal/kg
Calcium	1.15%

Holder

Protein	14.5%
Energy	2,820 kcal/kg
Calcium	1.1%

Breeder (female)

Protein	17%
Energy	2,880 kcal/kg
Calcium	2.4%

Feeding Breeder Turkey Toms

The growing and adult males for breeding populations of turkeys are kept separate from the females.

The diet for growing male breeder turkeys is restricted, using various feed restriction programs for the following reasons: to prevent obesity and to reduce feed costs.

The nutritional requirements vary with age. Examples of turkey breeder tom diets are as follows (same formulation as for the hens):

Starter	0–3 weeks
Grower 1	3–7 weeks
Grower 2	7–12 weeks
Developer	12–19 weeks
Holder	19 to end

(*Nicholas Technical Information*)

Breeder Management

Frequently, the failure of turkey breeding programs can be traced to problems of management.

Egg Handling and Incubation

Egg Handling

Proper egg handling is critically important. A fertile turkey egg contains a delicate living embryo that can be easily killed. A newly laid egg has a temperature of 106°F. Inside, the embryo is growing very rapidly. There is no air cell, the shell is damp, and at this stage the embryo is very susceptible to sudden temperature changes. As the egg cools, embryo growth slows and by the time the internal temperature gets down to 80°F, growth is arrested. Cooling eggs too fast reduces fertility and hatchability.

Eggs should be collected frequently, especially during cold weather. Dirty eggs usually show lowered hatchability and may transmit disease. Also, they are the main reason for exploders in the hatchery. Washing turkey eggs intended for hatching appears to be a debatable practice. Not only is washing ineffective in sanitizing the eggs, it may actually be harmful to fertility.

EGG FUMIGATION. Egg fumigation with formaldehyde kills bacteria on the shell and prevents disease outbreaks in the poults. Formaldehyde is a penetrating gas that is very deadly to bacteria. However, formaldehyde gas will not kill bacteria beneath the eggshell. For this reason, eggs should be fumigated soon after gathering before the bacteria has time to penetrate deep into the shell. The recommended fumigation procedure follows: Place the freshly gathered eggs in an airtight cabinet with heating and forced-air circulation. This cabinet should be large enough to hold one gathering of eggs. Eggs must be in wire-mesh trays, such as are used in setting hatching eggs. This ensures free circulation of the gas around the eggs. Tests have shown that gases are unable to circulate and penetrate when flaps or egg cases are used. The

BOX 16.1   What's Ahead for Turkeys

Tremendous changes have occurred in the U.S. turkey industry (see Figures 16.24 and 16.25), and more changes lie ahead. The future turkey industry will be characterized as follows:

Figure 16.24 Growing turkeys. *(Courtesy, Gretta Irwin, Iowa Turkey Federation)*

Figure 16.25 Young turkey poults. *(Courtesy, Gretta Irwin, Iowa Turkey Federation)*

- It is debatable whether per capita consumption in the United States will increase due to competition with other meats, particularly chicken.
- There is opportunity for increases in turkey export and consumption worldwide, particularly in Europe.
- Further increases in growth rate and feed efficiency will be achieved.
- There will be fewer infertile turkey eggs and embryonic deaths; poult mortality will be reduced.
- The trend for fewer but larger turkey producers and processors will continue.

temperature during fumigation in the cabinet should be at least 70°F.

EGG STORAGE.   Eggs should be allowed to cool slowly, then stored at 55° to 60°F with 80% relative humidity. A recommended procedure is to pack eggs in pre-cooled cases, small end down. Turkey eggs need not be turned while they are in storage prior to incubation if the storage time does not exceed 2 weeks. However, if the eggs are to be held for longer periods, they should be turned at least three times a day. Since both fertility and hatchability decline as the holding period increases, holding turkey eggs longer than 2 weeks is not recommended. Extreme care must be exercised so that the eggs are not overheated, chilled, or frozen during shipping. They must be packed small end down in cases, with the special turkey-size flap and filler or the new filler-flap being used to prevent breakage. Shipping will sometimes result in decreased fertility and hatchability but under proper conditions (temperature, humidity, and handling) this will be reduced.

Incubation

Turkey eggs are incubated at 99.5°F and a relative humidity of 80 to 85%. Also, proper ventilation is essen-

tial. The developing embryos require air containing 20% oxygen and not more than 1.5% carbon dioxide. Since the oxygen requirements of the embryos increase as they develop, it is evident that ventilation must be increased as the hatch progresses. The eggs should be turned at least five times daily, and preferably every 2 hours during incubation. With machines equipped with turning devices, this is easily accomplished. The incubation period for large-type turkey eggs is 28 days. Eggs from small-type turkeys tend to hatch in 27 to 27 1/2 days. Only clean eggs with sound shells should be placed in the incubator. Washing is not recommended for turkey eggs.

 USEFUL WEBSITES

British United Turkeys (BUT): http://www.but.co.uk

British United Turkeys of America (BUTA): http://www.BUTAInfo.com/

Hybrid Turkeys (Canada): http://www.hybridturkeys.com

Nicholas Turkey Breeding Farms: http://www.nicholas-turkey.com

FURTHER READING

Barbut, S. *Poultry Products Processing: An Industry Guide.* Boca Raton, FL: CRC press, 2002.

Diseases of Poultry. Edited by B. W. Calnek, H. John Barnes, Charles W. Beard, Larry R. McDougald, and Y. M. Saif. Ames, IA: Iowa State Press, 1997.

Pattison, M. *The Health of Poultry.* Blackwell Science, 1993.

Poultry Health and Management: Chickens, Ducks, Turkeys, Geese, and Quail. Sainsbury, D. Blackwell Science, 2000.

Sturkie's Avian Physiology, 5th ed. Edited by G. C. Whittow. San Diego, CA: Academic Press, 2000.

17

Marketing and Grading

Objectives

After studying this chapter, you should be able to:

1. Understand the changing consumption of poultry and eggs.
2. List the sizes of the major components of the poultry industry.
3. List changes in poultry marketing.
4. Define the grades of eggs and poultry meat.
5. Understand the role of candling.
6. List the major products of eggs and poultry meat.

Marketing is selling. Further processing is done to improve sales and profitability, be it packaging, improving shelf life, or new products. The changes in the marketing of poultry and eggs have been particularly marked. See Figure 17.1 for live turkeys and Figures 17.2 to 17.23 for examples of further processed poultry products. These changes in marketing encompass processing and distribution, together with farm production. Research-based advances in poultry production (breeding, nutrition, housing, animal health, etc.) have reduced costs. The cost of production of chicken meat has been reduced by more than 50% in the last 30 years. This has been due to a number of factors coming from research, together with innovation and continued improvements in management. The research-based factors are, in the order of importance:

1. Breeding and genetics
2. Improved nutrition
3. Animal health and disease prevention

The relative contribution of breeding and genetics versus nutrition is 4:1. Improved animal health has greatly reduced the mortality rates (number of deaths) from about 20% in the 1920s to about 5% in the 1970s and also the rate of morbidity (number of sick birds). Research has contributed to the poultry industry by reducing the cost of production and increasing the number

and variety of poultry products. (For details see the following sections.) The animal health industry has made a tremendous contribution by developing and manufacturing antibiotics, other drugs, and vaccines.

There have been organizational changes that have further lowered production costs. This has transformed poultry production into an integrated data-based industry. Thus, the poultry industry is the most efficient, vertically integrated, and technologically advanced of all livestock production. There are also small- and medium-sized niche producers (e.g., organic, free-range, etc.)

POULTRY AND EGG CONSUMPTION AND PRICES

Chicken and turkey consumption, on a per capita basis, has increased. Egg consumption per person was decreasing until the early 1990s but now has stabilized (see Chapter 1, Table 1.21). Consumers have increased their consumption of chicken and turkey meat due to the lower prices (relative to red meats), convenient tasty products (home-cooked and fast food), and perceptions of healthfulness. The consumption of eggs in the United States declined from its high in 1945 until the early 1990s. This reflected changes in lifestyle, with fewer home-cooked breakfasts and the health concerns about cholesterol in eggs.

Figure 17.1 Young turkeys. *(Courtesy Gretta Irwin, Iowa Turkey Federation)*

Figure 17.2 Chicken products include chicken and cheese mini tacos.

Figure 17.3 Chicken products include Buffalo wings.

Figure 17.4 Chicken products include chicken breast patties.

Figure 17.5 Chicken products include chicken and rice soup.

Figure 17.6 Chicken products include prepared chicken salad.

Figure 17.7 Chicken products include noodles as a prepared children's meal.

Figure 17.8 Chicken products include chicken and dumplings.

Figure 17.9 Chicken products include chicken in water.

Figure 17.10 Chicken products include chicken broth.

Figure 17.11 Chicken products include chicken breast stuffed with broccoli and cheese.

Figure 17.12 Chicken products include BBQ chicken pizza.

Figure 17.14 Chicken products include ready to eat spicy Thai chicken with noodles.

Figure 17.15 Chicken products include rapidly prepared chicken skillet dinners.

Figure 17.13 Chicken products include chicken patties.

Figure 17.16 Chicken products include chicken nuggets in "fun " shapes for children.

Figure 17.17   Rock Cornish game hen.

Figure 17.19   Cooked turkey. *(Courtesy, Hubbard ISA)*

Figure 17.18   Turkey products include Italian sausage.

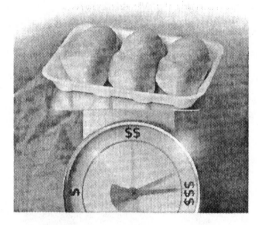

Figure 17.20   The importance of breast meat to profitability. *(Courtesy, Hubbard ISA)*

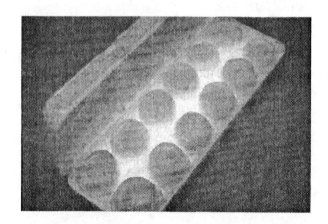

Figure 17.21   Kemin lutein eggs.

Figure 17.22 Ground turkey from range turkey.

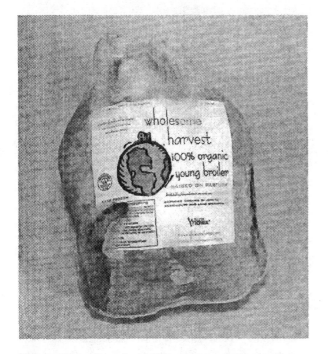

Figure 17.23 Organic chicken ready to cook.

MARKET VALUE OF POULTRY AND EGGS

U.S. poultry producers sold chicken, turkey, and eggs valued at a total of $22.2 billion in 1999. This represented 20% of the total livestock receipts and 10% of the total cash farm income that year. Of the combined total income from poultry and eggs, about $15.1 billion is derived from broilers, $4.3 bil-

lion from eggs, and $2.8 billion from turkeys. The farm price of poultry and eggs is like all commodities, determined by supply and demand.

Not only are production costs lower for poultry but also marketing costs for poultry products are much lower than for beef. This is due to larger operations, with lower costs of processing and efficiencies of scale.

FACTORS AFFECTING THE PRICE OF POULTRY PRODUCTS

Supply and demand dictates the prices of poultry and eggs. Producers have little control over prices in the short term. Longer term, they can affect aggregate demand with new products and/or exports. The nature of the industry is such that bird numbers can be increased or decreased fairly rapidly in response to price trends. A chick requires only about 6.5 weeks from conception to market, far less time than the 283 days required for gestation alone in cattle.

Demand for poultry meat continues to increase. The cost is low in comparison with pork or beef. There is some relationship between the prices of poultry and red meat. When red meat supplies are high, prices fall to a level where demand is increased. This can depress the demand for poultry, thereby decreasing prices. Conversely, when beef and pork prices are high, demand for poultry goes up, increasing poultry prices. If supplies of poultry are very high, not only is the price of poultry depressed but also that of beef and pork.

Impact of Holidays

Turkey is commonly eaten for holiday meals (e.g., Thanksgiving, Christmas). At this time, a large but temporary increase in demand occurs. Turkey consumption during the fourth quarter is now less than 40% of annual demand.

Impact of Disease Outbreaks

Widespread disease outbreaks with entire flocks lost from production due to depopulation can reduce supply and markedly increase prices.

Impact of Feed Costs

The largest single expense in the production of poultry products is feed. Thus, when feed grains and protein supplements are in short supply, the cost of feeding poultry may increase to an unprofitable level. In these circumstances, bird numbers are reduced. Conversely, when feed is cheap, bird numbers may be increased.

Impact of Media

Reports about the dangers of consuming poultry and eggs due to food-borne organisms and/or longer-term issues such as cholesterol can impact consumer demand. In Western Europe, scares about salmonella resulted in large, albeit temporary, decreases in egg consumption.

Impact of Imports and Exports

Poultry products leaving from or coming into the United States can affect poultry prices. The economic impact of this phase of marketing is discussed in Chapter 18.

MARKET CHANGES

Changes in marketing poultry and eggs have extended from the producer to the consumer involving technology, organization (with vertical coordination—for details see Chapter 18), and location. It is impossible to separate the changes from the forces behind them. Among the pertinent changes affecting the marketing of poultry and eggs are the following:

1. Poultry producers have become fewer in number and larger in size. These operations are large enough to command competent management "know-how," and to justify state-of-the-art automatic equipment and ventilated buildings.
2. Increased production efficiency, particularly feed efficiency.
3. Shifts in production areas (see Chapter 1 for details).
4. Hand in hand with increased vertical coordination, eggs, broilers, and turkeys have moved through shorter and more direct marketing channels. Increasingly, direct movement from packing plants to retailers is bypassing wholesale distributors, reducing storage time and increasing the shelf life for the consumer.
5. There has been a marked trend toward larger processors in the production areas. Because commercial egg production is more widely dispersed over the United States, the concentration of egg handling and processing is not as great in the market egg business as in poultry meat processing. Additional processing is greater with liquid and frozen egg products than with fresh eggs.
6. More new products. Throughout the poultry and egg production industry, renewed attention has been given to developing new poultry and egg products. In particular, more convenience foods have been developed.
7. There are larger and more efficient retail stores. For major grocery stores in the United States, see Table 17.1.

Marketing of Eggs

The most important product development has been the improvement in quality and freshness of eggs. New egg products include frozen omelets, egg sandwiches, and other breakfast egg dishes. Specialty eggs include:

- Hens fed a diet free of animal fat
- Certified organic, free-range
- Uncaged, unmedicated
- Cage-free vegetarian diet
- Naturally nested

Eggs from hens fed a diet free of animal fat have reduced levels of saturated fatty acids.

Marketing of Chicken and Turkey

In the past, chicken and turkeys were marketed as New York dressed (blood and feathers removed only). Today, chickens and turkeys are eviscerated and either sold as oven-ready or cut-up and/or further processed products (representing ~80% of chicken and turkey). The ever-increasing further processed products include (see Figures 17.2 to 17.17) chicken franks; boneless chicken in the form of breasts, patties, fillets, chunks, and nuggets; precooked fried chickens; chicken-filled rolled tortillas; and so on. Another development is the use of turkey and some chicken meat in the manufacture of such products as bologna, "low-fat hamburger," sausages, brats, and manufactured bacon (see Figures 17.18 to 17.20).

Another development in broiler marketing is an extensive network of fast-food (quick-service) restaurants and other carryout restaurants featuring chicken. Over 30% of the total broiler production is sold through these restaurants. It might be noted that individual fast-food chains have very specific requirements. For instance, Kentucky Fried Chicken prefers small and uniform broilers.

MARKET CHANNELS AND SELLING ARRANGEMENTS

Most eggs and poultry are marketed through retail food stores. The institutional markets, government purchases, and exports account for the remainder. In all cases, market channels have become more direct. Most eggs move from the producers to consumers either directly or via retail stores or restaurants. The marketing of poultry meat has

TABLE 17.1   Largest Grocery Retailers in the United States (2001 Sales)

Company	Stores	Sales (billion $)	% Market Share Supermarket	All retail
The Kroger Co.	3,211	50.7	12.8	10.0
Albertson's Inc.	2,573	38.3	9.7	7.6
Safeway Inc.	1,762	34.3	8.7	6.8
Ahold USA Inc.	1,446	34.3	8.7	6.8
Wal-Mart Supercenters	1,060	20.1	5.1	4.0
Sam's Club	498	18.4	4.6	3.6
Costco Wholesale Group	363	17.7	4.5	3.5
Delhaize America	1,406	15.2	3.8	3.0
Publix Super Markets Inc.	675	14.7	3.7	2.9
Winn-Dixie Stores Inc.	1,150	12.9	3.3	2.5
Great Atlantic & Pacific Tea Co.	822	10.5	2.7	2.1
H. E. Butt Grocery Co.	297	9.0	2.3	1.8
SuperValu Inc.	1,202	8.5	2.1	1.8
Meijer Inc.	152	6.1	1.5	1.2
Shaw's Supermarkets Inc.	187	4.6	1.2	1.0
Giant Eagle Inc.	250	4.4	1.1	0.9
Pathmart Stores Inc.	141	4.0	1.0	0.8
Hy-Vee Inc.	187	3.5	0.9	0.7
Raley's	149	3.0	0.8	0.6
Wegman's Food Markets Inc.	62	2.9	0.7	0.6
Aldi Inc.	596	2.9	0.7	0.6
Stater Brothers Markets	155	2.6	0.7	0.6
BJ's Wholesale Club Inc.	130	2.4	0.6	0.5
Harris Teeter Inc.	142	2.4	0.6	0.5
Flemming Companies Inc.	160	2.3	0.6	0.5

Note: Wal-Mart sales are for food only and do not reflect Sam's Club, which are shown separately. Aldi sales are fractionally lower than Wegman's, but both are shown as $2.9 billion due to rounding.

Source: Based on the 2000 rankings in the June 4, 2001 Feedstuffs company reports, Directory of Supermarket, Grocery &Convenience Store Chains, and FMI.

changed, with a substantial increase in further processed products. Broiler meat moves from producers to the processing plants and either to the store warehouses or directly to retail stores.

As integration and various forms of vertical coordination continue to spread within the poultry industry, fewer actual purchases and sales of products occur and fewer genuine negotiated prices are generated. Consequently, it is increasingly difficult for a poultry producer to determine what is the "going market," and markets are more vulnerable from the standpoint of manipulation. There are two major market news services: (1) the USDA Market News Service and (2) the Urner Barry report.

Market News Service

The Agricultural Marketing Service in USDA provides *Market News*. This is based on federal and state market news reporters covering 84 markets. Egg reports are made on frozen and dried eggs. These reports include data on supply and demand, movement, cold storage stocks, trading activity, price activities, and quality ranges. Prices and statistics are collected and disseminated nationwide on broilers/fryers, fowl, roasters, turkeys, ducks, and other miscellaneous poultry and rabbits.

Information on eggs and poultry is gathered from producers, major consuming centers, shipping points, and other sources. The information is available via the Web, fax, printed reports, trade publications, and the media. For poultry and egg news, see http://www.ams.usda.gov.

The Urner Barry Price Quotation

The Urner Barry Price Quotation provides a market-based wholesale price for shell eggs. Since 1858, Urner Barry has served the food industry by reporting and establishing price quotations for the poultry, egg, turkey, chicken, dairy, and beef markets. The company publishes several newsletters and directories. For more information, see http://www.urnerbarry.com.

MARKETING EGGS

In many respects, the poultry industry consists of two distinctly different industries: the egg industry and the poultry meat industry. Grading, pricing, and marketing structures are unique for each segment (see Boxes 17.1 to 17.5).

Problems Involved in the Marketing of Eggs

Eggs, like most agricultural commodities, are highly perishable. The shell is largely porous, thereby allowing gases, bacteria, and moisture to pass into and out of the egg. However, refrigeration minimizes these exchanges and bacterial growth and, consequently, prolongs the shelf life. Many of the consumer markets for eggs are often long distances from the areas of production. This necessitates refrigeration at all steps in the marketing and transportation phases of production.

A second problem incurred in the marketing of eggs involves shell breakage. Thus, packaging plays an extremely important role in egg marketing. Eggs that are cracked in processing can be salvaged through egg-breaking plants.

Finally, the greatest hurdle faced by the egg producer is lack of consumer knowledge. The egg is a highly nutritious food that is extremely versatile in the kitchen. It can be prepared as a main course or used to improve the cooking and organoleptic qualities of other foods.

Grading Eggs for Quality

If the egg industry is to remain viable and expand, only top-quality eggs can be marketed. A consumer is not likely to forget a bad experience of a rotten egg or an egg with blood or meat spots. Eggs are judged for quality through three sets of criteria: (1) external appearance, (2) candling, and (3) samples of eggs broken out to judge internal characteristics. Within the USDA, the Agricultural Marketing Service is the agency responsible for egg grading and standardization. Grading evaluates shell eggs (40%) on uniform official USDA quality standards. The USDA system for grading eggs is presented in Boxes 17.1 and 17.2.

The European Union has different rules on eggs. For instance, eggs marketed as Class A must be:

- Clean
- Normal and undamaged shell
- Not washed or cleaned prior to and after grading

External Appearance

The initial evaluation of eggs begins with the examination of the external appearance. The size and shape of the egg, as well as color and texture, are extremely important. The shells must be clean and free from cracks. Consumers show a preference for the color of eggshell. Consumers in the Unites States usually choose white shells, except in some areas (e.g., Boston) where they prefer brown shells.

BOX 17.1 U.S. Standards and Grades for Shell Eggs

QUALITY OF SHELL EGGS

AA Quality

The shell must be clean, unbroken, and practically normal. The air cell must not exceed 1/8 inch in depth, may show unlimited movement, and may be free or bubbly. The white must be clear and firm so that the yolk is only slightly defined when the egg is twirled before the candling light. The yolk must be practically free from apparent defects.

A Quality

The shell must be clean, unbroken, and practically normal. The air cell must not exceed 3/16 inch in depth, may show unlimited movement, and may be free or bubbly. The white must be clear and at least reasonably firm so that the yolk outline is only fairly well defined when the egg is twirled before the candling light. The yolk must be practically free from apparent defects.

B Quality

The shell must be unbroken, may be abnormal, and may have slightly stained areas. Moderately stained areas are permitted if they do not cover more than 1/32 of the shell surface if localized, or 1/16 of the shell surface if scattered. Eggs having shells with prominent stains or adhering dirt are not permitted. The air cell may be over 3/16 inch in depth, may show unlimited movement, and may be free or bubbly. The white may be weak and watery so that the yolk outline is plainly visible when the egg is twirled before the candling light. The yolk may appear dark, enlarged, and flattened, and may show clearly visible germ development but no blood due to such development. It may show other serious defects that do not render the egg inedible. Small blood spots or meat spots (aggregating not more than 1/8 inch in diameter) may be present.

(continued)

BOX 17.1   Continued

TERMS DESCRIBING THE EGGSHELL

a. Dirty. An individual egg that has an unbroken shell with adhering dirt or foreign material, prominent stains, or moderate stains covering more than 1/32 of the shell surface if localized, or 1/16 of the shell surface if scattered.

b. Check. An individual egg that has a broken shell or crack in the shell but with its shell membranes intact and without leaking contents. A "check" is considered to be lower in quality than a "dirty."

TERMS DESCRIBING THE EGG WHITE

a. Clear. A white that is free from discolorations or from any foreign bodies floating in it. (Prominent chalazas should not be confused with foreign bodies such as spots or blood clots.)

b. Firm (AA quality). A white that is sufficiently thick or viscous to prevent the yolk outline from being more than slightly defined or indistinctly indicated when the egg is twirled. With respect to a broken-out egg, a firm white has a Haugh unit value of 72 or higher when measured at a temperature between 45° and 60°F.

c. Reasonably firm (A quality). A white that is somewhat less thick or viscous than a firm white. A reasonably firm white permits the yolk to approach the shell more closely, which results in a fairly well-defined yolk outline when the egg is twirled. With respect to a broken-out egg, a reasonably firm white has a Haugh unit value of 60 to 72 when measured at a temperature between 45° and 60°F.

d. Weak and watery (B quality). A white that is weak, thin, and generally lacking in viscosity. A weak and watery white permits the yolk to approach the shell closely, thus causing the yolk outline to appear plainly visible and dark when the egg is twirled. With respect to a broken-out egg, a weak and watery white has a Haugh unit value lower than 60 when measured at a temperature between 45° and 60°F.

e. Blood spots or meat spots. Small blood spots or meat spots (aggregating not more than 1/8 inch in diameter) may be classified as B quality. If larger, or showing diffusion of blood into the white surrounding a blood spot, the egg shall be classified as Loss. Blood spots shall not be due to germ development. They may be on the yolk or in the white. Meat spots may be blood spots that have lost their characteristic red color or tissue from the reproductive organs.

f. Bloody white. An egg that has blood diffused through the white. Eggs with bloody whites are classed as Loss. Eggs with blood spots that show a slight diffusion into the white around the localized spot are not to be classed as bloody whites.

TERMS DESCRIBING THE YOLK

a. Outline slightly defined (AA quality). A yolk outline that is indistinctly indicated and appears to blend into the surrounding white as the egg is twirled.

b. Outline fairly well defined (A quality). A yolk outline that is discernible but not clearly outlined as the egg is twirled.

c. Outline plainly visible (B quality). A yolk outline that is clearly visible as a dark shadow when the egg is twirled.

d. Enlarged and flattened (B quality). A yolk in which the yolk membranes and tissues have weakened and/or moisture has been absorbed from the white to such an extent that the yolk appears definitely enlarged and flat.

e. Practically free from defects (AA or A quality). A yolk that shows no germ development but may show other very slight defects on its surface.

f. Serious defects (B quality). A yolk that shows well-developed spots or areas and other serious defects, such as olive yolks, which do not render the egg inedible.

g. Clearly visible germ development (B quality). A development of the germ spot on the yolk of a fertile egg that has progressed to a point where it is plainly visible as a definite circular area or spot with no blood in evidence.

h. Blood due to germ development. Blood caused by development of the germ in a fertile egg to the point where it is visible as definite lines or as a blood ring. Such an egg is classified as inedible.

GENERAL TERMS

a. Loss. An egg that is inedible, cooked, frozen, contaminated, or containing bloody whites, large blood spots, large unsightly meat spots, or other foreign material.

b. Inedible eggs. Eggs of the following descriptions are classed as inedible: black rots, yellow rots, white rots, mixed rots (addled eggs), sour eggs, eggs with green whites, eggs with stuck yolks, moldy eggs, musty eggs, eggs showing blood rings, eggs containing embryo chicks (at or beyond the blood ring state), and any eggs that are adulterated as such term is defined pursuant to the Federal Food, Drug, and Cosmetic Act.

c. Leaker. An individual egg that has a crack or break in the shell and shell membranes to the extent that the egg contents are exuding or free to exude through the shell.

From USDA Publication AMS 56.200.

BOX 17.2   U.S. Standards and Grade Classes for Shell Eggs

a. **U.S. Grade AA**

1. U.S. Consumer Grade AA (at origin) shall consist of eggs that are at least 87% AA quality. The maximum tolerance of 13% which may be below AA quality may consist of A or B quality in any combination, except that within the tolerance for B quality not more than 1% may be B quality due to air cells over 3/8 inch, blood spots (aggregating not more than 1/8 inch in diameter), or serious yolk defects. Not more than 5% (7% for Jumbo size) Checks are permitted and not more than 0.50% Leakers, Dirties, or Loss (due to meat or blood spots) in any combination, except that such Loss may not exceed 0.30%. Other types of Loss are not permitted.

2. U.S. Consumer Grade AA (destination) shall consist of eggs that are at least 72% AA quality. The remaining tolerance of 28% shall consist of at least 10% A quality and the remainder shall be B quality, except that within the tolerance for B quality not more than 1% may be B quality due to air cells over 3/8 inch, blood spots (aggregating not more than 1/8 inch in diameter), or serious yolk defects. Not more than 7% (9% for Jumbo size) Checks are permitted and not more than 1% Leakers, Dirties, or Loss (due to meat or blood spots) in any combination, except that such Loss may not exceed 0.30%. Other types of Loss are not permitted.

b. **U.S. Grade A**

1. U.S. Consumer Grade A (at origin) shall consist of eggs that are at least 87% A quality or better. Within the maximum tolerance of 13% which may be below A quality, not more than 1% may be B quality due to air cells over 3/8 inch, blood spots (aggregating not more than 1/8 inch in diameter), or serious yolk defects. Not more than 5% (7% for Jumbo size) Checks are permitted and not more than 0.50% Leakers, Dirties, or Loss (due to meat or blood spots) in any combination, except that such Loss may not exceed 0.30%. Other types of Loss are not permitted.

2. U.S. Consumer Grade A (destination) shall consist of eggs that are at least 82% A quality or better. Within the maximum tolerance of 18% which may be below A quality, not more than 1% may be B quality due to air cells over 3/8 inch, blood spots (aggregating not more than 1/8 inch in diameter), or serious yolk defects. Not more than 7% (9% for Jumbo size) Checks are permitted and not more than 1% Leakers, Dirties, or Loss (due to meat or blood spots) in any combination, except that such Loss may not exceed 0.30%. Other types of Loss are not permitted.

c. **U.S. Grade B**

1. U.S. Consumer Grade B (at origin) shall consist of eggs that are at least 90% B quality or better, not more than 10% may be Checks and not more than 0.50% Leakers, Dirties, or Loss (due to meat or blood spots) in any combination, except that such Loss may not exceed 0.30%. Other types of Loss are not permitted.

2. U.S. Consumer Grade B (destination) shall consist of eggs that are at least 90% B quality or better, not more than 10% may be Checks, and not more than 1% Leakers, Dirties, or Loss (due to meat or blood spots) in any combination, except that such Loss may not exceed 0.30%. Other types of Loss are not permitted.

d. **Additional tolerances**

1. In lots of two or more cases:

   i. For Grade AA—No individual case may exceed 10% less AA quality eggs than the minimum permitted for the lot average.

   ii. For Grade A—No individual case may exceed 10% less A quality eggs than the minimum permitted for the lot average.

   iii. For Grade B—No individual case may exceed 10% less B quality eggs than the minimum permitted for the lot average.

2. For Grades AA, A, and B, no lot shall be rejected or downgraded due to the quality of a single egg except for Loss other than blood or meat spots.

From USDA Publication AMS 56.200.

Candling

When eggs are graded for quality, they are routinely held against a light source (candling) so that the grader can determine certain quality characteristics of the internal parts of the egg without breaking the shell. When an egg is candled it is possible to evaluate:

- The shell for cracks and thickness.
- Size of the air cell.
- Firmness of the white—high-quality eggs have firm albumen.
- Detection of unsightly meat and blood spots.

BOX 17.3 U.S. Standards for Quality of Individual Shell Eggs

Quality Factor	AA Quality	A Quality	B Quality
Shell	Clean	Clean	Clean to slightly stained[1]
	Unbroken	Unbroken	Unbroken
	Practically normal	Practically normal	Abnormal
Air Cell	1/8 inch or less in depth	3/16 inch or less in depth	Over 3/16 inch in depth
	Unlimited movement	Unlimited movement and	Unlimited movement
	and free or bubbly	free or bubbly	and free or bubbly
White	Clear	Clear	Weak and watery
	Firm	Reasonably firm	Small blood and meat spots present[2]
Yolk	Outline slightly defined	Outline fairly well defined	Outline plainly visible
	Practically free from defects	Practically free from defects	Enlarged and flattened
			Clearly visible germ development but not blood
			Other serious defects

For eggs with dirty or broken shells, the standards of quality provide two additional qualities. They are:

Dirty	Check
Unbroken, adhering dirt or foreign material, prominent stains, moderate stained areas in excess of B quality.	Broken or cracked shell but membranes intact, not leaking.[3]

[1] *Moderately stained areas permitted (1/32 of surface if localized, or 1/16 if scattered).*

[2] *If they are small (aggregating not more than 1/8 inch in diameter).*

[3] *Leaker has broken or cracked shell membranes, and contents leaking or free to leak.*

From USDA Publication AMS 56.200.

Breakouts

Periodically, samples from a large batch of eggs are broken-out in order that the contents can be closely evaluated. Color of the white and yolk, odor, and general appearance can be readily evaluated.

The *yolk* of a high-quality egg must be round and firm and the *white* must be firm with a rather clear demarcation between the thin and thick albumen. *Egg albumen quality* is routinely measured in Haugh units.

Haugh units are determined by a micrometer measure of albumen height. The Haugh units are derived from a chart listing albumen heights, egg weights, and Haugh units. The pH (hydrogen concentration) of egg white is another means to estimate egg quality. The pH of a freshly laid egg is about 7.6 to 8.2. As the egg ages, carbon dioxide is lost and the pH increases up to 9.5.

Shell quality is evaluated by measuring the following:

- Breaking strength
- Thickness
- Specific gravity

Proper Care of Eggs at the Farm Level

The processing of quality eggs begins at the farm level. Producers must carefully select for egg production and egg quality in their breeding programs. Once a good flock of layers has been established, good feeding and management of birds is required to maximize the genetic potential of the birds.

Frequent or continuous gathering of eggs helps maintain quality because if eggs are exposed to ambient temperatures for extended periods, quality declines rapidly. Once collected, the eggs should be cooled rapidly to prevent spoilage. Washing is an extremely important part of producing quality eggs.

GRADING AND SIZING OF EGGS

The Agricultural Marketing Service is responsible for egg grading and standardization. The USDA system for grading and sizing of eggs is presented in Boxes 17.1 to 17.5. The grading of eggs involves their sorting according to weight and quality.

BOX 17.4 Tolerance for Individual Case Within a Lot

U.S. Consumer Grade	Case Quality	Origin (%)	Destination (%)
Grade AA	AA (min)	77	62
	A or B	13	28
	Check (max)	10	10
Grade A	A (min)	77	72
	B	13	18
	Check (max)	10	10
Grade B	B (min)	80	80
	Check (max)	20	20

From USDA Publication AMS 56.200.

BOX 17.5 U.S. Weight Classes for Consumer Grades for Shell Eggs

Size or Weight Class	Minimum Net Weight per Dozen (ounces)	Minimum Net Weight 30 per Dozen (pounds)	Minimum Net Weight for Individual Eggs at Rate per Dozen (ounces)
Jumbo	30	56	29
Extra-large	27	50	26
Large	24	45	23
Medium	21	39	20
Small	18	34	17
Peewee	15	28	—

The weight classes for U.S. Consumer Grades for Shell Eggs shall be as indicated in Table I of this section and shall apply to all consumer grades. From USDA Publication AMS 56.200.

Egg Quality

There are three grades for eggs that cover the range of eggs: AA, A, and B. The characteristics of AA, A, and B eggs together with dirty and check are shown in Boxes 17.1 to 17.4. These standards are based on:

- Interior quality factors of the white and yolk
- The size of the air cell
- Shell quality factors—cleanliness and soundness of the shell

Size

Eggs are also classified as jumbo, extra-large, large, medium, small, and peewee according to weight (or size) (see Box 17.5).

Products of the Egg-Breaking Industry

The egg-breaking segment of the egg industry involves a considerable portion of the total eggs marketed. This segment of the egg industry was originally developed as an outlet for soiled, cracked, and abnormal eggs unsuitable for marketing as shell eggs. Today, because of the high demand for broken eggs, many quality eggs are being used, especially peewee, small, and medium eggs. There are numerous advantages of processed eggs over shell eggs including:

1. Less storage space is involved.
2. Quality of frozen and dried eggs is preserved longer.
3. Packaging and processing is facilitated.
4. Processed eggs may be cheaper than shell eggs.
5. Less labor is involved in the egg-breaking industry.
6. One can choose specific parts of the egg for a particular need.

Egg-breaking operations are automated. After the eggs are mechanically broken, they are checked for abnormalities and odor. The broken eggs can then be processed whole or separated into yolks and whites. Whole eggs are mixed and strained before packaging. Yolks are also mixed and strained. Additionally, in frozen yolks, salt or sugar may be added to improve the rubbery consistency of the yolk after freezing and defrosting. Glycerine, molasses, or honey may also be used. Whites are strained to remove the chalazae, meat spots, blood spots, and broken shells, and may be passed through a chopper to homogenize the product. Of the three processed forms—frozen, liquid, and dried eggs— dried eggs have the longest shelf life and require the lowest expense for storage, since refrigeration is not needed. However, the vacuum spray processing of drying incurs additional costs.

Table 14.9 shows the quantities and kinds of egg products produced by the egg-breaking industry.

Pricing Eggs

The egg-pricing system in use today is still operating largely on the basis of wholesale price. Wholesalers, who formerly handled most of the eggs, have been largely displaced by packers. There is no satisfactory price barometer for eggs.

Packaging Material

The USDA has approved recommendations for new standard fiber cases. These include:

- Bursting strength
- Type of fiber
- Thickness of fiberboard
- Partitions
- Sealing

MARKETING POULTRY MEAT

The poultry meat industry encompasses primarily chickens, turkeys, and ducks. Geese, emus, ostriches, game birds, and pigeons are also raised for meat but might be considered as specialty or niche items.

Food Safety

Poultry and eggs are inspected for safety. In the United States, inspection occurs at all poultry-processing plants. The USDA determines that the poultry is fit for human consumption, ensures that the processing is done in a sanitary manner, prevents adulteration of poultry and poultry products, and requires that poultry and poultry products are properly labeled (for details see Chapter 12).

Market Classes and Grades of Poultry

The USDA has established specifications for different kinds, classes, and grades of poultry (see Boxes 17.6 and 17.7). Kind refers to species of poultry, such as chickens, turkeys, and ducks. Class refers to groups of poultry that are essentially of the same physical characteristics, such as broilers or hens. The classes do not necessarily reflect what is occurring in the poultry industry.

The factors determining the grade of carcass, or ready-to-cook poultry parts therefrom, are conformation, fleshing, fat covering, pinfeathers, exposed flesh, discoloration, disjointed bones, broken bones, missing parts, and freezing defects.

The USDA Agricultural Marketing Service is responsible for poultry grading, evaluating poultry (60% of chicken, 80% of turkey) on uniform official USDA quality standards. These quality grades are A, B, and C. There are also U.S. Procurement Grades for institutional use; these grades are U.S. Procurement Grade 1 and U.S. Procurement Grade 2. Procurement grades emphasize meat yield more than appearance (also see Box 17.8).

Marketing Chicken Meat Products

Most of the poultry meat marketed comes from the slaughter of broilers, although substantial quantities of meat from turkeys are marketed.

Marketing Broilers

In the United States, a major issue in marketing broilers is transportation from the area of production to the area of consumption. The geographical locations having the greatest shortages of broiler chickens are the New York tristate metropolitan area, California, the Upper Midwest, and New England. The locations having the greatest surpluses of broilers are the Southeast and the Delmarva peninsula. Broilers must be transported considerable distances to their targeted market.

Broilers are slaughtered at an average of 6.5 weeks old and at an average weight of 5 lb (see Chapter 15 for details). During slaughter, there are losses including:

- Blood—4% of liveweight
- Feathers—5% of liveweight

These, plus the losses of weight from the removal of offal, do not represent a complete economic loss. They can be sold for by-products (see Chapter 15 for details). Of the ready-to-cook product, the breakdown is as follows: breast, 26%; thigh, 17%; leg, 16%; back, 17%; wings, 12%; neck, 6%; and giblets, 6%.

The broiler can then be sold cut-up or whole and subsequently fried, roasted, barbecued, and so on. It can be deboned and used in such processed forms as sandwich rolls, frozen pies, and cold cuts. Fast-food chains (e.g., McDonald's, Burger King, Wendy's, Kentucky Fried Chicken) have a very

---

BOX 17.6   USDA Classification Of Chickens[1]

  a. Rock Cornish game hen or Cornish game hen. A Rock Cornish game hen or Cornish game hen is a young immature chicken (usually 5 to 6 weeks of age), weighing not more than 2 pounds ready-to-cook weight, which was prepared from a Cornish chicken or the progeny of a Cornish chicken crossed with another breed of chicken.

  b. Rock Cornish fryer, roaster, or hen. A Rock Cornish fryer, roaster, or hen is the progeny of a cross between a purebred Cornish and a purebred Rock chicken, without regard to the weight of the carcass involved; however, the term *fryer, roaster, or hen* shall apply only if the carcasses are from birds with ages and characteristics that qualify them for such designation under paragraphs (c) and (d) of this section.

  c. Broiler or fryer. A broiler or fryer is a young chicken (usually under 13 weeks of age), of either sex, that is tender-meated with soft, pliable, smooth-textured skin and flexible breastbone cartilage.

  d. Roaster or roasting chicken. A bird of this class is a young chicken (usually 3 to 5 months of age), of either sex, that is tender-meated with soft, pliable, smooth-textured skin and breastbone cartilage that may be somewhat less flexible than that of a broiler or fryer.

  e. Capon. A capon is a surgically unsexed male chicken (usually under 8 months of age) that is tender-meated with soft, pliable, smooth-textured skin.

  f. Hen, fowl, or baking or stewing chicken. A bird of this class is a mature female chicken (usually more than 10 months of age) with meat less tender than that of a roaster or roasting chicken and nonflexible breastbone tip.

  g. Cock or rooster. A cock or rooster is a mature male chicken with coarse skin, toughened and darkened meat, and hardened breastbone tip.

---

[1]Note that there are substantial differences between the USDA classification and what is occurring in the poultry industry—for instance:

  • Broiler chickens are marketed at about 6.5 weeks old at 5 lb.

  • Rock Cornish are less than 5 weeks old.

  • The term fryer is rarely used.

  • Roasters are much younger than 3 to 5 months.

  • Few capons, hens, or roosters are marketed to the general public.

From http://www.ams.usda.gov/poultry/standards/AMS-PYST-2002.htm#70-201.

BOX 17.7   USDA Classification of Turkey[1]

The following are the various classes of turkeys:

  a. Fryer-roaster turkey. A fryer-roaster turkey is a young immature turkey (usually under 16 weeks of age), of either sex, that is tender-meated with soft, pliable, smooth-textured skin, and flexible breastbone cartilage.

  b. Young turkey. A young turkey is a turkey (usually under 8 months of age) that is tender-meated with soft, pliable smooth-textured skin, and breastbone cartilage that is somewhat less flexible than in a fryer-roaster turkey. Sex designation is optional.

  c. Yearling turkey. A yearling turkey is a fully matured turkey (usually under 15 months of age) that is reasonably tender-meated and with reasonably smooth-textured skin. Sex designation is optional.

  d. Mature turkey or old turkey (hen or tom). A mature or old turkey is an old turkey of either sex (usually in excess of 15 months of age), with coarse skin and toughened flesh.

  e. For labeling purposes, the designation of sex within the class name is optional, and the two classes of young turkeys may be grouped and designated as "young turkeys."

Note that there are substantial differences between the USDA classification and what is occurring in the poultry industry—fryer-roasters predominating. From http://www.ams.usda.gov/poultry/standards/AMS-PYST-2002.htm#70-201.

---

significant impact on the marketing of broiler chickens. According to the USDA Agricultural Marketing Service, young chickens are classified by weight with different markets.

- < 4.2 lb food service/fast food (predominantly 3.6–4.00 lb)
- 4.21–5.25 lb retailer (dressed)
- >5.26 lb deboning

In 2002, these represented the following percentages of chickens slaughtered, respectively:

- < 4.2 lb, 29%
- 4.21–5.25 lb, 29%
- > 5.26, 42%

Larger broiler chickens for deboning are somewhat over 50% of chicken meat.

BOX 17.8 USDA Standards for Quality of Ready-to-Cook Poultry

A QUALITY

a. Conformation. The carcass or part is free of deformities that detract from its appearance or that affect the normal distribution of flesh. Slight deformities, such as slightly curved or dented breastbones and slightly curved backs, may be present.

b. Fleshing. The carcass has a well-developed covering of flesh considering the kind, class, and part.
   1. The breast is moderately long and deep, and has sufficient flesh to give it a rounded appearance with the flesh carrying well up to the crest of the breastbone along its entire length.
   2. The leg is well fleshed and moderately thick and wide at the knee and hip joint area, and has a well-rounded, plump appearance with the flesh carrying well down toward the hock and upward to the hip joint area.
   3. The drumstick is well fleshed and moderately thick and wide at the knee joint, and has a well-rounded, plump appearance with the flesh carrying well down toward the hock.
   4. The thigh is well to moderately fleshed.
   5. The wing is well to moderately fleshed.

c. Fat covering. The carcass or part, considering the kind, class, and part, has a well-developed layer of fat in the skin. The fat is well-distributed so that there is a noticeable amount of fat in the skin in the areas between the heavy feather tracts.

d. Defeathering. The carcass or part shall have a clean appearance, especially on the breast and legs, and shall be free of protruding feathers and hairs as defined in 7 CFR §70.1. A carcass or part shall be considered free from protruding feathers when it complies with the tolerances specified in the following table:

   1. The carcass may have lightly shaded areas of discoloration, provided the aggregate area of all discolorations does not exceed an area equivalent to the area of a circle of the diameter specified in the following table. Evidence of incomplete bleeding, such as more than an occasional slightly reddened feather follicle, is not permitted.

Carcass Weight		Maximum Aggregate Area Permitted	
Minimum	Maximum	Breast and Legs	Elsewhere
None	2 lb	3/4 in.	1 1/4 in.
Over 2 lb	6 lb	1 in.	2 in.
Over 6 lb	16 lb	1 1/2 in.	2 1/2 in.
Over 16 lb	None	2 in.	3 in.

   2. The carcass may have moderately shaded areas of discoloration and discolorations due to flesh bruising, provided:
      i. They are not on the breast or legs, except for the area adjacent to the hock joint;
      ii. They are free of clots; and
      iii. They may not exceed an aggregate area equivalent to the area of a circle of the diameter specified in the following table:

Carcass Weight		Maximum Aggregate Area Permitted	
Minimum	Maximum	Hock Area of Legs	Elsewhere
None	2 lb	1/4 in.	5/8 in.
Over 2 lb	6 lb	1/2 in.	1 in.
Over 6 lb	16 lb	3/4 in.	1 1/4 in.
Over 16 lb	None	1 in.	1 1/2 in.

   3. Parts, other than large carcass parts, may have lightly shaded areas of discoloration, provided the aggregate area of all discolorations does not exceed an area equivalent to the area of a circle of the diameter specified in the following table. Evidence of incomplete bleeding, such as more than an occasional slightly reddened feather follicle, is not permitted.

Carcass Weight		Maximum Aggregate Area Permitted
Minimum	Maximum	Parts
None	2 lb	1/2 in.
Over 2 lb	6 lb	3/4 in.
Over 6 lb	16 lb	1 in.
Over 16 lb	None	1 1/4 in.

(continued)

BOX 17.8 Continued

4. Parts, other than large carcass parts, may have moderately shaded areas of discoloration and discolorations due to flesh bruising, provided:
   i. They are not on the breast or legs, except for the area adjacent to the hock joint;
   ii. They are free of clots; and
   iii. They may not exceed an aggregate area equivalent to the area of a circle of the diameter specified in the following table:

Carcass Weight		Maximum Aggregate Area Permitted
Minimum	Maximum	Parts
None	2 lb	1/4 in.
Over 2 lb	6 lb	3/8 in.
Over 6 lb	16 lb	1/2 in.
Over 16 lb	None	5/8 in.

5. Large carcass parts, specifically halves, front halves, or rear halves, may have lightly shaded areas of discoloration, provided the aggregate area of all discolorations does not exceed an area equivalent to the area of a circle of the diameter specified in the following table:

Carcass Weight		Maximum Aggregate Area Permitted	
Minimum	Maximum	Breast and Legs	Elsewhere
None	2 lb	1/2 in.	1 in.
Over 2 lb	6 lb	3/4 in.	1 1/2 in.
Over 6 lb	16 lb	1 in.	2 in.
Over 16 lb	None	1 1/4 in.	2 1/2 in.

6. Large carcass parts, specifically halves, front halves, or rear halves, may have moderately shaded areas of discoloration and discolorations due to flesh bruising, provided:
   i. They are not on the breast or legs, except for the area adjacent to the hock joint;
   ii. They are free of clots; and
   iii. They may not exceed an aggregate area equivalent to the area of a circle of the diameter specified in the table above.

B QUALITY

a. **Conformation.** The carcass or part may have moderate deformities, such as a dented, curved, or crooked breast, crooked back, or misshapen legs or wings, which do not materially affect the distribution of flesh or the appearance of the carcass or part.

b. **Fleshing.** The carcass has a moderate covering of flesh considering the kind, class, and part.
   1. The breast has a substantial covering of flesh with the flesh carrying up to the crest of the breastbone sufficiently to prevent a thin appearance.
   2. The leg is fairly thick and wide at the knee and hip joint area, and has sufficient flesh to prevent a thin appearance.
   3. The drumstick has a sufficient amount of flesh to prevent a thin appearance with the flesh carrying fairly well down toward the hock.
   4. The thigh has a sufficient amount of flesh to prevent a thin appearance.
   5. The wing has a sufficient amount of flesh to prevent a thin appearance.

c. **Fat covering.** The carcass or part has sufficient fat in the skin to prevent a distinct appearance of the flesh through the skin, especially on the breast and legs.

d. **Defeathering.** The carcass or part may have a few scattered protruding feathers and hairs. A carcass or part shall be considered to have not more than a few scattered protruding feathers when it complies with the tolerances.

e. **Exposed flesh.** A carcass may have exposed flesh provided that no part on the carcass has more than one-third of the flesh exposed. A part may have no more than one-third of the flesh normally covered by skin exposed.

f. **Disjointed and broken bones, missing parts, and trimming.**
   1. Parts may be disjointed, but are free of broken bones. The carcass may have two disjointed bones, or one disjointed bone and one nonprotruding broken bone.
   2. Parts of the wing beyond the second joint may be removed at a joint. The tail may be removed at the base.

(continued)

PRICING BROILERS. For broilers and turkeys, there is a great deal of formula pricing. Originally, the typical broiler formula was based on live price divided by 73% (the approximate yield of ready-to-cook from liveweight) plus 3 to 7 cents to cover processing costs. However, as more and more broilers were produced under contract, fewer and fewer live broilers changed hands. As a result, the trend is away from farm-based calculations to ready-to-cook prices.

Marketing Spent Hens

Most of the mature chickens slaughtered are hens that have ceased egg production. They are known as spent hens. These can be further processed to soup, chicken salad, or hot dogs. The low amount of meat to be harvested, together with problems of bone fracturing and hence fragments (see Chapter 11), is reducing the demand for spent hens. Moreover, the consumer expects pieces of quality chicken breast in soups. Spent hens are increasingly rendered or composted.

Marketing Turkeys

Over 80% of the turkey meat produced today is cut-up or used for further processed products. Whole-bird processed turkeys, such as self-basting turkeys, comprise less than 20% of this total.

Turkey operations have become integrated; coordinated by the processor. Additionally, cooperatives continue to play a significant role in the marketing of turkeys in some parts of the United States.

Pricing Turkeys

The Urner Barry daily report of wholesale turkey prices can be used as a guide in price negotiations.

Marketing Ducks and Geese

While duck and goose consumption is low in the United States, increases are possible. White Pekin-type meat ducklings are ready for market between 7 and 8 weeks of age (see Chapter 19 for details).

CURRENT AND EMERGING ISSUES IN MARKETING (CONSUMER TRENDS)

There are many changes in the poultry meat and egg products being consumed. There have been tremendous changes in the products available. In addition, consumers are purchasing poultry meat at restaurants. Chicken can also be purchased as kabobs on a skewer ready to barbecue at home. Turkey has moved from seasonal (Thanksgiving, Christmas) to year-round consumption with turkey ham, sausage, bacon, and other products. Lifestyle changes such as people living alone or as couples without children have increased small packaging (e.g., a single turkey drumstick).

Designer or Fortified Products Including Eggs

It is possible to improve the nutritional value of eggs. Eggs are already being marketed with increased vitamin E and/or omega-3-polyunsaturated fatty acids. High vitamin E eggs have been approved by the Food and Drug Administration in the United States and are available to consumers in some markets (e.g., California). Eggs fortified with omega fatty acids (normally found in fish oils) and lutein are also available. The increase in content of specific nutrients is achieved by changing the feed the hen receives. This is a niche market with a premium to the producer. Enhancing the egg content of other nutrients is possible and likely. There are also other exciting potential developments:

- Researchers have shown that it is possible to reduce the cholesterol content in eggs. This is not done commercially because the drugs have not received FDA approval.
- It is likely that the egg contains specific nutrients that have health-promoting effects on people in the same way that milk does. As yet, there is very little information available on these zoonutrients. In the long term with the advances in human genomics, we may see a "personalized" egg meeting the nutritional needs of an individual.

All Natural

Cal-Maine Foods markets "all natural" eggs under the Egg Land, Best Egg (EB) label. The feed for the producing hens contains no animal products. The eggs have reduced saturated fatty acid content.

Free-Range and Organic Poultry Meat and Eggs

Some consumers prefer or, in some cases, only purchase free-range or organic poultry and eggs. These consumers are often more affluent with a high disposable income. Small breeders supply specialized markets.

As the name implies, free-range production is a system where the birds have access to the open at least during the day with housing for roosting and protection from weather. Free-range hens produce less eggs than those in battery cages. The reasons for the purchase of free-range poultry and eggs include:

- Concern for the welfare of the birds.
- Belief in the superior eating quality of free-range poultry and eggs.

In the United States, Cal-Maine Foods markets free-range eggs under the brand Egg Land's Farmhouse eggs. In the United Kingdom, free-range eggs have increased from about 2% to over 15%.

Organic production most frequently incorporates free-range conditions together with feeding organically raised grain. Feeding wheat can induce a biotin deficiency. The reasons for consumer preference of organic poultry and eggs include (in addition to the rationales for free-range):

- Concern for the environment.
- Belief that organic food is more healthy.

The USDA has established rules for organic production of poultry and eggs. The European Union has regulations on both free-range and organic poultry. These include breed, age going to market, feed, raising conditions in fields/paddocks, and welfare and veterinary care. In the UK and other countries, there have been marked increases in free-range and organic chickens and turkeys.

The French Ministry of Agriculture has a "Label Rouge" for chickens. This has captured >16% of the market for chicken. "Label Rouge" is a trademark of the French Ministry of Agriculture. There are requirements including the following:

- Access to open air
- Specific number per unit area
- Specific number per site
- Only certain genetics and feed allowed
- Birds must be slaughtered no earlier than 81 days old
- Farm to slaughter < 2 hr

This is the same as the European Union's specifics for free-range. Another example of the free-range chicken is the "Poulet d'Or" (gold chicken) sold by the supermarket chain, Waitrose. They use a relatively slow-growing French breed marketed at a minimum of 81 days. The birds are raised in fields with a wheat- and other cereal-based diet that is free of antibiotics and genetically modified grain. In the case of organic poultry meat, the feed must be organically grown and the birds treated in a manner consistent with regulations.

Similarly in Europe, organic and free-range eggs are gaining in popularity with more affluent consumers. Again, specific genetics is beginning to be used such as the Columbian Blacktail (a cross between Rhode Island Reds and other breeds) by the British supermarket chain, Waitrose. Irrespective of breed, the birds have movable housing on fields. They can get some nutrition from the land, but also receive a high-quality diet.

Pasture-Raised Poultry/Day-Range Poultry

Both chickens and turkeys can be raised on pasture with supplementary feed being provided. This can involve:

- Pastured pens. These are floorless with an area for shelter and can be moved regularly, even daily.
- Net or day-range.
- Yarding (allowed out during the day).
- Free-range with portable housing.

The pluses of these alternative poultry systems are the premium prices the meat may attract and increased profit if the producer "home" kills and processors. The negatives include the reduced growth together with problems with marketing, disease, and predators (foxes, hawks, etc.). Further information is available through the USDA Sustainable Agriculture Research and Education Program.

Nongenetically Modified (GM)/ Nongenetically Engineered (GE) Poultry

In both North and South America, many poultry are fed a diet of corn and soybeans. A significant proportion of the corn (>30%) and soybeans (60%) is transgenic, that is, it contains genes that have been introduced into the plant by the techniques of molecular biology or genetic engineering (GE). In the United States, approval of foods and drugs is based on the efficacy (Does it work?) and safety (Is it safe?) and not the process (How was it made?).

There is little doubt that those GE or genetically modified (GM) plants that have received approval have the attributes described (efficacy) and are safe for livestock and people (safety). However, there is a separate issue: Does the public have a right to know how their food is being produced? Should food from GM/GE crops be labeled? This discussion, particularly in Europe, has been extended to: Should poultry and other livestock products be labeled? This is despite the fact that it is not possible to distinguish poultry meat or eggs from birds fed GM or non-GM crops. Organic poultry must be fed non-GM crops.

TRENDS AHEAD IN MARKETING

- Vertical coordination dominates over 90% of egg, broiler chicken, and turkey production. Closer control is likely with more vertically integrated production, particularly in table eggs and turkeys. The motivating forces are the advantages in efficiency, management, quality control, marketing, and distribution. With increasing mechanization and lower labor requirements, more production units will be owned by integrators, rather than controlled on a contract basis.
- There will be greater consolidation in the poultry industry.

Other trends include the following:

- Information and data management will be increasingly used to improve the efficiency and profitability of production.
- More specialized products will be marketed.
- Marketing directly to retailers will increase, with specific quality requirements enforced in contracts. It is very possible that there will be considerable "scan-based marketing" (where the processor stocks the poultry counter in the grocery store and is paid a percentage/fixed amount when the products pass through the scanner).
- Retailers and restaurants will make specific requirements on animal welfare and perhaps also food safety (e.g., use of antibiotics).
- More niche products (natural, antibiotic-free, free-range, organic, etc.) will be marketed, particularly in the European Union followed by Canada and the United States. These will attract premium prices.
- In North America, duck meat consumption will increase at the high end (quality) of the market.
- Governments will increasingly encourage exports where there is a competitive advantage.

 USEFUL WEBSITES

Poultry and Egg Statistics

USDA National Agricultural Statistics Service
Agricultural Statistics for 2002: http://www.usda.gov/nass/pubs/agr02/acro02.htm

Agricultural Statistics for 2003: http://www.usda.gov/nass/pubs/agr03/acro03.htm

USDA, Economic Research Service, USDA, Washington, DC: http://www.ers.usda.gov/Publications

World Poultry and Egg Exports—FAO Database

FAO Stat Agriculture Data, Agriculture and Food Trade, Crops and Livestock Primary and Processed: http://apps.fao.org/page/collections?subset=agriculture

USA Poultry and Egg Exports

http://www.usda.gov/nass/pubs

Poultry Inspection and HACCP

USDA's FSIS: http://www.fsis.usda.gov

USDA Publication AMS 56.200: http://www.ams.usda.gov/poultry/standards/AMSEGGST.html

FURTHER READING

Barbut, S. *Poultry Products Processing: An Industry Guide.* Boca Raton, FL: CRC Press, 2002.

Bell, D. D, and W. D. Weaver. *Commercial Chicken Meat and Egg Production,* 5th ed. Norwell, MA: Kluwer Academic Publishers, 2002.

Martinez, Steve W. *Vertical Coordination of Marketing Systems: Lessons from the Poultry, Egg, and Pork Industries.* U.S. Department of Agriculture, Agricultural Economic Report No. 807, 2002.

18

Poultry Industry and Business

Objectives

After studying this chapter, you should be able to:

1. Define *vertical integration*.
2. List facilities and services provided by the producer in marketing and production contracts.
3. List facilities and services provided by the bird owner in marketing and production contracts.
4. List the advantages and disadvantages of vertical coordination.
5. List the three major types of business organizations.
6. List job opportunities in the poultry industry.
7. List the top exporters and importers of poultry products.

VERTICAL COORDINATION OF THE POULTRY INDUSTRY

Much of the poultry production involves the contracting of certain services by the farmer with highly integrated poultry companies. Therefore, it is imperative that the producer becomes well informed relative to contracts. Vertical coordination occurs between the stages of the poultry industry (see Figures 18.1 and 18.2). The levels of control from the contractors or integrator vary, from the least to the most, in the following manner:

1. Open production (open or spot market) least
2. Market contracting (or market-specific contract) ⇓
3. Production contracting (or resource providing the contract) ⇓
4. Ownership (vertical integration) most

Open Production

Open production is also known as an open or spot market. In this, there is very limited coordination.

What limited coordination that exists is provided by cash or spot prices.

Market Contracting

Market contracting involves the grower contracting with the processor for number, time, and quality characteristics, together with pricing method. (The processor has a limited involvement in some of the decision making of the grower.)

Production Contracting

Production contracting involves the processor or contractor having a major role in producer/grower decision making. The processor has ownership of various production inputs (e.g., feed) and acts as the market for farm products (poultry or eggs).

VERTICAL INTEGRATION

Vertical integration is a process in which a single company owns two or more stages in the supply or value chain (e.g., production and processing) with common management control.

309

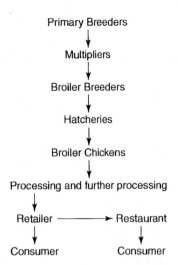

Figure 18.1 Vertical stages in broiler chicken industry.

Figure 18.2 Example of further processed chicken breast from Tyson, the largest producer of broiler chicken in the United States.

Vertical Coordination in the Broiler Industry

Vertical coordination has long dominated the broiler chicken industry (see Figure 18.1). Coordination was initially led by the feed industry but is now led by processors (e.g., Tyson Foods, Perdue Farms, etc.). There is a mixture of production contracts (~80%) and complete vertical integration (> 18%). Tyson Foods is the leading producer of broiler chickens and furthers processing (for example see Figure 18.2) in the United States. Contracts from Tyson Foods specify the use of company-supplied birds from Tyson's breeding company, Cobb-Vantress.

Production Contracts

Production contracts most frequently involve the processor providing:

- Baby chicks
- Feed
- Management and veterinary services

and the grower providing:

- Labor
- Chicken house

The processor receives broiler chickens at the desired weight and quality characteristics. The grower receives payment based on per pound (or kilogram) live broilers produced and premiums/deductions based on performance. In tournament contracts, payments are based on the relative performance of growers. There are some cooperatives that employ marketing contracts for organic and free-range broiler meat.

Consolidation in the Broiler Industry

Not only is the broiler industry highly vertically integrated but also there is increasing consolidation due at least in part to companies acquiring smaller companies. This is illustrated in Table 18.1.

Vertical Coordination in the Turkey Industry

Turkey production today is vertically coordinated (see Figure 18.3) with:

- ~55% production contracts
- >30% vertical integration
- ~10% sold on the spot market

Like the broiler industry, turkey production involves either production contracts (similar to those in the broiler industry) between growers and processors or vertical integration. (See Figures 18.4 to 18.6 for stages in turkey production). Marketing contracts via cooperatives continue, but their relative importance has declined. It is possible that niche products (free-range, organic, etc.) will be

TABLE 18.1 Increasing Consolidation of Broiler Production

	Percentage Market Share of Broilers Processed in United States		
	1980	*1990*	*2000*
Top 3 Companies	21	35	40
Top 10 Companies	50	63	70
Top 20 Companies	71	79	87

Source: Based on WATT Poultry USA.

Primary Breeders
↓
Multipliers
↓
Turkey Breeders
↓
Hatcheries
↓
Turkeys
↓
Processing and further processing
↓
Retailer ——————→ Restaurant
↓   ↓
Consumer   Consumer

Figure 18.3 Vertical stages in turkey industry.

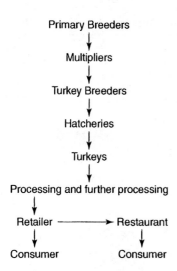

Figure 18.4 Young turkey poult. *(Courtesy, Gretta Irwin, Iowa Turkey Federation)*

Figure 18.5 Processing turkeys at West Liberty Foods. *(Courtesy, Gretta Irwin, Iowa Turkey Federation)*

Figure 18.6 Refrigerated storage of turkey hams. *(Courtesy, Gretta Irwin, Iowa Turkey Federation)*

merchandized via cooperatives. Jennie-O Foods is the leading processor of turkeys in the world.

WORLD TRADE IN POULTRY AND EGGS

World trade of poultry and eggs is increasing at a rapid rate. According to the UN's Food and Agriculture Organization (FAO), world chicken meat imports increased from about 2.2 million metric tons ($3.55 billion) in 1990 to about 5.9 million metric tons ($6.04 billion) in 2000. Similarly, world egg imports grew from about 6,600 metric tons ($39.1 million) in 1990 to about 0.91 billion metric tons ($0.99 billion) in 2000. Import and export of major poultry products are considered below, with trade in ducks discussed in the section World Trade in Duck and Goose Meat.

Exports of Poultry Meat

The United States is the leading exporter of poultry meat (see Table 18.2). Other major poultry exporting nations include China, Brazil, and France, although China is a net importer of poultry meat. Between 1990 and 2000 there were marked increases in chicken meat exports: United States, 4.4 fold; Brazil, 3.1 fold; and France, 71%.

Importance of Poultry Meat Exports

An indicator of the importance of exports to a country's economy can be gained from the following statistics for the United States for 1999 (from the USDA Economic Research Service). Exports of chicken meat were valued at $1.4 billion (4.92 billion lb) and represented 17% of production. The major markets were Russia and China. Exports of turkey meat by the United States were 378 million lb

TABLE 18.2   Poultry Meat Exports in 2000

Country	Meat Exports (million tons)
1. USA	2.72
2. China and Hong Kong	1.19 (China, 0.8; Hong Kong, 0.39)
3. Brazil	0.90
4. France	0.43[1]
5. Thailand	0.27
6. Netherlands	0.24[1]
7. Canada	0.12
8. Hungary	0.10
9. United Kingdom	0.09[1]
10. Germany	0.04[1]

[1]Excludes trade within the European Union.

Source: USDA—National Agricultural Statistics Service.

(172 thousand metric tons) ($198 million) in 2000 with Mexico being, by far, the major market.

Imports of Poultry Meat

The major importers of poultry meat include China and the Russian Federation (Russia) (see Table 18.3). If we look at net imports (imports minus exports), the world situation looks different. China is both a leading exporter and importer. Russia is still the major net importer of poultry.

Exports and Imports of Eggs

There is much trading in eggs. The major exporters include Belgium, China, and Germany (see Table 18.4). The United States exports significant numbers of eggs (earning $108.5 million in 2000). The major importers of eggs include countries in the European Union, with Germany being the largest importer and net importer of eggs together with Hong Kong (see Table 18.5).

TABLE 18.3   Poultry Meat Imports in 2000

Country	Meat Imports (million tons)
1. China	1.21
2. Hong Kong	1.12
3. Russian Federation	1.00
4. Japan	0.56
5. Saudi Arabia	0.37
6. Mexico	0.27
7. Canada	0.15
8. Germany	0.14
9. Ukraine	0.09
10. Republic of Korea	0.06

[1]Excludes trade within the European Union.

Source: USDA—National Agricultural Statistics Service.

TABLE 18.4   World Egg Exports in 2000

Country	Egg Exports Million lb (thousand metric tons)
1. Belgium	168 (76)
2. China	143 (65)
3. Germany	146 (66)
4. USA	143 (65)
5. India	127 (57)
6. Spain	112 (51)
7. France	110 (50)
8. Iran Islamic Republic	86 (39)
9. Malaysia	86 (39)
10. United Kingdom	84 (38)

Source: FAO database.

TABLE 18.5   World Egg Imports in 2000

Country	Imports Million lb (thousand metric tons) 2000	
1. Germany	493	(222)
2. Hong Kong	180	(81)
3. Italy	142	(64)
4. France	138	(62)
5. Netherlands	124	(56)
6. Singapore	89	(40)
7. United Kingdom	80	(36)
8. Switzerland	75	(34)
9. Belgium	64	(29)
10. Canada	51	(23)
11. Denmark	44	(20)
12. Georgia	42	(19)

USA 3576 Metric Tons (2000)

Source: FAO database.

Trends in World Trade in Poultry

Marked changes in poultry imports and exports have occurred. With the competitive advantages of low-cost locally produced corn and soybeans, Brazil increased its exports with a 3.1 fold increase between 1990 and 2000 as broiler production increased rapidly. It is likely that Argentina will also become a significant exporter of poultry meat because its production also increased at a rapid rate with analogous competitive advantages. The United States also has a competitive advantage. With the accession of China to the WTO (World Trade Organization) and likely reductions in tariffs, exports of U.S. poultry and eggs are also likely to increase. This is assuming that there are not artificial constraints to trade such as regulation not based on science.

World Trade in Duck and Goose Meat

World trade in duck and goose meat is substantial, amounting to more than $300 million in 2000. This includes high-value products such as duck and goose liver for pâté de foie gras. Major exporters are France, Hungary, Hong Kong, and Thailand with, respectively, $39, $35, $27, and $26 million in duck meat exports in 2000. Major importers are Hong Kong, Germany, Saudi Arabia, Japan, and the United Kingdom.

INTERNATIONAL COMPETITIVENESS

Competitiveness of the Poultry Industry in North and South America

The competitiveness of poultry production in North America stems from a number of factors including:

- Low cost, high-quality processed corn and soybeans for feed
- Infrastructure (roads, railroads, processing plants, supplies, etc.)
- Legal system with enforceable contracts and title to land
- Education for labor force (schools, community colleges, and land-grant universities)
- Technology and the research-extension base
- Efficient management and labor force in vertically integrated companies
- The availability of capital for loans (developed banking system)
- Relatively low taxes and transparent regulatory environment with few corrupt government officials

Argentina and Brazil also have the advantage of low-cost corn and soybeans. The loan deficiency payments for U.S. corn and soy products increase production. This tends to lower prices and acts as an indirect subsidy for producers of poultry, eggs, and other livestock products.

Competitive disadvantages for the United States, Canada, and Western Europe include the high cost of labor together with costs of environmental and welfare regulations.

Competitiveness of the Poultry Industry in the European Union

The success of the poultry industry in the European Union stems from a number of factors including:

- The common agricultural policy and other policies that restrict imports
- Consumer demand for specific products (compared to commodity products) including free-range and organic poultry and eggs
- Retailers supplying and promoting specific poultry products
- Supply chains tightly linking producers with consumers

Vertical Coordination in the Egg Industry

Egg production is vertically coordinated in the following manner:

- Producer contracts (~ 1/3)
- Vertical integration (~60%)
- Some marketing contracts (~3%)

Producer contracts involve the contractor who provides:

- Layers
- Feed
- Other supplies

The grower provides:

- Labor
- Housing and equipment

The eggs belong to the contractor and the grower is paid based on number of eggs, with incentive payments.

Vertical integration in the egg industry has been predominantly led by producers. They become integrators with ownership of:

- Laying hens and the facilities to produce them
- Feed mills
- Hatcheries
- Egg packing
- Marketing

Cooperatives using marketing contracts represent about 3% of the industry. Growth for this sector for niche products is possible.

Cal-Maine Foods is the leading egg company in the United States. Cal-Maine produces about 80% of its eggs via full vertical integration with the remainder via production contracts. The company

Figure 18.7 Ready to cook foods from eggs (Pour a Quiche).

Figure 18.8 The capital needs of the poultry industry have increased with size. This is an aerial view of an Arkansas broiler facility. *(Courtesy, Dr. Jerry Sell)*

sells the eggs as shell or table eggs. Another major egg processor is Michaels Foods. It employs a mix of production and marketing contracts. The company processes over 90% of eggs as further processed products (e.g., precooked omelets, reduced cholesterol products, etc.). (For more details of major companies, see Chapter 14.) An example of a further processed egg product is shown in Figure 18.7.

Advantages and Disadvantages of Vertical Coordination

The advantages of vertical coordination, particularly moving toward vertical integration, include:

- Reduction in risk and uncertainties (compared with oscillation in prices on the spot market)
- Increased profitability due to different points on the value/supply chain
- Increased access to capital at potentially lower interest rates (see Figure 18.8 for an example of the importance of capital)
- Reduced transaction costs
- Quality control
- Improved management
- Improved efficiency through specialization and elimination of redundancy
- Discounts for bulk purchase
- Economies of scale allowing the use of new technology in production and processing, and hence even greater efficiencies
- Ability to trademark and obtain premiums

It is said that a problem for farmers' profitability is that they "buy retail and sell wholesale" (many businesses buy wholesale and sell retail!). Greater vertical coordination leans more toward the prevailing system in business—"buy wholesale, sell retail."

Disadvantages of vertical coordination include:

- Reduced freedom and decision making for the grower
- Changes in the power structure of the grower-integrator relationship (while growers may experience a reduction in negotiating position, processors are experiencing a similar situation with retailers)

Long-term relationships between grower-processor-retailer require all to be profitable and expand, with the success of each being mutually advantageous.

Contracts

There are considerable advantages in contracts because they are legally enforceable. A poultry producer should read the agreement with care and consult a legal advisor to ensure that all of the terms are clearly spelled out and understood. A good contract must be clear, complete, and concise as to the duties and responsibilities of all parties. Actions for breach of contract and public laws (supported by State's Attorneys General) can protect growers.

Specifics of the Contract

The following specifics should be included in the contract:

- The contract should be specific as to starting and ending dates per brood or time basis. A time contract usually specifies four to six broods per year.
- Each party should retain the same right for continuing or closing the program. This could minimize hardships for growers who have used credit in providing housing and equipment.
- Cancellation. This should be specific and clearly understood, with equal rights and privileges.
- Management. The contract makes known who is responsible for decisions; and details of the management program should be spelled out.
- Payment or settlement. The contract should be clear as to the method of computing rate, time, incentives, penalties, and losses including condemnation.
- Arbitration. There may be procedures providing for binding settlement to avoid court proceedings.
- Legal relationship of contracting parties. It should be clearly stated whether or not the contract is a partnership, employer-employee situation, or arranged on an independent basis. This is important for social security and income tax purposes.
- Infrastructure (roads, railroads, processing plants, supplies, etc.)
- Legal system with enforceable contracts and title to land
- Education for labor force (schools, colleges, and universities)
- Technology and the research base
- Efficient management and labor force in vertically integrated companies
- The availability of capital for loans (developed banking system)

Competitive disadvantages for the poultry industry in Western Europe include the high cost of labor and feed ingredients together with costs of environmental and welfare regulations.

POULTRY BUSINESS ISSUES

Details of poultry business issues, particularly obtaining credit for small and moderate producers, are presented in Appendix I.

Types of Business Organizations

There are three major types of business organizations: (1) sole proprietorship, (2) partnership, and (3) corporation.

Sole Proprietorship

The sole proprietorship is a business owned and operated by one individual. This is the most common type of business organization in U.S. agriculture. Under this, one person controls the business, having sole management and control. This may be modified through contracts. The sole proprietors get all the profits or the losses.

There are two major limitations: (1) it may be more difficult to acquire new capital for expansion, and (2) not much can be done to provide for continuity with the passing of the owner.

General Partnership

A partnership is an association of two or more persons who, as co-owners, operate the business. About 13% of U.S. farms are partnerships. Most involve family members. For a partnership to be successful, the enterprise must be sufficiently large to utilize the abilities of the partners and to compensate them.

A partnership has advantages including combining resources that can increase efficiency, equitable management, tax savings (a partnership does not pay any tax), and flexibility. Partnerships may have disadvantages such as liability for debts, uncertainty of length of agreement, difficulty of determining value of a partner's interest, and limitations on management effectiveness.

Limited Partnership

A limited partnership is an arrangement in which two or more parties supply the capital (limited partner), but only one partner is involved in the management (general partner). The advantages include bringing in outside capital and limited liability. The disadvantages are that the general partner has unlimited liability and the limited partners have no voice in management.

Corporation

A corporation is a device for carrying out a farming enterprise as a legal entity entirely distinct from the persons who control it. It is restricted to doing only what is specified in its charter and must register in each state it does business in. Advantages include continuity, transfer of ownership, and limitation of the liability of shareholders to the value of their stock. A corporation may be family owned.

Capital Needs

Farmers have on average about 80% equity in their business and 20% borrowed money (debts).

Credit is an integral part of the poultry business including:

1. Short-term loans (1 year or less) are used for purchase of feed, supplies, etc.
2. Intermediate-term loans (1 to 7 years) are used for buying equipment or remodeling existing buildings.
3. Long-term loans (up to 40 years) are secured by mortgage on real estate and are used to buy land (or possibly to finance major new buildings).

RECORDS AND DATABASES

Records

The key to good business and management is records. Why keep records? The chief functions of records and accounts are:

- To provide information on which the management and business may be analyzed, with the strong and weak points ascertained. This assists more effective planning.
- Ultimately to provide profit and loss statements.
- Good records, properly analyzed and used, increase net earnings and are essential to good management.

Like any other business, today's successful poultry producers must have, and use, complete computerized records.

Databases

In the United States, the managers in the poultry industry have access to databases (e.g., the blue book from the Agri-Start database), covering detailed comparison of production in chicken production units and complexes. This allows comparison of an individual unit or its manager with others in the industry and with previous months and years. This is a tremendous tool to improve the quality of management and the efficiency of production.

Budgets

A budget is a projection of records and accounts and a plan for organizing and operating ahead for a specified period of time. A short-time budget is usually for 1 year, whereas a long-time budget is for a period of 3 to 5 years.

Enterprise Accounts

Where an enterprise is diversified, enterprise accounts should be kept—three different accounts for three different enterprises.

- They make it possible to determine which enterprises have been most profitable, and which least profitable.
- They facilitate comparison of a given enterprise with competing enterprises of like kind.
- They allow determination of profitableness at the margin (the last unit of production). This will give an indication as to whether to increase the size of a certain enterprise.

Tax Management, Estate Planning, and Wills

Tax management, estate planning, and wills are considered in detail in Appendix I and are mentioned here for completeness. Tax management and reporting consists of complying with the law but paying no more tax than required. The preparation of wills, trusts, and partnership agreements requires consideration of the effects of federal and state tax law. A will is a set of instructions drawn up by or for an individual that details how he or she wishes the estate to be handled after death. Despite the importance of a will, many farmers die without one. Always consult a CPA and attorney in these activities.

Liability

Poultry producers, as well as livestock producers, are vulnerable to damage suits. Moreover, the number of damage suits arising each year is increasing at an alarming rate, with astronomical damages being claimed. Comprehensive personal liability insurance protects a producer who is sued for alleged damages suffered from an accident involving his or her property. Workers' compensation insurance and employer's liability insurance protect farmers against claims and court awards from injured employees.

Workers' Compensation

Workers' compensation laws are in force throughout the United States. These cover on-the-job injuries and protect disabled workers regardless of whether their disabilities are temporary or permanent. Workers' compensation provides employees with assured payment for medical expenses or lost income due to injury on the job. Producers should seek the advice of paid agricultural business consultants and insurance agents experienced in workers' compensation and liability insurance.

Social Security

In the United States, Social Security is a system of retirement, health, and disability payments with both employees (e.g., agricultural workers) and employers (e.g., poultry producers) paying in on a regular basis. Employers pay their contribution and deduct each employee's contribution from his or her paycheck. Employers report cash wages for each employee by January 31 of each year. Agricultural workers admitted to the United States on a temporary basis are not eligible.

ALLIED POULTRY INDUSTRIES

The poultry industry leans heavily on the support of allied industries. These include:

- The animal health industry, which includes manufacturers of antibiotics and other drugs, together with vaccines (see Table 18.6).
- Feed and feed ingredient manufacturers.
- Facility and equipment designers and manufacturers.
- Waste disposal engineering and systems.
- Food manufacturers who use so much poultry meat and eggs.

TABLE 18.6   Animal Health and Nutrition Company Rankings (2001 sales)

Company	Sales (million $, U.S.)	% Change from 2000
Merial	1,636	2
Hills Pet Nutrition	1,200	6
Roche Vitamins	1,137	−3
Pfizer	1,022	−3
Intervet	956	1
Bayer	930	−1
BASF	930	5
Fort Dodge	776	−2
Schering-Plough	694	−4
Elanco	686	3
Novartis	550	−8
Degussa	513	16
Aventis	505	−6
Pharmacia	469	6
Royal Canin	420	15
Novus International	400	0
Idexx	386	5
Alpharma	335	11
Virbac SA	309	1
Lohmann	290	5
Ceva	168	26
Phibro	140	5
Vetoquinol	123	25
Phoenix	90	25
Norbrook	85	20
Virbac Corp.	61	14
Heska	48	−8
Embrex	45	15

Source: From Feedstuffs, 2002.

- Transportation companies (ranging from airlines carrying day-old chicks to truckers carrying poultry to processors and poultry products to railroads carrying feed ingredients to poultry-raising regions to international ports and shipping moving poultry products for export).

POULTRY RESEARCH

Poultry-related research is conducted in universities, government laboratories, and in private industry (integrators, food industry, animal health and feed companies, equipment manufacturers). Research has contributed to the poultry industry by the following:

- Reduction in the cost of production due to:
  - Improved growth rate and/or carcass yield or dressing percentage (for an example of improved carcass composition, see Figure 18.9).
  - Depressed feed required to achieve a unit of meat or eggs and reduced cost of feed.
  - Improved facilities designed for optimal production and/or reducing capital requirements (lower-cost construction and/or greater longevity).
  - Automation to reduce labor costs, etc.
- Decreases in the impact of environmental factors such as heat stress and diseases on production.
- Increases in both the variety of poultry products and their quality (eating properties)/healthfulness (nutritional and food safety characteristics).
- Improvements in the sustainability of the industry by addressing regulatory issues (such as environmental problems including the disposal of the carcasses of mortalities and animal waste) and welfare concerns (this being a regulatory, a consumer, and a social issue).

Figure 18.9   Type made the difference! *Left*: cross-section of new broad-breasted turkey carcass. *Right*: cross-section of old type turkey carcass. *(Courtesy, National Turkey Federation)*

CAREERS FOR POULTRY SCIENCE GRADUATES

The poultry industry requires well-educated employees. There are career opportunities for graduates of poultry science programs and animal science programs where students have had significant poultry experience.

Careers for poultry science graduates include the following:

1. Agricultural business and finance (loan officer, equipment design, and sales)
2. Agricultural sales and marketing (salesperson)
3. Biotechnology
4. Extension-4H, county agents, state specialist
5. Food industry (marketing, sales, new food development)
6. International agriculture (U.S. government or nongovernmental representation, international sales, and market development)
7. Meat science (USDA inspectors, plant supervisors, managers, etc.)
8. Nutrition and feed industry (feed-processing supervisors, managers, nutritional consultants)
9. Poultry breeding/genetics (unit supervisors, executives, managers, etc.)
10. Poultry health (sales, research, management)
11. Poultry production (unit supervisors, executives, managers, etc.) quality control, information analyst, human resources
12. Poultry promotion and information (article writer, marketer)
13. Research (laboratory technician, researcher, following M.S. or Ph.D. for industry, government, and universities)
14. Teaching/education (vocational agriculturists, community college, 4-year college, land-grant universities)
15. Veterinary medicine (following DVM, many opportunities)

FURTHER READING

Martinez, Steve W. *Vertical Coordination of Marketing Systems: Lessons from the Poultry, Egg, and Pork Industries.* U.S. Department of Agriculture, Agricultural Economic Report No. 807, April 2002.

19

Other Poultry—Ducks, Geese, Pigeons, Ratites (Ostriches Plus Emus and Rheas), Game and Ornamental Birds

---

Objectives

After studying this chapter, you should be able to:

1. List other types of poultry production besides broiler, turkey, and chicken egg production in the United States.
2. List several breeds for each of these types of poultry.
3. Briefly describe the production practices associated with these types of poultry.
4. List diseases important in duck production.
5. List the countries in which duck production and consumption are far more important than in the United States.

On a worldwide basis, and particularly in Asia, duck and goose production for meat is critically important. In the United States, ducks and geese are also economically important (see Figure 19.1 for an illustration of a day-old duckling). There is also commercial production of pigeons and the ratites (ostriches and emus), together with game and ornamental birds.

DUCKS AND GEESE

The aggregate world production of duck and goose meat together is approximately the same as the world production of turkey meat. The People's Republic of China is by far the major producer of duck and goose meat in the world (see Tables 19.1 and 19.2). Production in China is increasing at a rapid rate. Other major producers of duck meat include France and Thailand (see Table 19.1). Other major producers of goose meat include Egypt and Hungary (see Table 19.2). In the United States, produc-

tion of duck (with 116 million lb in 2000) and goose meat is of some importance, although production of chickens and turkeys dominates.

Duck and Goose Meat

Duck meat and goose meat are popular with many lovers of good food. Both goose and duck are high in fat. Examples of dishes include the following:

- Cooked and smoked meats (e.g., duck with plum sauce)
- Pekin duck
- Foie gras (the fatty liver of force-fed ducks or geese)—pâté
- Terrines and mousse
- Duck confit (meat roasted with herbs and possibly citrus fruit)
- Duck fat
- Cured duck gizzards in salads
- Duck or goose sausages

319

Figure 19.1   Day-old duckling. *(Courtesy, Dr. Jerry Sell)*

TABLE 19.1   World Production of Duck Meat (in Metric Tons)

Country	1990	2000	(% Increase)
1. China	0.59×10^6	1.99×10^6	(+239%)
2. France	0.11×10^6	0.24×10^6	(+116%)
3. Thailand	91×10^3	0.10×10^6	(+12.3%)
4. Vietnam	40×10^3	64×10^3	(+63%)
5. USA	44×10^3	53×10^3	(+19.2%)
6. Malaysia	36×10^3	51×10^3	(+39%)
7. Republic of Korea	14×10^3	45×10^3	(+219%)
8. Egypt	32×10^3	42×10^3	(+31%)
9. Germany	23×10^3	40×10^3	(+74%)
10. United Kingdom	25×10^3	40×10^3	(+61%)
11. Hungary	32×10^3	28×10^3	(−12%)
12. Myanmar	11×10^3	22×10^3	(+99%)
13. Philippines	12×10^3	21×10^3	(+81%)
14. Mexico	17×10^3	20×10^3	(+14%)
15. Indonesia	10×10^3	14×10^3	(+31%)
World	1.19×10^6	2.89×10^6	(+143%)

Source: FAO.

TABLE 19.2   World Production of Goose Meat (in Metric Tons)

Country	1990	2000	(% Increase)
China	47×10^6	1.85×10^6	(+290%)
Egypt	28×10^3	42×10^3	(+50%)
Hungary	55×10^3	39×10^3	(−29%)
Madagascar	8.8×10^3	13×10^3	(+48%)
Poland	9.0×10^3	9.2×10^3	(+2%)
World	0.62×10^6	2.00×10^6	(+220%)

Source: FAO.

Feathers

The feathers are used for insulation in winter clothes, for comforters (e.g., the eider goose down), and pillows.

Down

Down represents 10% of the feathers in ducks and geese. Down, the small fluffy feathers without shaft, are used in winter clothes and bedding (particularly the higher-priced quality products). They are used as insulation and for comfort. In ducks and geese, the air-filled down feathers are under the other plumage. They provide thermal protection. When feathers are sold as down, the product must be more than 80% down and less than 20% feathers. Feathers are washed (5–8 times) in soap to eliminate animal waste, but some oils should be left attached.

Ducks

Ducks are among the most versatile of animals. They will live under a wide range of climatic conditions. They are free from such common poultry diseases as leukosis, Marek's disease, and infectious bronchitis.

About 20 million ducks are marketed in the United States each year. Duck production was once centered on Long Island, New York, but ducks are now raised in Indiana, Michigan, Minnesota, and North Carolina. The largest producer of ducks in Europe is Cherry Valley Farms (UK).

Meat

Duck (either sex) for meat can be sold as:

- Broiler duckling (~7 weeks old)
- Roaster duckling (< 16 weeks old)
- Mature duck (> 6 months)

The broiler and roaster ducklings have tender meat while the mature duck has tougher meat. Broiler ducklings are marketed as ducklings at 7 weeks of age, at a liveweight of 9.5 lb (4.3 kg), which is equivalent to a dressed weight of 5.5 lb (~2.5 kg). They are generally frozen and ready to cook after thawing. The dressing percentage (carcass weight as a percentage of the body weight) for Pekin ducks is 58%.

An uncooked dressed duck has the following composition:

- ~49% water
- 39% fat
- 12% protein

While there is little demand for duck eggs in the United States, there are significant markets in both Europe and Asia.

Domestication

The domestic duck originated from the Mallard duck. The Mallard is found across Europe and Asia. There is disagreement on where ducks were domesticated. It has been speculated that Mallard ducks were separately domesticated in East Asia, leading to the Asian breeds (e.g., the Pekin) and close to Europe (in the Fertile Crescent), leading to the European breeds such as the Rouen and Aylesbury.

The Muscovy was domesticated in South America perhaps during the pre-Inca period in Colombia or Peru, well before contact with Europeans from 1492 onward. Muscovies were transported to Europe and Africa and later to Asia. The wild species of Muscovy is a waterfowl that is unusual because it is a perching bird. Its range is the South American rain forest.

Duck Classification

Ducks are classified as follows:

Phylum: chordata

Subphylum: Vertebrata

Class: Aves

Superorder: Carinatae

Order: Anseriformes

Species: *Anas platyrhynchos* (Mallard and domestic duck)

Cairina moschata (Muscovy)

Domestic ducks are either in the species *Anas platyrhynchos* (domestic ducks and Mallard ducks) or *Cairina moschata* (Muscovy "ducks"). Domestic ducks have been referred to in the past as *Anas domesticus*. While Muscovies are sometimes referred to as a breed of duck, they are in fact a separate genus and species. Strictly speaking, they should be referred to as ducklike rather than ducks. For the purpose of convenience, ducks and Muscovies will be considered together.

Breeding

The breeding and improvement of ducks has received less attention than the breeding of chickens and turkeys, reflecting their lesser economic importance in North America and Western Europe. Duck breeding in the United States is largely confined to breeding for meat production, with selection for more rapid and efficient gains, more lean meat, and higher egg production and hatchability. (For a diagram of the external anatomy of ducks see Figure 19.2.)

Breeds

Duck breeds are classified as either meat producers or egg producers

MEAT BREEDS. The Pekin duck is by far the predominant breed of duck for commercial production world wide. White Pekin and Aylesbury ducks together with Muscovies are excellent meat breeds. Rouen, Cayuga, Swedish, and Call ducks reach weights that would make them valuable as meat producers, but their poor egg production and colored plumage make them less satisfactory for commercial production.

White Pekin. White Pekin ducks are large white-feathered birds (see Figure 19.3). Adult drakes weigh 9 lb (~4 kg) and adult females weigh 8 lb (3.6 kg). They have orange-yellow bills, reddish yellow

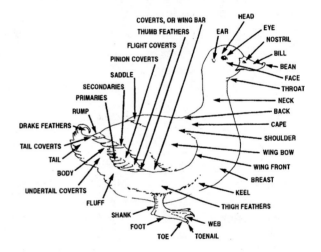

Figure 19.2 Diagram of the male duck. (The language describing and locating the different parts of the bird.) *(Original diagram from E. Ensminger)*

Figure 19.3 White Pekin ducks. *(Courtesy, Watt Publishing)*

shanks and feet, and yellow skin. Their eggs are tinted white. The breed originated in China and was introduced into the United States in the 1870s.

In the United States, commercial duck production relies solely on White Pekin ducks. A similar situation exists in much of Europe. They are ideally suited for meat production, producing excellent quality meat and reaching a market weight of 7 weeks. White Pekins are fairly good egg producers, averaging about 160 eggs/year. They are not good setters and they seldom successfully raise a brood. It is said that they should be treated gently to obtain maximum egg production.

Aylesbury Ducks. Aylesbury ducks have white feathers, white skin, flesh-colored bills, and light-orange legs and feet. Eggs are tinted white. Adult drakes weigh 9 lb (4.2 kg) and adult females weigh 8 lb (3.6 kg). The breed originated in England. They have excellent meat quality and reach market weight in about 8 weeks. Egg production is generally below that of White Pekins, but records of 300 eggs per year have been reported. Aylesburys show some lack of interest in setting. Until recently, the Aylesbury was the major breed in the UK but has been surpassed by the Pekin.

Muscovy. Numerous breeds of Muscovies exist (see Figure 19.4). The white variety is the most desirable for market purposes due to its excellent quality meat, provided they are marketed before 17 weeks of age. Their low egg production (40–45 eggs/year) makes them unsuitable for use on large commercial duck farms. Adult drakes weigh 10 lb (4.5 kg) and adult female ducks weigh 7 lb (3.2 kg). Muscovies are good setters, hatching and caring for approximately 30 ducklings annually.

Although they are not ideally suited to commercial production, Muscovies have excellent possibilities for small general farms that have special retail outlets. In Asia, hybrid or "mule" ducks are produced by mating female Muscovies with "Mallard-type" drakes. These produce satisfactory meat yields, but they are sterile.

EGG BREEDS. Khaki Campbell and Indian Runners are excellent egg-laying breeds, being the number one and number two egg producers per year, respectively. Where specialty duck egg markets exist, either breed would be a good choice.

Khaki Campbell. Khaki Campbell ducks are excellent layers with 365 eggs/year by some hens reported, but they are not valued tremendously for their meat. Young drakes and ducks weigh 3.5 to 4 lb (~1.75 kg) at 2 months of age; adult drakes and ducks weigh 4.5 lb (~2 kg). The breed originated in the United Kingdom from a cross of Fawn and White Runner, Rouen, and Mallard ducks. Males have brownish bronze lower backs, tail coverts, heads, and necks, with the rest of their plumage being khaki (see Figure 19.5). They have green bills and dark-orange legs and toes. Females have seal-brown heads and necks, with the rest of the plumage being khaki. They have greenish-black bills and brown legs and toes. The White Campbell is derived from the Khaki, but has not become a popular egg producer. These ducks have orange bills and legs.

Indian Runner. The Indian Runner originated in the East Indies (present-day Indonesia), with its egg-producing capability developed in Western Europe. Three Indian Runner varieties are recognized: White, Penciled, and Fawn. All three varieties have orange to reddish orange feet and shanks. Characteristically, the Runners stand erect. They weigh about the same as Khaki Campbells.

Figure 19.4 Colored Muscovy ducks. *(Courtesy, Watt Publishing)*

Figure 19.5 Khaki Campbell drake. *(Courtesy, USDA)*

Breeding Programs for Large-Scale Production

The growth of duck production, particularly in Asia, stems in part from genetic improvement. Breeding programs are making rapid progress in a manner analogous to that of chickens and turkeys. Cherry Valley Farms (UK) is a major source of breeding stock.

CARE OF BREEDERS. Birds can be brought into full production at about 7 months of age by providing a long-day length (14 hours of light daily).

Incubation

Duck eggs and Muscovy eggs require, respectively, 28 and 35 days of incubation. Incubators designed for hatching duck eggs are available. On large commercial farms, ducklings are frequently taken from the machines as they hatch. However, great care must be taken to prevent chilling of newly hatched ducklings. It may be wise for small producers to keep the machine closed until hatching is completed.

Brooding and Rearing

Ducklings should be moved from the hatcher to comfortable brooding quarters as quickly as possible. Prevent chilling and do not overcrowd the birds during transit. Provide feed and water for ducklings as soon as they are placed in the brooder.

Buildings of practically any type can be used to brood ducklings as long as the birds are kept warm, dry, and free of drafts. Ventilation systems and windows should be designed so that fresh air can be brought into the building without chilling the ducklings. Litter flooring may be used with chopped straw, wood shavings, and peat moss. Ducklings need supplementary heat for about 4 weeks after they hatch. Ducklings need clean drinking water at all times. It may be supplied in hand-filled water fountains or by automatic waterers.

Feeding

Maximum efficiency for growth and reproduction can be obtained by using commercially prepared diets as pellets. Four diets are recommended: starter, grower, finisher, and breeder (see Chapter 7 for details).

Management

Ducks respond to management; good management = success, poor management = trouble.

Both breeder and market ducks lend themselves to environmentally controlled buildings and automation. The largest duck farm in Europe is Cherry Valley Farms in Rothwell, England. At this company, Pekin ducks are raised in barns with straw on the floors and a predominantly wheat diet. They have water available for drinking and preening.

Diseases

On large commercial duck farms, the risk of disease can be minimized by (1) a closed flock policy and (2) controls on the entry of vehicles and visitors. In case of disease, consult with a veterinarian. Examples of common diseases are presented.

- *Botulism* occurs in both young and adult ducks. It is caused by the bacterium *Clostridium botulinum*, which grows in decaying organic material. Ducks ingesting toxins produced by the bacteria lose control of their neck muscles. Maintaining clean facilities will prevent this disease.
- *Fowl cholera* can cause high mortality. Strict sanitation will help control fowl cholera. Burn or bury dead birds. Vaccines can control losses from fowl cholera.
- *Necrotic enteritis*, caused by *C. perfringens*, is found in breeding stock. Breeder houses and yards must be free of wet litter. Mortality may be sporadic over a long period of time. Some antibiotics are effective for treatment and prevention.
- *Reproductive disorders.* Paralysis of the intromittent organ of drakes can be observed early in the mating season. Females can have prolapse of the vagina, impacted oviducts, and egg yolk peritonitis. The birds rendered incapable of reproduction should be culled.
- *Viral hepatitis* outbreaks can cause 80 to 90% mortality in flocks of ducklings (1–5 weeks) (see Figure 19.6). This highly contagious disease

Figure 19.6 Duckling killed by infection with duck hepatic virus (DHV). *(Reprinted from* Diseases of Poultry, *10th edition, edited by B.W. Calnek, with permission from Iowa State Press)*

strikes swiftly without warning. Vaccination may be recommended against duck viral hepatitis. Antibodies produced by the laying ducks are passed through the egg to the young ducklings. This gives passive immunity to protect during the first 3 weeks. Antibody therapy at the time of initial loss is an effective flock treatment.

- *Brooder pneumonia* is caused by fungi in the litter. Good litter and a dry brooder house help prevent brooder pneumonia.
- *New duck disease* (infectious serositis) is one of the most serious diseases affecting ducklings, with losses up to 75%. It is a bacterial disease caused by *Moraxella anatipestifer*. Symptoms resemble those of chronic respiratory disease of chickens. The first signs of the disease are sneezing and loss of balance. Afflicted ducklings fall over on their sides and backs. Antibiotics have been used with some success.
- *Coccidiosis* is not as severe as in chickens. It can cause mortality and morbidity and hence poor performance. The causal organism in ducks is different from that in chickens.
- *Duck virus enteritis* (duck plague) is an acute viral disease affecting ducks and geese. It is transmissible by contact, swimming water, and from migratory waterfowl. Symptoms include watery diarrhea and nasal discharge. General droopiness develops in about 5 days after exposure. Symptoms last for 3 to 4 days, frequently ending in death. Postmortem examination shows multiple hemorrhages in body organs. Strict sanitation will help control the disease. A modified chicken embryo-adapted vaccine has been developed and used successfully.

Marketing and Processing Ducks

Feed should be withheld from the birds 8 to 10 hours before slaughter, but water may be provided up to the time of killing. Clean, uncrowded rearing facilities will help to prevent bruising, cutting, and other factors that cause poor product. Ducklings can be transported in crates or trailers to the slaughterhouse.

In large commercial slaughterhouses, ducklings are dipped through a molten wax after slaughtering. When the wax hardens—by immersion in cold water—it can be peeled free to remove any feathers that remain. Wax can be reused if it is remelted and the feathers are screened out. Small quantities of wax can be purchased for use in small farm slaughtering operations. Wax is highly combustible; hence, care must be taken to prevent contact with an open flame.

Birds are then eviscerated. Wash the eviscerated bird thoroughly. If birds are to be marketed

Figure 19.7 Processed ready to cook duckling with orange sauce.

frozen, package them in shrinkable plastic bags. There is further processing of duck (for example see Figure 19.7). If special—New York Dressed—markets are available, chill the ducklings and sell them uneviscerated.

The federal grades of ducks are U.S. Grade A, U.S. Grade B, and U.S. Grade C and are essentially the same for all poultry (see Chapter 17). The grading program is voluntary.

Geese

While goose production in the United States is modest (< 1 million birds), world production of goose meat is large. The People's Republic of China is the major producer of goose meat in the world, and production is increasing rapidly (see Table 19.2).

Geese are very hardy and can live almost entirely on good pasture. Yet, the production of geese for meat purposes has never enjoyed the popularity in the United States that it has in some European and Asian countries. Geese are produced commercially in Missouri, Iowa, South Dakota, Minnesota, Wisconsin, Ohio, Indiana, California, and Washington. In the United States, the number of farms selling geese has decreased. In addition, geese can be raised as a hobby for ornamental and exhibition purposes. Geese are not susceptible to many poultry diseases.

Canada geese will not be considered even though they are a common wild goose of North America. There are major restrictions on hunting or holding Canada geese in the United States. Before

Canada geese can be sold or transferred to another person, a permit must be obtained from the Fish and Wildlife Service (Department of the Interior).

Geese Classification

Geese are classified as follows:

Phylum: Chordata

Subphylum: Vertebrata

Class: Aves

Superorder: Carinatae

Order: Anseriformes

Species: *Anser anser* (North American and European breeds of domestic geese)
Anser cygnoides (Asian and African breeds of domestic geese)

Domestication

The North American and European breeds and Asian and African breeds of domestic geese were domesticated from different species of waterfowl and in different geographical locations. The North American and European breeds are descendants of the Greylag goose. Asian and African breeds are descendants of the swan goose. Domestication of geese occurred at least 5,000 years ago, independently in the Fertile Crescent, probably in Egypt and in China. This is analogous to the domestication of pigs with both a Middle East and East Asian domestication of different populations of Eurasian wild boars.

Meat

Goose (either sex) is sold as either young or mature goose. The young goose has tender meat. In the mature goose, the meat is tougher. Young geese are marketed at 12 to 16 weeks old, with a dressed weight of 11 lb (5 kg). An uncooked dressed young goose has the following composition:

- ~50% water
- 34% fat
- 16% protein

Much of the fat is under the skin, and removal of the skin reduces fat to 7%.

Breeding

The breeding and improvement of geese in North America has received far less attention than in Europe and Asia.

BREEDS. Toulouse, Emden, and African geese are heavy breeds. These are the most popular breeds in

Figure 19.8 Emden (*left*) and Toulouse geese. *(Courtesy, Watt Publishing)*

the United States for meat production. Other common breeds are Chinese, Buff, Pilgrim, Sebastopol, and Egyptian. There are considerable differences in breeds of geese. In choosing a breed, therefore, one should consider the purpose for which they are to be used. Geese are raised for meat and/or eggs, as show birds, or even guard animals. When choosing a breed, one should determine the market requirements such as size (market body weight) and plumage color (white is generally preferred). Strains that lay the most eggs produce goslings at lowest cost. For examples of goose breeds, see Figure 19.8.

African Goose. The African goose has a distinctive knob or protuberance on its head. Its carriage is erect. The head is light brown, the knob and bill are black, and the eyes are dark brown. The plumage is ash brown on the wings and back and light ash brown on the neck, breast, and underside of the body. The African goose is a good layer, grows rapidly, and matures early. It is not as popular for market production as the Emden or the Toulouse because of its dark beak and pinfeathers.

Buff. The Buff has only fair economic qualities as a market goose, with a limited number raised for market. The color varies from dark buff on the back to a very light buff on the breast and from light buff to almost white on the underpart(see Figure 19.9).

Chinese Goose. The Chinese goose grows rapidly, is attractive, and makes a desirable medium-size market goose. It is very popular as an exhibition and ornamental breed. There are two standard varieties: the Brown and the White. It is smaller than other standard breeds and more "swanlike" in appearance. Both varieties mature early and are better layers than other breeds, usually averaging from 40 to 65 eggs per bird per year.

Figure 19.9 Breeding flock of Buff geese. *(Courtesy, USDA)*

Egyptian.   The Egyptian is a long-legged, but very small, goose kept primarily for ornamental or exhibition purposes. Its coloring is mostly gray and black, with touches of white, reddish brown, and buff.

Emden.   The Emden was one of the first breeds imported into the United States. This breed was known at first as Bremen, after the German city from which it was initially exported. The present name is after Emden (Germany) from which it was exported to England.

The Emden is pure white and "sprightly." It is much tighter feathered than the Toulouse and appears more erect (see Figure 19.8). The Emden is a fairly good layer (35–40 eggs per goose per year), but production depends on the breeding and selection of the flock. It is usually a better setter than the Toulouse and is one of the most popular breeds for marketing. It grows rapidly and matures early.

Pilgrim.   The Pilgrim is a medium-size goose that is good for marketing. A unique feature of this breed is that males and females may be distinguished by color. In day-old goslings the male is creamy white and the female gray. The adult male remains all white and has blue eyes; the adult female is gray and white and has dark hazel eyes.

Sebastopol.   The Sebastopol is a white ornamental goose that is very attractive because of its soft plume-like feathering. This breed has long, curved, profuse feathers on its back and sides and short, curled feathers on the lower part of the body.

Toulouse.   The Toulouse goose derives its name from the city of Toulouse in southern France, which is noted for its geese. This breed has a broad, deep body and is loose-feathered, a characteristic that gives it a massive appearance. The plumage is dark gray on the back, gradually shading to light gray edged with white on the breast and to white on the abdomen. The eyes are dark brown or hazel, the bill pale orange, and the shanks and toes are deep reddish orange (see Figure 19.8).

Incubation

Eggs should be washed soon after gathering. Store eggs at 55°F and a relative humidity of 75% until set for hatching. If eggs are held for more than a couple of days, turn them daily to increase the percentage of hatch. Hatchability decreases fairly rapidly after a 6- to 7-day holding period. Eggs properly stored can be held 10 to 14 days with fair results. The incubation period for most goose eggs is 29 to 31 days. For the Egyptian goose, it is 35 days.

Incubators, either still-air or forced-draft, can be used to hatch goose eggs. However, artificial incubation of goose eggs is much more difficult than with chicken eggs because more time and higher humidity are required.

Feeding

Goslings should have drinking water and feed when they are started under the brooder or hen. Supply plenty of watering space. For the first 3 weeks, feed goslings 20 to 22% protein goose starter pellets. After 3 weeks, feed 15% protein goose grower pellets. Geese can go on pasture as early as the first week, but they will only receive significant nutrition from forage at 5 to 6 weeks of age. This may be provided as silage. Geese are very selective and tend to pick out the palatable forages. They will reject alfalfa and narrow-leaved tough grasses and select the more succulent clovers and grasses. Geese should not be raised on dried-out mature pasture. Suitable stocking densities are 20 to 40 birds per acre (~75 birds per hectare) depending on the size of the geese and the quality of the pasture.

Management

Breeder geese should be fed a pelleted breeder ration at least a month before egg production is desired. They do much better and waste less feed on pellets than on mash. A chicken-breeder ration may be used if special feeds for geese are not available. Provide oystershells (or other calcium sources), grit, and plenty of clean, fresh drinking water at all times. Lights in the breeder house can be used to stimulate earlier egg production.

Diseases

Geese are resistant to diseases. If there are signs of disease, a poultry pathologist should be consulted immediately (also see Chapter 11).

Marketing

Following slaughter, geese can be scalded in a commercial scalder. Water temperature should be 145°

to 155°F (~66°C) and the length of the scald should be from 1½ to 3 minutes. Detergent should be added to the water to hasten thorough wetting of the feathers.

After scalding, the birds may be "rough-picked" by hand, picked on some type of conventional rubber-fingered picking machine, or placed in a spinner-type picker. After "rough-picking," it is difficult to remove the remaining pinfeathers and down. Because of the difficulty of handpicking, it is common practice to finish the "rough-picked" birds by dipping each bird in melted, specially formulated wax.

In large-scale operations, the birds are waxed in an on-the-line process with a wax temperature of 140° to 220°F (60–100°C). After waxing, the birds are exposed to a cold water spray or dipped in a tank of cold water to cool and harden the wax to a "tacky" state. The wax is then removed by hand, resulting in a clean, attractive carcass. The wax is reused by melting and straining out the feathers, pins, and so forth.

The U.S. standards of quality are essentially the same for all poultry. Geese are graded for conformation, fleshing, and fattening. Defects, such as missing skin and bruises, are also considered in establishing quality.

FEATHERS. Goose feathers are valuable for the bedding and clothing industries due to their properties (see the discussion on down feathers).

THE RATITES—OSTRICHES, EMUS, AND RHEAS

The ratites (walking birds) are flightless birds that include the emu (*Dromaius novae-hollandiae*) from Australia, the kiwi (from New Zealand), the ostrich (*Struthio camelus*) from Africa (see Figure 19.10), and the rhea (*Rhea americana*) from South America. The ratites are often thought to be primitive birds. For instance, the feathers are primitive and are not interlocking as with other birds. Of the ratites, the ostrich is raised commercially, with some emu farming/ranching and experimentation with rhea production.

Evolution

The ratites evolved from birds that could fly. It is mostly likely that a common ancestor lived in Gondwanaland. This was a supercontinent made up of Africa, India, Australia, South America, and Antarctica in the Mesozoic era. With the breakup of Gondwanaland, the population of ratite ancestors were isolated and evolved into ostriches (Africa), emus (Australia), and rheas (South America). Using molecular clocks, it has been estimated

Figure 19.10 Ostrich. *(Courtesy, Jerry Sell, Iowa State University)*

that the ancestors of the different ratites of South America, Africa, and Australia diverged about 51 million years ago after the complete breakup of Gondwanaland and its components separated by continental drift.

Classification

Ratites are classified as follows:

Phylum: Chordata

Subphylum: Vertebrata

Class: Aves

Superorder: Ratatae

Order: Struthioniformes

Species: *Dromaius novae-hollandiae* (emu)
Struthio camelus (ostrich)
Rhea americana (rhea)

Domestication

Ratites have been domesticated in the past 150 years. Ostriches and emus are produced commercially in Southern Africa, Australia, and North America and Europe. Some rheas are also farmed. There has been a significant growth of emu and ostrich production in both the United States and countries in the European Union. Until recently, the meat was essentially a by-product, and ostriches were raised for hide for leather making. Feathers (ostriches) and oil (emus) are other important by-products. Emu oil is used, for instance, in cosmetics. Advantages of emus and ostriches are the quality of hide, the feathers, the meat, and the rapid growth rate.

With the possible exception of sales of founder breeding stock, profitability in the last 15 years in the ratite industry has been low or nonexistent. Prices for a breeding pair of ostriches are about $2,000 to $4,000. Some consider that ostrich production is more economically viable. This can be compared to $400 to $450 for slaughter birds. Costs for raising emus and ostriches are relatively high. These include:

- Bird housing and land
- Feed
- Incubation facilities
- Veterinary care

The Ostrich (*Struthio camelus*)

The ostrich (*Struthio camelus*) was named as the camel bird. The adults of these, the largest birds, are about 7 1/2 ft high (~2.75 meters) and weigh 300 lb (140 kg). The ostrich cannot fly, but it can run fast, attaining a speed of 40 miles (~65 kilometers) per hour. Ostriches can live to be 70 years of age.

Ostriches probably evolved in Africa but extended their range over geological history to include Eurasia. Wild ostriches are found in much of Africa and until about 100 years ago in Arabia and the Middle East. Based on the latter, it is not surprising that ostriches are mentioned in the Bible.

Subspecies

There are four subspecies of the single ostrich species. These are limited to distinct geographical regions in Africa:

Struthio camelus camelus (North Africa)

S.c. molybdophanes (Africa/Horn of Africa)

S.c. massaicus (East Africa—Kenya and Tanzania)

S.c. australis (South Africa)

History of Ostrich Farming/Ranching

Ostriches have been farmed in South Africa since about 1857 (see Figure 19.11 for ostriches being farmed). Farming was initiated to enhance production of feathers, which were used extensively in fashionable hats, and so on. There was a peak in farmed ostriches in 1914 with about 800,000. With the onset of World War I, demand for feathers plummeted as did ostrich numbers. The farmed ostrich (some-

Figure 19.11   Ostriches in pen. *(Courtesy, Jerry Sell, Iowa State University)*

times referred to as *S.c. domesticus* or the black) is predominantly derived from chicks from wild populations of *S.c. massaicus* (red-necks) and *S.c. australis* (blue-necks).

Size of Ostrich Production in the World

The cumulative total of ostriches slaughtered in the world is 340,000 per year, resulting in about 80 million lb (35 thousand metric tons) of meat. South Africa's ostrich production is presently about 270,000 birds slaughtered per year. Sales of South African ostrich products include the following:

- Hides (~$40 million per year)
- Meat (~$10 million/year)
- Feathers (~$5 million per year)

Production also occurs in the United States, Europe, Australia, and Israel.

Biology of Ostriches

GASTROINTESTINAL PHYSIOLOGY/ANATOMY AND NUTRITION.   The ostrich gastrointestinal tract resembles on a larger scale that of the chicken but without a crop. Ostriches swallow stones to aid gizzard grinding of feed. Moreover, they consume significant amounts of soil, about 2% of feed consumed. This can affect mineral nutrition either as a source of minerals or, by chelation, decreasing absorption of specific minerals.

In the wild, ostriches are selective grazers but do consume some dry bones. Ostriches are able to digest plant fiber such as hemicellulose and cellulose due to microbial fermentation in the large intestine and ceca, with the production of volatile fatty acids (e.g., acetate). But transmit time is long (1.5–2 days). It is not surprising that feeds have higher total metabolizable energy (TME_n) in turkeys than chickens.

Figure 19.12   Young ostriches. *(Courtesy, Jerry Sell, Iowa State University)*

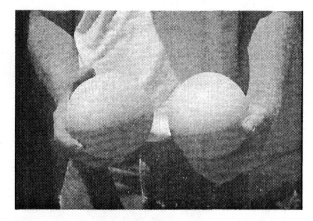

Figure 19.13   Ostrich eggs. *(Courtesy, Jerry Sell, Iowa State University)*

Digestibility of neutral digestible fiber is higher than in chickens. It increases in ostrich chicks up to about 60% at 17 weeks old. Similarly, digestibilities of individual amino acids are consistently higher than in chickens.

WATER BALANCE.   As the name camel bird might imply, the ostrich has relatively low water needs. Studies indicate when water is available, an ostrich will gain water from drinking (7.9 liters/day), metabolism (0.4 liters/day), and food (0.2 liters/day), and lose water in feces and urine (5.4 liters/day) and by evaporation (for cooling) (3.1 liters/day).

HORMONES.   Ostriches and emus have the full repertoire of avian hormones. However, ratites appear to have abnormal thyroid function, resulting in their neotenic or "infantile" appearance.

GROWTH.   Growth rates in young ostriches vary considerably (see Figure 19.12 for young ostriches). On average, female ostriches grow faster than males, gaining 1.6 kg (3.5 lb) per week compared to 1.2 kg (2.6 lb) per week in males. Average growth rates of 1 lb/day (0.44 kg/d) have been reported in ostriches (at 100–150 lb or 50–70 kg).

REPRODUCTION.   Both wild and farmed ostriches are seasonal breeders. The breeding season varies with the country and geography where the ostriches are raised. For instance in South Africa, the breeding season is between June and February (midwinter through late summer). In Australia, it is between July to March (midwinter through late summer). In contrast in the United States, the breeding season is May to September (late spring to the end of summer). However, close to the equator, ostriches may breed throughout the year.

The anatomy and physiology of the reproductive system of the ostrich is essentially very similar to that in the chicken (see Chapter 2), except in scale. For instance, each testes of the adult male is 6 in. (15

cm) long and 8 in. (20 cm) in circumference. In the wild, and with many farmed ostriches, the female (or hen) ostriches will incubate about 20 eggs at a time when with other females. The "major" or "dominant" female incubates the eggs frequently, discarding eggs that are not her own (see Figure 19.13 for an ostrich egg illustration).

PRACTICAL ASPECTS OF REPRODUCTION.   Adult breeding ostriches are farmed either in pairs (one male and one female) or trios (one male and two females). The only reliable method of determining the sex of young chicks is examining the sex organs. It is advisable to keep the males and females separated prior to pairing for mating. At the time of pairing, the ostriches should be kept in a pen or paddock fenced with smooth wire about 6 ft high (~2 m).

Farmed ostriches show much lower reproductive efficiency and fertility than chickens. In a single breeding season, a mature female ostrich will lay about 50 eggs (compared to more than 300 per year in a chicken). However, about half of the adult hens do not produce any eggs. Nutrition is critical to reproduction, but requirements are still not adequately documented.

Infertility.   Female infertility includes vaginal prolapse, egg retention, and oviductal infections. Male infertility may be age related (male too young) or due to exhaustion of spermatozoa supplies due to repeated mating during prolonged egg laying in females.

Artificial Insemination.   Artificial insemination is beginning to be employed. Ostrich semen can be collected manually into a warmed collection tube (see Figure 19.14 for a diagram of ostrich spermatozoan). Injections of oxytocin have been found to increase semen volume.

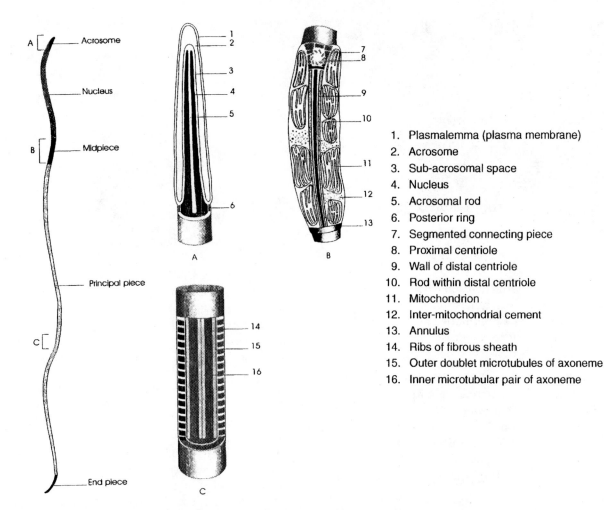

1. Plasmalemma (plasma membrane)
2. Acrosome
3. Sub-acrosomal space
4. Nucleus
5. Acrosomal rod
6. Posterior ring
7. Segmented connecting piece
8. Proximal centriole
9. Wall of distal centriole
10. Rod within distal centriole
11. Mitochondrion
12. Inter-mitochondrial cement
13. Annulus
14. Ribs of fibrous sheath
15. Outer doublet microtubules of axoneme
16. Inner microtubular pair of axoneme

Figure 19.14   Schematic diagram of the structure of the ostrich spermatozoan. *(Reprinted from* The Ostrich. *(1999). D.C. Deeming (ed.) with permission from CABI Publishing, Wallingford, UK)*

CHARACTERISTICS OF OSTRICH EGGS.   Ostrich eggs have a similar composition to chicken eggs.

1. Egg production per hen per breeding season: ~50 eggs.
2. Fertility of eggs is low: 75% in South Africa and even lower elsewhere.
3. Hatchability of fertile egg about 70%.
4. Ostrich eggs are ~2.2 to 4.4 lb (~1–2 kg) (average 3.3 lb or 1.5 kg) (see Figure 19.13).
5. Incubation temperature: 97.5°F (36.5°C).
6. Length of incubation: 42 days.
7. Mass of ostrich chick at hatching: ~66% of initial egg mass.
8. Bacterial contamination in eggs: ~13%.

EGG STORAGE AND INCUBATION.   As opposed to the placing of eggs in nests by many other species of birds, in ostriches the eggs are buried. Thus, the techniques of egg storage and incubation by the industry must mimic the natural conditions. It is critically important to store ostrich eggs under con-ditions optimal for viability during incubation, hatching, and posthatching growth. Eggs should be collected within 1 day of laying and stored at ~71°F (~22°C). Incubation temperature is thought to be optimal at 97.5°F (36.4°C). Eggs should be positioned with the large end up at a 45-degree angle. The eggs should be turned twice daily.

BROODING AND REARING.   The brooding period of young ostriches (chicks) is very critical. Chicks should be brooded at a temperature of about 90°F (32.2°C) for 14 days. Following this, the temperature is reduced a few degrees each week.

Ostrich Production

Ostriches are kept in paddocks or large fenced areas. The stocking density varies with the ability of the land to produce adequate forage. For instance in South Africa, the stocking density on the velt is between 2.5 and 25 acres per bird (1–10 hectare/bird).

Limitations To Ostrich Production

The efficiency of production is low for a number of reasons. There is a lack of knowledge of optimal nutrition, disease prevention, and environmental conditions (e.g., ventilation). Genetically improved emus and ostriches have not been developed due to the forming of breeding pairs. Until recently artificial insemination had not been developed. This slowed development of selective breeding programs (to improve growth rate, disease resistance, etc.) Moreover, there are technical problems as indicated by the low hatchability of eggs and relatively high posthatch mortality.

With commercial ratite production around the world, the birds are being exposed to diseases that they have not previously seen. Mortalities of over 25% have been reported in the first 3 months of life for ostriches. Skewed sex ratios (more males) are due to higher mortality and incubation failures in females.

At this time, there has been little consolidation or vertical integration with concomitant economies of scale. In summary, the major biological limitations to ostrich farming are the following:

1. Low egg fertility
2. Low hatchability
3. Relatively high chick mortality

Slaughter

In many countries such as the United States, Australia, and Europe, ostriches go to market at 9 months of age (190 lb or ~85–90 kg). This optimizes the economics of ostrich meat production with feed: gain declining after this age. In South Africa, ostriches are slaughtered at 14 months since this results in the best hides for leather. Feathers are harvested on the farm. Ostriches are transported to abattoirs for processing plants. Birds are slaughtered by electrical stunning or captive bolt. Following bleeding out, the plucking is completed, the head is removed, and the hide is removed. This is followed by evisceration and, finally, removal of meat cuts.

Characteristics of Ostrich Muscle/Meat

Ostrich muscle has the following characteristics:

- Very low in fat (~0.5%)
- Relatively low in cholesterol (~60 mg per 100 g)
- High protein (21%)
- Water (77%)

The low-fat content is due to there being little intramuscular fat. The fatty acid profile differs from most meats being 37% saturated fatty acids, 32% monounsaturated fatty acids, and 32% polyunsaturated fatty acids.

Ostrich meat is less juicy than beef or chicken due in part to the low fat. This problem is accentuated if the meat is overcooked (it should not be cooked too well done!). Ostrich meat can be used in most processed meats, particularly as a replacement for beef. The high pH value indicates that it should not be used for raw hams.

Emus

There is even less information about emu production than there is for ostriches. In general, emu husbandry is assumed to be similar to that of ostriches. For illustrations of emus and emu farming see Figures 19.15 and 19.16.

Slaughter

Slaughter of emus can employ electrical stunning prior to slaughter or a modified captive bolt.

Reproduction

Reproduction is seasonal in emus. Like sheep, they are considered short-day breeders (requiring short daily photoperiods to come into reproductively mature/active condition). As emus form pair bonds,

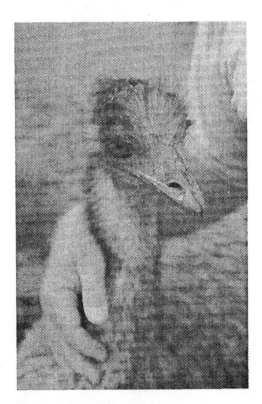

Figure 19.15 Emu. *(Courtesy, USDA)*

Figure 19.16   Emus in pen. *(Courtesy, USDA)*

genetic improvement is difficult. With the development of artificial insemination, it will be possible to use superior males for genetic improvement.

PIGEONS

Pigeons are used for the following purposes (see Figure 19.17):

- The sport of racing pigeons
- Flyers and performers
- Showing fancy pigeons
- Meat production

Pigeon Meat

Pigeon meat is sold as either squab (young, immature pigeon) or pigeon (adult male or female). Squab is an exceptionally tender meat while pigeon

Figure 19.17   Pigeon. *(Original diagram from E. Ensminger)*

is a tougher meat. An uncooked dressed pigeon is about 57% water, 24% fat, and 19% protein.

Classification

Pigeons are classified as follows:

> **Phylum:** Chordata
> **Subphylum:** Vertebrata
> **Class:** Aves
> **Superorder:** Carinatae
> **Order:** Columbiformes
> **Species:** *Columbia livia* (pigeon)

Domestic pigeons have been referred to in the past as *Columbia domesticus*.

Domestication

Pigeons were probably domesticated about 5,000 years ago. Paintings and carvings showing these birds have been found in the ancient cultures of Europe, Asia, and the Near East. Pigeons and doves have been traditionally associated with peace and love.

Breeds

Through the centuries, selective breeding of pigeons has resulted in about 200 different breeds, each with distinct behavior, size, shape, stance, feather form, and colors. The Homer, White King, and Swiss Mondaines are the most popular.

Breeding

Pigeons are ready to mate at about 4 to 5 months of age. They mate in pairs and usually remain with their mates throughout life. The pigeon hen lays an .egg, generally skips a day, then lays again. The incubation period is about 17 days. Both males and females sit on or incubate eggs.

Both parents care for the young, called *squabs* (the newly hatched chicks are sometimes referred to as *squeakers*). They feed them by regurgitating a thick, creamy mixture, called pigeon milk, into the open mouths of the young. This milk is produced by the crop when stimulated by the hormone, prolactin. The "milk" consists of fat and protein from crop epithelial cells that are sloughed off. Pigeons grow rapidly. Squabs exceed the normal adult weight at the time they are ready to leave the nest, about 30 days of age.

FEEDING AND WATERING.   Pigeons are grain eaters. They prefer a variety or mixture of grains or commercial pigeon pellets. Example rations are given in Chapter 7.

GAME AND ORNAMENTAL BIRDS

Most game and ornamental birds are of the order Galliformes(see Figures 19.18 to 19.21 for examples). Game birds are raised for sale to game preserves or for shooting preserves or for meat. Some producers raise them for profit, on a full-time or part-time basis. They also may be raised for pleasure. There is a market for the sale of ornamental birds.

Feeding and Watering

Unless otherwise stated, these game and ornamental birds should receive a commercial game bird diet (see Chapter 7 for details). If this is not available, turkey feed can be used. Water should always be available.

Diseases

The birds are subject to many poultry diseases and respond to the same treatments as chickens and turkeys.

Bobwhite Quail (*Colinus virginianus*)

The name, Bobwhite, comes from its call. In the spring, cocks not yet paired sound their loud, tuneful "bobwhite" call. The Bobwhite quail lives on fallow fields, meadowlands rich in bushes, and open woodlands from the Canadian border to Mexico and Cuba. Bobwhites are also raised on quail farms for hunting purposes.

Guinea Fowl

The fowls derive their name from Guinea, a part of the West Coast of Africa (see Figure 19.18 for an illustration of guinea fowl).

Guinea Fowl Meat

Some restaurants serve guineas at banquet and club dinners as a special delicacy; as a substitute for game

Figure 19.18 Guinea fowl. *(Courtesy, USDA)*

birds—grouse, partridge, quail, and pheasant. Guinea fowl are marketed at 12 weeks old with dressed carcass, 1.5 kg.

Breeds

There are three principal varieties of domesticated guinea fowl in the United States: Pearl, White, and Lavender.

Reproduction

Hens can lay more than 100 eggs per year. The incubation period is 26 to 28 days.

Feeding and Watering

A commercial chicken or turkey breeder mash (22–24% protein) should be fed to layers. Young guineas should have a growing chicken diet.

Partridge

Both the Hungarian partridge of Europe and the Chukar of Asia have been introduced into the United States. Hungarian partridges, sometimes referred to as gray cannonballs, were introduced into the Western United States early in the 20th century. Today, most partridges are found on the Canadian plains and in the North Central and Northwestern United States.

Coturnix Quail (Japanese and European Quail—*Coturnix coturnix coturnix* and *Coturnix coturnix japonica*)

Coturnix quail are divided into two subspecies: the European quail (*Coturnix coturnix coturnix*) and the Asian or Japanese quail (*Coturnix coturnix japonica*) (see Figures 19.20 and 19.21). These quail are also known as coturnix quail, pharoah's quail, stubble quail, and eastern quail. The Japanese quail is smaller than the Bobwhite quail, but it produces a larger egg. Japanese quail were either domesticated in Japan about the 11th century or brought to Japan from China about that time. They were first raised as pets and singing birds. By 1900 in Japan, these quail had become widely used for meat and egg production. Japanese quail were imported into the United States from Japan. Japanese quail are used for eggs, meat, hunting dog training, and research (they are a useful model for poultry).

Breeds

The following varieties (breeds) of Japanese quail are known: Manchurian Golden, British Range, English White, and Tuxedo.

Figure 19.19   Pheasant. *(Courtesy, USDA)*

Figure 19.21   Adult Japanese quail.

Figure 19.20   Day-old Japanese quail. *(ISU photo by Bob Elbert)*

Peafowl

The peafowl belongs to the same family as pheasants and chickens, differing particularly in plumage. The peafowl is native to India and South Asia. About 300 B.C., Alexander the Great introduced the birds to Greece.

Breeds

There are three varieties of peafowl: Indian Blue, Java Green, and Congo. The Indian Blue is most common.

Pheasants

These are generally classed as game birds (see Figure 19.19). Pheasants originated in Eurasia. They were first brought to North America by Benjamin Franklin's son-in-law, but the venture was unsuccessful. Later, the U.S. Consul in Shanghai sent 28 Chinese pheasants to Oregon. This latter introduction met with success.

Breeds

Pheasants are classed as (1) game breeds or (2) ornamental breeds. The game breeds are Blackneck Pheasant, Chinese Ringneck Pheasant, English Ringneck Pheasant, Formosan Pheasant, and Melanistic Mutant Mongolian Pheasant. The ornamental breeds are Amherst's Pheasant, Golden Pheasant, and Reeves' Pheasant. The Chinese Ringneck pheasant is the most popular variety.

 USEFUL WEBSITES

Cherry Valley Farms ducks
http://www.cherryvalley.co.uk/

FURTHER READING

Barbut, S. *Poultry Products Processing: An Industry Guide.* Boca Raton, FL: CRC Press, 2002.

Deeming, D. C., ed. *The Ostrich: Biology, Production, and Health,* 1–358. New York: CABI Publishing, 1999.

Diseases of Poultry. Edited by B. W. Calnek, H. John Barnes, Charles W. Beard, Larry R. McDougald, and Y. M. Saif. Ames, IA: Iowa State Press, 1997.

Pattison, M. *The Health of Poultry.* Harlow, England: Longman, 1993.

Reproduction in Farm Animals. Edited by B. Hafez and E. S. E. Hafez. Philadelphia: Lippincott, Williams and Wilkins, 2000.

Sturkie's Avian Physiology, 5th ed. Edited by G. C. Whittow. San Diego, CA: Academic Press, 2000.

Business Suggestions for Small- and Moderate-Sized Poultry Producers

TYPES OF BUSINESS ORGANIZATIONS

The success of your poultry enterprise is dependent upon the type of business organization. Four major types of business organizations are found in poultry and other agricultural enterprises: (1) sole proprietorship, (2) the partnership, (3) the corporation, and (4) contracts.

Proprietorship (Individual)

The sole proprietorship is a business owned and operated by one individual. This is the most common type of business organization in U.S. farming as a whole. Under the sole proprietorship, one person controls the business, having sole management and control. This may be modified through contracts. The sole proprietors get all the profits of the business; likewise, they must absorb all the losses.

There are two major limitations: (1) It may be more difficult to acquire new capital for expansion; and (2) not much can be done to provide for continuity with the passing of the owner.

Partnership (General)

A partnership is an association of two or more persons who, as co-owners, operate the business. About

USEFUL WEBSITES

for Range Production of Poultry

http://www.wisc.edu/cias/research/livestoc.html#poultry

13% of U.S. farms are partnerships. Most agricultural partnerships involve family members who have pooled land, buildings/equipment, their labor, and management to operate a larger business. It is a good way to bring a son or daughter, who is usually short on capital, into the business. Although there are financial risks to each member of such a partnership and potential conflicts in management decisions, the existence of family ties tends to minimize such problems.

In order for a partnership to be successful, the enterprise must be sufficiently large to utilize the abilities and skills of the partners and to compensate the partners adequately in keeping with their respective contributions to the business. A partnership has the following advantages:

- **Combining resources can increase efficiency.** It is very important that the partners agree on the value of each person's contribution to the business, and that this be clearly spelled out in the partnership agreement.
- **Equitable management.** Unless otherwise agreed upon, all partners have equal rights, regardless of financial interest. Any limitations, such as voting rights proportionate to investments, should be a written part of the agreement.
- **Tax savings.** A partnership does not pay any tax on its income, but it must file an informational return. The tax is paid as part of the individual tax returns of the respective partners, usually at lower tax rates.
- **Flexibility.** Usually, the partnership does not need outside approval to change its structure or operation—the vote of the partners suffices.

337

Partnerships may have the following disadvantages:

- **Liability for debts.** In a partnership, each partner is liable for all the debts and obligations of the partnership.
- **Uncertainty of length of agreement.** A partnership ceases with the death or withdrawal of any partner, unless the agreement provides for continuation by the remaining partners.
- Difficulty of determining the value of partner's interest.
- Limitations of management effectiveness.

Limited Partnership

A limited partnership is an arrangement in which two or more parties supply the capital, but only one partner is involved in the management. This is a special type of partnership with one or more "general partners" and one or more "limited partners."

The limited partnership avoids many of the problems inherent in a general partnership, and it has become the chief legal device for attracting outside investor capital into farm ventures. As the term implies, the financial liability of each partner is limited to his or her original investment. The partnership does not require, and in fact prohibits, direct involvement of the limited partners in management. A limited partner is in a similar position to a stockholder in a corporation.

The advantages of a limited partnership are:

- It facilitates bringing in outside capital.
- It need not dissolve with the loss of a partner.
- Interests may be sold or transferred.
- The business is taxed as a partnership.
- Liability is limited.
- It may be used as a tax shelter.

The disadvantages of a limited partnership are:

- The general partner has unlimited liability.
- The limited partners have no voice in management.

Corporations

A corporation is a device for carrying out a farming enterprise as an entity entirely distinct from the persons who are interested in and control it. Each state authorizes the existence of corporations. As long as the corporation complies with the provisions of the law, it continues to exist—irrespective of changes in its membership. There has been increased interest in the use of corporations for the conducting of farm business.

From an operational standpoint, a corporation possesses many of the privileges and responsibilities of a real person. It can own property; it can hire labor; it can sue and be sued; and it pays taxes.

Separation of ownership and management is a unique feature of corporations. The owners' equity interest in a corporation is represented by shares of stock. The shareholders elect the board of directors who, in turn, elects the officers. The officers are responsible for the day-to-day operation of the business. Of course, in a close family corporation, shareholders, directors, and officers can be the same people.

The major advantages of a corporate structure are:

- It provides continuity despite the death of a stockholder.
- It facilitates transfer of ownership.
- It limits the liability of shareholders to the value of their stock.
- It may make for some savings in income taxes.

The major disadvantages of a corporation are:

- It is restricted to doing only what is specified in its charter.
- It must register in each state.
- It must comply with stipulated regulations, involving considerable paperwork and expense.
- It is subject to the hazard of higher taxes.
- It is possible to lose control.

Another type of corporation is family owned (privately owned). It enjoys most of the advantages of its generally larger outside investor counterpart, with few of the disadvantages. The chief advantages of the family owned corporation over a partnership arrangement are that it alleviates unlimited liability; a lawsuit cannot financially destroy the entire business and the individual partners with it; and it facilitates estate planning and ownership transfer.

CAPITAL NEEDS

Farmers have on average about 80% equity in their business and 20% borrowed money (debts).

CREDIT

Credit is an integral part of today's poultry business. Wise use of it can be profitable, and unwise use disastrous. Accordingly, poultry producers should know more about credit. They need to know something about the lending agencies available to them, the types of credit, and how to go about obtaining a loan.

Types of Credit or Loans

It is important that the best available source of the loan be secured. Getting the needed credit through the right kind of loan is an important part of sound financial poultry management. The following three general types of agricultural credit are available, based on length of life and type of collateral needed:

1. **Short-term loans.** This type of loan is made for operating expenses and is usually for 1 year or less. It may be used for the purchase of birds, feed, and for operating expenses; and it is repaid when eggs or birds are sold. Security, such as a chattel mortgage on the birds, may be required by the lender.
2. **Intermediate-term loans.** These loans are used for buying equipment, for making land improvements, and for remodeling existing buildings. They are paid back in 1 to 7 years. Generally, they are secured by a chattel mortgage.
3. **Long-term loans.** These loans are secured by mortgage on real estate and are used to buy land or make major improvements to farmland and buildings, or to finance construction of new buildings. They may be for as long as 40 years. Usually they are paid off in regular annual or semiannual payments. Among the best sources for long-term loans are an insurance company, the Federal Land Bank, the Farm Home Administration, or an individual.

Credit Sources

Farmers borrow from banks and the Farm Credit Service. But agricultural financing is changing. Today, farmers are tapping the vast supply of equity or risk capital that is constantly seeking investment opportunities.

Credit Factors Considered and Evaluated by Lenders

Potential money borrowers sometimes make their first big mistake by going in "cold" to see a lender, without adequate facts and figures, with the result that they have two strikes against getting the loan.

When considering and reviewing loan requests, the lender tries to arrive at the repayment ability of the potential borrower. Likewise, the borrower has no reason to obtain money unless it will make money. Lenders need certain basic information including:

- Lenders are impressed with borrowers who have a feasibility study showing where they are, where they are going, and how they expect to get there. This gives an assurance of the necessary management skills. An analysis of the present and future projections is imperative.
- It is the obligation of borrowers to present the following information to the lender:
  - Name of applicant and spouse; age of applicant
  - Number of children (minors; legal age)
  - Years in geographical area
  - References
  - Owner or tenant
  - Location—legal description and county
  - Type of enterprise
  - Financial statement—the borrower's financial record and current financial position + assets and liabilities

Borrowers should always have sufficient slack and/or insurance to absorb reasonable losses due to such unforeseen circumstances as floods, storms, diseases, and poor markets; thereby permitting lenders to stay with them in adversity and to give them a chance to recoup their losses in the future. The financial statement should include the following:

1. Current assets:
   - Number of birds.
   - Feed.
   - Machinery.
   - Cash. (There should be reasonable cash reserves to provide a cushion against emergencies.)
   - Bonds or other investments.
   - Cash value of life insurance.
   - Fixed assets—land/property with estimated value.
   - Long-term contracts.
2. Current liabilities:
   - Mortgages.
   - Contracts.
   - Open accounts—to whom owed.
   - Cosigner or guarantor on notes.
   - Any taxes due.
   - Current portion of real estate indebtedness due.
3. Fixed liabilities—amount and nature of real estate debt:
   - Date due.
   - Interest rate.
   - To whom payable.
   - Contract or mortgage.

Shrewd lenders usually ferret out many things. Most lenders recognize that the potential borrower is the most important part of the loan.

Lenders consider the borrower's character, integrity, experience, credit rating, health, and so forth.

Production records. This refers to a good set of records showing efficiency of production. These should show egg production and feed efficiency of layers and rate of gain and feed efficiency of meat birds plus mortalities. Lenders insist on good records and good performance.

Progress with previous loans. Has the borrower paid back previous loans plus interest; has the borrower reduced the amount of the loan, thereby giving evidence of progress?

Profit and loss (P & L) statement. This serves as a valuable guide to the potential ahead. Preferably, this should cover the previous 3 years. Also, most lenders prefer that this be on an accrual basis (even if the poultry producer is on a cash basis in reporting to the Internal Revenue Service).

- Is it an economic unit with good buildings, recently purchased equipment, and adequate water?
- Are the right poultry being produced, and is there a market?
- Is there adequate collateral (or security) to cover the loan, with a margin?

The loan request. Poultry producers are in competition for money from other businesses. Hence, it is important that their request for a loan be well presented and supported. The potential borrower should tell the purpose of the loan; how much money is needed, when it is needed, and what it is needed for; the soundness of the venture; and the repayment schedule.

Credit Factors Considered by Borrowers

Credit is a two-way street; it must be good for both borrowers and lenders. If borrowers are the right kind of people and on a sound basis, more than one lender will want their business. Thus, it is usually well that borrowers shop around. There are basic differences in length and type of loan, repayment schedules, services provided with the loan, interest rate, and the ability and willingness of lenders to stick by the borrower in emergencies and times of adversity. Thus, interest rates and willingness to loan are only two of the several factors to consider.

Helpful Hints for Building and Maintaining a Good Credit Rating

Poultry producers who wish to build and maintain good credit are admonished to do the following:

1. Keep credit in one place, or in a few places. Generally, lenders frown upon "split financing."
2. Borrowers should shop around for a creditor (a) who is able, willing, and interested in extending the kind and amount of credit needed; and (b) who will lend at a reasonable rate of interest; then stay with the borrower.
3. Get the right kind of credit. Do not use short-term credit to finance long-term improvements or other capital investments. Also, use the credit for the purpose intended.
4. Be frank with the lender. Be completely open and aboveboard. Mutual confidence and esteem should prevail between borrower and lender.
5. Complete and accurate records should be kept by all enterprises. By knowing the cost of doing business, decision making can be on a sound basis.
6. Take an annual inventory for the purpose of showing progress made during the year.
7. Repay loans when due. Borrowers should work out a repayment schedule on each loan, then meet payments when due. Sale proceeds should be promptly applied on loans.
8. Plan ahead—analyze the next year's operation and project ahead.

Borrow Money to Make Money

Poultry producers should never borrow money unless they are reasonably certain that it will make or save money. With this in mind, borrowers should ask, "How much should I borrow?" rather than, "How much will you lend me?"

Calculating Interest

The charge for the use of money is called interest. The basic charge is strongly influenced by the following:

- The basic cost of money in the money market.
- The servicing costs of making, handling, collecting, and keeping necessary records on loans.
- The risk of loss.
- Competition between lenders.

Interest rates vary among lenders and can be quoted and applied in several different ways. The quoted rate is not always the basis for proper comparison and analysis of credit costs. Even though several lenders may quote the same interest rate, the effective or simple annual rate of interest may vary widely. The more common procedures for determining the actual annual interest rate, or the equivalent of simple interest on the unpaid balance, follow. (Of course, the going rate of interest should be substituted for the 12% figure used herein.)

If borrowing $12,000,

Interest = .12 × $12,000 = $1,440 per year

Points

Some lenders charge points. A point is 1% of the face value of the loan. Thus, if 4 points are being charged on a $12,000 loan, $480 dollars will be deducted and the borrower will receive only $11,520. But the borrower will have to repay the full $12,000. Obviously, this means that the actual interest rate will be more than the stated rate. But how much more?

Assume that a $12,000 loan is for 1 year and the annual rate of interest is 12%. Then the payment by the borrower of 4 points would make the actual interest rate as follows:

Interest = .12 × $12,000 = $1,440

Average use of money is $11,520 for one year

Effective rate of interest is $1,440 (interest)/
$11,520 = 12.5%

RECORDS AND ACCOUNTS

The key to good business and management is records. The historian, Santayana, put it this way, "Those who are ignorant of the past are condemned to repeat it." Also, good records help the poultry producer to overcome the banker's traditional fear of feathers.

Why Keep Records?

The chief functions of records and accounts are:

1. To provide information by which your poultry business may be analyzed, with its strong and weak points ascertained. From these facts, you may adjust current operations and develop more effective planning.
2. To provide profit and loss statements.
3. To provide a net worth statement, showing financial progress during the year.
4. To provide cash flow projections, showing when money is needed, and showing loan repayability.
5. To furnish an accurate, but simple, net income statement for use in filing tax returns.
6. To keep production records on birds.
7. To aid in making a credit statement when a loan is needed.
8. To keep a complete historical record of financial transactions for future reference.

Good records, properly analyzed and used, will increase net earnings and serve as a basis for sound management.

Kind of Record or Account Book

The record forms will differ somewhat according to the type of enterprise. For example, with layers, cost per dozen eggs is the important thing, whereas in broiler chicken, turkey, and duck production, it is cost per pound of bird. Net returns are important, but it is also necessary that records show all the items of cost and income—egg production, feed consumption, and mortality of layers; rate of growth, pounds of chicken per 100 lb of feed, mortality, and quality of broilers produced.

It is recommended that you obtain a copy of a record book, particularly one prepared for and adapted to your business. Also, some commercial companies distribute very acceptable record books without charge.

Kind of Records to Keep

Your records should be easy to keep and should give the information desired to make a valuable analysis of the business. You should keep the following kinds of records.

Annual Inventory

The annual inventory is the most valuable record that poultry producers can keep. It should include a list and value of real estate, poultry, equipment, feed, supplies, and all other property, including cash on hand, notes, and bills receivable. Also, it should include a list of mortgages, notes, and bills payable. It shows the producers what they own and what they owe; whether they are getting ahead or going behind. The following pointers may be helpful relative to the annual inventory:

1. The inventory should be taken at the beginning of the account year; usually this means December 31 or January 1.
2. It is important that each item be properly and separately listed.

METHOD OF ARRIVING AT INVENTORY VALUES.   It is difficult to set up any hard and fast rule to follow in estimating values when taking inventories. Perhaps the following guides are as good as any:

- Estimating the value of farm real estate is, without doubt, the most difficult of all. It is suggested that the owner use either (a) the cost of the farm, (b) the present sale value of the farm, or (c) the capitalized rent value according to its productive ability with an average operator.
- Buildings are generally inventoried on the basis of cost less observed depreciation and obsolescence. Once the original value of a building is arrived at, it is usually best to take

depreciation on a straight-line basis by dividing the original value by the estimated life in terms of years. Usually 4% or more depreciation is charged off each year for income tax purposes (consult your CPA for details).

- Birds are usually not too difficult to inventory because there are generally sufficient current sales to serve as a reliable estimate of value.
- The inventory value of equipment is usually arrived at by either of two methods: (a) the original cost less a reasonable allowance for depreciation each year, or (b) the probable price that it would bring at a well-attended auction. Under conditions of ordinary wear and reasonable care, it can be assumed that the general run of equipment will last about 5 years. Thus, with new equipment, the annual depreciation will be the original cost divided by 5.
- The value of feed and supplies can be based on market price.

Two further points are important. Whatever method is used in arriving at inventory value (a) should be followed at both the beginning and the end of the year, and (b) should reflect the operator's opinion of the value of the property involved.

Record of Receipts and Expenses

These records are essential to any type of well-managed business. To be most useful, these entries should not only record the amount of the transaction but should also give the source of the income or the purpose of the expense, as the case may be. In other words, they should show the producer from what sources the income is derived and for what it is spent.

The following kinds and arrangements of farm record books are commonly used for recording receipts and expenditures:

1. Those that devote a separate page to each enterprise; that is, a separate page is used for the layer enterprise, another for replacement pullets, still another for crops, and so on.
2. Those that provide for a record of receipts and expenses on the same page, using one column for receipts and another for expenses. This type is easy to keep, but very difficult to analyze from the standpoint of any particular enterprise.
3. Those that combine the features of both "1" and "2." This type is more difficult to keep than the others, and may be confusing to the person keeping the record.

Household and personal accounts should be kept, but should be handled entirely separate from the poultry enterprise accounts because they are not farming expenses as such.

Summarizing and Analyzing the Records

At the end of the year, the second or closing inventory should be taken, using the same method as was followed in taking the initial inventory. The final summary should then be made, following which the records should be analyzed. In the latter connection, you should remember that the purpose of the analysis is not to prove that you have or have not been prosperous. You probably know the answer to this question already. Rather, the analysis should show actual conditions on the farm and point out ways in which these conditions may be improved.

Although you can summarize and analyze your own records, there are many advantages in having the services of a specialist for this purpose. Such a specialist is in a better position to make a "cold" appraisal without prejudice, and to compare enterprises with those of other similar operators. Thus, the specialist may discover that in comparison with other operators, the broilers on a given farm are requiring too much feed to make a pound of gain, or that the layer enterprise is much less profitable than others have experienced. The local county agent can either render or recommend such specialized assistance. In some areas, it may consist of joining a cooperative farm record group or engaging the services of a consultant; in some states, such service is provided by the extension agent.

Computers and Record Keeping

Accurate and up-to-the-minute records have taken on increasing importance in the poultry business. There is hardly any limit to what computers can do if fed the proper information. Among the difficult questions that they can answer for a specific poultry enterprise are:

- How is the entire operation doing so far? It is possible to obtain a quarterly or monthly progress report; often making it possible to spot trouble before it is too late.
- What enterprises are making money; which ones are losing? By keeping records for each enterprise, it is possible to determine strengths and weaknesses; then either to rectify the situation or shift labor and capital to a more profitable operation.
- Is each enterprise yielding maximum returns? By having profit, or performance, indicators in each enterprise, it is possible to compare these (a) with the historical average of the same poul-

try farm or (b) with the same indicators of other similar establishments.

- How to plan ahead? By using projected prices and costs, computers can show what moves to make for the future. They can be used to determine when to buy feed, purchase chicks, market meat birds, and so forth. They are a powerful planning tool.
- How can income taxes be cut to the legal minimum? By keeping accurate records of expenses and depreciation, it is possible to cut tapes.

BUDGETS IN THE POULTRY BUSINESS

A budget is a projection of records and accounts and a plan for organizing and operating ahead for a specified period of time. A short-time budget is usually for 1 year, whereas a long-time budget is for a period of years. The principal value of a budget is that it provides a working plan through which the operation can be coordinated. Changes in prices, droughts, and other factors make adjustments necessary. But these adjustments are more simply and wisely made if there is a written budget to use as a reference.

How to Set Up a Budget

It is important (1) that a budget is kept, (2) that it be on a monthly basis, and (3) that you be comfortable with whatever forms or system is used. No budget is perfect. But it should be as good an estimate as can be made—despite the fact that it will be affected by unpredictable events such as diseases, markets, and others.

How to Figure Net Income

When calculating net income, there are other expenses that must be taken care of before net profit is determined, namely:

1. Depreciation on buildings and equipment. It is suggested that the useful life of buildings and equipment be as follows, with depreciation accordingly: buildings, about 31.5 years; and machinery and equipment, 5 years. Sometimes, a higher depreciation, or amortization, is desirable because it produces tax savings, and is protection against obsolescence due to scientific and technological developments.
2. Interest on owner's money invested in farm and equipment. This should be computed at the going rate in the area, say 12%.

   Here is an example of how the above works.

   Let's assume that on your poultry enterprise, there was an annual gross income of $150,000 and a gross expense of $80,000 or a surplus of $70,000.

Let's further assume that there are $20,000 worth of equipment, $50,000 worth of buildings, and $175,000 of the owner's money invested in farm and equipment. Here is the result:

Gross profit =	$70,000

Depreciation:

Equipment	$20,000 × 0.2 (20%) = $4,000	
Buildings	$50,000 × 0.032 (3.2%) = $1,600	
	Subtotal = $5,600	
Interest	$175,000 × (0.12) 12% = $21,000	
		$26,600
Return on labor and management		$43,400

Some people prefer to measure management by return on invested capital, and not wages. This approach may be accomplished by paying management wages first, then figuring return on investment.

Enterprise Accounts

Where a poultry enterprise is diversified (for example, a farm having layers, replacement pullets, and crops), enterprise accounts should be kept—three different accounts for three different enterprises. The reasons for keeping enterprise accounts are:

1. You can determine which enterprises are most profitable, and which are least profitable.
2. You can compare a given enterprise with competing enterprises of like kind.
3. You can determine the profitableness at the margin (the last unit of production). This will give an indication as to whether to increase the size of the enterprise.

ESTATE PLANNING

The preparation of wills, trusts, and partnership agreements requires consideration of the effects of federal and state tax laws. Always use the services of a CPA and attorney in estate planning.

Special-Use Valuation

Owners of farms (including poultry) and small businesses have an estate planning advantage by means of what is called special-use valuation. Under this concept, a farm can escape valuation for estate tax purposes at the highest and best use. Thus, a poultry farm located in an area undergoing development may be considerably more valuable to developers than it is as a farm. Nevertheless, if the family is willing to continue the farming use for 10 years, the farm can be included in the estate at its value as a farm. Though the procedures are clear as to how special-use valuation is elected, the frequency with which mistakes are made indicates the importance of having a competent tax attorney or CPA firm prepare the estate tax return.

Gift Tax Exclusion

It is possible to reduce estate taxes by gifting. The nontaxable gift tax exclusion is $10,000 per donee per year. A husband and wife who elect gift-splitting may jointly give $20,000 per recipient per year. These gifts may be interests in a farming operation.

Unlimited Marital Deduction

An unlimited deduction is permitted for the value of all property included in the gross estate that passes to the decedent's surviving spouse in the specified manner. Always consult a professional (CPA and attorney).

Wills

A will is a set of instructions drawn up by or for an individual that details how he or she wishes the estate to be handled after death. Despite the importance of a will in distributing property in keeping with the individual's wishes, many farmers die without a will. State law determines property distribution in such cases.

All producers should have a will drawn up by an attorney. By so doing, (1) the property will be distributed in keeping with their wishes, (2) they can name the executor of the estate, and (3) sizable tax savings can be made by the way in which the property is distributed. Wills can be changed and updated from time to time. This can be done by either a properly drawn-up codicil (formal amendment to a will) or a completely new will revoking the old one. The same attorney should prepare both the husband's and wife's wills so that a common disaster clause can be incorporated, and the estate planning of each can be coordinated.

Trusts

A trust is a written agreement by which an owner of property (the trustor) transfers title to a trustee for the benefit of persons called beneficiaries. Both real and personal property may be placed in trust. A trust should be written by your attorney.

The trustee may be an individual(s), bank, or corporation, or a combination of two or three of these. Management skill should be considered carefully in choosing a trustee. A trust can continue for any period of time set by the owner—for a lifetime, until the youngest child reaches age 21, and so forth.

LIABILITY

Most poultry producers are in such a financial position that they are vulnerable to damage suits. Moreover, the number of damage suits arising each year is increasing at an almost alarming rate, and astronomical damages are being claimed. Over 90% of the court cases involving injury result in damages being awarded. Comprehensive personal liability insurance protects a farm operator who is sued for alleged damages suffered from an accident involving his or her property or family.

Both workers' compensation insurance and employer's liability insurance protect farmers against claims or court awards resulting from injury to employees. Workers' compensation usually costs slightly more than straight employer's liability insurance, but it carries more benefits to the worker. An injured employee must prove negligence by his or her employer before the company will pay a claim under employer's liability insurance, whereas workers' compensation benefits are established by state law, and settlements are made by the insurance company without regard to who was negligent in causing the injury.

Workers' Compensation

Workers' compensation laws are in full force throughout the United States. These cover on-the-job injuries and protect disabled workers regardless of whether their disabilities are temporary or permanent. Although differences exist among the individual states in their workers' compensation laws, principally in their benefit provisions, all statutes follow a definite pattern as to employment covered, benefits, insurance, and the like. Workers' compensation provides employees with assured payment for medical expenses or lost income due to injury on the job. Whenever an employment-related injury results in death, compensation benefits are paid to the worker's dependents.

Generally, all employment is covered by workers' compensation. A few states provide exemptions for farm labor or exempt farm employers of fewer than 10 full-time employees. Farm employers in these states may elect workers' compensation protection. Producers in these states may wish to consider coverage as a financial protection strategy since the upper limits for settlement of lawsuits are set by state law under workers' compensation. This required employee benefit is costly for producers and usually requires purchase of insurance. Costs vary among insurance. Some states have a quasi-government provider of workers' compensation to assure availability of coverage for small businesses. Producers should seek the advice of paid agricultural business consultants, area extension farm management specialists, and insurance agents experienced in workers' compensation and liability insurance.

SOCIAL SECURITY

In the United States, Social Security is a system of retirement, health, and disability payments with both employees (e.g., agricultural workers) and employers (e.g., poultry producers) paying in on a regular basis. Employers pay their contribution and deduct the employee's contribution from each employee's paycheck. Employers report cash wages for each employee by January 31 of each year. Agricultural workers admitted to the United States on a temporary basis from a foreign country are not eligible.

CULLING

Profitability of a small farm (including for specific niche markets) depends on sound management. A critical decision is whether to cull birds or not. For useful guidelines see the table below.

Useful Culling Chart

Separating Layers from Nonlayers

Character	Layer	Nonlayer
Comb	Large, smooth, bright red, glossy	Dull, dry, shriveled, scaly
Face	Bright red	Yellowish tint
Vent	Large, smooth, moist	Shrunken, puckered, dry
Pubic bones	Thin, pliable, spread apart	Blunt, rigid, close together
Abdomen	Full, soft, pliable	Contracted, hard, fleshy
Skin	Soft, loose	Thick, underlaid with fat

Separating High from Low Producers

Character	High Producer (continuous laying)	Low Producer (brief laying)
Vent	Bluish white	Yellow
Eye-ring	White	Yellow
Earlobe	White	Yellow
Beak	White	Yellow
Shanks	White, flattened	Yellow, round
Plumage	Worn, soiled	Not much worn
Molting	Late, rapid	Early, slow

Raising Ducks and Geese in Free-Range Conditions

DUCKS

Management

Ducks respond to management; good management = success, poor management = trouble! Ducks are less subject to diseases than chickens or turkeys. If disease occurs, it may be due to unsanitary surroundings or genetic weaknesses.

Breeding Programs for Small-Scale Production

Section of Breeders

For small-scale production, breeding stock can be selected from ducklings hatched in the spring. Potential breeders are selected when the birds are 6 to 7 weeks of age. The distinctly different voices of males and females make it easy to separate the birds—females honk and males belch. Care should be taken to select the proper sex ratio—one drake to about six ducks. A few extra drakes and ducks are selected to allow for mortality and further culling during the conditioning period. Drakes should come from the earlier hatched flocks to ensure their readiness for mating by the beginning of the following year.

Breeders for meat breeds should be selected that are vigorous and have good weight, conformation, feathering, and growth rate. A "rule of thumb" is that drakes that weigh 5.5 lb (2.5 kg) at 6 weeks should weigh 7.5 lb (3.4 kg) by 8 weeks; (female) ducks that weigh 5.5 lb (2.5 kg) at 6 weeks should weigh 7 lb (3.2 kg) by 8 weeks. Ducklings should be selected from ducks having high fertility, hatchability, and egg production by use of trap nests and family or progeny testing. Fertility, egg production, and hatchability are economically as important to the producer of market ducks as body weight and conformation. For breeders for egg breeds, egg production is paramount for selection.

Breeding Facilities

There is little need for expensive breeder facilities. A simple shed or house can be used. As most duck eggs are laid at night, breeder ducks and drakes are confined to the laying houses at night. Up to 500 breeders can be housed in one flock. Smaller flocks (50–60 breeders) may show higher egg production and lower mortality. Provide 2.5 sq ft (0.23 m) of floor space per bird in the breeder house plus yard.

Breeder houses must be one story because domestic ducks (except Muscovies) cannot fly well. Keep breeder houses clean, dry, and well ventilated to prevent overheating during summer and to reduce dust, noxious gases, and so forth. Care should be taken to prevent the entry of rain and snow through the ventilating system. Supplementary heat need not be supplied to breeding stock.

Straw makes good bedding material as do peat moss, peanut shells, or wood shavings. Dry litter should be added frequently. Remove damp litter before it starts to mold. While ducks can make nests in the litter, simple nest boxes can be provided in long rows along the wall. Nests should be 12 in. wide, 18 in. deep, and 12 in. high (30 × 45 × 30 cm) and placed at floor level.

The use of outside drinking facilities instead of water fountains within the building has the major advantage of preventing wet pens. Ducks can be left without water during night confinement provided they have no access to feed. If it is necessary to provide watering facilities within the house, they should be placed above wire flooring or a screened drain.

Yards should slope gently away from the breeder houses to provide good drainage. Failure to provide enough open, well-drained yard space will cause dirty runs and increase the danger of disease.

Although not necessary for the production of fertile eggs, many commercial producers provide swimming water for breeding flocks. Swimming water can be provided in concrete troughs, which should be about 3 ft wide (1 m) and 8 to 12 in. (~25 cm) deep. Locate the troughs at the end of the yard opposite the breeder house.

Ducks may appear nervous and even run in circles if disturbed. All-night lights (one 15-watt lamp per 200 sq ft or 18 m) may be used in the house if this is a problem.

Care of Breeders

Do not bring birds into full production before 7 months of age, because of small eggs and low hatchability. Ducks are brought into full production by giving them 14 hours of light daily. This long-day length is proved for females for 3 weeks before the start of egg production, and to males 4 to 5 weeks before the start of egg production, so they are ready for mating. The 15-watt lamps used to reduce stampeding will stimulate egg production, but 40- to 60-watt lamps give a better response.

Egg production increases rapidly once sexual maturity is reached. The flock should be laying 90% or more of full production within 5 to 6 weeks. Daily egg production will remain above 50% for about 5 months in meat-type breeds. High-producing egg-type breeds will have greater persistency.

Producers have varying opinions about the value of keeping breeding stock once the level of egg production sinks below 50%. Some find it economical to force-molt the birds and bring them back into production in 8 to 10 weeks. Others obtain all possible eggs until production reaches 30% then they either force-molt them for an additional period of egg production or sell them for meat.

Levels of fertility and hatchability generally parallel those for egg production. They are highest when egg production is high and then taper off. Failure to obtain satisfactory levels of fertility and hatchability calls for a thorough inspection and evaluation of all phases of management.

Most duck eggs are laid before 7:00 A.M. It is advisable to gather them at this time because prompt collection lessens the problem of soiled and cracked eggs. The breeding stock should be let out of the house at the time of starting the first collection. If some ducks are laying, let them remain on the nest and make a second collection several hours later.

Carefully wash soiled eggs immediately after collection using warm water at 110° to 115°F (~40°C) containing egg sanitizer. Do not use water colder than the eggs because it will cause contraction of the egg contents, with the result that dirt, bacteria, and mold spores may be drawn through the pores of the shell. Egg sanitizers can be purchased at farm supply stores. Always follow the manufacturer's directions. Cracked, misshapen, or abnormally small or very large eggs should not be saved for incubation because they have very low hatchability.

Store hatching eggs at 55°F (13°C) at a relative humidity of 75%. Eggs do not need turning if weekly settings are made. Hatching eggs may be stored for 2 weeks without marked reduction in hatchability, but they should be turned daily after storage of >1 week. Eggs should be stored small end down. Precautions should be taken to prevent the flow of air from the cooling unit passing directly over the eggs. Excessive evaporation and enlargement of the air cell decrease hatchability.

Artificial Incubation

Incubators designed for hatching duck eggs are available. When only a few eggs are to be hatched a regular chicken-egg machine may be used. For best results, follow the manufacturer's directions.

Start the incubator a day or two in advance of the first setting. This will allow time in which to bring the machine into correct incubator condition before the eggs are set. Remove duck eggs from the storage area 5 to 6 hours before being set. This gives them time to warm to room temperature and lessens the drop in temperature of the incubator when the eggs are set. Place the eggs in the incubator small end down. Most incubators are equipped with automatic turning devices. These should be set to turn eggs every 3 hours. If manual turning is necessary, it should be done at least 3 times daily.

Candle the eggs after 7 to 10 days of incubation. This can be done by passing each egg over a small hand candler (light source) or by placing a tray of eggs above a bright light. The living embryo of a fertile egg will appear as a dark spot in the large end of the egg near the air cell. Blood vessels radiating from this spot give the appearance of a spider floating within the egg. An embryo that has died before candling will appear as a spot stuck to the shell membranes without clearly radiating blood vessels. Infertile eggs will appear clear. Remove cracked eggs together with infertile eggs and eggs containing dead embryos from the incubator. Eggs are frequently candled again after 25 days of incubation (32 days for Muscovy eggs). At this time the bills of normally developing ducklings can be seen within the air cells. Considerable movement can also be observed.

On large commercial farms, ducklings are frequently taken from the machines as they hatch. However, great care must be taken to prevent chilling of newly hatched ducklings. It may be wise for small producers to keep the machine closed until hatching is completed. Ducklings that must be helped from the shell should not be saved for breeding stock. Hereditary factors may be partially responsible for this condition.

Natural Incubation

Natural methods of incubation can be used on small farms. This is especially true for muscovies since these are excellent setters and incubate eggs readily. Other breeds do not set regularly. Eggs from these can be hatched under broody bantam chicken hens.

Clean, dry nesting facilities must be provided for setting hens and ducks. Feed and water should be within close proximity because the female must obtain her daily requirements within short periods of time. Delay in finding feed and water will result in undue chilling of the eggs. This situation is more critical with broody hens because they will have to keep their nests for 1 week longer than if they were setting on chicken eggs.

Brooding and Rearing

Ducklings should be moved from the hatcher to comfortable brooding quarters as quickly as possible. Prevent chilling and do not overcrowd the birds during transit. Provide feed and water for ducklings as soon as they are placed in the brooder.

Buildings of practically any type can be used to brood ducklings as long as the birds are kept warm, dry, and free of drafts. Ventilation systems and windows should be designed so fresh air can be brought onto the building without chilling the ducklings. Floors can be either wire or litter. If the expense can be justified, welded wire (>¾ in. or 2 cm) about 4 in. (2 cm) above concrete is the most satisfactory type of flooring. It may be used over the entire floor space or over only part of the floor. Wire flooring has the major advantages of keeping ducklings away from manure and dampness and of being washed down daily if adequate floor drains are present. Litter flooring may be more practical for most small producers. Straw, wood shavings, and peat moss make good litter. Make sure litter is free of mold. Moldy litter can cause high mortality in young stock.

Ducklings grow rapidly. Make sure they have adequate floor space. For 3-week-old ducklings, allow 0.5 sq ft ($0.045m^2$) of space per bird on wire and 1 sq ft ($0.09m^2$) per bird on litter. If confinement rearing is practiced, increase the floor space to 2 sq ft ($0.19 m^2$) per bird by 7 weeks of age.

Ducklings need supplementary heat for about 4 weeks after they hatch. In hot summer weather, heat may be needed for only the first 2 or 3 weeks. Electric, gas, coal, or wood-burning brooder units can be used for small operations. For larger operations, forced hot-air or hot-water systems are more efficient because they require less labor and fuel. The brooder temperature should be kept at 85° to 90°F (~31°C) the first week, then reduced by 5°F (2.75°C) per week during succeeding weeks. By 4 weeks the ducklings will be feathered enough to venture outdoors in all but extremely cold weather. If hovers are used the first week, use brooding guards to keep the ducklings confined to the comfort zone.

Ducklings need clean drinking water at all times. It may be supplied in hand-filled water fountains or by automatic waterers. To prevent wet litter, place the water supply above wire flooring or on a screened drain. Clean waterers daily.

Ducklings should be given access to outside yards when they are old enough to tolerate weather conditions. The young birds will manage nicely on grassy areas that have adequate shade. Although they are waterfowl, ducklings cannot tolerate chilling rains until they are about 4 weeks old. Young ducklings should be given shelter at the first signs of precipitation. This precaution can usually be disregarded when the ducklings are 4 weeks old. Birds 5 to 8 weeks of age need shelter only in extreme winter conditions. Yards should slope gently away from the houses to provide drainage. Locate watering facilities at the far end—lowest point—of the yard. Most commercial growers provide swimming water for the ducklings at 5 weeks of age. However, ducklings can be raised without swimming water. Stagnant pools of water should be prevented because they are sources of disease.

Manure will build up in yards after a few weeks. The rate will depend on the density of the birds. If the yards are located on light sandy soil, it will be relatively easy to scrape off the top surface. Periodic cleaning of yards should be part of the planned work schedule.

Feather pulling beginning at 4 weeks of age may be a sign of overcrowded conditions in the yards and houses. Steps should be taken to stop this vice as soon as it starts. The birds should be given additional space. If feather pulling continues, it may be necessary to trim the bill.

Night lights may reduce running and ensure that birds find feed and water. Use one reflected 15-watt lamp per 200 sq ft (19 m^2).

Feeding

Maximum efficiency for growth and reproduction can be obtained by using commercially prepared diets (see Chapters 7 and 19) as pellets. These are recommended over mash because they are easier to consume, they reduce waste and dust, and feed conversion is usually superior. Lack of pelleted feed should not discourage those who wish to produce ducklings on a small scale. Satisfactory results are possible with mash.

Feed ducklings the starter diet the first 2 weeks after they hatch. To encourage early consumption, place the feed in baby-chick-sized hoppers and locate them close to the water supply. When ducklings reach 2 weeks of age, switch them to the grower diet and feed this diet until they are ready for market.

Potential breeders should be fed a holding diet containing less energy. When fed in restricted amounts, the diet will keep the breeders from putting on excess fat but provide the nutrients needed. For 100 breeders provide 45 lb (20 kg) feed per day. Feed half in the morning and half in the late afternoon. The pellets should be scattered over a large area so all birds get a chance to obtain their daily requirements.

Increased nutritional requirements for reproduction make it essential to feed breeding stock a breeder diet. The breeders should be switched to the breeder diet about 1 month prior to the date of anticipated egg production. To ensure good eggshell quality, the breeders should be given crushed oystershells in separate hoppers or a high-calcium feed.

Many types of feeders can be used for ducks. Ordinary hoppers used in commercial chicken production work well provided they are arranged at floor level. Since ducks grow at fairly rapid rates and consume large quantities of feed in short periods of time, it is advantageous to use hoppers that hold large quantities of feed. Small feeders can be used until the ducklings are 2 weeks old. Larger feeders should be used for older market ducklings and breeding stock. Feed hoppers that are used outdoors should have lids that fit securely.

Provide water whenever feed is available. This is especially important when breeders are confined during the night. If hoppers are within the building and water supplies outdoors, hoppers should be closed overnight to prevent the breeders from choking on dry feed. Open the hoppers in the morning as soon as the birds have access to water. Breeders will adjust to this routine and egg production will not be affected.

Diseases

Ducks raised in small numbers and in relative isolation suffer little from diseases. Moreover, proper sanitation and biosecurity (e.g., reducing contact with visitors and wild birds) minimize the risks of disease. In case of disease, consult with a veterinarian. Duck diseases are presented in Chapter 19. Those that especially affect small or free-range producers include:

- *Fowl cholera* can lead to deaths of many ducks. Sanitation is important to control the disease. Do not allow "mudholes" and slimy areas to form. Burn or bury dead birds. Vaccines can control losses from fowl cholera.
- *Necrotic enteritis* is found in breeding stock. Breeder houses and yards must be free of wet litter and mudholes.
- *Reproductive disorders* in birds should result in them being culled.
- *Brooder pneumonia* is caused by fungi in the litter. Good litter and a dry brooder house help prevent this.
- *New duck disease* is a serious disease of ducklings. The first symptoms are sneezing and loss of balance. Afflicted ducklings fall over on their sides and backs.
- *Duck virus enteritis* is an acute viral disease. It is transmissible by contact, swimming water, and from migratory waterfowl. Symptoms include watery diarrhea and nasal discharge, with general droopiness developing in about 5 days. Symptoms last for 3 to 4 days, frequently ending in death.

Marketing and Processing Ducks

Feed should be withheld from the birds 8 to 10 hours before slaughter, but water may be provided up to the time of killing. Clean, uncrowded rearing facilities will help to prevent bruising, cutting, and other factors that cause poor product. Ducklings can be transported in crates or trailers to the slaughterhouse. Construction of elaborate slaughtering facilities is justified only for large commercial operations.

Ducks may be sold live or can be processed by using facilities similar to those used for small chicken flocks. For slaughter, hang ducklings by the feet or place them in special slaughtering funnels. Take a long, thin, sharp knife and draw it across the outside of the throat high up on the neck just under the lower bill. This will sever the jugular vein and allow swift, complete bleeding. When bleeding has ceased, birds can be scalded and picked or they can be dry picked. Dry picking has the advantage of producing exceptionally attractive carcasses, but it is slower and there is greater danger of tearing the skin. For scalding, immerse the ducklings for 3 minutes in hot water at 140°F (60°C). Pick immediately after scalding and remove all remaining pinfeathers;

grasp the pinfeathers between the thumb and a dull knife.

After slaughtering, ducklings can be dipped through a molten wax. When the wax hardens—by immersion in cold water—it can be peeled free to remove any feathers that remain. Wax can be reused if it is remelted and the feathers are screened out. Wax is highly combustible; hence, care must be taken to prevent contact with an open flame. Birds are then eviscerated. Wash the eviscerated bird thoroughly.

GEESE

Breeding

Selection of Breeders

Breeder geese should be selected for size, prolificacy, and vigor. Medium-size birds of each breed make the best breeders. Breeders should be selected from stock that has been tested and found to be superior in desired traits such as market body weight, egg production, and so on. Trapnesting is essential for identifying eggs from the better birds.

Breeder Facilities

Breeding geese are thought to prefer to be outdoors. Except in extremely cold weather or in storms, mature geese seldom use a house. Colony poultry houses, open sheds, or barns are provided for shelter in the North.

Geese make nests on the floor of the house or in coops, boxes, or barrels provided in the yard. Very crude nests are used in the open for many farm flocks of geese. Straw or grass hay is used for outside nests as well as for the nests on the floor of a house. One nest should be provided for every three females and the geese should be allowed to select their own nests. Inside nests should be separated by partitions, while outside nests should be placed some distance apart to reduce fighting.

Mating

Geese should be mated at least 1 month prior to the breeding season. For larger breeds of geese, a ratio of one male to three or four females may be used. Ganders of some of the lighter breeds will mate satisfactorily with four or five females. A gander may refuse to mate with some females. Geese matings should not be changed from year to year except when they prove unsatisfactory. Geese are very slow to mate with new birds, so it is difficult to make changes or to introduce new stock. If matings are changed, it is advisable to keep previously mated geese as far apart as possible.

It is difficult to distinguish male and female geese, except the Pilgrim. For all breeds, the gender can be determined by examination of the reproductive organs. This is done as follows: Lift the goose by the neck and lay it on its back with the tail pointed away from you. Move the tail end of the bird out over the edge so it can be readily bent downwards. Then insert your pointer finger (sometimes it helps to have a little Vaseline on it) into the cloaca about ½ in. (1 cm) and move it around in a circular manner several times to enlarge and relax the sphincter muscle, which closes the opening. Next apply some pressure directly below and on the sides of the vent to evert or expose the sex organs.

Incubation

Gather eggs twice daily, especially during cold weather. Store them at 55°F (12°C) and a relative humidity of 75% until set for hatching. Goose eggs are washed just like chicken eggs. Wash soiled eggs in warm [100°–150°F (~40°C)] water and a detergent sanitizer. Eggs should be washed soon after gathering, then dried and stored until ready for the incubator. If eggs are held for more than a couple of days, turn them daily to increase the percentage of hatch. Hatchability decreases fairly rapidly after a 6- or 7-day holding period. Eggs properly stored can be held 10 to 14 days with fair results. The incubation period for most geese eggs is 29 to 31 days. For Egyptian geese, it is 35 days.

Small, inexpensive electric incubators, either still-air or forced-draft, can be used to hatch goose eggs. However, artificial incubation of goose eggs is much more difficult than with chicken eggs because more time and higher humidity are required. Breeders should gain experience with chicken eggs before attempting artificial incubation of goose eggs. When using an incubator, always follow the manufacturer's instructions.

Many goose breeders prefer to set eggs under turkeys or Muscovies to allow the geese to continue to lay. If the setting hen does not turn the eggs, mark them with crayon or pencil and turn them daily by hand. Moisture is needed where chicken or turkey hens are used for setting. Sprinkle the eggs during the incubation period and have the nest and straw on the ground or on grass-covered turf. Some growers report better hatchability if they sprinkle the eggs lightly or dip them in water daily during the last half of the incubation period, but eggs need no additional moisture if the setting goose or Muscovy has water for bathing.

Remove goslings from the nest as they hatch and keep them in a warm place until the youngest is several hours old. If this is not done, the setting hen may leave the nest along with the hatched goslings before all the eggs are hatched.

Brooding and Rearing

A special brooder building is not required for brooding small numbers of geese. Any small building or a corner of a garage or barn can be used as a brooding area for a small flock if it is dry, reasonably well lighted and ventilated, and free from drafts. The building must also be protected against wild predators, dogs, cats, and rats. For brooding large numbers of geese, provide a barn, large poultry house, or regular broiler house. Allow at least 0.5 sq ft (0.045 m²) of floor space per bird at the start of the brooding period and gradually increase the space to 1.25 sq ft (0.11 m²) at the end of 2 weeks. If the birds are confined longer because of inclement weather, provide additional space as they increase in size. Cover the floor with 4 in. (10 cm) of such absorbent litter as wood shavings, chopped straw, or peat moss. To maintain good litter, stir frequently, remove wet spots, and periodically add clean, dry litter. Be sure litter is free from mold.

Goslings can be successfully brooded by broody chicken hens and most breeds of geese. If the young birds were not hatched by the brooding female, place them under her at night. Be certain broody birds are free of lice and mites. One hen can raise five goslings. In mild weather, the hens may only need to brood the goslings for 10 to 14 days after which they can get along without heat.

Goslings are artificially brooded in many types of heated brooders. Infrared lamps are a convenient and satisfactory source of heat, provided enough of them are used to furnish heat for the lowest temperatures expected. When using hover-type brooders, brood only about one-third as many goslings as the brooder's chick capacity. Because goslings are large in size, raise the hover 3 to 4 in. (~9 cm) higher than for baby chicks. Fence in the brooding area for the first few days with a corrugated paper or wire mesh fence.

At the start set the temperature of the hover at 85° to 90°F (~31°C). Reduce the temperature 5° to 10°F (~4°C) per week until 70°F (21°C) is reached. The behavior of the goslings will indicate their comfort. If they are cold, they will huddle together under the lamps. If they are too warm, they will move away from the heat source. In warm weather, the goslings can go outdoors as early as 2 weeks, but they will need frequent attention until they learn to go back into the coop or brooder when it rains. They must be kept dry to prevent chilling that can result in piling and smothering. Houses are usually not needed after the geese are 6 to 8 weeks of age.

Feeding

Goslings should have drinking water and feed when they are started under the brooder or hen. Supply plenty of watering space. Use waterers that the birds cannot get into, but that are wide and deep enough for them to dip both bill and head. Start with two automatic cup-type waterers for each 100 to 200 goslings, depending on the environmental temperature. Increase the number of waterers as the birds grow. Watering jars or a trough with wire guard and running water are also suitable for young goslings. If troughs are installed, figure on 8 ft (2.4 m) of trough space for 500 goslings for the first 2 weeks of age; then, as needed, increase the space up to 20 ft (6 m). On range, the waterer can consist of a barrel or large tank rigged to an automatic float in a watering trough. If waterers are indoors, they should be kept on wire platforms with underdrainage to help keep the litter dry.

For the first few days of feeding use shallow pans or small feed hoppers in addition to the regular feeders. For each 100 confined goslings on full feed, provide either two hanging tube feeders with pans that are 50 in. (125 cm) in circumference or 8 ft. (2.4 m) of trough space. Increase the feeding space as the birds grow. When feed intake is being restricted, provide enough space so that all geese can eat at one time. For geese raised on range for market, use two wooden hoppers or two turkey range feeders for each 250 birds. The hoppers should be large enough so that they will need to be filled only once or twice weekly. Construct the hoppers so that feed is protected from rain, sun, and wind. Mechanical feeders are used for large-scale production. Geese may be fed pellets, mash, or whole grains. For the first 3 weeks, feed goslings 20 to 22% protein goose starter pellets. After 3 weeks, feed 15% protein goose grower pellets.

Although geese can go on pasture as early as the first week, they will receive significant nutrition from forage at 5 to 6 weeks of age. This may be provided as silage. Geese are very selective and tend to pick out the palatable forages. They will reject alfalfa and narrow-leaved tough grasses and select the more succulent clovers and grasses. Geese should not be raised on dried-out mature pasture. Suitable stocking densities are 20 to 40 birds per acre depending on the size of the geese and the quality of the pasture. A 3 ft (1-m)-woven-wire fence will confine the geese to the grazing area. Be sure that the pasture areas and green feed have not been treated with pesticide that may be harmful to the birds.

If pasture is plentiful and of good quality, the amount of pellets may be restricted to about 1 to 2 lb per goose per week until the birds are 12 weeks of age. However, for maximum growth, increase the amount of feed as the supply of young, tender grass decreases or when the geese reduce their consumption of grass.

From 12 weeks to market, offer pellets on a free-choice basis, even when on range. Mash or whole grains can be fed alone or they can be mixed at a 50:50 mash-to-grain ratio. At 3 weeks of age, use a mash-to-grain mix of approximately 60:40. Change this ratio gradually during the growing period until at market age the geese are receiving a 40:60 ratio of mash to grain. Depending on the quality and quantity of available pasture, adjust these ratios up or down slightly.

Wheat, oats, barley, and corn may be used as the whole grains in various mixtures, such as equal parts of wheat and oats. All-corn can be substituted when the goslings are 6 weeks old. For maximum growth, it is important that mash-and-grain mixtures provide similar nutrient intake (15%) as the all-mash diets. Grower-size insoluble grit should be freely available to geese throughout the growing period.

Management

Geese generally start laying in February or March and often continue to lay until early summer. However, the Chinese breed may start laying early in the winter.

Breeder geese should be fed a pelleted breeder ration at least a month before egg production is desired. They do much better, and waste less feed, on pellets than on mash. A chicken-breeder ration may be used if special feeds for geese are not available. Provide oystershells (or other calcium sources), grit, and plenty of clean, fresh drinking water at all times.

Lights in the breeder house can be used to stimulate earlier egg production. In commercial flocks, artificial methods of hatching and rearing are also used. To maintain egg production, feed chicken layer pellets or mash, confine broody geese away from but in sight of their mates, and gather eggs several times each day to break up broodiness. Young ganders make good breeders, but both sexes usually give the best breeding results when they are 2 to 5 years old. Good fertility may be obtained in eggs from young birds, but these eggs may not hatch well. Although young flocks are considered more profitable, females will lay until they are about 10 years of age. Ganders may be kept for more than 5 years.

GEESE AS WEEDERS. Weeder geese are said to control and eradicate troublesome grass and certain weeds in a great variety of crops and plantings, including cotton, hops, onions, garlic, strawberries, nurseries, corn, orchards, groves, and vineyards. The geese eat grass and young weeds but are said to not touch certain cultivated plants.

Diseases

When geese are managed properly, they seldom get a disease. Nevertheless, the producer should be alert to the first signs of trouble. If there is suspicion of disease from the appearance of the birds or deaths, a veterinarian should be consulted immediately. Keep an accurate record on the number and dates of deaths (also see Chapter 11).

Marketing and Processing Geese

Geese bring the highest prices at Thanksgiving and Christmas. Geese should be fasted for about 12 hours before killing, but they should have access to water. To kill, geese should be placed in funnels or hung by the legs in shackles, then the throat should be cut at the base of the beak to sever the jugular vein and carotid artery. Geese can be scalded or picked dry. The dry method, if well done, results in an attractive carcass but is considered too slow and laborious to be economical. There is also more danger of tearing the skin in dry picking.

Geese can be scalded in a commercial scalder, or they can be hand-scalded in a small operation. Water temperature should be 145° to 155°F (~65°C) and the length of the scaled should be from 1½ to 3 minutes. Detergent should be added to the water to hasten thorough wetting of the feathers. To hand-scald, grasp the goose firmly by the bill with one hand and by the legs with the other, then submerge its body (breast down) in the scalding water. Pull the bird repeatedly through the water against the lay of the feathers; this serves to force the water through the feathers to the skin. The sparser feathering on the back needs lighter scalding than the heavier and denser feathering on the breast.

After scalding, the birds may be "rough-picked" by hand, picked on some type of conventional rubber-fingered picking machine, or placed in a spinner-type picker. After "rough-picking," it is difficult to remove remaining pinfeathers and down. This can be done by grasping the pinfeathers between the thumb and a dull knife. Because of the difficulty of handpicking, it is common practice to finish the "rough-picked" birds by dipping each bird in melted specially formulated wax.

In small operations, dry off the "rough-picked" geese just enough to take the wax; then dip them several times in wax held at 150° to 160°F (65–70°C) to build up a heavy enough layer of wax to supply good pulling power. A better job occurs with the use of two tanks of wax, one held at 160° to 170°F (~75°C), and the second at about 150°F (65°C). The hotter wax is used for penetration and the cooler wax for buildup. After waxing, the birds are dipped in a tank of cold water to cool and harden the wax to a "tacky" state. The wax is then removed by hand, resulting in a clean, attractive carcass. The wax can be reused by melting and straining out the feathers. The U.S. standards of quality are essentially the same for all poultry. Geese are graded for conformation, fleshing, and fattening. Defects, such as missing skin and bruises, are also considered in establishing quality. For further information, see Chapter 17.

Feathers

Goose feathers are a source of extra income. Three geese usually yield 1 lb dry feathers. Producers can wash and dry the feathers on their own premises. To wash feathers, use soft, lukewarm water to which has been added either detergent or borax and washing soda. Rinse the feathers, then wring and spread them out to dry.

Raising Ostriches

OSTRICH PRODUCTION

Ostriches are kept in paddocks, or large fenced areas. The stocking density varies with the ability of the land to produce adequate forage.

PRACTICAL ASPECTS OF REPRODUCTION

Adult breeding ostriches are farmed either in pairs (one male and one female) or trios (one male and two females). The only reliable method of determining the sex of young chicks is examining the sex organs. It is advisable to keep the males and females separated prior to pairing for mating. At the time of paring, the ostriches should be kept in a pen or paddock fenced with smooth wire about 6 ft high (~2 m).

Farmed ostriches show much lower reproductive efficiency and fertility than chickens. In a single breeding season, a mature female ostrich will lay about 50 eggs (compared to more than 300 per year in a chicken). However, about half of the adult hens do not produce any eggs. Nutrition is critical to reproduction, but requirements are still not adequately documented.

Egg Storage and Incubation

As opposed to placing of eggs in nests with other species of birds, in ostriches the eggs are buried. The techniques of egg storage and incubation must mimic the natural conditions.

It is critically important to store ostrich eggs under conditions optimal for viability during incubation, hatching, and posthatching growth. Eggs should be collected within 1 day of laying and stored at about 71°F (21°C). Incubation temperatures are thought to be optimal at 97.5°F (36.4°C). Eggs should be positioned with the large end up at a 45-degree angle. The eggs should be turned twice daily.

Brooding and Rearing

The brooding period of young ostriches (chicks) is very critical. Chicks should be brooded at a temperature of about 90°F (32.2°C) for about 14 days. Following this, the temperature is reduced a few degrees each week.

Raising Pigeons

BREEDING FACILITIES

Pigeon houses are called lofts. The quarters should be dry, well ventilated, and provided with plenty of daylight. A loft 7 ft (~2 m) wide and 10 to 12 ft (3.5 to 4 m) long will provide ample room for 15 pairs of birds; that is, about 5 sq ft (0.4 m²) per pair. Breeder houses should be equipped with nests, bowls, feed hoppers, bathing pans, and a rack for nesting material.

REPRODUCTION

Care of Breeders

Pigeons are ready to mate at about 4 to 5 months of age. They mate in pairs and usually remain with their mates throughout life. Pairs may be changed if desired by placing the male and female in a coop together and leaving them there for 6 to 14 days. No more than 10 to 15 pairs of mated birds should be kept in one loft.

Eggs

The pigeon hen lays an egg, generally skips a day, then lays another egg.

Incubation

The incubation period is about 17 days. Both males and females sit on eggs. The male generally sits on the eggs during the middle of the day and the female the remainder of the time.

Brooding and Rearing

Both parents care for the young, called squabs (the newly hatched chicks are sometimes referred to as squeakers). They feed them by regurgitating a thick, creamy mixture, called pigeon milk, into the open mouths of the young. This milk is produced by the crop when stimulated by the hormone, prolactin. The milk consists of fat and protein from crop epithelial cells that are sloughed off. Pigeons grow rapidly. Squabs exceed the normal adult weight when they are ready to leave the nest, about 30 days of age. Young should be banded when about 7 days of age for record keeping.

FEEDING AND WATERING

Pigeons are grain eaters. They prefer a variety, or mixture, of grains or commercial pigeon pellets. Provide clean, fresh drinking water (also see Chapter 7).

CARE AND MANAGEMENT

Diseases

A clean, well-kept loft will contribute to the health of the birds. Pigeons, quarters, and nests should be checked frequently for lice and mites.

MARKETING, RELEASING

There is a demand for squabs. Squabs should be slaughtered at about 30 days old. Squabs are killed and dressed like other poultry. If the young are to be kept as breeders or for show, allow them a few days of flight in the pen with their parents before moving them to a separate pen for young unmated birds.

Raising Game and Ornamental Birds

BOBWHITE QUAIL

Those raising quail should remember that quail are living beings, that they have been stressed by taking them out of their natural environment and keeping them in confinement, and that they are 100% dependent on the caretaker.

Reproduction

When selecting breeders, there are a few issues. The larger varieties should be selected when growing birds for meat purposes. The small- and moderate-sized varieties generally are desired when growing birds for hunting preserves because they usually fly better and faster than the larger birds. Birds or hatching eggs should be disease-free, preferably tested for and free of pullorum and typhoid. Select stock with a history of good egg production, hatchability, and livability.

Caging and Facilities

Breeders may be kept in cages or in floor pens. Cages can be either for an individual pair or a colony cage. Caged breeders require about 1 sq ft (0.1 m^2) per pair. A solid partition between cages is desirable to prevent fighting. Floor pens should provide a minimum of 1 sq ft (0.1 m^2) per bird. The advantages of cages over floor pens are that the eggs are cleaner, there is less exposure of the birds to diseases and parasites, and there is less fighting.

The major advantage of the floor system is that more margin of error in feeding and watering is possible, since the birds can move about and choose. Year-round production can be obtained by providing up to 17 hours per day of lighting. Bobwhite hens will begin egg production at 16 to 24 weeks of age. In a 6-month breeding season, a hen should lay about 90 eggs. Eggs should be stored at 55° to 65°F (~15°C) and a relative humidity of 75 to 80%.

Incubation

The period of incubation is 23 to 24 days.

> Incubation for 20 days at 99.75°F (37.5°C), humidity— wet-bulb setting 84° to 86°F (~29.5°C)
>
> Hatcher temperature, 98.75°F (37.1°C) [still-air 100°F(37°C)], humidity–wet-bulb hatching 87° to 98°F (30.5–36.7°C)

Brooding

Brooding is that period from hatching to 5 or 6 weeks of age. Three general types of brooding facilities and equipment are used: (1) batteries, (2) floor pens with litter, and (3) raised wire floors. The major advantage of batteries and raised wire floors is that less space is required. The floor pen with hover or infrared brooder is the simplest method.

Nutrition

Water and commercial game bird diets should be available to the birds at all times (also see Chapters 6 and 7).

Care and Management

Future breeders should be beak trimmed at hatching so as to reduce cannibalism. The incidence of cannibalism is increased by crowding, bright lights, insufficient feeders and water space, too high temperatures, and placing different types of birds together. Quail seem to be more sensitive to management than chickens.

Diseases

Bobwhite quail are susceptible to diseases and parasites that affect poultry including:

- Ulcerative enteritis, or quail disease (caused by an anaerobic bacterium)
- Blackhead
- Coccidiosis
- Quail bronchitis (caused by a virus)
- Botulism
- Pox
- Chronic respiratory disease (CRD)
- Pullorum

Occasionally infectious coryza, fowl cholera, staphylococcosis, and trichomoniasis are reported. Internal parasites include large roundworms, cecal worms, capillary worms, gapeworms, and tapeworms. External parasites include fleas, lice, mites, and ticks. Disease prevention programs are similar to those with chickens.

Marketing and Releasing

Few thrills are as exciting as the sudden whir of a Bobwhite covey rise, especially to a hunter. Coveys lie in a circle with their heads pointing outwards. Thus, they can quickly spot an enemy, then fly off in all directions. Most producers of birds for hunting preserves condition them in flight pens for about a month before release. Flight pens are usually outside wire pens about 12 to 15 ft (4–5 meters) wide and 100 to 150 ft (30–50 m) long, with an enclosed shelter at one end. Many producers contract a year or two ahead for the sale of their birds and eggs. Bobwhite quail are generally considered mature at 16 weeks of age, at which time they can be used for meat or hunting preserve purposes.

GUINEA FOWL

Guineas might be more popular were it not for their harsh and seemingly never-ending cry and their bad disposition.

Reproduction

Selection of Breeders

In selecting breeders, size and uniform color are most important. At maturity, both males and females range from 3 to 3.5 lb (~1.5 kg) in weight. Birds can be sexed by the cry, and by larger helmet and wattles together with the coarser head of the male.

Breeding Facilities

As the breeding season approaches, wild mated pairs range off in the fields in search of hidden nesting places in which it is difficult to find the eggs. Domesticated guinea breeding birds are usually allowed free-range. However, on some farms the breeders are kept confined during the laying period in houses equipped with wire-floored sun porches.

Care of Breeders

In the wild state, guinea fowl mate in pairs. However, under domestic conditions, one male is usually kept for every four or five females. Guinea breeders are usually allowed free-range. They are difficult to confine in open poultry yards unless their wings are pinioned or one wing is clipped.

Eggs

A hen of good stock that is properly managed will lay 100 or more eggs per year. Guinea eggs are smaller than chicken eggs; about 1.4 oz (40 g) versus 2 oz (60 g) for a chicken egg.

Incubation

The incubation period is 26 to 28 days. Chickens can be used for hatching guinea eggs in small flocks. For large flocks, incubators are better. Forced-draft incubators should be operated at about 99.5° or 99.7°F (37.5°–37.6°C) and 57 to 58% humidity.

Brooding and Rearing

Guinea chicks are known as baby keets. Guinea keets may be raised in the same kind of brooder houses and brooders as baby chicks or poults. Hovers should be started at 95°F (35°C) for the first 2 weeks, then lowered by 5°F (2.75°C) per week thereafter.

Feeding and Watering

A good commercial chicken or turkey breeder mash, containing 22 to 24% protein, should be fed to layers. Young guineas should have a growing diet (also see Chapters 6 and 7).

Care and Management

Most Guineas are raised in small flocks. Chicken hens make the best mothers for guinea keets. Guinea hens are likely to take their keets through wet grass and lead them too far from home.

Diseases

Guineas are subject to the same diseases and parasites as other poultry and respond to the same treatments.

Marketing and Releasing

The normal marketing season is late summer and throughout the fall. The demand is for young birds weighing 1.75 to 2.5 lb (~1 kg) liveweight. Many hotels and restaurants in large cities serve guineas at banquet and club dinners as a special delicacy or as a substitute for other game birds. Guineas are killed and dressed in the same way as chickens.

HUNGARIAN OR GRAY PARTRIDGE

Reproduction

Selection of Breeders

The sexes can be accurately separated by an experienced person on the basis of the secondary wing covert feathers. Birds with physical defects such as eye defects, twisted beaks, deformed feet, or bumped or twisted backs, or lacking in fleshing should be rejected.

Breeding Facilities

Recommended breeding facilities consist of mating pens with doors facing a central courtyard.

Care of Breeders

The Gray partridge is monogamous and intolerant of density, particularly during the breeding season. For these reasons, it is important that partridges be isolated in pairs as soon as premating signs are evident in the spring. The latter is accomplished by releasing several pairs of partridges in a central courtyard surrounded by individual mating pens with the doors open. The females will select males of their choice and entice them into a pen, following which the doors may be closed.

Eggs

Egg laying occurs in the spring and summer. They are erratic in egg production. Hens will average about 30 eggs per year, with a range from 0 to 60. Eggs should be gathered frequently and stored for (not to exceed) 10 days at 55°F (12.5°C) and 90% relative humidity.

Incubation

The incubation period is 24 to 25 days. The incubator temperature should be 99° to 100°F (37.2°–37.8°C) for forced-air and 101°F (38.3°C) for still-air-type incubators. A wet-bulb reading of 84° to 85°F (28.9°–29.4°C) is recommended (about 60% relative humidity).

Brooding and Rearing

Hungarian partridges can be successfully raised in commercial brooder batteries, preferably with no more than 30 chicks per group because of their pugnacious nature even as chicks. The temperature of the brooder should be 95°F (35°C) at shoulder height of the chicks, with the temperature reduced 5°F (2.75°C) each week following the first week. At 5 to 6 weeks of age the birds can be moved from the brooder to wire-covered outside runs. However, during nights they should be herded back under brooders until they are 10 weeks of age.

Feeding and Watering

Water and commercial game bird diets should be available to the birds at all times. Example rations for Hungarian or Gray partridge are given in Chapter 7.

Care and Management

The wire-covered runs (used following the brooder stage) should provide 5 to 6 sq ft (~0.5 m^2) of space per bird. Pens should be constructed of 1 in. (2.5 cm) wire mesh.

Diseases

Hungarian partridges are susceptible to poultry diseases and parasites.

Marketing and Releasing

Birds are usually released when 16 to 20 weeks of age. Merely dumping the birds in a release site will defeat the intended purpose. Portable release pens approximately 10 ft × 10 ft × 4 ft high (3 × 3 × 1.2 m), covered with 1 in. (2.5 cm) wire mesh, should be moved to the area in which the release is planned. Groups of 12 to 15 birds should be maintained in each pen for a few days, with feed and water provided. Once the birds have become used to the area, one end of the pen can be opened, allowing the birds to leave and drift away. Release pens should be 200 yd (200 m) apart.

COTURNIX QUAIL

Coturnix quail can be either Japanese quail (*Coturnix coturnix japonica*) or European quail (*Coturnix coturnix coturnix*). The former has been domesticated.

Reproduction

Selection of Breeders

Quail can be separated by their feather color difference when they reach about 3 weeks of age. Adult males should weigh 4 to 5 oz (120–150 g), and adult females 4.5 to 6 oz (125–180 g).

Breeding Facilities

Quail can be raised on the floor. If they are raised for breeding or for egg or meat production, they will perform better in cages. **Pedigree cages** 5 in. × 8 in. × 10 in. (12.5 cm × 20 cm × 25 cm) will hold a pair of quail. **Colony cages** 2 ft × 2 ft × 10 in. (60 cm × 60 cm × 25 cm) will accommodate up to 25 quail, while a 2 ft × 4 ft × 10 in. (60 cm × 120 cm × 25 cm) cage will accommodate up to 50 quail. Adult quail will perform better if given 16 to 25 sq in. (100–160 cm^2) of floor space per bird. They need 0.5 to 1 in. (1.25–2.5 cm) of feeder space per bird and 0.25 in. (0.6 cm) of trough space for water.

Care of Breeders

Quail may start laying as early as the sixth week. When males are sexually mature, a large glandular or bulbous structure appears above the cloacal opening. If this gland is pressed, it will emit a foamy secretion. Adult female hens require 14 to 18 hours of light per day to maintain maximum egg production and fertility.

Eggs

The eggs weigh ~1/3 oz (10 g), about 8% of the female body weight. The hens lay an egg every day for 8 to 12 months. The eggs are edible and can be prepared in the same manner as chicken eggs. For incubation, collect eggs two to three times daily, and store at 55°F (12.8°C) and 70% humidity.

Incubation

The incubation period is 14 to 17 days. Domesticated Japanese quail have lost the broodiness trait; hence, eggs must be incubated under a hen or artificially.

The recommended temperature and humidity for single-stage forced-draft incubation is:

1 to 14 days temperature: *dry-bulb*, 99.5°F (37.5°C) and 60% humidity; or wet-bulb, 87°F (30.5°C).

14 days to hatching temperature: *dry-bulb*, 99°F (37.2°C) and 70% humidity; or wet-bulb, 90°F (32.2°C).

Brooding and Rearing

Brooder temperature should be 95°F (35°C) at head level of chicks for first week. Thereafter, decrease the temperature 5°F (2.75°C) every week until after the fourth week. Beak-trim the chicks to prevent pecking and cannibalism. Young birds can be transferred from brooder to cages or floor around the fourth week.

Feeding and Watering

Provide water and commercial game or turkey feed. Also see Chapter 7 for suggested rations for quail.

Care and Management

Quail may be managed much the same as chickens or turkeys, except for size. Care should be taken with small quail to prevent drowning in water troughs. Feather picking or other forms of cannibalism occur frequently when Japanese quail are kept on wire. Beak trimming is the best preventive. Other steps to lessen cannibalism are reducing light intensity and increasing dietary fiber or grit.

Diseases

Although the quail is a hardy bird compared to other poultry species, it can be affected by most of the common poultry diseases. Sanitation is the best preventive measure, including the control of rats, mice, and fleas. Eggs should be fumigated shortly after they are collected or at least within 12 hours after they have been placed in the incubator.

PEAFOWL

Reproduction

Selection of Breeders

Select breeders that have the best color. Breeders should be healthy and free from leg weaknesses and crooked toes.

Breeding Facilities

Peafowl are not inclined to wander far. Their choice of a roost is a tall tree or on top of a building.

Care of Breeders

A good male can be mated to as many as five hens.

Eggs

The hen is usually 2 years old before starting to lay; and she will usually lay 10 to 12 eggs, although some hens will lay up to 35 eggs.

Incubation

Incubation may be (1) by setting 6 to 10 eggs under a peahen, (2) by setting 6 to 10 eggs under a broody chicken, or (3) by artificial incubation.

Brooding and Rearing

If the birds are brooded artificially, they may be treated the same as baby chicks or turkey poults. Hold the hover temperature at 95°F (35°C) for the first week, then decrease the temperature by 5°F (2.75°C) per week thereafter.

Feeding and Watering

The feeding of peafowl is very simple. A 30% protein turkey prestarter can be used for the first 4 to 5 weeks, after which a 26% protein turkey starter mash or game bird starter mash may be used (also see Chapters 6 and 7). Provide fresh clean water.

Care and Management

Beak trimming may be necessary to prevent cannibalism. Provision of adequate space and a good diet can reduce cannibalism. Peafowl have a very raucous voice that may annoy the neighbors. If a male in full plumage is shipped, the tail should be wrapped and sewed with clean muslin next to the feathers and burlap on the outside.

Diseases

Peafowl are subject to most of the diseases associated with chickens and turkeys, with the two most common being blackhead and coccidiosis.

Marketing and Releasing

Peafowl are usually sold in pairs, as ornamental birds, at about ~$200 (US) per pair. Peafowl are edible and are regarded as a delicacy for special occasions.

PHEASANTS

Pheasants are generally raised for the purpose of stocking farms reserved for hunting.

Reproduction

Selection of Breeders

Select from earliest hatches, from the sixth week on. Choose birds for their size, weight, and feathering, without visible defects. Select birds that feather rapidly in true plumage color. Test all breeders for pullorum disease each season before egg production starts.

Breeding Facilities

Cold is not a factor. But the birds need protection from winds, such as can be provided by brush or cornstalks. Hobbyists sometimes use an open-front shed. Pheasants should be moved into laying pens prior to February 15; individual pens enclosing 1 cock and about 7 hens, or community pens holding about 140 hens and 20 cocks. Provide nesting boxes under brush or under other covering.

Care of Breeders

Pheasants are polygamous. Egg production can be increased beginning January 1 by artificial lighting according to one of the following:

- Constant light. Increase light to 14 to 15 hours daily and hold constant throughout breeding season.
- Increase light to 14 hours and hold until peak production is reached, then increase 15 minutes per week until maximum of 17 hours light per day is reached.

Eggs

Pheasants in the wild lay a clutch of 10 to 12 eggs. In early spring, gather eggs several times each day. Hold eggs at 55° to 60°F (13° to 16°C) and humidity of 70 to 75%. But incubate as soon as possible.

Incubation

The incubation period is 23 to 24 days. Most incubator manufacturers specify the temperature and humidity. If they do not, the following recommendations may be used:

INCUBATOR TYPE

Forced-Air		Still-Air	
°F	(°C)	°F	(°C)
99.75	(37.6)	102.25	(39.0)
99	(37.2)	102	(30.0)
86–88	(30.0–31.1)		
92–95	(33.3–35.0)		

Brooding and Rearing

Brooding practices for pheasants are much the same as for chickens. Most commercial operators use battery brooders for the first 7 to 9 days, then move into brooder houses. Smaller producers usually raise chicks in brooder houses from the start.

- The temperature should be 95° to 97°F (35°–36°C) for the first week, then reduced 5°F (2.75°C) each week.
- Birds should be moved from brooder houses and placed in flight pens at 7 weeks of age.
- Provide water in 1-gal (4-l) fountains.
- Start chicks by feeding on cardboard flats. Then progress to feeders.
- Provide 1 linear in. (2.5 cm) feeder/chick until 6 weeks of age, then 2 to 3 linear in. (5–7.5 cm) from 6 weeks on.

Feeding and Watering

Commercial game bird feeds are available in most areas. Also provide fresh clean water. Also see Chapters 6 and 7 for nutritional requirements.

Care and Management

Beak trim birds when moving them to the laying pen. Pheasant chicks are usually beak trimmed at 7 to 9 days of age to prevent cannibalism. Also, blunt cocks' spurs and trim and blunt inside toenails. If pheasants are to be raised in open-top pens, they must be pinioned (have the first section of the wing removed) or have their feathers clipped. To catch birds, drive them into a catching pen, then trap them with a nylon net similar to a fish net.

Diseases

Pheasants are subject to many of the diseases and parasites of chickens and usually respond to the same treatments. The most common diseases are coccidiosis, botulism, pullorum, and gapes.

Marketing and Releasing

For hunting, pheasants can usually be released at the tenth to fourteenth week of age. They should not be released until they are fully feathered and able to care for themselves. Pheasant is also a specialty meat item. They should be processed at about 22 weeks of age and packaged in an attractive form. Frozen birds are sold individually vacuum-packed.

INDEX

A

Account book, kind of, 341
Additives
  antifungal, 110–111
  new developments, 120
  nonnutritive, 108–112
Adipose tissue, 246, 247
  growth, 247
  impact of, on the poultry
    industry, 247
Adrenocorticotropic hormone
    (ACTH), 40, 42
Advertisers, 18
Aflatoxin, 110
African goose, 325
Age, nutrition and, 77–78
Aggressive behavior, 157
Agonistic behavior, 157
Agriculture, U.S. Department of
  Agricultural Marketing Service, 295
  classification of chickens, 302
  classification of turkeys, 302
  Food Safety and Inspection
    Service, 195, 196–197
  standards, for quality of ready-to-
    cook poultry, 303–305
Agri-Start database, 316
Airflow, 209
Albumen, 220
Alfalfa products, 108
Allergies, 9
Allied poultry industries, 317
All natural eggs, 306
All-pullet flocks, 227
Alternative replacement
    programs, 227
Amino acids, 32
  true digestibility, 84–85
Amylase, 30
Anabolism, 78
Animal behavior, 153–159
Animal fat, 9–10
Animal proteins, 102–107
Annual inventory, 341–342
Anorexia, 92
Anthelmintics, 112
Antibiotics, 111, 188–189
  for broilers, 255
  food safety and, 198

judicious use, 186
for turkeys, 273
Anticoagulants, 144
Anticoccidiais, 112
Anticoccidial agents, 176
Antifungal additives, 110–111
Antioxidants, 109
Antiparasitic drugs, 112
Apparent metabolizable energy
    (AME), 82
Arachidonic acids, 80
Arachnids, 177
Archaeopteryx, 23
Arizonosis, 166–167
Arsenicals, 111–112
Artificial incubation of ducks,
    347–348
Artificial insemination
  of broiler breeders, 259–260
  of chickens, 70
  of ostriches, 329
  of turkeys, 282–283
Ascites, 186–187
Aspergillosis, 164, 166–167, 172
Aspergillus flavus, 110
Association of American Feed Control
    Officials (AAFCO), 106
Atherosclerosis, 9
Automation, 209–210
Autonomic nervous system, 39
Auxillary power in broiler
    production, 251
Aviagen Group, 68
Avian diseases, 163
Avian encephalomyelitis, 163,
    166–167
Avian influenza, 164, 165
Avian leukosis, 165
Avian pneumovirus, 165
Avian pox, 165
Avian sperm, 34
Aylesbury ducks, 322

B

Bacterial diseases, 165, 172
Barley, 97
Barred Plymouth Rock, 3
Beak trimming, 56–57, 137, 237,
    255, 277

Beetles, 178–179
Behavior
  aggressive or agonistic, 157
  feeding and, 155–156
  norms, 157–158
  parental, 156–157
  problems, 188, 238
  sexual, 156
  studies, 20
Beltsville Small White, 72, 270
Bentonites, 109
Bile, 31
Biological tests of poultry feeds,
    113–114
Biology, 22–44
  circulatory system, 26–27
  digestive system, 28–33
  egg production, 223
  endocrine system, 39–40
  excretory system, 33
  integument, 23–25
  nervous system, 39
  reproductive system, 33–39
  respiratory system, 27–28
  special senses, 39
  structural systems, 25–26
Biosecurity, 139, 182
Bioterrorism, 188
Biotin, 93
Birds, evolution and classification,
    22–23
Blackhead, 173, 174–175
Blood cells, 26
Blood groups, 63
Blood meal, 103
Blood spots, 38
Bluecomb disease, 166–167
Bobwhite quail (*Colinus virginianus*),
    333, 356–357
  brooding, 356
  care and management, 356–357
  diseases, 357
  feeding, 132
  incubation, 356
  marketing and releasing, 357
  nutrition, 356
  reproduction, 356
Body conformation, 63
Bomb calorimetry, 113

Bone diseases, 187
Bone fracturing, 187
Botanics/botanicals, 96, 189
Botulism
    in ducks, 323
    in poultry, 164, 166–167
Breakouts, 299
Breast blisters, 263–264
Breeding
    of broilers, 257
    changes in methods, 5
    of chickens
        breeds and varieties, 66–68
        business organization, 68–69
        methods of mating, 70
        selecting and culling, 70–72
        types and classes, 65–66
    of ducks, 321
    of geese, 325, 350–351
    genetics and, 226
    of pigeons, 332
    of turkeys, 65, 72
        management, 283–284
        nutrition, 284–285
Breeding flocks, 17
Brewers' and distillers' grains, 98, 100
British Thermal Unit (BTU), 81
Broad-Breasted Bronze, 4, 72, 269
Broad-Breasted White, 72, 270
Broiler breeders
    feeding, 128–129
    females, 258–259
    males, 259–260
    nutrition, 260
    photostimulation, 258
    reproduction, 257–260
Broiler chickens, 63
    brooding, 256–257
Broiler contract, analyzing outcome
        of, 248–250
Broiler genetics, measurement to
        improve, 71
Broiler industry, consolidation, 310
Broiler production chickens, 65–66
Broilers
    business aspects, 248–250
    commercial production, 247–255
    cost of producing chicken
        meat, 248
    economics, 249
    feeding, 128
    financing, 248
    harvesting and processing,
        260–263
    health, 253–255
    houses and equipment, 216
    housing and equipment, 250–252
    labor requirements, 248
    lighting requirements, 206,
        251, 256
    management, 255–257

marketing, 301–302
nutrition and feeding, 252–253
pricing, 305
production costs, 135
raising males and females
        separately, 255–256
Bronchitis, infectious, 37
Bronze turkey, 72
Brooder houses, 140, 215, 277
    cleaning, 276
Brooder pneumonia, 324, 349
Broodiness, 238
Brooding, 57, 140
    of bobwhite quail, 356
    brooder houses and
        equipment, 140
    of ducks, 323
    equipment, 215
    of geese, 351
    management schedule, 140–141
    moisture control, 140
    for pigeons, 355
    temperature control, 140
    in turkey management, 274–275
    ventilation, 140
Brooding turkey poults, 275–278
Brown eggs, 240
Buckwheat, 98
Budgets in the poultry business,
        316, 343
Buff, 325
Burial in open-bottom pits, 151
Business aspects of egg production
    costs and returns, 225–226
    egg contracts, 224–225
    financing, 224
    labor requirements, 224
    location, 224
    records and accounting, 225
Business issues, 315
    budgets, 316
    capital needs, 315–316
    databases, 316
    enterprise accounts, 316
    liability, 316
    records, 316
    social security, 316–317
    tax management, estate planning,
        and wills, 316
    types of business organizations, 315
    workers' compensation, 316
Business organization
    of poultry breeding, 68–69
    types of, 337–338
Butylated hydroxyanisole (BHA), 109
Butylated hydroxytoluene (BHT), 109

C

Caged layer fatigue, 187
Cage system, 213–214, 229–230
Calcium (C), 85

Calcium supplements, 107
Cal-Maine Foods, 313–314
Calorie, 80
Campylobacter jejuni, 192, 196
Canada geese, 324–325
Candida albicans, 110
Candling in grading eggs, 298
Canker, 173
Cannibalism, 137, 158, 188, 238
Capital needs of poultry business,
        315–316, 338
Caponizing, 33, 136–137
Capons, 33, 264
Carbohydrates, 32, 79–80
Carotenoids, 109–110
Casein, 105
Catabolism, 78
Catching turkeys, 278
Ceca, 33
Central nervous system (CNS), 39
Cereal milling by-product feeds, 98
Cervical vertebrae, 25
Chemical analysis of poultry
        feeds, 113
Chemical senses, 154
Cherry Valley Farms, 69
Chicken(s)
    artificial insemination, 70
    consumption, 288
    domestication and early use, 1–3
    feeding, 126–129
    marketing, 294
    mating behavior, 156
    selecting and culling, 70–72
    types of classes, 65–66
    USDA classification, 302
    vaccination programs, 183
Chicken genome, 60
Chicken meat
    changes in production, 6
    composition, 245–246
    future, 264
    marketing, 242–245, 301–302, 305
Chicks
    brooder houses and
        equipment, 215
    delivering, 57
    separating male from female, 226
    troubleshooting, 138
Chinese goose, 325
Cholecalciferol (D₃), 92
Cholesterol, 9–10, 38
Choline, 93
Chromosomes, 59–60
Chronic respiratory disease (CRD),
        48–49, 166–167
Circulatory system, 26–27
Cleaning, 183
    equipment for, 213
Cloaca, 33, 39
Clostridium perfringens, 192

Clutches, 44
Cobalamins, 94
Cobb-Vantness, 68
Coccidiosis, 173, 174–175, 176,
  255, 324
Coccidiostats, 112
  for broilers, 255
  for turkeys, 273
Cockfighting, 157
Coconut meal, 101
Cold stress, 188
  adaptation to, 44
Colon, 33
Color sexing, 62–63
Commercial breeding of poultry, 58
Commercial feeds. *See also* Feed(s)
  buying, 115
  farm-mixed versus complete, 115
  selecting, 116
Commercial production of broilers,
  247–255
Communication methods, 139,
  154–155
Composting, 151
Computers, 139
  in diet formulation, 125–126
  in record keeping, 342–343
Condensed fish solubles, 106
Conduction, 44
Confinement production, 121
Consolidation in the broiler
  industry, 310
Consumer trends in marketing,
  305–307
Contour feathers, 24
Contracts, 314–315
  broiler, 248–250
  egg, 224–225
  production, 310
  specifics of, 315
  turkey, 268
Convection, 44
Copper, 89
Copra meal, 101
Coracoids, 25
Corn, 97
Cornish Rock, 3
Corporations, 315, 338
Corticotropin-releasing hormone
  (CRH), 40, 42
Coryza, 166–167
Costs
  of poultry feeds, 114
  of producing eggs, 225
Cottonseed meal, 101
Coturnix quail, 333–334, 358–359
  care and management, 359
  diseases, 359
  feeding and watering, 359
  reproduction, 359
Crab meal, 106

Credit, 338–341
Critical Control Points (CCPs),
  196–197
Crop, 28, 30
Crossbreeding, 48
Crumbles, 116
Cryptosporidiosis, 173
Culling, 72, 238, 345
Curled-toe paralysis, 94

D

Dairy products, 104–105
Dark meat, 26
Databases of the poultry business, 316
Day-old chicks, 226
Day-range poultry, 307
Dead birds, disposal, 256
Deglutition, 28
7-dehydrocholesterol, 92
Dented eggshells, 38
Deoxyribonucleic Acid (DNA), 59
Designer products marketing, 306
Desnooding, 277–278
Diabetes mellitus, 9
Diac muscle, 25–26
Diagnostic laboratory, 186
Diets. *See also* Commercial feeds;
  Feed(s)
  balancing, 123–126
  factors in formulating, 122
Digestible energy, 81
Digestion
  physiology, 28–33
  process, 28
Digestive system, 28–33
Disaccharides, 79
Disease. *See also specific*
  of coturnix quail, 359
  of ducks, 323–324, 349
  of geese, 326
  genetic resistance, 186
  of Hungarian or gray partridge, 358
  impact on poultry pricing, 293
  of peafowl, 360
  of pheasants, 361
  treatment, 186
Disinfectants, 183, 184–185
Disinfection for broilers, 254
DNA, mitochondrial, 59
Domestication
  of ducks, 321
  of geese, 325
  of pigeons, 332
  of ratites, 327–328
Dominant genes, 61–62
Double-yolked eggs, 38
Dried buttermilk, 104
Dried fish solubles, 106
Dried hydrolyzed casein, 105
Dried hydrolyzed whey, 104
Dried milk albumen, 105

Dried milk protein, 105
Dried skimmed milk, 104
Dried whey, 104
Dried whey solubles, 105
Drugs, 111–112
Dry spreading, 213
Dubbing, 57, 238
Duck plague, 166–167
Ducks, 319, 320–324
  artificial incubation, 347–348
  breeding, 321, 323, 346–348
  breeds, 321–322
  brooding and rearing, 323, 348
  classification, 321
  diseases, 323–324, 349
  domestication, 321
  eggs, 218
  feathers, 320
  feeding, 130, 323, 349
  incubation, 323
  management, 323, 346
  marketing, 305, 324, 349–350
  meat, 319–320, 320–321
  natural incubation, 348
  processing, 324, 349–350
  world trade in meat, 313
Duck viral hepatitis (DVH), 163,
  166–167
Duck virus enteritis, 324, 349
Dust bathing, 158

E

Educational programs in food
  safety, 193
Efficiency, improving, 135–136
Egg-breaking industry, 18
  products, 300–301
Egg contracts, 224–225
Egg-handling equipment, 232
Egg industry, vertical coordination
  in, 313–314
Egg-layer genetics, measurement to
  improve, 71
Egg laying, 25, 39
Egg production, 63
  business aspects of, 224–226
  changes in, 5–6
  management, 234–239
  turkey, 285–286
Egg production chickens, 66
Eggs, 35–36
  abnormalities, 38–39
  albumen quality, 299
  biology of the production, 223
  brown, 240
  clean, 55
  color, 38
  commercial production, 218,
    223–234, 239
  composition, 38, 218–223
  consumption, 6, 288

Eggs, *(continued)*
    of coturnix quail, 359
    eating, 238
    exports and imports, 312
    future issues, 241
    grading and sizing, 296, 299–301
    guinea fowl, 357
    handling hatching, 49–50, 239
    incubation, 259
    infertile, 36
    interior quality, 49
    marketing, 240, 294, 296–299
    market value, 293
    nutrition and, 8–9
    partridge, 358
    peafowl, 359
    pheasant, 360
    pigeon, 355
    pricing, 301
    producing quality, 238–240
    production of turkey, 281
    proper care of, at the farm
        level, 299
    Salmonella and, 194–195
    sanitizing, 55–56
    selection, 49
    shape, 37–38, 49
    size, 36–37, 49, 259
    storage, 259
    structure, 38
    troubleshooting, 138–139
    white, 240
    world trade, 311–313
Egg shell, 223
    composition, 223
Egg white, 223
    composition, 223
Egg yolk, 218, 223
    composition, 220–223
Egyptian goose, 326
Embryo communication, 51–52
Embryonated eggs, 239
Embryonic development, 19, 51
Emden goose, 326
Emergency warning systems, 213
Emerging diseases, 165
Emus, 327–328, 331–332
    reproduction, 331–332
    slaughter, 331
Encephalomalacia, 92–93
Encephalomyelitis, 165
Endocrine system, 39–401
Energy, 78–82
    biological measure of utilization,
        114
    digestible, 81
    gross, 81
    metabolizable, 81
    net, 81
    saving, 141–142
    true metabolizable, 81–82
Energy feedstuffs, 97–100

Enterprise accounts, 316, 343
Environment
    for broilers, 253–254
    effect of, on poultry, 151–152
Environmental controls, 205
Environmentally controlled
        buildings, 203–206
Environmental quality, 233–234
Environmental stress, 187–188
Enzymes, 28, 110
Epidemic tremor, 166–167
Epididymis, 33
Erysipelas, 273
Erythrocytes, 27
Escherichia coli, 191, 192
Esophagus, 28, 29–30
Estate planning, 316, 343
    gift tax exclusion, 344
    special-use valuation, 343–344
    trusts, 344
    unlimited marital deduction, 344
    wills, 344
Ethology, 153
Ethoxyquin, 109
Europe, welfare and poultry
        production in, 159–160
European Union
    competitiveness of the poultry
        industry, 313
    welfare and poultry
        production, 160
Excreta, facilities and equipment for
        handling, 213
Excretory system, 33
Exercise, lack of, 9
Exhibition chickens, 66
Expenses, record of, 342
Exports
    of eggs, 312
    poultry meat, 311–312
    poultry pricing and, 294
External parasites, 177–178
Exudative diathesis, 92–93

F

Family selection, 71
Farm level, proper care of eggs at,
        299
Fats, 80, 100
Fat-soluble vitamins, 89, 92–93
Fatty acid synthesis, 247
Feather meal, 104
Feather picking, 158
Feathers, 24–25, 263
    as animal feed, 263
    colors, 64
    contour, 24
    development, 62
    goose, 353
Fecundity, 36
Federal Water Pollution Control Act
    Amendments, 149

Feed(s), 97–112. *See also*
        Commercial feeds; Nutrients
    buying, 114–116
    conversion, 233
    efficiency, 8, 63–65
    evaluating, 112–114
    feathers as, 263
    impact on costs, on poultry
        pricing, 293
    new developments, 120
    preparation, 116
    regulation of intake, 42–43
    standards, 121–132
Feedback
    negative, 40, 42
    positive, 42
Feeders, 216
    for broilers, 252
    for layers, 232
    for pullets, 231
    for turkeys, 270, 272
Feeding, 126–129, 232–233
    behaviors associated with, 155–156
    of coturnix quail, 359
    of ducks, 130, 323, 349
    equipment for, 210
    of game birds, 131–132
    of geese, 130–131, 326, 351–352
    of guinea fowl, 132, 333, 357
    of Hungarian or gray partridge, 358
    of ostriches, 131
    of peafowl, 132
    of pheasants, 361
    of pigeons, 131
    standards for, 122–123
    of turkeys, 129–130
Feed laws, 116
Feedstuffs, physical evaluation of, 112
Feed substitutions, 116
Female reproductive system, 34–39
Fertility, 48
Fertilizers, precautions when using
        manure as, 147–148
Fiber, 80
Fighting in chickens, 157
Filoplume, 25
Fish liver and glandular meal, 106
Fish meal, 105–106
Fish protein concentrate, 106
Fish residue meal, 106
Flavoring agents, 109
Flies, 151, 179
Flightiness, 158
Flock health, 180–182, 233–234
Flock mating, 70
Floor system, 230
Fluke, 176
Folacin (folic acid), 93
Follicle, 35
Food, Drug, and Cosmetic Act
    (1938), 116
Food and Drug Act (1906), 116

Food industry standards for the welfare of poultry, 161
Food safety, 18, 191–198
  antibiotics and, 198
  educational programs, 193
  eggs and poultry meat, 193–196
  government and, 191–193
  irradiation and, 197–198
  marketing of poultry meat and, 301
  pathogens and microorganisms and, 193
  role of USDA, 196–197
Food Safety and Inspection Service (FSIS), 196–197
Forced molting, 228, 284
Forced resting, 136, 228
Fortified products, marketing, 306
Fowl cholera, 163, 166–167, 273, 323, 349
Fowl pox, 164, 166–167, 273
Fowl typhoid, 168–169
Free-range meat and eggs, marketing, 306–307
Fructose, 79
FSH, 44
Fungal or fungal product diseases, 172
Fusarium, 110

G

Galactose, 79
Gallbladder, 31
Game birds, 333
  diseases in, 333
  feeding, 131–132, 333
  sport, and hobby operations, 17
  watering, 333
Geese, 319, 324–327
  breeding, 325–326, 350–351
  classification, 325
  diseases, 326
  domestication, 325
  feathers, 320
  feeding, 130–131, 326, 351–352
  incubation, 326, 350
  management, 326
  marketing, 305, 327, 352–353
  meat, 319–320, 325
  processing, 352–353
  world trade in meat, 313
Gene knockout, 20
General partnership, 315
Genes, 58–59
  lethal, 48, 65
  semilethal, 48
Genetically modified organisms (GMOs), 60
Genetics, 19–20, 48, 58–65
  breeding and, 226
  breeding turkeys, 65
  chromosomes, 59–60
  DNA and, 59
  dominant and recessive genes, 61–62

genome, 59–60
heterosis, 65
inbreeding, 65
inheritance of economically important traits in poultry, 62–65
lethal genes and abnormalities of development, 65
multiple gene inheritance, 61
mutations, 65
resistance to disease, 186
simple gene inheritance, 61
variation and selection, 60
Genome, 59–60
Genomics, 20
Gift tax exclusion, 344
Gizzard, 28, 30, 31
Glandular meal, 103
Glucagon, 30
Glucose, 79
Glycerol-3-hosphate, 247
Glycogen, 79
Gonadotropin-releasing factor (GnRH), 42, 44
Gossypol, 89, 101
Government, food safety and, 191–193
Grading of eggs, 296, 299–301
Grains, 97–100
Gray partridge. *See* Hungarian or gray partridge
Grit, 110
Gross energy, 81
Grow-out operations, 17
Growth, role of nutrition in, 77–78
Growth hormone (GH), 246
Guinea fowl, 4, 333, 357–358
  care and management, 357
  diseases, 358
  feeding, 132, 357
  marketing and releasing, 358
  reproduction, 357
  watering, 357
Gumboro, 168–169

H

Habituation, 153
HACCP-based inspection, 196–197
Halal slaughtering, 262
Hatchability, 47–50
Hatcheries, 17, 52–57, 69
  building, 54–55
  by-products, 104
  changes, 5
  equipment, 55
  management and sanitation, 55–56, 182
Hatchery manager, 54
Haugh units, 299
Head picking, 238
Health
  broiler, 253–255
  flock, 180–182

layers, 233–234
poultry, 9, 162–190
turkey, 272–273
Hearing, 154
Heart disease, 9–10
Heat production, 204, 207–208
Heat stress, 187–188
  adaptation to, 43–44
Hemorrhagic enteritis, 168–169, 273
Hendrix Poultry Breeding, 68–69
Herring meal, 105
Heterosis, 58, 65
Hexamitiasis, 173, 174–175
Histomoniasas, 173
Holidays, impact on poultry pricing, 293
Hormones, 39–40, 41–42, 246
House mouse (*Mus musculus*), 143
Hubbard-ISA, 69
Hungarian or gray partridge
  care and management, 358
  diseases, 358
  feeding and watering, 358
  marketing and releasing, 358
  reproduction, 358
Hybrid vigor, 58
Hy-Line International, 69
Hypertension, 9
Hypervitaminosis, 92
Hysteria/fright, 188
Hy-Vac, 69

I

Immunity, 179–180
Immunization, 183, 186. *See also* Vaccinations
Immunoglobulins, 180
Imports
  of eggs, 312
  of poultry meat, 312
  poultry pricing and, 294
Imprinting, 153
Inbreeding, 48, 65
Incinerators, 213
Incrossbreeding, 48
Incubation, 46–57
  of bobwhite quail, 356
  of ducks, 323, 347–348
  of geese, 326
  hatchability, 47–50
  hatcheries, 52–57
  incubator operation, 50–52
  for pigeons, 355
  of turkey eggs, 286
Indian Runner, 322
Induced molting, 136, 259
Infectious bronchitis, 37, 164, 168–169
Infectious bursar disease, 165
Infectious disease, 163–172
  prevention and control, 180–186
  spread, 178
Infectious sinusitis, 168–169, 272

Infectious synovitis, 168–169
Infertile eggs, 36
Infertility for ostriches, 329
Inorganic elements, toxic levels of, 119
In ovo feeding, 96, 253
In ovo vaccination, 188
Insects, 177
Insulation, 208–209
Insulin, 30
Insulin-like growth factor-I
    (IGF-I), 246
Integrated Pest Management
    (IPM), 177–178
Integration of body processes, 39–44
Integument, 23–25
Internal parasites, 172–173, 176
Iodine, 89
Iron, 89
Irradiation, food safety and, 197–198

J

Japanese quail, feeding, 132
Joule, 81

K

Khaki campbell, 322
Kidneys, 33
Kilocalorie, 80–81
Kosher slaughter, 262

L

Lagoon, 213
Large intestine, 33
Large roundworms, 174–175
Laryngotracheitis, 164, 168–169
Layers, 223–239
    feeding, 126–127
    future issues, 241
    health program, 233–234
    housing and equipment,
        213–215, 231–232
    lighting requirements, 205
    management, 237–238
    nutrition, 232–233
    production costs, 135
    troubleshooting, 138
Learning, 153
Leg weakness, 187
Lethal genes, 48
    abnormalities of development, 65
Leucocytozoonosis, 173, 174–175
Leukosis, 165, 168–169
Liability, 316, 344–345
Lice, 174–175
Lighting, 141
    for breeder turkeys, 282
    for broilers, 206, 251, 256
    for brooding broiler chickens, 257
    layers, 205
    management, 139–140
    in poultry houses, 202–203,
        205–206, 237

    for pullets, 205
    sexual maturity and reproduction
        and, 44
    for turkey breeders, 282
    for turkeys, 206
Limited partnership, 315, 338
Linear programming in balancing
    diet, 125–126
Linoleic acids, 80
Linseed meal, 102
Lipases, 30
Lipids, 32
Lipogenesis, 247
Liquid eggs, 240
Listeria, 195
Listeria monocytogenes, 191, 192
Litter, 144–146, 150
    for broilers, 256, 257
    reducing need of, 145–146
Livability, 64–65
Liver, 31
Liver meal, 103
Lohmann-Wesjohann Group, 69
Lorenz, Konrad, 20
Low atmospheric pressure, 188
Low-density lipoprotein (LDL),
    9–10
Luteinizing hormone (LH), 42, 44

M

Macrochromosomes, 59–60
Macrominerals, 85
Magnesium, 85, 88
Maintenance, role of nutrition in, 77
Male reproductive system, 33–34
Management, 237–238
    bobwhite quail, 356–357
    broilers, 255–257
    Coturnix quail, 359
    ducks, 323
    geese, 326
    Hungarian or gray partridge, 358
    layers, 237–238
    pheasants, 361
    pigeons, 355
    schedule in brooding, 140–141
    turkey, 274–278, 286
Manganese, 89
Mannose, 79
Manure, precautions when using, as
    a fertilizer, 147–148
Marek's disease, 165, 168–169
Marine by-products, 105–106
Market channels and selling
    arrangements, 294–295
Market egg-producing operations, 17
Market forecasters, 18
Marketing, 288
    of chicken, 294
    of chicken meat, 242–245
    of chicken meat products,
        301–302, 305

    current and emerging issues in,
        305–307
    of ducks, 305, 324, 349–350
    of eggs, 294, 296–299
    of geese, 305, 327, 352–353
    of Hungarian or gray partridge, 358
    of peafowl, 360
    of pheasants, 361
    of poultry meat, 301–305
    trends ahead in, 307
    of turkey meat, 265–267
    of turkeys, 294, 305
Marketing contracting, 307
Marketing specialists, 18
Market News, 295
Market value of poultry and
    eggs, 293
Mash, 116
Mass mating, 70
Mating, methods of, 70
Mating behavior
    in chickens, 156
    in geese, 350
    in turkeys, 156
Meat meal, 103
Meat packing by-products, 103
Meat production, biology of,
    246–247
Meat spots, 38
Meckel's diverticulum, 31–32
Media, impact of, on poultry
    pricing, 294
Medicine, poultry in, 20
Medullary bone, 25
Megacalorie, 81
Meleagris gallopavo gallopavo, 4
Meleagris gallopavo silvestris, 4
Memory, 153
Menadione dimethylpyrimidol
    (MPB), 93
Menadione sodium bisulfate
    (MSB), 93
Menadione sodium bisulfite
    complex (MSBC), 93
Menhaden, 105
Mesotocin, 39
Metabolic diseases, 186–187
Metabolizable energy, 81
Methlonine, 96
Micelles, 31
Michaels Foods, 314
Microchromosomes, 59–60
Microminerals, 85
Milk protein products, 105
Millet, 98
Minerals, 32, 85–89
Mineral supplements, 107
Mite infestation, 174–175
Mitochondrial DNA, 59
Moisture control in brooding, 140
Molasses, 100
Mold inhibitors, 110–111

Molting, 136, 227–228
  forced, 228, 284
  forced rest, 228
  induced, 259
Monoglycerides, 32
Monosaccharides, 79
Moraxella anatipestifer, 324
Morbidity, 162
Mortality, 162, 213
Motion, pendular, 31
Multiple gene inheritance
        (quantitative traits), 61
Muscle, 245
  development, 246–247
  disease in and myopathy, 187
  growth, 247
Muscovy, 322
Muscular dystrophy, 89
Muscular system, 25–26
Mutations, 65
Mycoplasma gallisepticum, 48–49, 55
Mycoplasma synovial, 48, 55
Mycoplasmosis, 168–169
Mycosis, 168–169
Mycotoxicoses, 170–171
Mycotoxins, 110, 172
Myoblasts, 247
Myofibril proteins, 245
Myoglobin, 245
Myotubes, 247

N

National Human Genome Research
        Institute, 20
National Poultry Improvement Plan
        (USA), 71–72
National Research Council (NRC)
        requirements for poultry, 123
Natural incubation of ducks, 348
Naturally ventilated buildings, 203
Necrotic enteritis in ducks, 323, 349
Negative feedback, 40, 42
Nematodes, 176
Nervous system, 39
Nesting, 238
Net energy, 81
Newcastle disease, 37, 164–165,
        170–171, 273
New duck disease, 324
New Hampshire, 3
Niacin (Nicotinic
        acid/nicotinamine), 93
Nickerson Group, 69
Nitrofurans, 111–112
Nitrogen-free extract (NFE), 79
Nitrogen from poultry waste, 149–150
Nongenetically modified
        (Gm)/nongenetically
        engineered (Ge) poultry, 307
Nonnutritive additives, 108–112
Normal mortality and morbidity
        losses, 162

North America, competitiveness of
        the poultry industry in, 313
Norway rat *(Rattus norvegicus),* 143
Nutrients. *See also* Commercial feeds;
        Diets; Feed(s)
  classification of, 76
  determination of requirements, 122
  functions of, 76–78
Nutrition, 48, 96
  of bobwhite quail, 356
  of breeding turkeys, 284–285
  broiler breeder, 260
  in broiler production, 252–253
  in flock health, 233
  of laying hens, 232–233
  of pullets, 227
  studies, 19
  for turkeys, 272–274
Nutritional muscular dystrophy, 92–93

O

Oats, 97
Offal and bone, 263
Oilseed meals, 100
Olfaction, 154
Open market, 307
Open production, 307
Organic poultry meat and eggs,
        marketing, 306–307
Ornamental birds, 333
  diseases, 333
  feeding and watering, 333
Oropharynx, 29–30
Osteoporosis, 187
Ostrich, 327–328, 328
  biology, 328–330
  brooding and rearing, 354
  characteristics of eggs, 330
  characteristics of muscle/meat, 331
  egg storage and incubation, 354
  farming/ranching, 328
  feeding, 131
  history, 328
  practical aspects of
        reproduction, 354
  production, 328, 330–331, 354
  slaughter, 331
  subspecies, 328
Ovary, 34
Oviduct, 35
Oviposition/egg laying, behaviors
        associated with, 158

P

Pancreas, 30
Pancreatic polypeptide, 30
Pantothenic acid, 93
Parasites, 172
  control for broilers, 255
  external, 177–178
  internal, 172–173, 176
  turkeys and, 273–274

Paratyphoid, 170–171
Parental behavior, 156–157
Parthenogenesis, 47
Partnership, 337–338
  general, 315
  limited, 315, 338
Partridge, 333. *See also* Hungarian or
        gray partridge
Pasteurization, 193–194
Pasture-raised poultry, 307
Pathogenic diseases, 163–172
Pathogen Reduction and Hazard
        Analysis and Critical Control
        Point (HACCP) System,
        196–197
Pathogens and microorganisms, and
        food safety, 193
Peafowl, 334, 359–360
  breeds, 334
  diseases, 360
  feeding, 132
  marketing and releasing, 360
  reproduction, 359
Peanut meal, 102
Peck order, 20, 157
Pedigree selection, 71
Pekin ducks, 69
Pellet binders, 109
Pellets, 116
Pendular motion, 31
Pen mating, 70
Per capita consumption, of poultry
        products, 18
Peripheral nervous system (PNS), 39
Peristalsis, 29–30, 31
Perosis, 89
Persecution, 188
Pests, control, 142
Pet food, 262–263
Phase feeding, 233
Pheasants, 334, 360
  breeds, 334
  care and management, 361
  diseases, 361
  feeding, 132, 361
  marketing and releasing, 361
  reproduction, 360–361
  watering, 361
Phosphorus, 85, 150
Phosphorus supplements, 107
Photostimulation, 282
  of broiler breeders, 258
Physical evaluation of feedstuffs, 112
Pigeons, 332, 355
  breeding, 332
  breeding facilities, 355
  breeds, 332
  care and management, 355
  classification, 332
  domestication, 332
  feeding, 131, 355
  marketing, releasing, 355

Pigeons, *(continued)*
  meat, 332
    reproduction, 355
    watering, 355
Pilchard meal, 105
Pilgrim, 326
Plant proteins, 100–102
Plasma, 26
Plumage color, 62
Plumules, 25
Pneumatic bones, 25
Pneumonia/respiratory diseases,
        272–273
Pollution laws and regulations,
        148–149
Polyneuritis, 94
Polysaccharides, 79
Positive feedback, 42
Postslaughter, 262
Potassium, 88
Poultry
  commercial breeding, 58
  consumption, 288
  defined, 1
  development of American,
        production, 5–8
  effect of environment, 151–152
  effect of light, 44
  feeds, 97–112
  market classes and grades, 301
  market value, 293
  in medicine, 20
  relative importance of different
        species, 15–16
  in scientific research, 18–20
  word origins related to, 3
  world distribution, 12–13
  world trade, 311–313
Poultry blood, 26–27
Poultry by-product meal, 104
Poultry enterprise, 133–134
  choice of species, and specialized
        versus integrated, 133
  location, 133
  size, 133–134
Poultry equipment
  requirements of, 209–212
  types of, 210–212
Poultry facilities, siting, 151
Poultry farms, changes in the
        number and size of, 7
Poultry health, 162–190
  causes of avian diseases, 163
  glossary of, terms, 189–190
  for layers, 234
  monitoring, 162
  new issues, 188–189
  pathogenic or infectious diseases,
        163–172
Poultry houses, 199–217, 229–233
  animal waste management, 203
  building materials, 207

cage types, 207
  design and construction, 206–207
  environmental controls, 203–206
  environmental factors, 207–209
  floor types, 206–207
  housing systems, 229–230
  humidity, 202
  insulation, 202, 208–209
  lighting, 202–203, 205–206
  location, 199–200
  prefabricated houses, 209
  requirements, 201–203, 229
  space requirements of building
        and equipment, 200–212
  temperature, 202
  vapor barrier, 202
  ventilation, 202, 203, 206
Poultry industry, 317
  competitiveness
    in the European Union, 313
    in North America, 313
  components, 16–18
  future, 21
  impact of adipose tissue, 247
  in United States., 3, 13–20
  vertical coordination, 307
Poultry management, 134–135
Poultry meat
  exports of, 311–312
  imports of, 312
  marketing, 301–305
  nutrition and, 9
  Salmonella and, 194
Poultry mortalities, 151
Poultry nutrition, 75–96
  classification of nutrients, 76
  functions of nutrients, 76–78
  nutrient composition of poultry
        and eggs, 75–76
Poultry operations, changes in
        ownership and
        organization, 8
Poultry producers, business
        suggestions for small- and
        moderate-sized, 337–345
Poultry products
  basics for good nutrition, 8–9
  factors affecting the price of,
        293–294
  to feed the hungry, 10–12
  health issues and, 9
  in industry, 20
  per capita consumption of, 18
Poultry reproduction, 46–47
Poultry research, 317
Poultry science, careers for,
        graduates, 318
Poultry wastes, 103–104, 147–152,
        237–238, 256
  management, 203
  processing, 150–151
Prebiotics, 96, 189

Predator control for turkeys, 278
Prehension, 28
Pricing
  broilers, 305
  eggs, 301
  turkeys, 305
Primary breeders, 68–69, 257
  of turkeys, 268–269
Prions, 172
Probiotics, 96, 189
Procenzymes, 28
Processing
  of ducks, 324, 349–350
  of geese, 352–353
Processing industry, 18
Processing plant prior to
        slaughter, 262
Production contracting, 307, 310
Production-promoting drugs, 111–112
Product quality, presence of
        substances affecting, 119
Proprietorship, 337
  sole, 315
Proteins, 32, 82–85, 245
  animal, 102–107
Protein supplements, 100–107
Protein utilization, biological
        measure of, 113–114
Protozoan diseases, 172, 176
Proventriculus, 28, 30
Public relations and education, 18
Pullets
  housing and equipment for,
        230–231
    replacement, 215–216
  lighting requirements, 205
  nutrition, 227
Pullorum disease, 48, 170–171
Pyridoxal, 94
Pyridoxamine, 94
Pyridoxine, 94

Q

Quail. *See also* Bobwhite quail;
        Coturnix quail
  eggs of, 218
Qualitative traits, 61
Quality
  grading eggs for, 296
  of shell eggs, 296–299
Quantitative trait loci (QTL), 60

R

Random Sample Performance
        Tests, 71
Range turkeys, 278
Rapeseed/canola meal, 102
Ratites, 327
  classification, 327
  domestication, 327–328
  evolution, 327
Ready-to-lay pullets, 226–227

Receipts, record of, 342
Recessive genes, 61–62
Records
  computers and, 342–343
  kinds, 341
  of the poultry business, 316
  reasons for keeping, 341
  summarizing and analyzing, 342
  type to keep, 341–342
Red blood cells, 27
Red fibers, 26
Red jungle fowl *(Gallus gallus),* 2
Refrigeration of eggs, 232
Refuse Act (1899), 148–149
Relative humidity in broiler
    production, 251
Rendering, 96
Replacement pullets, 226–227
  feeding, 127
Reproduction
  of bobwhite quail, 356
  of coturnix quail, 359
  of emus, 331–332
  of guinea fowl, 333, 357
  of Hungarian or gray
      partridge, 358
  of peafowl, 359
  of pheasants, 360–361
  role of nutrition in, 78
Reproductive diseases, 187, 349
Reproductive system, 33–39
  female, 34–39
  male, 33–34
Research, poultry, 317
Research and development
    scientists, 18
Respiratory diseases, 164–165
Respiratory system, 27–28
Rheas, 327–328
Rhinotracheitis, 165
Riboflavin (Vitamin B$_2$), 93–94
Rice, 97
Roasters, 263–264
  breast blisters, 263–264
Rodenticides, 144
Rodents, 142–144, 151, 179
  controls, 143–144
  observation, 143
Roof rat *(Rattus rattus),* 143
Roundworms, 176
Rye, 98

S

Safflower meal, 102
Salivary glands, 28
Salmonella, 55, 192, 194–195
Salmonella serotype *Enteriditis,*
    191, 195
Salmonella serotype
    Typhimurium, 191
Salmon meal, 105
Salt (sodium Chloride/NaCl), 85

Sanitation
  egg, 236–237
  in flock health, 233
  in hatchery, 55–56
Sarcoplasmic soluble proteins, 245
Sardine meal, 105
Satellite cells, 247
Scientific research, poultry in, 18–20
Screw-processed cottonseed meal, 119
Scrotum, 33
Sebastopol, 326
Segmentation contractions, 31
Selenium, 89
Semen quality, 260
Semilethal genes, 48
Seminal fluid, 34
Senses, 154
  special, 39
Sepalm produces, 69
Sesame meal, 102
Sex, nutrition and, 78
Sexing, 259
  color, 62–63
Sexing chicks, 56
Sex-linked dwarfism, 64
Sexual behavior, 156
Sexual maturity and reproduction,
    effect of light on, 44
Shank color, 62
Shell, 220
  quality of, 49, 299
Shell eggs, 240
  quality of, 296–299
Shrimp meal, 106
Simple gene inheritance, 61
Single-Comb White Leghorn hens, 57
Single nucleotide polymorphisms
    (SNPs), 60
Sizing of eggs, 299–301
Skeletal system, 25
Skin, 23–24
  color, 62
Slaughter, 261, 262
  of emus, 331
  halal, 262
  kosher, 262
  of ostriches, 331
  turkey, 279–280
Sleeping, 158
Small intestine, 31–32
  digestion and absorption in the,
      31–32
Smith incubator, 46
Smoking, 9
Social order, 20
Social relationships, 158–159
Social security, 316–317, 345
Soft-shelled eggs, 38–39
Somatic nervous system, 39
Somatostatin, 30
Sorghum, 97
Soybean meal, 100–101

Special senses, 39
Specific Pathogen Free (SPF)
    program, 186
Spent hens, marketing, 305
Spermatogenesis, 33
Spermatozoa, 59
Spot market, 307
Spraying equipment, 213
Squabs, 332
Squeakers, 332
Standard operating procedures, 134
Staphylococcus aureus, 192
Stereotypies, 158
Stress, 159
  heat, 187–188
Stromal proteins, 245
Sunflower seed meal, 102
Swellhead syndrome, 165

T

Table eggs, handling, 239
Tankage, 103
Tapeworms, 174–175, 176
Taste, 154
Tax management, 316
Telomere, 60
Temperature
  for broilers, 256
  for brooding broiler chickens, 257
  control of, in brooding, 140
Territorial behavior, 155
Testes, 33
Thermoregulation, 43–44, 157–158
Thiamin, 94
Thyroxine, 246
Tibial dyschondroplasia, 187
Toe clipping, 278
Toe picking, 238
Tongue, 28
Toulouse goose, 326
Toxic levels of inorganic elements, 119
Toxicology, 19
Trace minerals, 88–89
Traits
  inheritance of economically
      important, 62–65
  qualitative, 61
  quantitative, 61
Transgenics, 20, 60
Transportation/hauling, 261
Trapping, 144
Trichomoniasis, 173, 174–175
Trigeminal chemoreceptors, 154
Triglyceride breakdown by adipose
      tissue, 247
Triglyceride synthesis by adipose
      tissue, 247
Triticale, 98
Troubleshooting, 137–139
  chicks, 138
  eggs, 138–139
  layers, 138

True metabolizable energy, 81–82
Trusts, 344
Tryglycerides, 32
Trypsinogen, 30
Tuberculosis, 170–171
Turkey(s) (*Meleagris gallopavo*), 4
   abnormalities and injuries, 274
   aggressive behavior, 157
   artificial insemination, 282–283
   breast blisters, 274
   breeder management, 286
   breeding and genetics, 72,
      268–272
   breeds and varieties, 269–270
   calluses, 274
   cannibalism, 274
   changes in, 6
   combating disease, 273
   commercial production, 267–268
   consumption of, 288
   domestication and early use, 4
   egg handling, 285–286
   facilities for breeder, 217
   facilities for market, 217
   feeding, 129–130
   further processing, 265–267
   future, 286
   grading, 280
   handling, 278
   health, 272–273
   incubation of eggs, 286
   inspection, 279–280
   lighting requirements, 206
   marketing, 294, 305
   mating behavior, 156
   moving, 278
   natural mating, 283
   nutrition and feeding, 272–274
   parasites and, 273–274
   pendulous or drop crop, 274
   predator control, 278
   pricing, 305
   processing, 278–280
   production costs, 135
   products and by-products, 280
   rickets, 274
   semen, 282–283
   slaughtering and processing,
      279–280
   stampeding, 274
   USDA classification of, 302
   vaccinations, 183, 273
Turkey breeders, 280
   egg production, 281
   lighting, 282
   management, 286
   reproduction and, 280–281
Turkey-cock, 4
Turkey contracts, 268
Turkey coryza, 170–171
Turkey genome, 60

Turkey industry
   breeding operators and systems, 270
   business aspects, 267–268
   feeders, 270, 272
   financing, 267–268
   housing and equipment, 270–272
   labor requirements, 268
   litter, 270
   vertical coordination in the, 310–311
   waterers, 272
Turkey management, 274
   brooding, 274–275
   range turkeys, 278
   turkey poults, 275–278
Turkey meat
   biology of meat production, 267
   composition, 267
   marketing, 265–267
Turkey poults, 275–278
   facilities for brooding, 217
Turkey venereal, 170–171

U

Ulcerative enteritis, 170–171
Undesired behavior, 158
United Kingdom, welfare and poultry
      production in, 159–160
United States, welfare and poultry
      production in, 160–161
Unlimited marital deduction, 344
Urine, 33
Urner barry price quotation, 295
Uterus, 39

V

Vaccinations, 57, 237
   for broilers, 254
   for brooding broiler chickens, 256
   for chickens, 183, 186
   for *Eimeria tenella*, 176
   for turkeys, 273
Vagina, 39
Vapor barrier, 209
Vapor production of poultry, 205
Vas deferens, 33
Vasotocin, 39
Ventilation
   in broiler production, 250–251
   in brooding, 140, 141
   for layers, 231
   in poultry houses, 202, 203, 206
Vent picking, 238
Vertical coordination
   advantages and disadvantages
      of, 314
   in the broiler industry, 310
   in the egg industry, 313–314
   in the poultry industry, 307
   in the turkey industry, 310–311
Vertical integration, 309–311
Veterinarian, 186

Vibrionic hepatitis, 170–171
Viral diseases, 163–165
Viral hepatitis of ducks, 323–324
Vision, 154
Visual displays of poultry, 154–155
Vitamin A, 89, 92
Vitamin B$_1$, 94
Vitamin B$_3$, 93
Vitamin B$_6$, 94
Vitamin B$_{12}$, 94
Vitamin D, 92, 144
Vitamin E, 92–93
Vitamin K, 93
Vitamins, 32, 89
   fat-soluble, 89, 92–93
   water-soluble, 93–94
Vitamin supplements, 107–108
Vocal communication, 154

W

Water, 94–95
Waterers, 212, 216
   for broilers, 252
   for brooding broiler chickens, 256
   for layers, 232
   for pullets, 231
   for turkeys, 270, 272
Watering
   of coturnix quail, 359
   of guinea fowl, 333, 357
   of Hungarian or gray partridge, 358
   of pheasants, 361
Water quality for broilers, 253
Water-soluble vitamins, 93–94
Welfare, 159–161
Wet spreading, 213
Whale meal, 106
Wheat, 97
Whey, 104–105
White eggs, 240
White fish meal, 105
White gizzard disease, 89
White Pekin ducks, 321–322
White Plymouth Rock, 3
Wild birds, 179
Wills, 316, 344
Workers' compensation, 316, 344–345
World trade
   in duck and goose meat, 313
   in poultry and eggs, 311–313

X

Xanthophylls, 109–110

Y

Yeasts, 106–107
Yolkless eggs, 38

Z

Zinc, 89
Zymogens, 28